Epigenetics in Health and Disease

Epigenetics in Health and Disease

Igor Kovalchuk, Ph.D., MD
Olga Kovalchuk, Ph.D., MD

Vice President, Publisher: Tim Moore
Associate Publisher and Director of Marketing: Amy Neidlinger
Editorial Assistant: Pamela Boland
Development Editor: Russ Hall
Operations Specialist: Jodi Kemper
Assistant Marketing Manager: Megan Graue
Cover Designer: Alan Clements
Managing Editor: Kristy Hart
Senior Project Editor: Lori Lyons
Copy Editor: Charlotte Kughen
Proofreader: Sarah Kearns
Indexer: Angela Martin
Senior Compositor: Gloria Schurick
Manufacturing Buyer: Dan Uhrig

Publishing as FT Press
Upper Saddle River, New Jersey 07458

FT Press offers excellent discounts on this book when ordered in quantity for bulk purchases or special sales. For more information, please contact U.S. Corporate and Government Sales, 1-800-382-3419, corpsales@pearsontechgroup.com. For sales outside the U.S., please contact International Sales at international@pearsoned.com.

Printed in the United States of America

First Printing May 2012

ISBN-10: 0-13-259708-X
ISBN-13: 978-0-13-259708-1

Pearson Education LTD.
Pearson Education Australia PTY, Limited.
Pearson Education Singapore, Pte. Ltd.
Pearson Education Asia, Ltd.
Pearson Education Canada, Ltd.
Pearson Educación de Mexico, S.A. de C.V.
Pearson Education—Japan
Pearson Education Malaysia, Pte. Ltd.

Library of Congress Cataloging-in-Publication Data

Kovalchuk, Igor.
Epigenetics in health and disease / Igor Kovalchuk, Olga Kovalchuk. — 1st ed.
p. cm.
Includes index.
ISBN 978-0-13-259708-1 (hardcover : alk. paper)
1. Epigenetics—History. 2. Epigenetics—Health aspects. 3. Pathology, Cellular. 4. Cells—Morphology. I. Kovalchuk, Olga, MD. II. Title.
QH430.K68 2012
571.9'36—dc23

2012007714

Dedicated to Anna.

Contents

Acknowledgments

We would like to thank Zoe Migicovsky, Corinne Sidler, Alena Babenko, Melanie Kalischuk, Andriy Bilichak, Stephanie Wickersham, Joel Stimson, Aki Matsuoka, and Munima Alam for helping us in writing this book, collecting literature, and checking chapters for comprehension. Special thanks to Valentina Titova for the tremendous help in proofreading the book.

About the Authors

Igor Kovalchuk, Ph.D., MD, is Professor and Board of Governors Research Chair at the Department of Biological Sciences, University of Lethbridge (Alberta, Canada). He edits *Frontiers in Plant Microbe Interaction*, *Frontiers in Epigenomics*, and other journals. As principal investigator in the university's Plant Biotechnology laboratory, he studies genetic and epigenetic regulation of plant response to stress, including the transgenerational effects of stress and microevolution of plant stress tolerance/resistance.

Olga Kovalchuk, Ph.D., MD is Professor and Board of Governors Research Chair and CIHR Chair in Gender and Health at the University of Lethbridge and a member of the editorial boards of *Mutation Research—Fundamental and Molecular Mechanisms of Mutagenesis* and *Environmental and Molecular Mutagenesis*. She researches the role of epigenetic dysregulation in carcinogenesis; epigenetic regulation of cancer treatment responses; radiation epigenetics and role of epigenetic changes in genome stability and carcinogenesis; radiation-induced oncogenic signaling; and radiation-induced DNA damage, repair, and recombination.

1

Historical perspective

Genetics can be broadly defined as the science studying the mechanisms of inheritance in general and genes in particular. *Epigenetics* can be, in part, defined as the branch of biology dealing with the mechanisms of inheritance. In contrast to genetics, epigenetics involves the control of gene expression that is not accompanied by any changes in DNA sequence. Epigenetics deals with the mechanisms of heredity which do not involve modifications of DNA sequence and are reversible in nature.

Success of a certain population depends on the fine balance between the ability to retain a given genotype in the stable environment and the ability to evolve by modification in response to substantial environmental changes. Changes in the genome can be dual in nature; they might deal with stable physical changes in DNA sequence leading to mutations and reversible chemical modifications of nucleotides or chromatin structure leading to epimutations. Mutations are the basis of genetic changes.

This book introduces you to the concept of epigenetics and epigenetic regulation. The book discusses processes of evolution in light of current understanding of the role of epigenetics and describes the role of epigenetic regulations in the growth and development of somatic cells, tissue differentiation, and the maintenance of epigenetic states in various cells of the same organisms. Furthermore, the book provides an introduction to an in-depth understanding of the role of epigenetics in the mechanisms of inheritance and interaction with the environment. The chapters also describe the role of epigenetics in health and disease. Finally, the book introduces you to the concepts of silencing, co-suppression, and paramutations, and discusses the role of epigenetics in these processes.

This book is aimed primarily at students beginning to study epigenetics, whether at the undergraduate or graduate level. It may also be essential reading for research scientists in the field of epigenetics, genome stability, stress tolerance and adaptation, transgeneration effects, genome evolution, and other related fields, as well as anyone who simply wishes to know more about the field of epigenetics.

The mechanisms of environmental influences on the phenotypic appearance of organisms and inheritance were developed nearly two centuries ago and represented a prominent part of the descriptive work performed by Jean-Baptiste Lamarck and Charles Darwin. Although their ideas were often viewed as too preliminary and naïve, it was those ideas that laid a solid and important foundation for the development of the field of epigenetics. Epigenetics has a lot to do with an organism's interaction with the environment; therefore, it is important to review how our understanding of the interactions between the organism's genome, surroundings, and phenotype has developed over time.

The French biologist Jean-Baptiste Lamarck (1744–1829), who is credited with the first use of the word "biology," was the first scientist who proposed a theory of evolution. He used the term *transformation* rather than *evolution* to suggest that organisms change and transform as the result of "a new need that continues to make itself felt." His first reference to evolution as a process of less complex species becoming more complex appeared in 1800 in his Floreal lecture. Within next 20 years, Lamarck published three important works (*Recherches sur l'organisation des corps vivants*, 1802; *Philosophie Zoologique*, 1809; *Histoire naturelle des animaux sans vertèbres* (in seven volumes, 1815–1822) in which he developed his ideas of evolution and formulated the laws that described evolution as a process. Lamarck writes:

> **Law 1:** Life, by its own forces, continually tends to increase the volume of every body which possesses it and to enlarge the size of its parts up to a limit which it brings about itself. **Law 2:** The production of a new organ in an animal body results from the appearance of a new want or need, which continues to make itself felt, and from a new movement which this want gives birth to and maintains. **Law 3:** The

development of the organs and their strength of action are constantly in proportion to the use of these organs. **Law 4:** All that has been acquired, impressed upon, or changed in the organization of individuals during the course of their life is preserved by generation and transmitted to the new individuals that come from those which have undergone those changes.

Lamarck used these laws to explain the two forces he saw as comprising evolution; a force driving animals from simple to complex forms, and a force adapting animals to their local environments and differentiating them from each other.

Lamarck is remembered primarily for his belief in the inheritance of acquired characteristics and the **use and disuse** model by which, according to Lamarck, organisms develop their characteristics. The theory of evolution developed by Lamarck is frequently referred to as Lamarckism or Lamarckian evolution. This theory is also often referred to as **soft inheritance**. The term was first suggested by Ernst Mayr to explain the ideas of Lamarck and Étienne Geoffroy Saint-Hilaire (1772–1844) and to contrast those ideas with the modern idea of inheritance, which Mayr referred to as **hard inheritance**. Geoffroy, a French naturalist and a colleague of Lamarck, defended Lamarck's idea of the influence of the environment on species evolution. He further developed Lamarck's idea suggesting that the environment causes a direct induction of organic change that is the transmutation of species in time.

Perhaps the first attempt at rejection of soft inheritance was made by the English surgeon William Lawrence (1783–1867) in 1819. He stated that "The offspring inherit only connate peculiarities and not any of the acquired qualities" (Lawrence and William, 1819). The inheritance of acquired characteristics was also rejected by the German biologist August Weismann (1834–1914). In the 1880s, he performed an experiment in which he cut off the tails of 22 generations of mice, thus proving that the loss of tail cannot be inherited. Furthermore, in 1893, Weismann proposed his own theory of inheritance. He discovered that the cells that produce the **germ plasm** (now known as gametes) separate from somatic cells at an early stage of organismal development. Weissman could not understand how

somatic and **gametic cells** communicated with each other, and therefore, he declared that the inheritance of acquired characteristics was impossible. He further suggested that the organism's body (the **somatoplasm**) exists for only one generation, whereas the hereditary material, which he called germ plasm, is immortal and passed from generation to generation. Although being rather naïve and futuristic, this view led to an important suggestion: Nothing that happens to somatic cells may be passed on with the germ plasm. Thus, this model underlies the modern understanding of inheritance in which germlines are main cells passing hereditary information from one generation to another. At the same time, because this model suggested that the germ plasm is a self-sufficient substance that is not influenced by the environment, it represented a unilateral understanding of evolutionary processes.

Another theory of evolution was synthesized and described by the English biologist and social philosopher Herbert Spencer. In 1857, he published his theory of evolution in his essay "Progress: Its Law and Cause." Spencer characterized the process of evolution as "evolution of complexity"; he suggested that evolution was a process in which simple organisms always evolved into more complex ones, therefore, evolution itself was progressive in nature. Currently, this view of evolution is considered to be misleading; it is generally accepted that species evolve in response to the environment in the process of natural selection that does not have directionality. The absence of the logical explanation of natural selection did not allow Spencer's theory of evolution to become more prominent. At the same time, it was Spencer who popularized the term **evolution** itself. Moreover, after reading Darwin's *The Origin of Species*, published just two years after Spencer's essay, Spencer attempted to use Darwin's theory for explanation of the role of evolution in society. Moreover, he also tried to incorporate it into his own theory of evolution, coining the now-common phrase **survival of the fittest**.

In 1859, Charles Darwin published the work *"On the Origin of Species by Means of Natural Selection, or the Preservation of Favoured Races in the Struggle for Life"* (Darwin, 1859) (commonly known as *The Origin of Species*) that became a foundation of evolutionary biology and a reason for plenty of scientific discussions.

Darwin's theory suggested that species in the population evolve through a process of natural selection. His book offered multiple examples of how the diversity of life on our planet arose by common descent with modification through a branching pattern of evolution. Darwin proposed that within a certain species, individuals that are less fit for their particular environment are less likely to survive and reproduce compared to those that are well-adapted and have better survival and reproductive potential. The more successful individuals leave more progeny, and thus pass their heritable traits to the next generation. As a result, a certain part of population adapts to the changed environment and eventually might become a separate species. One important thing to note here is that the environment plays a critical role in shaping species' evolution. Darwin accepted a version of the inheritance of acquired characteristics proposed earlier by Lamarck.

Later on, Darwin set forth his provisional hypothesis describing the mechanisms of heredity. In 1868, he presented this idea in his work *The Variation of Animals and Plants under Domestication* (Darwin, 1868). The theory of pangenesis suggests that each individual cell of an organism not only experiences environmental changes and responds to them but also generates molecules that accumulate in germ cells. Darwin believed that these molecules, which he called **gemmules** (Darwin, 1868; Darwin, 1971), are capable of contributing to the development of new traits and organisms. Nowadays, it is a striking fact that small non-coding RNAs such as **microRNAs** (**miRNAs**) and **small interfering RNAs** (**siRNAs**) generated by somatic cells are indeed able to travel within the organism reaching the gametes and potentially influencing the phenotypic appearance of progeny.

1-1. Ontogeny and phylogenetics

Ontogeny (ontogenesis, morphogenesis) is a branch of science describing the development of an organism from the fertilized egg to its adult form.

Phylogenetics (phylogenesis) is the study of evolutionary relatedness among various groups of organisms.

Neo-Darwinism is a comprehensive theory of evolution, frequently called the Modern Synthesis, that combines Mendelian genetics with Darwinian natural selection as a major factor in evolution and population genetics. The term Neo-Darwinism was first used by George Romanes (1848–1894) in 1895 to explain that evolution occurs solely through natural selection as it was proposed by Alfred Russel Wallace (1823–1913) and August Weismann. Neo-Darwinism suggests that evolution occurs without mechanisms involving the inheritance of acquired characteristics based upon interactions with the environment. Thus, this modernized Darwinism accepted some ideas developed by Darwin's original theory of evolution via natural selection, but at the same time it separated them from Darwin's hypothesis of **pangenesis** and the Lamarckian view of inheritance.

Historically, many scientists tried to prove or disprove Darwin's theory of pangenesis. Francis Galton (1822–1911), a cousin of Darwin, conducted many experiments that led him to refute the pangenesis theory. Initially, he accepted the theory, and, in consultation with Darwin, he tried to detect how gemmules were transported in the blood. In his very simple hypothesis, he suggested that if gemmules were transferred to gametic cells though the blood then blood transfusion between various breeds of animals would generate new traits in progeny. In a long series of experiments initiated around 1870, he transfused the blood between dissimilar breeds of rabbits and found no evidence of characteristics transmitted by blood transfusion.

Darwin challenged the validity of Galton's experiment. He wrote in 1871:

> Now, in the chapter on Pangenesis in my "Variation of Animals and Plants under Domestication," I have not said one word about the blood, or about any fluid proper to any circulating system. It is, indeed, obvious that the presence of gemmules in the blood can form no necessary part of my hypothesis; for I refer in illustration of it to the lowest animals, such as the Protozoa, which do not possess blood or any vessels; and I refer to plants in which the fluid, when present in the vessels, cannot be considered as true blood.

Until the end of the nineteenth century, Darwin's theory of pangenesis was accepted by many scientists. The work of Gregor Johann

Mendel on plant hybridization fundamentally changed scientists' understanding of the mechanism of inheritance. Although Mendel published his work in 1866, it was not until 1900 that his ideas were re-examined. Upon re-discovering the significance of Mendel's work, a new era of Mendelian genetics began in which scientists completely rejected the possibility of the transmission of information from somatic cells to gametes and thus to progeny. It was a real pushback for Lamarck's theory of evolution.

Many scientists still considered the possibility of environmentally induced heritable changes. The Russian scientist Ivan Michurin (1855–1935), one of the founders of scientific agricultural selection, also assumed that genotypes could change upon environmental pressure. He worked on hybridization of plants of similar and different origins, developing strategies for overcoming species incompatibility upon hybridization and cultivating new methods in connection with the natural course of ontogenesis (1922–1934). He was also interested in directing the process of predominance, evaluation, and selection and in working out methods of acceleration of selection processes. In the early twentieth century, he proved that the dominant traits in generation of hybrids depend on heredity, **ontogenesis,** and **phylogenesis** of the initial cell structure as well as on individual features of hybrids. Michurin was a true follower of Lamarck and Darwin, and he firmly believed that natural selection could be influenced by external factors, with man being the most influential one.

In the not-too-distant past, the ideas of Lamarck and Michurin seemed to be pseudo-scientific and impossible to believe in. But recently, a breakthrough publication describing changes in the genetic make-up of grafted plants appeared that became an eye-opener, suggesting many new possibilities for transmission of genetic material. Sandra Stegemann and Ralph Bock showed that transfer of genetic material from stock to scion is possible upon grafting of tobacco plants (Stegemann and Bock, 2009). The results of the study demonstrated that recipient plants acquired tolerance to an antibiotic in the same manner as donor plants, and they also confirmed the transfer of genetic material from a donor to a recipient. Although it is still unclear whether the acquisition of antibiotic resistance occurs via plastid transfer through plasmodesmata or via the transfer of a large portion of the plastid genome from a donor cell to a recipient cell, it can be definitely considered as an example of changes not only in phenotypic appearance but also in the genetic make-up of a grafted plant.

The emergence of epigenetics as science was closely linked to the study of evolution and development. Nowadays, we know that chromosomes are associated with both genetic and epigenetic regulation, thus driving the developmental processes. Despite the early discovery of chromosomes by Walther Flemming (1843–1905), the founder of cytogenetics, in 1879, it took many more experiments to links chromosomes to function, phenotypes, and developmental programming. The experiments by Edmund Wilson (1856–1939), Theodor Boveri (1862–1915), Walter Sutton (1877–1916), and later on Thomas Hunt Morgan (1866–1944) provided several evidences that chromosomes were indeed involved in developmental processes, and changes in chromosomes resulted in changes in phenotype. The **Boveri-Sutton chromosome theory** suggested that Mendelian laws of inheritance could be applied to chromosomes and chromosomes might thus be units of inheritance. Morgan's work in *Drosophila* showed that the inheritance of many genes was linked to the X chromosome; among them were genes coding for eye color. This and other works enabled him to become the first scientist to receive a Nobel Prize (1933) for his work in genetics. The report by Watson and Crick (1953) describing DNA structure and proposing the mode of DNA replication further reinforced the notion that DNA is the cell's genetic material. Although the studies of chromosome morphology indicated that somatic cells contained all of the chromosomes, it was not clear why the somatic cells of different tissues had different phenotypic appearance, raising doubts whether somatic cells actually did carry all the genes and not only those that were necessary for their growth and development.

Although investigations of epigenetic regulation of an organism's development and cell fate were being actively pursued throughout the twentieth century, the actual name "epigenetics" did not emerge until 1942 when Conrad Hal Waddington (1905–1975) used it to describe how genes might interact with their surroundings to produce a phenotype. Waddington described several essential concepts, including **canalization**, **genetic assimilation**, and **epigenetic landscape**. The concept of canalization in Waddington's understanding was the capacity of the organisms of a given population to produce the same phenotype regardless of the extent of genetic and environmental variations. He assumed that this robustness came as a result of evolution, shaping the developmental processes to perfection. Waddington's idea of genetic assimilation suggested that an organism responds to the environment in such a way that the acquired phenotype would become part of the developmental process of the organism.

To demonstrate that the phenomenon exists, Waddington induced an extreme environmental reaction in the developing embryos of the fruit fly *Drosophila*. When exposed to ether vapor, a small percentage of the *Drosophila* embryos developed a second thorax. It was obvious that bithorax embryos represent an abnormal phenotype. Waddington continued selection of bithorax mutant embryos, and after about 20 generations of selection, he obtained *Drosophila* flies that developed bithorax without being exposed to ether vapor. Waddington suggested that in this particular case, selection led to the production of the desired effect, which became canalized, and, as a result, bithorax appeared regardless of environmental conditions. Thus, Waddington's experiments demonstrated that Lamarckian ideas of inheritance of acquired characteristics could, at least in principle, be true. Finally, the epigenetic landscape, as suggested by Waddington, represents is a programmed cell fate where developmental changes would occur with increasing irreversibility, much like marbles rolling down a small-ridged slope toward the lowest elevation point. Nowadays, the term **epigenetic landscape** refers to the certain area of a chromatin in the cell with specific cytosine methylation and histone modifications involved.

During the past 50 years, the scientific community has witnessed a lot of rises and falls in an interest in epigenetics. Perhaps, the next important discovery in the area of epigenetics was Alexander Brink's report on the phenomenon of **paramutation**. In 1956, Brink described a somewhat puzzling and controversial phenomenon of the inheritance phenotype associated with the of *Red 1* (*r1*) locus in maize (Brink, 1956). It was observed that the spotted seed allele (*R-st*) was able to transform the *R-r* (purple color seeds) phenotype allele into a colorless seed phenotype in subsequent generations. As a result of the cross, all of the F_2 generation plants showed reduced anthocyanin in seeds, which was contrary to the expected segregation ratios according to the Mendelian law. The phenomenon that he proposed to be called paramutation involved heritable transmission of epigenetically regulated expression states from one homologous sequence to another. For a more detailed description of paramutations in plants and animals, see Chapter 13, "Paramutation, Transactivation, Transvection, and Cosuppression: Silencing of Homologuous Sequences."

In her early work, Barbara McClintock (1902–1992) also suggested that the chromosomal position effect might influence on the behavior of mutable loci in maize. She assumed that the observed difference in mutability ratios of suppressor elements in maize had the mechanism similar to the earlier described phenomenon of position-effect variegation. The latter term was first brought up by Hermann Joseph Muller (1890–1967) and was meant to describe the effect of chromosomal position on gene expression. Having observed gross chromosomal rearrangements, Muller noted changes in gene expression, and the genes that were brought into the area of heterochromatin expressed poorly. McClintock noted that some controlling elements, such as *Spm,* would suppress gene expression rather than mutate a gene; she also noticed that the suppression of gene expression would take place not only at the locus where the elements had been inserted but also at the neighboring loci.

The work of David Nanney (published in 1958) showed that the cytoplasmic history of conjugating parents had an impact upon the mating-type determination of resulting progeny in *Tetrahymena* (Nanney, 1958). This phenomenon was suggested to be of epigenetic nature.

In the early 1960s, Mary Lyon and Walter Nance presented the mechanism of another epigenetically regulated process, **X-chromosome inactivation**. It was suggested that inactivation of the mammalian female X chromosome occurred before the 32-cell stage of the embryo. However, there was no clear assumption that this process was indeed of epigenetic nature. The fact that no changes were observed at the level of DNA allowed Riggs (1975) and Holliday and Pugh (1975) to propose that DNA methylation could be a mechanism of X-chromosome inactivation.

In the 1970s, Hal Weintraub's work on the expression of globin genes revealed an influence of chromosomal location on the transcriptional activity. His observations were the source from which the suggestion came that the chromatin structure might regulate gene expression.

In the early 1980s, it became clear that there was an apparent correlation between the level of cytosine methylation at GpG DNA sequences and the level of gene transcription. Moreover, the mitotic heritability of **DNA methylation patterns** was also shown. Later on, by the mid-1980s, the influence of nuclear content on the genetic/phenotypic make-up of the organism was also revealed. It was found out that not only the DNA sequence of paternal or maternal alleles had an effect on the phenotype, but the origin of a particular chromosome itself could influence the phenotype. Thus, it was suggested that besides the DNA sequence, the chromosome also carried additional information.

In the 1990s, scientists presented more discoveries in the area of epigenetics, coming from studies of various organisms including protozoa, fungi, *Drosophila*, plants, and animals. In plants, it was found that the transgene coding for chalcone synthase (*Chs*) had various degrees of suppression of expression gene expression. It was perhaps the first well-documented event of **gene silencing** (Napoli et al., 1990).

In trypanosomes, it was discovered that silencing of the group of *Variable Surface antigen Genes* (*VSG*) is maintained by the incorporation of a novel base, β-D-glucosylhydroxymethyluracil (Borst et al. 1993). Because trypanosomes do not have the mechanism of cytosine methylation, it was suggested that the insertion of the modified base would also serve as a gene-silencing mechanism. Significant progress was made in understanding the mechanisms of X inactivation. A portion of the human X chromosome was identified to function as the X chromosome inactivation center; later on, the gene *Xist* was identified that appeared to be coding for a **non-coding RNA** expressed only in an inactive X chromosome (Willard et al., 1993). The analysis of the expression of the neighboring gene *Igf2* and *H19* pair provided a further understanding of the mechanism underlying chromosomal imprinting. The genes were mutually exclusively expressed depending on the maternal or paternal origin of the chromosome; if the *Igf2* gene was expressed from the paternal chromosome, then the *H19* gene was repressed, whereas if the *H19* gene was expressed from the maternal chromosome, the *Igf2* gene was repressed. Methylation analysis of the locus identified high frequency of occurrence of methylated CpGs. Therefore, it was proposed that methylation controlled the access to the enhancer element that functioned mutually exclusively for both genes. Indeed, in mice, it was found that a mutant impaired in the function of a 5-methyl-cytosine DNA methyltransferase lost the imprinting of the gene pair in ES cells.

The role of epigenetic regulation in control over gene expression was also demonstrated by the experiments on fungi. Gene duplication in *Neurospora crassa* often resulted in the occurrence of two events: frequent mutations and hypermethylation of both gene copies, a phenomenon known as **repeat-induced point mutation (RIP)**. Furthermore, for the first time, it was shown that cytosine methylation in *Neurospora* could occur at non-CpG sites. A similar phenomenon was observed in *Drosophila*; the duplication of the brown gene translocated near heterochromatin increased the level of repression in the active copy. Because in *Drosophila,* cytosine methylation is not used as a process of gene expression regulation, there should be a different repression mechanism. The research in this direction resulted in the development of the concept of chromosomal **boundary elements**, the areas of the chromosome that contained a 300 bp nuclease-resistant core surrounded by nuclease hypersensitive sites that were first described in *Drosophila*. It was suggested that such elements allow the separation of a chromatin domain along the chromosome, thus leading to differential areas of chromosome compaction and gene expression. In yeasts, the Sir2, Sir3, and Sir4 proteins (silent information regulator proteins) were identified that were proposed to control repressive states near heterochromatic regions. The evidence that Sir3 and Sir4 interacted with the tails of histones H3 and H4 further confirmed the importance of both these proteins and histones in the maintenance of the chromatin state. By the end of the 1990s, histone-modifying enzymes such as acetylases and deacetylases were identified, and the MeCP2 protein complex that was able to bind to methylated DNA and histone deacetylases were described.

One more important discovery made in the late 1990s was the description of the phenomenon of **RNA interference**. A series of work by Craig Mello and Andrew Fire culminated in the famous publication in *Nature* (Fire et al., 1998) that described the ability of double-stranded RNA molecules to inactivate the expression of genes in *C. elegans*; the effect of interference was evident in both injected animals and progeny. Because just a few molecules per cell were sufficient to trigger the effect, the authors suggested the existence of an amplification component, currently known as a mechanism that involves the function of RNA-dependent RNA polymerase. The importance of this work for studying organism development, the therapy of various human diseases, as well as for the development of biotechnology and basic science was recognized with the Nobel Prize awarded to Mello and Fire in 2006.

In 1990, Robin Holliday defined epigenetics as "the study of the mechanisms of temporal and spatial control of gene activity during the development of complex organisms" (Holliday, 1990). Today, the definition of epigenetics has been changed; it is now described as the study of the mechanisms of inheritance and control of gene expression that do not involve permanent changes in the DNA sequence. Such changes occur during somatic cell division and sometimes can be transmitted transgenerationally through the germline.

The last ten years were marked by the most prominent achievements in epigenetic research. In 2000, it was discovered that the Sir2 protein of yeasts was in fact a histone deacetylase. Studies of the heterochromatin states, replication processes, the activity of the Sir3 protein in yeast, and the heterochromatin protein (HP1) in mammals showed that heterochromatin was not in a solid inert stage reversible only during replication but rather in an active equilibrium stage of protein exchange between the nuclear soluble compartment and heterochromatin itself, regardless of cell cycle status.

By the early 2000s, most of the histone modifications and the enzymes that catalyze them were discovered, and it was believed that besides histone methylation, all other modifications such as acetylation, phosphorylation, ubiquitination, and so on were reversible. Thus, various histone methylation states were regarded as a permanent epigenetic mark of chromatin status and were reversible only during replication. The results of studies by Cuthbert et al. (2004) and Henikoff et al. (2004) raised the possibility that histone methylation could be reversible, and their suggestions were met with true enthusiasm. In his works, Cuthbert demonstrated that peptidylarginine deaminase was able to remove single methylation events at the arginine amino acid of histone H3 (Cuthbert et al., 2004). Henikoff et al. (2004) showed that H3.3, a histone H3 variant, was able to replace histone H3 in a transcription-dependent and replication-independent manner, opening the possibility for more flexible regulation of methylation after the process transcription was over.

Another breakthrough was the discovery that nuclear organization and silencing at telomeres were not necessarily completely interrelated. The experiment showed that if telomeres and the associated silencing complex were released from the periphery of the nucleus and were able to move throughout the nucleus, the silencing at telomeres was established with similar efficiency (Gasser et al., 2004). This is truly exciting—it suggests that chromatin compartmentalization and gene silencing processes are not rigid and predefined states; there indeed exists an active exchange between the nuclear pools of proteins and small RNAs that are able to establish a certain chromatin state at any given locus regardless of its nuclear location.

A curious phenomenon was reported for *Arabidopsis*; it was on the borderline of epigenetic regulation and described a non-Mendelian inheritance. An *hth* mutant is homozygous for the mutation in the *HOTHEAD* (*HTH*) gene that encodes a flavin adenine dinucleotide-containing oxidoreductase involved in the creation of the carpel during the formation of flowers. It was reported that in the progeny of the *hth* mutant, the percentage of the frequency of appearance of the *HTH* phenotype and *HTH* genomic sequence was ~15% (Lolle et al., 2005). It was first proposed that reversion was triggered by RNA synthesized by *HTH/hth* parents and stored in the progeny *hth/hth* plants. Four alternative explanations have been proposed: Two of them were in part similar to an original explanation made by Lolle et al. (2005) and dealt with template-directed gene conversion; the third one offered the process of mutation accumulation followed by selection; and the fourth one involved **chimerism**. Later on, two publications seemed to put everything in place: Peng et al. (2006) and Mercier et al. (2008) reported that the *hth* mutant showed a tendency toward outcrossing and recovered a normal genetic behavior when grown in isolation. Despite the fact that

Arabidopsis is an extreme self-pollinator (less than 0.1% of outcrossing), in the *hth* plants the frequency of outcrossing among neighboring plants was ~12%. This can be an excellent alternative explanation for the apparent genetic instability of *hothead* mutants.

Now that the genome sequences of model organisms such as *C. elegans*, *Drosophila*, *Arabidopsis*, human, mice, rice, and so on are available, more and more investigators have attempted to understand the organization of the genome and chromatin and explain the mechanisms of inheritance, maintenance of genome stability, and regulation of gene expression. What has become clear is that these mechanisms are both genetic and epigenetic in nature. As it was recently put by Daniel E. Gottschling ("Epigenetics: from phenomenon to field" in *Epigenetics*; eds. C.D. Allis, T. Jenuwein, D. Reinberg), it was time to move "above genetics"—a literal meaning of epigenetics as several important genomes have already been sequenced.

There are multiple examples of the influence of environment on the genetic and epigenetic make-up of the organism. The phenomena of stress-induced transposon activation, non-targeted mutagenesis, stress-induced communication between cells and organisms, and evidences of transgenerational changes induced by stress are just some representations of epigenetic effects of the environment on the organism.

The non-linear response to DNA damaging agents is one of the most interesting examples of an epigenetically controlled process. It has already been known that a higher dose of mutagen does not necessarily result in a higher level of damage to DNA. In fact, low doses of ionizing radiation often lead to disproportionally high levels of DNA damage. Doses of ionizing radiation that are believed to have a negligible effect on a cell often exert dramatic influence on DNA damage and cell viability.

In the past, cell-to-cell communication between neighboring cells as well as communication between cells of different tissues and organs in multicellular organisms were considered Lamarckian/Darwinian and thus improbable. There are multiple examples of physiological cell-to-cell communications in simple and complex organisms involving hormonal signaling, neurotransmission, and so on. Moreover, it is believed that damaged tissues are able to communicate with non-damaged tissues—a phenomenon known as **bystander effect**. The phenomenon of bystander effect has also been observed between whole living organisms.

Can organisms communicate memory of stress across generations? According to Darwin, organisms evolve from the pool of individuals with spontaneous changes/mutations through the process of natural selection. The process of mutagenesis is believed to be random, and the majority of mutations are deleterious. The rare mutations that become beneficial under certain environmental conditions have a chance to be fixed in a population. Because mutagenesis does not occur frequently, the fixation of desired traits would take place very rarely. In contrast, processes of acclimation and adaptation are rapid ones that allow organisms to acquire protection against stress in a single generation after stress exposure. These processes cannot be explained by the laws of Mendelian genetics. In this book, you find multiple examples demonstrating the inheritance of stress memory in various organisms across generations.

This chapter attempted to explain what epigenetics is, how it is involved in the regulation of growth and development of the organism, how it controls interactions of the organism with the environment, and what roles epigenetics plays in the mechanisms of inheritance and evolutionary processes.

There have been many more important discoveries in the field of epigenetics, and we apologize to all those authors whose work, though relevant, is not mentioned in this chapter because of limitations of space.

References

Borst et al. (1993) Control of antigenic variation in African trypanosomes. *Cold Spring Harb Symp Quant Biol.* 58:105-14.

Brink RA. (1956) A genetic change associated with the *R* locus in maize which is directed and potentially reversible. *Genetics* 41:872-879.

Cuthbert et al. (2004) Histone deimination antagonizes arginine methylation. *Cell* 118:545-53.

Darwin CR. (1871) Pangenesis. *Nature. A Weekly Illustrated Journal of Science* 3:502-503.

Darwin CR. (1859) *On the Origin of Species by Means of Natural Selection, or the Preservation of Favoured Races in the Struggle for Life* (1st ed.), London: John Murray.

Darwin CR. (1868) The variation of animals and plants under domestication. London: John Murray. 1st ed., 1st issue.

Fire et al. (1998) Potent and specific genetic interference by double-stranded RNA in Caenorhabditis elegans. *Nature* 391:806-11.

Gasser et al. (2004) The function of telomere clustering in yeast: the circe effect. *Cold Spring Harb Symp Quant Biol.* 69:327-37.

Henikoff et al. (2004) Epigenetics, histone H3 variants, and the inheritance of chromatin states. *Cold Spring Harb Symp Quant Biol.* 69:235-43.

Holliday R. (1990) Mechanisms for the control of gene activity during development. *Biol Rev Camb Philos Soc*, 65:431-471.

Lawrence, William FRS. (1819) *Lectures on physiology, zoology and the natural history of man.* London: J. Callow, p. 579.

Lolle et al. (2005) Genome-wide non-mendelian inheritance of extra-genomic information in *Arabidopsis*. *Nature* 434:505-9.

Mercier et al. (2008) Outcrossing as an explanation of the apparent unconventional genetic behavior of *Arabidopsis thaliana* hth mutants. *Genetics* 180:2295-7.

Nanney DL. (1958) Epigenetic factors affecting mating type expression in certain ciliates. *Cold Spring Harb Symp Quant Biol.* 23:327-35.

Napoli et al. (1990) Introduction of a Chimeric Chalcone Synthase Gene into Petunia Results in Reversible Co-Suppression of Homologous Genes in trans. *Plant Cell* 2:279-289.

Peng et al. (2006) Plant genetics: increased outcrossing in hothead mutants. *Nature* 443:E8.

Stegemann S, Bock R. (2009) Exchange of genetic material between cells in plant tissue grafts. *Science* 324:649-651.

2

Chromatin dynamics and chromatin remodeling in animals

A proper function of eukaryotic cells requires the packaging of the long DNA chain into a small nucleus. In mammals, approximately 2 meters of DNA has to fit into the nucleus, which is about 5 micrometers in size. Even a more complex task is to package DNA so that it allows for the development- and tissue-specific expression pattern. Chromatin compaction is achieved through the activity of small basic proteins called histones. DNA is wrapped twice around the histone octamer, thus taking the basic unit of chromatin assembly—the **nucleosome**, which compacts the DNA length about sevenfold—from 2 nm to roughly 11 nm. The binding of histone H1 between two nucleosomes triggers further compaction of nucleosomes into a 30 nm chromatin fiber. The binding of chromatin fibers to a protein scaffold generates a looped domain structure of approximately 300 nm in width.

Most chromatin in eukaryotic non-dividing somatic cells is present in undefined, compact structures that allow local decondensation in response to internal and external factors. Mitosis results in an additional temporary compaction of these chromatin loops into an approximately 700 nm-wide metaphase chromosome arm. Chromatin is not a rigid structure, and dynamic changes are crucial for most DNA-dependent processes, such as replication, transcription, and DNA repair. Chromatin modifications change chromatin structure by altering physical properties of individual nucleosomes. This occurs mainly by the addition or removal of a charge to or from target proteins or nucleic acids. The interaction between the charge of residues and chromatin structure is rarely direct, as suggested in the case of histone acetylation. However, in most cases, the chromatin status is defined by complex interactions of modifications of many specialized sets of nuclear proteins.

Chromatin structure is regulated through processes such as DNA methylation, ATP-dependent remodeling of nucleosome cores, covalent modifications of histone tails, the replacement of core histones by their variants, and nucleosome eviction. **Chromatin remodelers** are multiprotein complexes that alter histone—DNA interactions to form, disrupt, or reposition nucleosomes using the energy derived from ATP hydrolysis. This chapter focuses on chromatin architecture, the role of ATP-dependent remodeling factors, and chromatin-binding proteins in various nuclear processes.

Epigenetics of nuclear architecture

During the interphase, chromosomes in the nucleus of a cell are located in specific regions called **chromosome territories (CTs)**, with the regions around them called **interchromatin compartments (ICs)**. The positions of CTs and ICs in the nucleus allow the efficient regulation of gene expression. The nuclear architecture of the majority of eukaryotic cells is defined by gene-rich CTs that occupy the interior regions of the nucleus, whereas gene-poor CTs are located at the periphery of the nucleus (Croft et al., 1999). A similar correlation is found among transcriptionally active and inactive genes. Whereas the former are primarily located closer to the center of the nucleus, the latter reside farther from the nuclear interior. The nuclear positioning of CTs and ICs is a dynamic process during which chromatin organization changes according to the need of a specific type of cells/tissues. During development, CTs containing transcriptionally active genes form loops of chromatin that extend toward the center of the nucleus (reviewed in Bártová et al., 2008). The specific positioning of active decondensed chromatin loops defines the active expression of entire gene clusters, such as the major histocompatibility cluster (MHC) and the epidermal differentiation complex (EDC). Although it is most common for inactive genes to be located at the nuclear periphery, there are exceptions from this rule—many active genes are found at the periphery of the nucleus.

Several models describing the possible arrangement of CTs and ICs in the nucleus have been proposed. The first model, called the **interchromatin compartment model (CT-IC)**, suggests a sponge-like CT architecture (see Figure 2-1). One of the first reports that possibly inspired the development of this model was the report showing peripheral invaginations of chromosome territories, which position transcriptionally active chromatin into the interior of the interphase nucleus (Verschure et al., 1999). Another model, named the **interchromosome domain (ICD)** model, proposes that CTs are smooth domains that exist as separate units of chromatin. The ICD model was proved to be conceptually unsustainable after the discovery of chromosome intermingling (Branco and Pombo, 2006). Thus, the CT-IC model is more likely a model that is able to explain the occurrence of frequent chromosome translocations and gene-gene interactions than the ICD model.

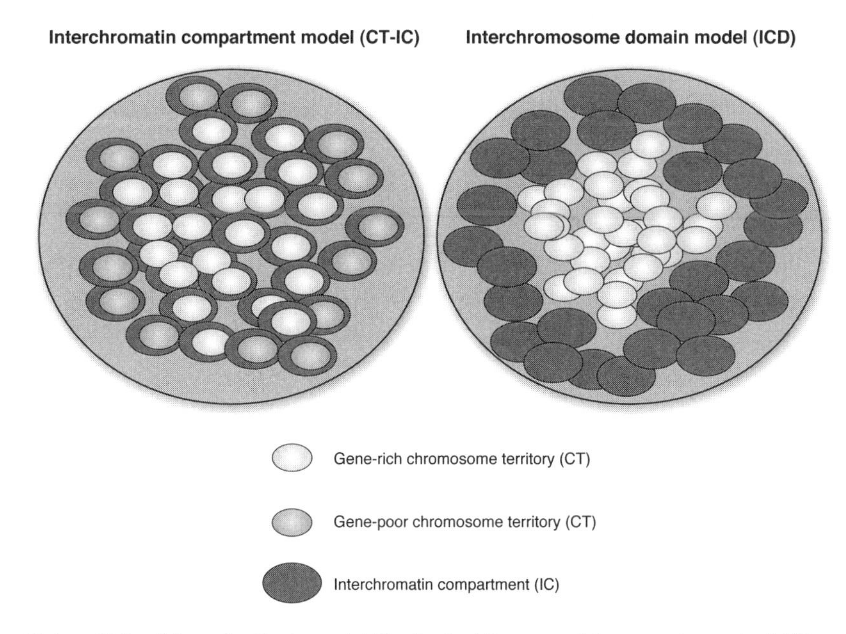

Figure 2-1 A hypothetical model illustrating the possible arrangement of CT and IC regions.

Transcription from the active CTs located on decondensed chromatin loops takes place in nuclear compartments known as transcription factories with the help of RNA polymerase II (reviewed in

Bártová et al., 2008). In contrast, the positioning of chromatin at the nuclear periphery helps establish and maintain transcriptional gene silencing. A perfect example of a correlation between the nuclear positioning and transcriptional activity is represented by **X chromosome inactivation (Xi)**.

A condensed X chromosome is localized in the most peripheral region of the interphase nucleus and is highly enriched in H3K27me3 and H3K9me2 heterochromatin marks. Furthermore, the positioning of genes on the chromosome is also very specific; all genes of the X chromosome are located at the outer rim of the active X and Xi territories (Clemson et al., 2006). The structure of active and inactive chromatin loops is not rigid, and the acquisition of active chromatin marks such as H3K9ac or H3K4me2 and passive chromatin marks such as deacetylated histone 3 and H3K9me2 results in changes in chromatin condensation that allow to form and separate out small active and passive chromatin regions with opposite chromatin activity in the genomic regions.

Thus, the nuclear dynamics plays a crucial role in chromatin positioning and the formation of micro- and macro-loops of active/passive transcription. It has been suggested that changes in the shape of the nuclear envelope through lamins might alter the nuclear shape and thus modify chromatin conformation. The nuclear envelope consists of an inner and an outer nuclear membrane separated by an intermembrane lumen. Lamins are proteins that form intermediate filaments. Those that are located at the inner nuclear membrane are called **peripheral lamins**, and those located at the inner layer of the nucleus are termed **internal lamins**. Lamins that are intimately associated with the inner membrane provide the mechanical stability for the nuclear membrane.

There are two groups of lamins, A-type and B-type; and A-type lamins are divided into two types of proteins, lamins A and C, through alternative splicing. Lamins A, B, and C are phosphorylated during late G2/early mitosis by a mitosis-promoting factor. This results in depolymerization of lamin intermediates followed by the disintegration of the nuclear lamina. Besides playing a role in the organization of the nuclear envelope, lamins are also involved in chromatin organization and possibly transcription. Peripheral and internal A-type lamins associate with nuclear actin and c-fos, and this interaction influences transcriptional activity. In Drosophila, genes that interact with B-type lamins have a peripheral nuclear localization and are transcriptionally silent (Pickersgill et al., 2006).

The nuclear lamina directly interacts with the nuclear envelope through many **lamin-associated polypeptides (LAPs)**. These proteins may provide a link between lamins and the nuclear envelope as well as between lamins and peripheral chromatin. A functional interaction between lamins A and C and DNA/nucleosomes occurs in a sequence-independent manner (reviewed in Bártová et al., 2008). Interactions between LAPs and histones/chromatin are not passive as they influence histone modification, chromatin organization, and even gene expression (summarized by Schirmer and Foisner, 2007). A functional relationship between A-type lamins and specifically modified histones was demonstrated for diseases associated with mutations in lamins (reviewed in Bártová et al., 2008). Scaffidi and Misteli (2006) showed that the levels of repressive chromatin marks H3K9me3, H3K27me3, and the heterochromatin-associated protein 1 isoform γ (HP1γ) are reduced in the cells of Hutchinson-Gilford progeria syndrome (HGPS), a disease associated with mutations in the lamin A gene (LMNA). These changes were particularly noticeable in the inactive X chromosome. The repressive chromatin marks and HP1β were not only depleted in these cells but were found to be rearranged and dispersed into other chromatin regions, resulting in the formation of unusual hetero- and euchromatin.

The role of nuclear actin-related proteins (ARPs) in chromatin remodeling

Actin-related proteins (ARPs) were first discovered in the early 1990s, but their essential role as a chromatin remodeler has been only recently suggested. ARPs are involved in epigenetic control of such processes as DNA repair, chromosome segregation, and gene expression (reviewed in Meagher et al., 2009).

ARPs show a limited sequence identity with conventional actins, although like actin proteins, they maintain the capacity to fold forming a potential nucleotide-binding pocket. Most eukaryotes contain eight to ten classes of ARPs involved in protein complexes. All chromatin-remodeling complexes that contain a Swi2 (SWITCH)-related

DNA-dependent ATPase subunit also include one or more nuclear ARPs. Whereas in eukaryotes, yeast homologs of Arp4, Arp5, Arp6, and Arp8 function in the nucleus, Arp2 and Arp3 participate in cytoskeletal functions in the cytoplasm (reviewed in Meagher et al., 2009). Nuclear ARPs do not form filaments; instead, they act as essential subunits of macromolecular machines controlling chromatin dynamics.

Nuclear ARPs function as subunits of nucleosome remodeling complexes responsible for nucleosome phasing and repositioning as well as the exchange of histone variants within nucleosomes (see Figure 2-2). The complexes containing ARPs direct acetylation or methylation of lysine residues of histones. Although these chromatin activities are global in nature and are capable of opening or closing chromatin on large stretches of the genome, ARPs may also be involved in a more precise epigenetic control over the expression of particular regulatory genes that are critical for both the development and the response to environmental stress (reviewed in Meagher et al., 2009). Nuclear ARPs are also involved in the epigenetic control of chromosome segregation and DNA repair. Human ARP8 has been shown to be associated almost exclusively with mitotic chromatin in a dividing cell, and it is also essential for mitotic alignment of chromosomes. ARP4 has been found weakly localized to the surface of mitotic chromatin in cultured mouse and human cells. In addition, yeast ARP4 is also associated with centromeric DNA during metaphase (reviewed in Meagher et al., 2009).

Nuclear ARPs bind Swi2-related DNA-dependent ATPases as part of chromatin remodeling (SWI/SNF, SWR1, RSC, INO80, p400, see the section "The developmental role of SWI/SNF proteins: BAP in *Drosophila* and BAF in mammals" for details) and chromatin modifying complexes (NuA4 HAT). Besides Swi2 factors, the complexes consist of one ARP and an actin subunit. In mutant human cells deficient for the Swi2-related Brg1 subunit, the Brahma-associated factor (BAF) complex (see section "The developmental role of SWI/SNF proteins: BAP in *Drosophila* and BAF in mammals" for details) is also missing the ARP4 and actin subunits (reviewed in Meagher et al., 2009). Similarly, the binding of actin and ARP4 in the SWR1 complex and the binding of ARP8 in the INO80 complex require N-terminal regions of the Swi2-related Swr1 and Ino80 subunit, respectively.

More details on the possible role of ARPs in the formation of chromatin-remodeling complexes can be found in the review by Meagher et al. (2009). However, the exact role of ARPs in these complexes still remains to be established.

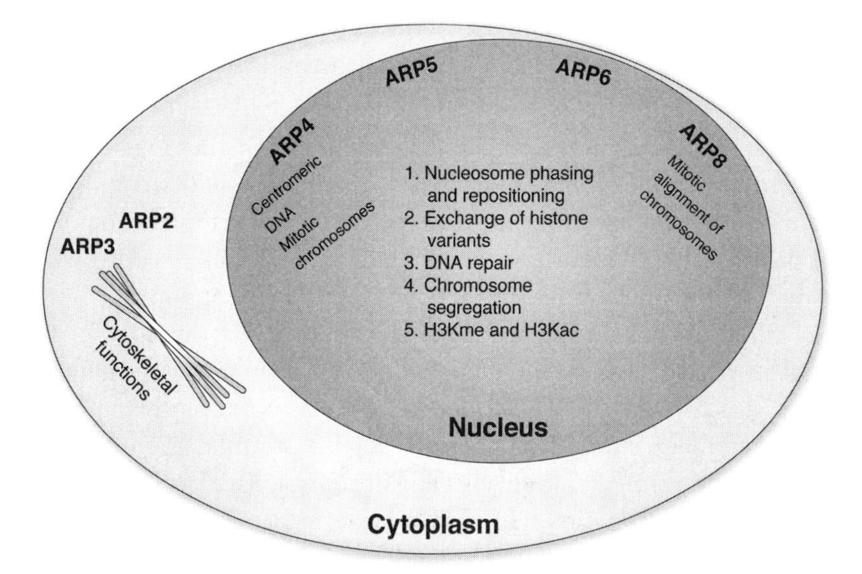

Figure 2-2 The functions of ARPs in cells.

Chromatin-modifying and chromatin-remodeling proteins

This chapter describes three major groups of proteins involved in chromatin modifications: **chromatin-remodeling complexes**, **effector proteins**, and **insulator proteins**.

Chromatin-remodeling complexes

Chromatin-remodeling complexes are energy-driven, multi-protein machinery that allows an access to specific DNA regions or histones by altering nucleosomal positions, histone-DNA interactions, and histone octamer positions. These chromatin remodelers contain a catalytic ATPase for inducing changes in local chromatin structure covering one or two nucleosomes. In chromatin-remodeling complexes, ATPases are grouped into four subfamilies: the **SWI/SNF**

(SWITCH/SUCROSE NONFERMENTING) ATPases, the **imitation switch (ISWI) ATPases**, the **chromodomain and helicase-like domain (CHD) ATPases**, and **INO80 ATPases**. Many of the currently known SWI/SNF-like ATPases do not fit any of these classes and require further characterization. Several recent reviews have summarized the current understanding of diversity and specialization of chromatin-remodeling complexes and modulation of remodeling activity by nucleosome modifications (Ho and Crabtree, 2010).

Because the main role of ATP-dependent chromatin remodelers is to increase the nucleosome mobility, it is surprising that there is such a great variety of non-redundant ATPases. Two possible explanations are proposed. First, Ho and Crabtree (2010) suggest that besides being important for nucleosome repositioning and transcription, ATPases play other functional roles in chromatin structure. For example, ISWI complexes have been shown to be required for maintaining the higher-order structure of the *D. melanogaster* male X chromosome, and INO80 complexes are involved in the regulation of telomere length, chromosome segregation, and checkpoint control as well as DNA replication during cell division (Ho and Crabtree, 2010). Second, it appears that ATP-dependent chromatin remodelers are not only simple activators of transcription, but they alternate between being activators and repressors, as was observed for BAF complexes. The switch between functioning as activators and repressors mode is context-dependent because the tissue-specific BAF complexes have been reported to interact with a variety of transcription factors in different cell types, depending on available interaction partners.

More progress in understanding the function of various ATPases comes from studying their role in animal development. Details on the functional role of chromatin remodelers in animal development have been covered by Ho and Crabtree (2010) in their review.

The developmental role of SWI/SNF proteins: BAP in Drosophila and BAF in mammals

In *Drosophila melanogaster*, the determination of body segmental identity requires the proper patterns of homeotic gene expression of both the *Bithorax* complex and the *Antennapedia* complex. The action of the gap and pair-rule genes is required for the establishment of correct gene expression patterns of the *Antennapedia* and *Bithorax* complexes. The expression of these genes is then maintained by TrxG and PcG proteins that positively and negatively regulate transcription, respectively. The Drosophila SWI/SNF proteins encoded by the trxG genes *brahma* (*brm*), *osa*, and *moira* (*mor*) function in a multisubunit complex known as **Brahma-associated proteins (BAP)**. Other subunits of the BAP complex that are required to antagonize the function of PcG proteins are SNR1 (Snf5-related protein 1; a homolog of mammalian BAF47 and yeast Snf5) and SAYP (supporter of activation of yellow protein—also known as E(Y)3; a homolog of mammalian BAF45-family members). The recruitment of the BAP complex to chromatin might be counteracted by PcG proteins; light microscopy showed that these two groups of proteins are mutually exclusive on salivary-gland polytene chromosomes (reviewed in Ho and Crabtree, 2010). The role of BAP as an antagonist of PcG during early Drosophila development apparently extends to other functions; depletion of BRM or other proteins of the BAP complex in the zygote leads to multiple defects in organ and gamete formation and causes lethality at the later stages of embryo development.

In mammals, the SWI/SNF protein complex is known as the BAF complex. Although BAF plays a role in development, currently there is no direct evidence showing that BAF may be involved in the specification of segment identity in vertebrates. The composition of the BAF complex is highly dynamic as progressive changes in subunit composition occur during the transition from a pluripotent stem cell to a multipotent neuronal progenitor cell to a committed neuron (reviewed in Ho and Crabtree, 2010). The ATPase subunit is one of 12 subunits of the BAF complex encoded by two homologs, *Brm* and *Brg1* (brahma-related gene 1). Many BAF subunits are encoded by gene families, thus allowing a diversity of stable assemblies in different cell types. For example, BRM and BRG1 are encoded by one gene family, and each BAF complex consists of only one of these proteins. Thus, it is not surprising that mice deficient in either BMR or BRG1 have different phenotypes. Maternal *Brg1* is required for the activation of genome expression at the two-cell stage of embryo development. Similarly, zygotic *Brg1* is essential for both the survival and proliferation of cells of the inner cell mass and the trophoblast. In contrast, *Brm* does not seem to be required for normal mouse development.

2-1. Antagonistic functions of TrxG and PcG proteins

The antagonistic function of trithorax and polycomb proteins was first described in *Drosophila melanogaster*. PcG and trxG genes were first discovered in Drosophila as master regulators of homeotic (Hox) gene expression (reviewed in Kim et al., 2009). Whereas the polycomb complexes function as repressors of target genes, the trithorax proteins work as activators targeting the identical DNA regulatory elements. These elements—the PcG or trxG response elements (PREs/TREs)—recruit PcG or TrxG proteins to form multimeric complexes on PREs/TREs. The interactions of PcG with PREs result in epigenetic inheritance of silent chromatin states, whereas the interactions of TrxG with TREs lead to the propagation of active chromatin. These PcG and TrxG complexes are not required for the initial establishment of expression patterns of homeotic genes (genes that control patterns of body formation), but they are essential for the maintenance of the established state throughout the rest of development (reviewed in Schuettengruber et al., 2007). PcG and trxG genes are also found in mammals and play a crucial role in cell lineage specification and stem cell maintenance. In contrast, PREs and TREs have only been reported in *Drosophila*. Analysis of the regulatory function of PcGs in mice and humans showed that PcG targets in ES cells are regulators of differentiation pathways. It can be hypothesized that PcG proteins suppress cell fate-specific genes allowing ES cells to remain in a pluripotent state. It is possible that the process of cell differentiation occurs through the removal of PcG proteins from their targets and the replacement of PcG proteins with TrxG proteins; this switch allows the activation of pathways involved in cell fate decision.

In *Drosophila*, PcG proteins form two separate multi-protein complexes, the Polycomb repressive complex 1 (PRC1) and the polycomb repressive complex 2 (PRC2). The PRC2 complex is formed by four core proteins: Enhancer of Zeste (E(z)), Extra Sex Comb (ESC), p55, and Suppressor of zeste 12 (Su(z)12). In contrast, mammals have two additional PRC2-related complexes. PRC2, trimethylates histone H3 at the position of lysine 27 through its catalytically active component EZH2; the inactivation of EZH2 mediates PRC2-mediated silencing. The H3K27 methylation mark recruits PRC1, and the PRC2 chromodomain-containing components may play an important role in this process. In *Drosophila*, the ubiquitylation of histone H2A is mediated by PRC1, and this process is essential for silencing of the Hox gene. In mammals, EZH2 directly recruits DNA methyltransferases to target genes, stabilizing the robust PcG-mediated repression (reviewed in Kim et al., 2009). In cancer cells, H3K27 methylation by PcG predisposes the marked genes to aberrant silencing through *de novo* methylation (Schlesinger et al., 2007). Similar to *Drosophila*, in mammals, the PcG-mediated repression involves other mechanisms besides DNA methylation and histone modifications. PcG complexes are involved in gene silencing during X-chromosome inactivation and genomic imprinting. Because ncRNAs play a crucial role during both of the aforementioned silencing processes, it is possible that PcG proteins also interact with the small RNA machinery. Detailed information on the composition and function of PcG complexes in different organisms is provided in the previously published reviews (Schwartz and Pirrotta, 2007).

TrxG proteins that act as antagonists of PcG proteins also form several protein complexes. One of the components, MLL1, catalyzes histone H3K4 trimethylation. BPTF, a subunit of the NURF chromatin-remodeling complex, recognizes H3K4me3 and facilitates the repositioning of nucleosomes on the Hox promoter and other genes involved in developmental regulation, allowing them to be expressed (reviewed in Kim et al., 2009).

Mouse **embryonic stem cells (ESCs)** produce a complex called esBAF that contains BRG1 and BAF155 but not BRM and BAF170. The ESC-specific BAF complexes regulate transcriptional networks of mouse ESCs allowing **pluripotency**. On the contrary, no other cell types have BAF complexes that contain BAF155 but not BAF170. Additional evidence that esBAF is required for the establishment of pluripotency in the early embryo comes from the observation that deletion of either BAF47 or BAF155 is lethal to the embryo before it has implanted. esBAF complexes may also be essential for transition of ESCs from a pluripotent state to different cell lineages. Indeed, it has been shown that RNAi-mediated depletion of either BAF57 or BAF155 prevents silencing of *Nanog*, a master regulatory transcription factor, and also hinders chromatin compaction and hetero-chromatin formation during differentiation (reviewed in Ho and Crabtree, 2010) (see Figure 2-3).

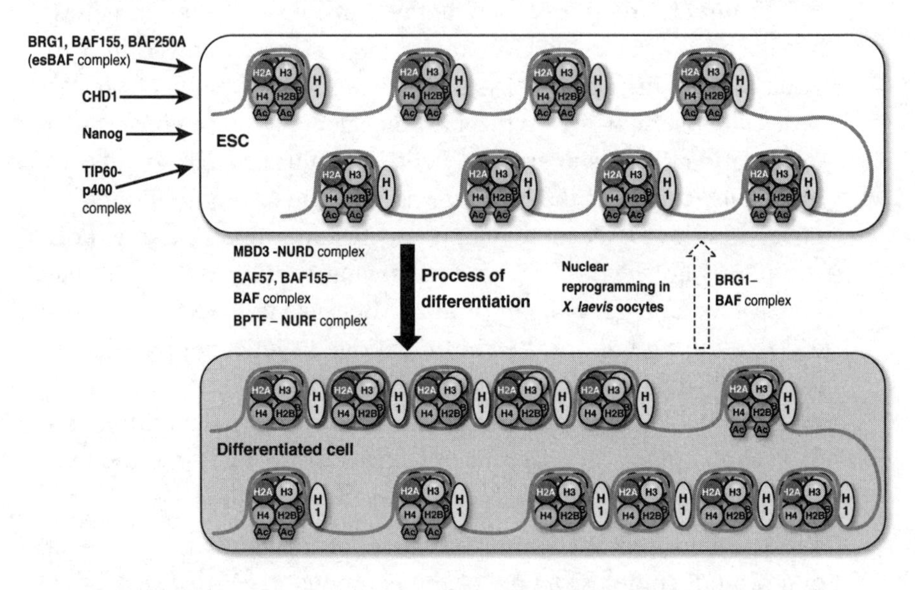

Figure 2-3 The role of chromatin-remodeling complexes in the maintenance of pluripotency.

Upon differentiation of ES cells into neuronal progenitors, several esBAF subunits are exchanged; the complex incorporates BRM and BAF60C and excludes BAF60B.

BRG1 is essential for the self-renewal of neuronal progenitors and for the normal differentiation of neurons from these progenitors (Lessard et al., 2007). Neuronal progenitors deficient in BRG1 misexpress key components of the sonic-hedgehog- and NOTCH-signaling pathways that direct neurogenesis. BAF45A, another protein of the BAF complex, is sufficient to induce the proliferation of neuronal progenitors at the point of their mitotic exit. Neuronal progenitor-specific BAF (npBAF) complexes in neuronal progenitors located in stem cell niches in the subventricular zone contain BAF45A and BAF53A proteins. If progenitor cells leave their stem cell niche and exit from mitosis, BAF45A and BAF53A proteins are replaced by

BAF45B and BAF53B, and the BAF complex changes its name to a neuron-specific BAF (nBAF) complex.

The coordination between tissue-specific BAF complexes and transcription factors is evident from studies of heart, skeletal muscle, and T-cell development. You can find more details of the role of tissue-specific BAF complexes in cell- and tissue-specific transcription in the review by Ho and Crabtree (2010).

Pluripotency of ESCs is maintained by the action of esBAF and TIP60-p400 complexes that repress inappropriate gene expression as well as by the CHD1 activity that apparently prevents chromatin compaction. Cell differentiation is triggered by silencing of pluripotency genes by BAF and NURD complexes and suppression of signaling molecules by the NURF complex, resulting in chromatin compaction. The reversal of this state can be observed in *Xenopus laevis* oocytes.

The developmental role of ISWI complexes

The ISWI family is the second well-characterized family of SWI-like ATP-dependent chromatin-remodeling complexes that were originally identified in *Drosophila melanogaster*. *Drosophila* has a single ISWI ATPase—the ISWI protein that is part of the core component of three types of ISWI complex: the **nucleosome-remodeling factor (NURF)**, the **chromatin-assembly factor (ACF)**, and the **chromatin accessibility complex (CHRAC)**. ISWI plays a critical role in cell viability and division. Loss-of-function mutations in *Iswi* are lethal during late pupal or larval development in *Drosophila*, whereas the expression of an ATPase-dead, dominant-negative allele of *Iswi* results in defects in organogenesis and at the molecular level in global decondensation of mitotic chromosomes (reviewed in Ho and Crabtree, 2010). In mutants, chromosome decondensation suggests that ISWI complexes may regulate higher-order chromatin structures. The fact that ISWI is required for incorporating linker histone protein H1 into chromatin supports this hypothesis.

In *Drosophila*, females deficient in ISWI are completely sterile due to the misregulation of bone morphogenetic proteins (BMPs). The deregulation of BMP-mediated gene expression in germinal stem cells leads to a rapid loss of self-renewal of these stem cells. The NURF complex might be the predominant among the three aforementioned complexes because deletion of *nurf301* (a component of the NURF complex) results in the development of the aforementioned phenotypes of the *Iswi* mutant.

In mice, cells contain two core ATPases of the ISWI class with non-overlapping protein expression patterns, SNF2H and SNF2L (Dirscherl and Krebs, 2004).

These ATPases are functionally distinct and can be found in different complexes. SNF2H is present in several chromatin-remodeling complexes, such as ACF, human CHRAC, **NoRC (nucleolar-remodeling complex)**, and the **WICH** complex **(WSTF (Williams–Beuren syndrome transcription factor)-ISWI chromatin remodeling)**, whereas SNF2L can be found in the NURF and **CERF (CECR2-containing remodeling factor)** complexes. It is curious that both ATPases may be involved either in transcriptional activation or repression, depending on the complex they are part of. SNF2L in the NURF complex and SNF2H in the NoRC complex are involved in transcriptional activation and repression. In contrast, SNF2H being part of the ACF, CHRAC, and WICH complexes is involved in various epigenetic processes, such as nucleosome assembly and spacing, the regulation of heterchromatin upon DNA replication and chromosome segregation (Dirscherl and Krebs, 2004).

Although there are no reports of *Snf2l*-null mice, the disruption of the bromodomain PHD-finger transcription factor (*BPTF*) gene encoding the largest subunit of mammalian NURF complexes causes embryonic lethality between embryonic days E7.5 and E8.5. Although ESCs in BPTF-null mice are viable, they are unable to give rise to mesodermal and endodermal cells. Similar to Drosophila, in mouse embryos, BPTF also interacts with transcription factors of the SMAD family and regulates BMP-mediated signalling during the establishment of germ layers in the embryo (reviewed in Ho and Crabtree, 2010). Nevertheless, it is unclear whether mammalian NURF complexes are required for proper oogenesis as they are in *Drosophila melanogaster*. In mice, ISWI proteins function beyond early embryonic development. As part of the CERF complex, SNF2L is required later in embryogenesis because deletion of the gene encoding the CERF-specific subunit CECR2 disrupts cranium formation.

In contrast to SNF2L mutants, mice deficient in SNF2H die shortly after embryonic implantation, which makes it impossible to characterize the function of SNF2H. However, the lethality of early embryos in *Snf2h*-null mice suggests that the complexes containing SNF2H are critical for embryo development. Mutations in other accessory subunits that form complexes with SNF2H are not necessarily lethal, but they cause severe developmental abnormalities. Mutations of WSTF (part of the WICH complex) results in the genetic disorder Williams–Beuren syndrome in humans. The disorder is characterized by mental retardation combined with growth deficiency and congenital cardiovascular disease. This phenotype has been confirmed in *Wstf*-haploinsufficient mice (reviewed in Ho and Crabtree, 2010).

Developmental roles of CHD complexes

The CHD family of SWI-like ATPases contains nine chromodomain-containing members broadly classified into three subfamilies based on their constituent domains: subfamily I (CHD1 and CHD2), subfamily II (CHD3 and CHD4), and subfamily III (CHD5, CHD6, CHD7, CHD8, and CHD9).

The **subfamily I member CHD1** was initially thought to be required for the activation of transcription; in humans, the tandem chromodomains of CHD1 bind H3K4me3, a well-known mark of open chromatin, and recruit post-transcriptional initiation and pre-messenger-RNA splicing factors (reviewed in Ho and Crabtree, 2010). More recent work in *Drosophila* did not support this role of CHD1 because *Chd1*-mutant *Drosophila melanogaster* zygotes are viable and display only a mild notched-wing phenotype. *Drosophila* CHD1 seems to play a far greater role in gametogenesis as *Chd1*-null females and males in *Drosophila* are sterile. It has been shown that *Chd1*-mutant females, if mated to wild-type males, lay fertilized eggs that die before hatching. Maternal CHD1 is also required for the incorporation of the histone variant H3.3 into the male pronucleus during decondensation after fertilization. After mating of *Chd1*-mutant females and normal males, the paternal genome is unable to participate in mitosis in the zygote, thus resulting in non-viable haploid embryos (Konev et al., 2007). A similar phenotype can be achieved when both genes encoding H3.3 are knocked out; viable adults are completely sterile. It is hypothesized that the chromatin remodeler CDH1 is absolutely required for the incorporation of the histone variant H3.3 into the paternal genome after fertilization (reviewed in Ho and Crabtree, 2010). Unfortunately, it is not clear whether CHD1 plays the same role in mammals as to date no *Chd1*-null mouse has been reported. Analysis of CHD1 depletion via RNAi in ESCs shows that these cells lose their hyperdynamic and euchromatic state, thus losing pluripotency (refer to Figure 2-3). The inconsistency between the phenotypes in *Drosophila* and mouse will be resolved when a *Chd1*-null mouse is developed.

In mammals, **the subfamily II members**, **CHD3 and CHD4**, are subunits of the **nucleosome-remodeling and histone deacetylase (NURD)** complexes that contain histone deacetylases (HDACs) and function as transcriptional repressors. The exact composition of NURD complexes is different in different cell types and can also vary in response to signals within a tissue. Thus, mammalian NURD complexes are similar to BAF complexes because they achieve diversity in regulatory functions through combinatorial assembly. The NURD complex typically consists of core ATPases and three main accessory subunits. The core ATPases are either CHD3 or CHD4, whereas the accessory subunits are represented by three gene families: MTA (metastasis-associated), MBD (a methyl-CpG-binding domain), and RbBP (the retinoblastoma-associated-binding protein). The MTA family is represented by three MTA proteins—MTA1, MTA2, and MTA3—and each complex contains only one protein type. Each NURD complex also contains MBD2 or MBD3 as well as RbBP4 and/or RbBP7. More information on the function of MBD proteins and the involvement of the NURD complex in other functional complexes is given below.

In humans, mutations in **CHD7, a member of the CHD subfamily III**, result in CHARGE syndrome characterized by multiple developmental abnormalities such as heart defects, severe retardation of growth and development, and genital abnormalities. A similar phenotype was achieved in *Chd7+/–* heterozygous mice. During differentiation, CHD7 is involved in transcriptional activation of tissue-specific genes. Also in *Drosophila* , the counterpart of *Chd7*, *kismet*, is required for transcription elongation and for the counteraction to PcG protein-mediated repression of transcription by recruiting the histone methyltransferases ASH1 and TRX to chromatin during development (reviewed in Ho and Crabtree, 2010). Severe developmental defects observed in patients with CHARGE syndrome suggests that in humans and Drosophila, CHD7 functions in gene transcription and cellular development may be very similar.

Developmental roles of INO80 complexes

INO80 also belongs to the family of SWI-like ATP-dependent chromatin-remodelers. In mammals, INO80 complexes contain two ATPases: INO80 and SWR1. Besides the core ATPases, INO80 and SWR1 complexes consist of several other accessory proteins. The complexes possess *in vitro* nucleosome remodeling activity and might function as transcriptional regulators (Ho and Crabtree, 2010). Another complex that is possibly involved in the INO80-related activities is the TIP60-p400 complex.

In yeast, the SWR1 complex is responsible for the replacement of the H2A histone with the histone variant Htz1 (H2AZ in mammals). In mammals, the homologs of yeast SWR1, p400, and SRCAP are involved in distinct complexes that share some common subunits. SRCAP-containing complexes mediate the ATP-dependent exchange of histone H2AZ/H2B dimers for nucleosomal H2A/H2B. This results in transcriptional activation of selected genes by chromatin remodeling. The histone variant H2AZ is conserved in eukaryotic organisms. In mice and Drosophila, it appears to be essential for their development. For example, the incorporation of H2AZ into chromatin of differentiating mouse ESCs regulates a proper lineage commitment. It has to be demonstrated whether SWR1 complexes are involved in the incorporation of H2AZ during ESC differentiation. In mammals, the TIP60-p400 complex consists of the p400 ATPase, the histone acetyltransferase TIP60, and about 16 accessory subunits. This complex is involved in transcriptional activation and DNA damage repair (reviewed in Ho and Crabtree, 2010). RNAi of the components of the TIP60-p400 complex, including TIP60 and p400, showed the premature differentiation and arrest of mouse ESCs. Knockout of one of the components of the complex, TRRAP, results in the embryo's death at the peri-implantation stage (Herceg et al., 2001).

Thus, four ATP-dependent chromatin remodelers—CHD1, NURD complexes, the TIP60-p400 complex, and the esBAF complex—seem to have non-redundant roles in pluripotency (refer to Figure 2-3), raising the question of whether each has a programmatic, specialized, non-overlapping role in maintaining the "landscape" of pluripotent chromatin.

Effector proteins

Effector proteins are proteins that read and implement modification-encoded biological messages. Chromatin modifications serve as recognition sites for the recruitment of these effector modules to specific chromatin domains (Kim et al., 2009). Effector proteins contain specific binding modules such as chromodomains, bromodomains, Tudor domains, PHD (Plant Homeodomain) domains, and others (see Table 2-1). These binding domains allow effector proteins to associate with one or more specific histone modifications or DNA methylation. Thus, certain histone modifications may either recruit or occlude effector proteins, further reinforcing a particular chromatin structure. Effector proteins such as HP1 and Polycomb group proteins may bind to two or more nucleosomes influencing large stretches of chromatin (reviewed in Kim et al., 2009). Alternatively, effector proteins were found to function as adaptors that attract additional chromatin-remodeling complexes. One such example is the HP1 protein.

Table 2-1 Effector proteins that interact with specifically modified histones

Effector protein	Binding modules	Target histone	Histone modification
HP1	Chromodomain	H3K9	Methylation
PC	Chromodomain	H3K27	Methylation
CHD1	Chromodomain	H3K4	Methylation
JMJD2A	Tudor	H3K4	Methylation
L3MBTL1	MBT	H1bK26, H4K20	Methylation
BPTF	PHD	H3K4	Methylation
ING2	PHD	H3K4	Methylation
UHRF1	SRA	H3K9	Acetylation
Rsc4	Bromodomain	H3K14	Acetylation

Table 2-1 Effector proteins that interact with specifically modified histones

Effector protein	Binding modules	Target histone	Histone modification
Bdf1	Bromodomain	H4K8	Acetylation
Taf1	Bromodomain	H4K16	Acetylation
14-3-3 protein	14-3-3	H3S10	Phosphorylation

The mammalian HP1 protein is a non-histone protein that regulates chromatin remodeling and transcription via the interaction with other histone and non-histone proteins. There are three isoforms of HP1, HP1α, HP1β, and HP1γ. Trimethylation of histone 3 at lysine 9 serves as a signal and the attachment site for HP1 binding. H3K9 is methylated by histone methyltransferases Suv39H1 and G9a, and the former one recruits HP1 to H3K9me3. Such an example can be found in HP1 binding to trimethylated H3K9 and DNMT1 (reviewed in Kim et al., 2009). The association of HP1 with SUV39H1 and/or DNMT1 promotes further H3K9 methylation, HP1 binding, and extensive DNA methylation, thus further stabilizing the repressed chromatin structure (see Figure 2-4). Histone acetylation also influences the association of HP1 with chromatin; the inhibition of histone deacetylases that results in decondensation of heterochromatin leads to reorganization of HP1 foci in the nucleus.

HP1 isoforms often co-localize with each other and with various chromatin fractions. The isoforms HP1α and HP1β co-localize in 3T3 and HeLa cells, whereas all three isoforms associate with mitotic chromosomes. Both HP1α and HP1β associate with heterochromatin, whereas HP1γ associates with either euchromatin or heterochromatin domains at a relatively similar rate. It is hypothesized that co-localization of HP1γ with euchromatin leads to silencing of euchromatic loci (reviewed in Bártová et al., 2008). Localization of the HP1 isoforms changes with cell cycle progression. During the interphase, HP1β is mainly bound to centromeric heterochromatin, whereas HP1α and HP1γ are associated with the nuclear compartment. During the metaphase of mitosis, centromers are only associated with HP1α but not with other isoforms. Upon differentiation of endodermal cells (Bártová et al., 2008), HP1 interacts with the transcriptional intermediary factor 1β (TIF1β). It is hypothesized that the TIF1β-HP1 interaction is probably required for induction of the endodermal pathway. In euchromatin of undifferentiated cells, TIF1β interacts only with the HP1β and HP1γ subtypes. In euchromatin of cells undergoing differentiation, TIF1β exclusively interacts with HP1γ, whereas in heterochromatin of these cells, it associates with HP1β. Indeed, dynamic nuclear repositioning of HP1 in differentiated cells results in colocalization of TIF1β and HP1 at centromeric regions. In contrast, the interaction of TIF1β with HP1α was not found in both undifferentiated and differentiated cells. It is hypothesized that HP1β is important for the transition of TIF1β from euchromatin to heterochromatin occurring during cell differentiation (reviewed in Bártová et al., 2008).

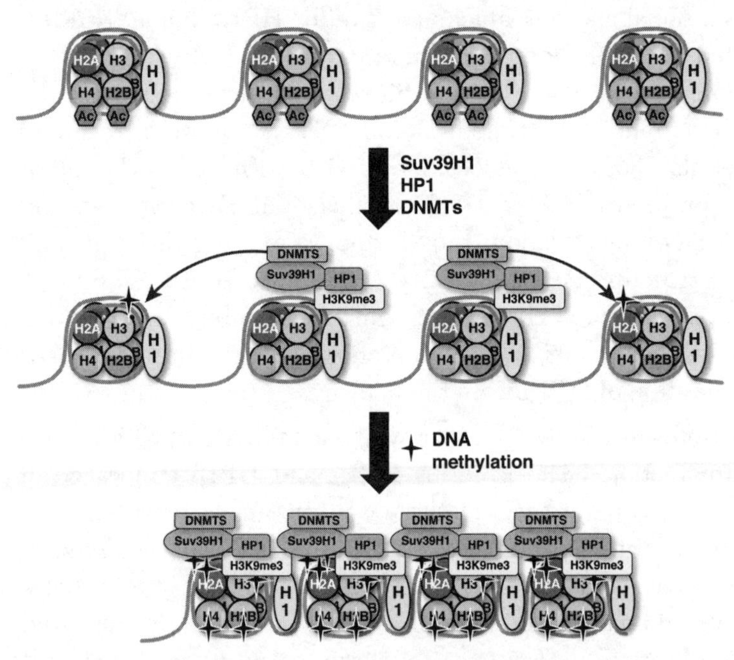

Figure 2-4 Specific histone modifications are recognized by effector proteins.

Another effector protein, BRTF, a component of the NURF chromatin-remodeling complex, recognizes trimethylated H3K4 through PHD fingers bringing the remodeler to a particular chromatin region. Other effector proteins possess enzymatic activities that also allow them to act as remodeling factors. For example, the CHD1 remodeling ATPase binds H3K4me3 and mediates subsequent recruitment of post-transcriptional initiation and pre-messenger-RNA splicing factors.

Another group of effector proteins is represented by proteins that bind to methylated DNA. Several such proteins have been identified during the past decade, including MeCP1, MeCP2, MBD1, MBD2, MBD4, and Kaiso. These proteins have one common ability: to bind methylated DNA and interpret DNA methylation marks in different biological contexts.

The MeCP1 protein complex consists of MBD2 and the chromatin-remodeling complex NURD/Mi2 containing the histone deacetylases HDAC1, HDAC2, and the histone-binding proteins RbA p46 and RbA p48, and several other proteins (reviewed in Kim et al., 2009). The MBD2 structure closely resembles MBD3, but it contains an additional 140 amino acid-long N-terminus. MBD2 plays a critical role in cell growth and is involved in the repression of several tumor suppressor genes. Also, it appears that in certain cancer lines, MBD2 is responsible for high levels of expression of human telomerase reverse transcriptase. Similarly, MBD2 was shown to demethylate promoters of SV40 and GL2T genes, which leads to their activation in cancer cells.

MeCP2 contains two main protein domains: a **methylbinding domain (MBD)** that allows MeCP2 to directly interact with DNA and a **transcriptional-repression domain (TRD)**. MeCP2 and MBD1/2 proteins have their own DNA-binding recognition sequence. Whereas MeCP2 preferentially binds the methylated CpGs flanked by a run of four A/T bases, for MBD1/2, such requirement does not exist. Also, MeCP2 is required in the maintenance of DNA methylation in the complex with DNMT1. MeCP2 has a higher affinity to methylated DNA than MeCP1 as MeCP2 can bind to a single methylated CpG, whereas MeCP1 prefers clustered CpG regions. As a result, more stable gene silencing and closed chromatin are formed if MeCP2 is involved. It is not surprising that mutations in MeCP2 are associated with various abnormalities, including disease syndromes. One of the most common causes of mental retardation in females characterized by a progressive neurological impairment is associated with mutations in MeCP2 (reviewed in Lan et al., 2010). It is speculated that the lack of expression of MeCP2 results in the overexpression of four neural development-related genes (Schwartz and Pirrotta, 2007).

MBD1 also plays an essential role in transcription regulation through binding methylated DNA. In humans, there are at least four isoforms of MBD1: MBD1v1, MBD1v2, MBD1v3, and MBD1v4

(Schlesinger et al., 2007). In mice deficient in CG-methylation, MBD1 remains to be associated with heterochromatic regions, whereas other MBD proteins do not do so. It was demonstrated that in mice, one of the major isoforms, MBD1a, contains a CXXC domain (CXXC-3) that binds specifically to non-methylated CpG, thus suggesting that MBD1 might also act as a transcription regulator in the non-methylated DNA sequence.

In contrast to other MBDs, MBD3 does not bind methylated DNA, presumably due to the insertion of two amino acids, His30 and Phe34, into its MBD domain. Nevertheless, MBD3 plays a crucial role in transcriptional regulation by interacting with the NURD/Mi2 complex. Through its MBD domain, MBD3 interacts with HDAC1 and MTA2. Although it is unable to directly bind DNA, the NURD/Mi2-MBD3 complex participates in the regulation of transcription. It is assumed that other specific DNA-binding proteins may recruit the NURD/Mi2-MBD3 complex to DNA. In mice, *Mbd3* mutation is lethal. The inner cell mass of *Mbd3*-null mice fails to develop into a mature epiblast, and therefore, it cannot develop the proper embryonic and extra-embryonic tissues after implantation. In contrast, *Mbd3*-null ESCs are viable and can initiate differentiation in culture, although they are unable to develop proper cell lineages due to inefficient silencing of pluripotency genes. The assembly of NURD complexes is severely impaired if *Mbd3* is mutated. It was also found that the overexpression of MBD3 might lead to global genome demethylation, possibly indicating a bipartite role of MBD3 in methylation and demethylation processes.

MBD4, also known as MED1, is a homolog of bacterial DNA repair glycosylases/lyases that are responsible for the removal of damaged methylated cytosines at CpG sites. Mutations in MBD4 result in high levels of mutations at the CpG sites and are frequently accompanied by malignant transformations. Thus, MBD4 may be considered to be a tumor suppressor gene (reviewed in Kim et al., 2009).

Kaiso is one of the most recently described proteins that are known to bind methylated DNA. Kaiso binds two symmetrically methylated CpG sites in a preferred sequence context CGCG. Similar to MBD1, Kaiso is also able to bind the non-methylated DNA sequence TNGCAGGA, although with a much lower affinity than it can bind the methylated regions. Being a global repressor of transcription, Kaiso is important for early embryo development (reviewed in Lan et al., 2010).

Recently, it has been shown that a multi-domain protein UHRF1 (ubiquitin-like, containing PHD and RING finger domains 1), another methyl CpG-binding protein, is required for CpG maintenance methylation at replication forks. UHRF1 harbors at least five recognizable functional domains: an ubiquitin-like domain (UBL) at the N terminus followed by a tandem Tudor domain, a plant homeodomain (PHD), a SET and RING associated (SRA) domain, and a really interesting new gene (RING) domain at the C terminus (reviewed in Hashimoto et al., 2009). Recent data demonstrated that the SRA domain of UHRF1 binds hemimethylated CpG dinucleotides and flips 5-methylcytosine out of the DNA helix, whereas its Tudor domain and PHD domain bind the tail of histone H3 in a highly methylation sensitive manner. It is suggested that the SRA domain of UHRF1 anchors the protein to hemimethylated DNA, thus recruiting DNMT1 to chromatin to facilitate the faithful inheritance of genomic methylation patterns. At the same time, the Tudor and PHD domains interact with histone silencing marks, such as H3K9me3, thus reinforcing a transcriptionally prohibitive chromatin state. UHRF1 might also be important for DNA repair/replication as reflected by the fact that murine UHRF1-null cells show an enhanced susceptibility to DNA replication arrest and DNA damaging agents.

The role of protein insulators

Insulators are DNA elements that can protect a gene from neighboring transcriptional influences to prevent inappropriate activation or repression of the gene. Insulators have two well-known functions that are represented by **enhancer blocking** and **barrier insulators** activities, so that they can either prevent distal enhancers from activating a promoter or block heterochromatin spreading that may lead to silencing of neighboring genes (see Figure 2-5). A single insulator element can control the expression of a single gene or several genes. Likewise, the expression of a single gene can be controlled by several distinct regulatory elements. Again, a functional constraint for the appropriate regulator/gene interaction needs to be achieved. In

Drosophila, five different insulator binding proteins have been identified, Zw5, BEAF-32, GAGA factor, Su(Hw) (Suppressor of Hairy-wing) and dCTCF (CCCTC-binding factor). Only dCTCF has a known conserved counterpart in vertebrates. Such difference might be explained by the fact that the *Drosophila* genome is much smaller and the distance between genes is shorter than in vertebrates. Therefore, the need for differential transcriptional control over closely located genes in *Drosophila* is much greater.

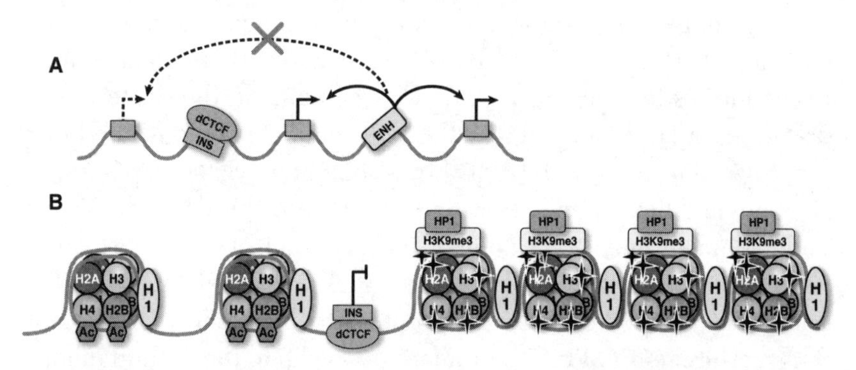

Figure 2-5 Two major functions of insulators: **A.** Preventing the regulation of promoter activity by an enhancer (ENH) element. **B.** Preventing spreading of repressive chromatin into active chromatin. dCTCF—an insulator protein commonly found in animals.

Enhancer blocking insulators are best exemplified by the *gypsy* element in *Drosophila* and the CTCF-binding sites that were originally identified in vertebrates and later on in *Drosophila* (Wallace and Felsenfeld, 2007). CTCF is able to interact both with itself and with other proteins to allow the formation of clusters. The interaction of CTCF with *gypsy* sequences leads to the formation of discrete domains. Indeed, CTCF molecules bound to distant sites, even on different chromosomes, can interact with one another *in vivo*.

Physical interactions between *cis*-regulatory elements and the possible role of CTCF in vertebrates are well demonstrated in studies of the mouse *β-globin* locus. In cells within which globin genes are expressed, the distant locus control regions (LCRs) cluster together with the developmentally appropriate promoters to form an **active chromatin hub (ACH)**. In contrast, in the progenitor cells that do not express globin genes, ACH is not formed. Three CTCF-binding sites located upstream of the mouse β-globin locus and one located downstream are in contact with one another and establish a compact domain structure that is dependent on CTCF expression. At the later developmental stages, neither CTCF nor its downstream binding site is required for the establishment of ACH and globin gene expression (reviewed in Wallace and Felsenfeld, 2007).

The mechanism that prevents enhancer-promoter interaction must be directional: Enhancer blocking occurs only if the insulator is physically present between a promoter and an enhancer. A processive model is suggested by the studies using the 5'-HS4 chicken *β-globin* insulator as a positional enhancer blocker on chromatinized episomes in human cells. The insertion of the insulator element between an enhancer and a promoter blocks the HS2 enhancer of the human *β-globin* locus control region from activating a downstream ε-*globin* gene. This leads to transcription inhibition, the accumulation of RNA pol II 5' of the insulator, and the interruption of the spreading of histone H3 and H4 acetylation (reviewed in Wallace and Felsenfeld, 2007). The inhibitory action of insulators is not absolute. Strong enhancers can overcome insulation, and the promoter targeting sequences (PTS) found in *Drosophila* can nullify insulator action and facilitate long-range enhancer-promoter interactions.

In vertebrates, a chromatin insulator protein CTCF plays an important role in several epigenetic processes including genomic imprinting, X-chromosome inactivation, transcription of non-coding RNAs at repetitive elements, and long-range chromatin interactions (reviewed in Filippova, 2008). Thus, CTCF-binding sites establish epigenetic boundaries by which correct gene expression is ensured during development and also contribute to higher-order genome organization within the nucleus.

In *Drosophila*, many genes that are expressed in a cell- and tissue-specific manner are located within the same cluster and thus require insulation. In the *Drosophila* bithorax complex (BX-C), three homeotic genes—Ultrabithorax (Ubx), Abdominal A (Abd-A), and Abdominal-B (Abd-B)—are regulated by a 300 kb-long *cis*-regulatory region. The special and temporal expression of these genes is controlled by nine distinct regulatory domains with eight chromatin boundaries between them. The unique feature of this regulatory region is the colinearity of the regulatory domains that control gene expression from the thoracic segment T3 through the abdominal segments A1–A9 (reviewed in Mohan et al., 2007). Three of the eight boundaries, Miscadastral pigmentation (Mcp), Frontabdominal-7 (Fab-7), and Frontabdominal-8 (Fab-8), located between the nine aforementioned regulatory domains have been functionally identified as chromatin insulators with enhancer-blocking activity (Maeda and Karch, 2006). Whereas dCTCF is associated with all but one of the known or predicted insulators of BX-C and was shown to be required for the function of Fab-8, the GAGA factor was shown to be required for the enhancer-blocking function of Fab-7.

2-2. Gypsy insulators

The Drosophila *gypsy* retrotransposon contains an insulator composed of 12 degenerate binding sites for the Su(Hw) protein that are separated by AT-rich DNA possessing sequence motifs common to matrix/scaffold attachment regions (MARs/SARs). *Gypsy* sequences are often located between the enhancer and promoter regions, thus preventing the activation of gene expression. Random insertions of *gypsy* retrotransposons may often result in random genetic and epigenetic mutations in the Drosophila genome. The analysis showed that a 350-bp DNA sequence of the *gypsy* insulator is recognized by a protein complex consisting of at least three components, Su(Hw), Mod(mdg4)67.2, and CP190. Whereas Su(Hw) and its interacting factor CP190 bind DNA directly via their zinc-finger domains, Mod(mdg4)-67.2 is recruited to the *gypsy* insulator sequence through physical interactions with Su(Hw) and CP190 (reviewed in Mohan et al., 2007). The Drosophila genome contains several hundred endogenous binding sites for Su(Hw). Multiple *gypsy* sites and their associated proteins cluster together to form "insulator bodies," with the ultimate effect of organizing the nearby chromatin into loop domains.

Conclusion

The exact organization of chromatin and its spatial arrangement in the nucleus in still unclear. The interchromatin compartment model (CT-IC) suggests that chromosomes are organized into chromosome territories of gene-rich and gene-poor regions, located at the nuclear interior and nuclear periphery, respectively. Various proteins, including lamins and lamin-associated proteins, SWR/SNF chromatin-remodeling proteins, and effector proteins are involved in regulation of chromatin positioning and chromatin compactness. At the same time, insulator proteins ensure that a particular chromatin structure does not spread from one region of the chromosome to another and that transcription in one region is not influenced by an enhancer from a different chromatin region. All these proteins are especially critical during the development of the organism, and two particular complexes—trxG and PcG—have antagonistic effect on chromatin structure and gene expression.

Exercises and discussion topics

1. Define chromatin remodelers.
2. Describe the models of chromosome arrangements in the nucleus.
3. What are the lamins and lamin-associated proteins? What is their role in chromosome organization?
4. What are the ARPs and what is their role in chromatin remodeling?
5. What is the difference between chromatin-remodeling complexes and effector proteins?
6. Describe the developmental role of SWR/SNF factors.
7. Explain what the TrxG and PcG proteins are and describe their antagonistic role on chromatin structure.
8. Describe the role of PREs/TREs in function of TrxG and PcG proteins.
9. Describe the structure of PRC1 and PRC2 in Drosophila. List the homologs of the proteins involved in these complexes in mammals.
10. Explain the role of chromatin-remodeling complexes in the maintenance of pluripotency.

11. Describe the developmental role of ISWI complexes in Drosophila.
12. Describe the role of ISWI homologs in mice. Explain what happens in knockout animals.
13. Describe the developmental role of the CHD family of SWI-like ATPases.
14. Describe the role of INO80 complexes in development.
15. What are the effector proteins? What is their role in chromatin maintenance?
16. Describe the role of insulator proteins as enhancer blockers and barrier insulators.

References

Bártová et al. (2008) Histone modifications and nuclear architecture: a review. *J Histochem Cytochem* 56:711-721.

Branco MR, Pombo A. (2006) Intermingling of chromosome territories in interphase suggests role in translocations and transcription-dependent associations. *PLoS Biol* 4:e138.

Clemson et al. (2006) The X chromosome is organized into a gene-rich outer rim and an internal core containing silenced nongenic sequences. *Proc Natl Acad Sci USA* 103:7688-7693.

Croft et al. (1999) Differences in the localization and morphology of chromosomes in the human nucleus. *J Cell Biol* 145:1119-1131.

Dirscherl SS, Krebs JE. (2004) Functional diversity of ISWI complexes. *Biochem Cell Biol* 82:482-489.

Filippova GN. (2008) Genetics and epigenetics of the multifunctional protein CTCF. *Curr Top Dev Biol* 80:337-360.

Hashimoto et al. (2009) UHRF1, a modular multi-domain protein, regulates replication-coupled crosstalk between DNA methylation and histone modifications. *Epigenetics* 4:8-14.

Herceg et al. (2001) Disruption of Trrap causes early embryonic lethality and defects in cell cycle progression. *Nature Genet* 29:206-211.

Ho L, Crabtree GR. (2010) Chromatin remodelling during development. *Nature* 463:474-484.

Kim et al. (2009) Epigenetic mechanisms in mammals. *Cell Mol Life Sci* 66:596-612.

Konev et al. (2007) CHD1 motor protein is required for deposition of histone variant H3.3 into chromatin in vivo. *Science* 317:1087-1090.

Lan et al. (2010) DNA methyltransferases and methyl-binding proteins of mammals. *Acta Biochim Biophys Sin (Shanghai)* 42:243-252.

Lessard et al. (2007) An essential switch in subunit composition of a chromatin remodeling complex during neural development. *Neuron* 55:201-215.

Maeda RK, Karch F (2006) The ABC of the BX-C: the bithorax complex explained. *Development* 133:1413-1422.

Meagher et al. (2009) Chapter 5. Nuclear actin-related proteins in epigenetic control. *Int Rev Cell Mol Biol* 277:157-215.

Mohan et al. (2007) The Drosophila insulator proteins CTCF and CP190 link enhancer blocking to body patterning. *EMBO J* 26:4203-4214.

Pickersgill et al. (2006) Characterization of the Drosophila melanogaster genome at the nuclear lamina. *Nat Genet* 38:1005-1014.

Scaffidi P, Misteli T (2006) Lamin A-dependent nuclear defects in human aging. *Science* 312:1059-1063.

Schirmer EC, Foisner R (2007) Proteins that associate with lamins: many faces, many functions. *Exp Cell Res* 313:2167-2179.

Schlesinger et al. (2007) Polycomb-mediated methylation on Lys27 of histone H3 pre-marks genes for de novo methylation in cancer. *Nat Genet* 39:232-236.

Schuettengruber et al. (2007) Genome regulation by Polycomb and trithorax proteins. *Cell* 128:735-745.

Schwartz YB and Pirrotta V (2007) Polycomb silencing mechanisms and the management of genomic programmes. *Nat Rev Genet* 8:9-22.

Verschure et al. (1999) Spatial relationship between transcription sites and chromosome territories. *J Cell Biol* 147:13-24.

Wallace JA, Felsenfeld G. (2007) We gather together: insulators and genome organization. *Curr Opin Genet Dev* 17:400-407.

3

Chromatin dynamics and chromatin remodeling in plants

Chromatin remodeling plays a crucial role throughout plant development, including embryonic and post-embryonic developmental processes such as seed maturation, germination, organogenesis, and flowering. Plant developmental plasticity is regulated by a number of epigenetic regulatory switches operating at the embryo stage as well as during the transition to postembryonic development, in sharp contrast to animals where epigenetic states are established early during embryonic development and cell/tissue differentiation. For this reason, plants might require a higher degree of flexibility in chromatin remodeling. Both shoot and root apical meristems (SAM and RAM) contain groups of pluripotent stem cells, which help maintain continuous growth and development of the aerial and underground structures of the plant as well as allowing the formation of gametic cells for plant reproduction. Thus, epigenetic regulation of the fate of the SAM cells is critically important for plant development.

Epigenetic regulation of seed development

Seed development is controlled by a complex function of positive and negative regulators of gene expression. During dormancy the majority of genes are repressed, whereas germination activates a specific set of genes. In Arabidopsis, the genes that function as positive regulators of the seeds development include *ABSCISIC ACID INSENSITIVE3* (*ABI3*), *BABY BOOM* (*BBM*), *LEAFY COTYLEDON* (*LEC*), *AGAMOUSLIKE15* (*AGL15*), *WUSCHEL* (*WUS*), *SOMATIC EMBRYOGENESIS RECEPTOR-LIKE KINASE* (*SERK*), and several others (reviewed in Ahmad et al., 2010). Three groups of proteins repress the seed transcriptome:

- Histone modifiers such as PcG and HISTONE DEACETYLASE6/19 (HDA6/HDA19)
- Chromatin remodelers BRAHMA (AtBRM1) and PICKLE (PKL)
- Transcription factors VP1/ABSCISIC ACID INSENSITIVE 3-LIKE (VAL) (Zhang and Ogas, 2009) (see Figure 3-1)

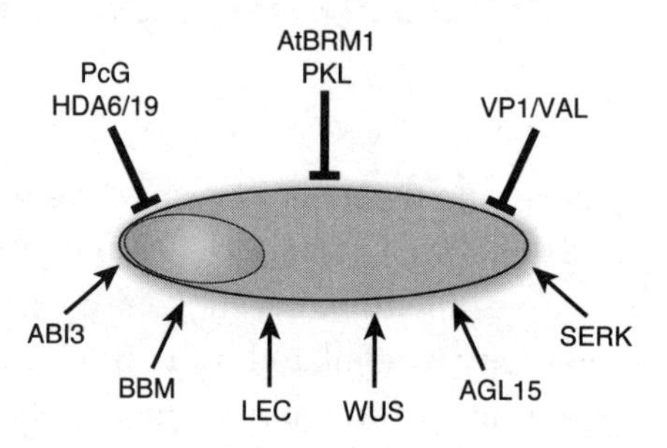

Figure 3-1 Positive and negative regulation of transcription in seeds.

The exact mechanism of silencing genes involved in seed germination and the role of PKL in deposition of H3K27me3 are still under investigation. Many questions remain unanswered, including the following:

- How is the restricted expression of seed genes released for seeds to germinate?
- What role does PcG play in the repressive process?
- How is timing of the events controlled during seed development?

Chromatin remodeling in shoot and root apical meristem

In the SAM, stem cells occupy the central region of the meristem, and their pluripotency is maintained through signals generated by cells of the organizing center (cells that surround the meristem cells and prevent their differentiation). How are the stem cells simultaneously maintained in a pluripotent state and able to generate progeny in which differentiation pathways are rapidly activated? Mutations at various genes encoding chromatin factors that maintain certain epigenetic marks through cell division alter expression patterns of meristem-specific genes and affect meristem size and organization, suggesting that proper chromatin structure is critically important for regulation of this dynamic process (reviewed in Jarillo et al., 2009). For example, mutants deficient for Chromatin-Assembly Factor-1 (CAF1), a histone-binding protein complex involved in *de novo* nucleosome assembly during DNA replication, are impaired in meristem organization (Ramírez-Parra and Gutiérrez, 2007). In Arabidopsis, CAF1 consists of three protein subunits encoded by FASCIATA1 (FAS1), FAS2, and MULTICOPY SUPPRESSOR OF IRA1 (MSI1). CAF1 is thought to be responsible for controlling stable inheritance, as well as maintenance, of epigenetic states during DNA replication.

Mutation in *FAS1*, a gene encoding one of the subunits of CAF1, results in spontaneous release of **transcriptional gene silencing (TGS)**, overexpression of genes located in heterochromatic genomic regions, and alteration in the expression of genes such as homeobox gene *WUSCHEL (WUS)* primarily expressed in cells forming the organizing center (see Figure 3-2; reviewed in Jarillo et al., 2009). *WUS* expression is normally suppressed by CAF1, and in *fas* mutants *WUS* transcription occurs ectopically in the SAM. Consequently, *fas* mutants display increased proliferation and fasciation in SAM and aberrant organization of the SAM. Curiously, mutations in the Arabidopsis gene *BRUSHY1 (BRU1)*, also known as *MGOUN3 (MGO3)/TONSOKU (TSK)*, result in phenotypes resembling *fas1*. Exact function of BRU1 is unclear, but it might also be involved in chromatin organization as BRU1 and FAS1 have common genomic targets.

Organization and maintenance of RAM requires proteins encoded by Arabidopsis homologues of the Drosophila *NUCLEOSOME ASSEMBLY PROTEIN1 (NAP1)* gene: *NAP1-RELATED PROTEIN1* and *2* (*NRP1* and *NRP2*). NRP1 and NPR2 bind histones H2A and H2B and associate with chromatin *in vivo*. Double mutants (*nrp1/nrp2*) are impaired in root growth and elongation due to the arrest of cell cycle progression at G2/M and disordered cellular organization in root tips. Similar to *fas1* mutant, *nrp1/nrp2* also exhibit release of TGS and altered expression of genes involved in root proliferation and patterning.

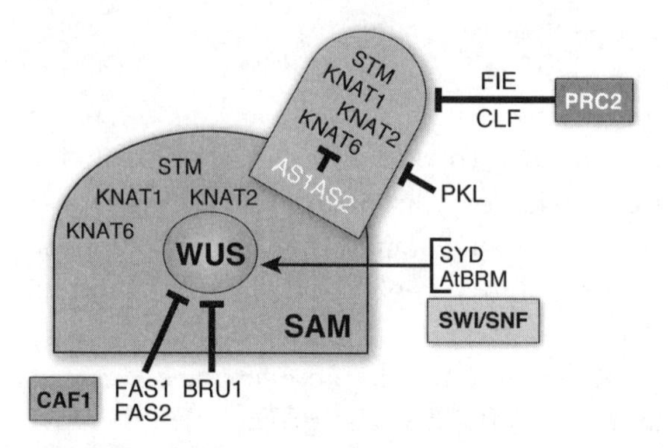

Figure 3-2 Regulation of expression of meristem-specific genes (modified with permission from Jarillo et al. (2009) Chromatin remodeling in plant development Int. J. Dev. Biol. 53: 1581-1596). The large knob is SAM; the small knob is newly formed primordia. See text for details.

Whereas the CAF1 complex restricts expression of *WUS* to the organizing center, SWI/SNF ATP-dependent chromatin-remodeling proteins ensure appropriate levels of *WUS* transcript in the other tissues. The SWI/SNF protein encoded by *SPLAYED (SYD)* binds the *WUS* promoter and controls its transcription thus directly affecting number of cells in SAM (refer to Figure 3-2). Plants impaired in *SYD* exhibit decreased *WUS* expression and premature termination of the SAM (Wagner and Meyerowitz, 2002). The SWI/SNF chromatin-remodeling ATPase, BRAHMA (AtBRM) might also be required for SAM maintenance. AtBRM is involved in the formation and/or maintenance of boundary cells during embryogenesis as well as being expressed in meristems and proliferating tissue. *AtBRM* mutants show a decrease in inflorescence meristem size and number of flowers (refer to Figure 3-2; reviewed in Jarillo et al., 2009). These observations confirm the importance of chromatin remodeling in the regulation of meristem activity, including maintenance of the pluripotent state and initiation of cell division and differentiation.

Role of PcG and trxG complexes in regulation of plant organ development

Control over the cell division and differentiation at the peripheral zone of the meristem also occurs with the help of chromatin-remodeling complexes. To achieve sequential production of lateral organs, it is important to keep meristem genes stably repressed outside the SAM. Also, the correct identity of developing organs must be maintained by the balance of expression of genes required for differentiation in developing tissues and organs and repression of these genes in the tissues where they have to be silenced. PcG and trxG complexes are essential components in the regulation of repressive and activating chromatin states, respectively (Pien and Grossniklaus, 2007). PcG proteins act as memory factors and stabilize the repressive states of homeotic gene expression whereas TrxG proteins establish transcriptionally active states of differentiation genes. SWI/SNF chromatin remodelers also play key roles in controlling expression of genes involved in organ initiation and cell differentiation.

Establishment and maintenance of repressive states by PRC complexes

Plants have multiple PRC2 complexes that regulate diverse developmental programs. For example, in Arabidopsis the *ESC* homologue is encoded by a single copy gene *FERTILIZATION INDEPENDENT ENDOSPERM (FIE)*, whereas the other PRC2 subunits have more than one homologue: EMBRYONIC FLOWER 2 (EMF2), FERTILIZATION-INDEPENDENT SEED2 (FIS2), VERNALIZATION 2 (VRN2), and VEFL36 are homologues of Su(z)12; Drosophila E(z) protein also has three homologues in Arabidopsis—MEDEA (MEA), CURLY LEAF (CLF), and SWINGER (SWN)—responsible for the HMTase activity of PRC2 complexes. Finally, the p55 homologue is encoded by *MSI1*, and although the *MSI1* gene family consists of five members, only MSI1 has been suggested as part of the PRC2 complex (reviewed in Jarillo et al., 2009). VRN2, EMF2, and FIS2 have been implicated in vernalization-mediated flowering, vegetative development, and seed development, respectively (reviewed in Ahmad et al., 2010).

During organ initiation in the newly formed primordia, two PRC2 subunits, CLF and FIE, maintain the repressed state of the meristem-specific genes *STM*, *KNAT1*, *KNAT2*, and *KNAT6* (refer to Figure 3-2). However, in *clf* mutants only *STM* and *KNAT2* are derepressed, whereas in knockdown *fie* mutants, all four genes are ectopically expressed in leaves (reviewed in Jarillo et al., 2009). CLF is able to bind to the *STM* locus, thereby promoting H3K27 trimethylation of the *STM* gene. CLF and its homologue SWN also methylate histone 3 to repress the floral homeotic gene AGAMOUS (AG) in vegetative tissues. Partial depletion of FIE activity showed that FIE is also involved in the maintenance of root identity and in the repression of floral homeotic genes such as AG, APETALA3 (AP3), and PISTILLATA (PI) (reviewed in Jarillo et al., 2009).

FERTILIZATION INDEPENDENT SEED (*FIS*) genes are another PRC2 complex required for regulation of gametophyte, embryo, and endosperm developing and include *MEA*, *FIE*, *MSI1*, and *FIS2*. This PRC2 complex is referred to as PRC2-MEA complex to distinguish it from other PRC2 complexes. The main known function of PRC2-MEA complex is repression of genes involved in early stages of endosperm development, such as *PHERES1* (*PHE1*), which is a MADS box gene (mostly transcription factors) (Pien and Grossniklaus, 2007). However, both aforementioned PRC2 complexes might be involved in *PHE1* repression, as besides MEA-dependent repression, CLF and SWN-dependent repression of *PHE1* was also shown. Various PRCs complexes act together to provide repressive H3K27me3 and H3K27me2 marks to specific homeobox genes (Pien and Grossniklaus, 2007). Although the subunit composition of these complexes might vary, *FIE* appears to be present in all of them.

Establishment of repressive states depends on PRC2, whereas maintenance of these states requires PRC1 in animals. However, in Arabidopsis no homologues of PRC1 complex have been identified, suggesting that plants either use different mechanisms of maintenance of silent states or that the mutants of such homologues may be lethal. The role of PRC1 in maintenance of heterochromatic states may be performed by LIKE HETEROCHROMATIN PROTEIN 1 (LHP1)/TERMINAL FLOWER 2 (TFL2), a plant homologue of HP1 (Turck et al., 2007). In animals, HP1 binds methylated H3K9 and associates with constitutive heterochromatin. However, LHP1 is not involved in the maintenance of constitutive heterochromatin in Arabidopsis but instead regulates expression of euchromatic genes, such as *FLC* (reviewed in Jarillo et al., 2009). The role in PRC1-mediated silencing was suggested for LHP1 because of its capacity to specifically associate with genes carrying the H3K27me3 mark (Turck et al., 2007).

PcG response elements (PREs) recruit PRC1 and PRC2 to specific genomic areas in animals (see Chapter 2, "Chromatin Dynamics and Chromatin Remodeling in Animals"). PREs have not been identified in *Arabidopsis*, so it is unclear how PRCs target specific genes. However, it is possible that this specificity is caused by PcG-interacting proteins. In human cells, the tumor suppressor Retinoblastoma protein (Rb) is required for maintenance of H3K27 methylation and for the ability of PRC2 and PRC1 to bind the promoter of the CDK inhibitor p16. The homologue of human Rb in plants—the RETINOBLASTOMA-RELATED (RBR) protein—interacts with the components of PRC2, FIE, and MSI1, suggesting that RBR may recruit PRC2 to the promoters of the target genes (reviewed in Desvoyes et al., 2010).

Establishment of active chromatin states

Similarly to animals, developmental stages in Arabidopsis are promoted by antagonistic action of TrxG homologues against PcG homologues (Pien and Grossniklaus, 2007). TrxG homologue identified in *Arabidopsis* is TRITHORAX 1 (ATX-1), which contains a SET domain and exhibits H3K4 methyl transferase activity *in vitro* (Alvarez-Venegas et al., 2003). Decreased levels of genomic H3K4 methylation *atx1* mutants provide further evidence ATX1 may be responsible for methylation of H3K4, although other TrxG proteins might also be involved. ATX1 might also be required for activation of floral homeotic genes repressed by PcG proteins such as FLC, AG, AP3, and PI (Alvarez-Venegas et al., 2003). Based on limited information, it has been suggested that plant Trx proteins might antagonize the repressor activity of PcG complexes. No other Trx homologues are known in plants. No Polycomb or Trithorax Responsive Elements (PRE/TREs) have been identified in plants, and therefore it is unclear how trxG and PcG complexes are directed to target genes.

Epigenetic control of flowering

Natural populations of Arabidopsis can be broadly divided into summer and winter annuals. Whereas summer annuals germinate in the spring and flower in spring or summer, winter annuals germinate in the fall and flower in spring. Flowering of winter annuals requires vernalization—a prolonged exposure to cold. On the molecular level, FRIGIDA (FRI) promotes high levels of expression of the floral repressor gene *FLC*, coding for a MADS-box protein that in turn inhibits the expression of the floral integrators *FLOWERING LOCUS T* (*FT*) and *SUPPRESSOR OF OVEREXPRESSION OF CO* (*SOC1*), whose expression levels determine the exact time of flowering. Vernalization causes repression of *FLC* expression and the release of repression on *FT* and *SOC1*; this state is mitotically stable and thus maintained for the rest of the plant life cycle. Activation and repression of *FLC* expression is regulated through a number of chromatin-remodeling processes (reviewed in Schmitz and Amasino, 2007).

Repression of *FT* requires a PRC2-like complex containing CLF, EMF2, and FIE, as well as two additional remodeling factors LHP1/TFL2 and EARLY BOLTING IN SHORT DAYS (EBS). Mutations in genes encoding any of these five proteins result in early flowering and FT upregulation, especially under short day conditions, suggesting they establish non-active chromatin conformation at *FT* chromatin under short day photoperiods (reviewed in Imaizumi and Kay, 2006). In addition to targeting the promoter and transcribed regions of *FT*, LHP1 represses the genes involved in vegetative expression of floral organ identity. EBS is a plant-specific protein containing a PHD Zn finger and a BAH (Bromo Adjacent Homology) domains, which are characteristics of chromatin-remodeling factors. Like LHP1, the role of EBS is not limited to floral transition and includes regulation of the expression of floral organ identity genes and the repression of germination during the dormancy period.

Floral integrator *SOC1* is regulated through the activity of one of the components of PCR2 complex, MSI1. Reduction of MSI1 levels results in late flowering and delayed induction of *SOC1* expression. Activation of *SOC1* by MSI1 does not require FLC; it correlates with increased levels of permissive chromatin marks such as H3K4me2 and H3K9ac in the *SOC1* chromatin.

Establishment of winter-annual state

Arabidopsis winter-annuals have a high *FLC* expression level that blocks flowering before winter. Resetting the *FLC* switch is necessary for perpetuation of the vernalization requirement to successive generations (reviewed in Schmitz and Amasino, 2007). Vernalization suppresses the expression of *FLC* through several histone modifications of the local chromatin, including deacetylation of histone H3 at the position of lysine 9 (K9), followed by methylation of histone H3 at lysine 9 (K9) and lysine 27 (K27). H3K9 deacetylation requires *VERNALIZATION INSENSITIVE 3* (*VIN3*), which encodes a PHD-domain protein. H3K9 and H3K27 methylations are dependent on the activity of *VIN3* and *VERNALIZATION 1* (*VRN1*), which encodes Myb-related DNA-binding protein and *VRN2*, which encodes a Polycomb group protein (Sung and Amasino, 2004; Bastow et al., 2004).

The mechanism of activation of *FLC* in winter annuals was deciphered by studying early flowering mutants, which exhibited low levels of *FLC* expression. Identified mutants were also deficient in genes encoding proteins belonging to the putative chromatin-remodeling complexes, the Polymerase Associated Factor 1 (PAF1) and SWR1 (Figure 3-3A).

Components of the PAF1 complex are encoded by genes such as *vernalization independent 4* (*vip4*), early flowering 7 (*elf7*), and *elf8* (He et al., 2004; Oh et al., 2004, 2008). In yeast, PAF1 complex interacts with SET1 HMTase, which methylates H3 at lysine 4 and SET2 HMTases responsible for methylation of H3K36 (Krogan et al., 2003; Ng et al., 2003). Mutation in any PAF1 complex homologues in Arabidopsis cause complete suppression of *FLC*, and thus mutants exhibit early flowering in winter annuals (Oh et al., 2004; He et al., 2004). Consequently, there is a correlation between *FLC* expression and high levels of H3K4 methylation surrounding *FLC* chromatin (He et al., 2004; Kim et al., 2005; Martin-Trillo et al., 2006). Evidence exists that SUPPRESSOR OF FRIGIDA 4 (SUF4) is an intermediate protein responsible for recruiting the PAF1 complex to *FLC* chromatin (see Figure 3-3A) (Kim and Michaels, 2006; Kim et al., 2006).

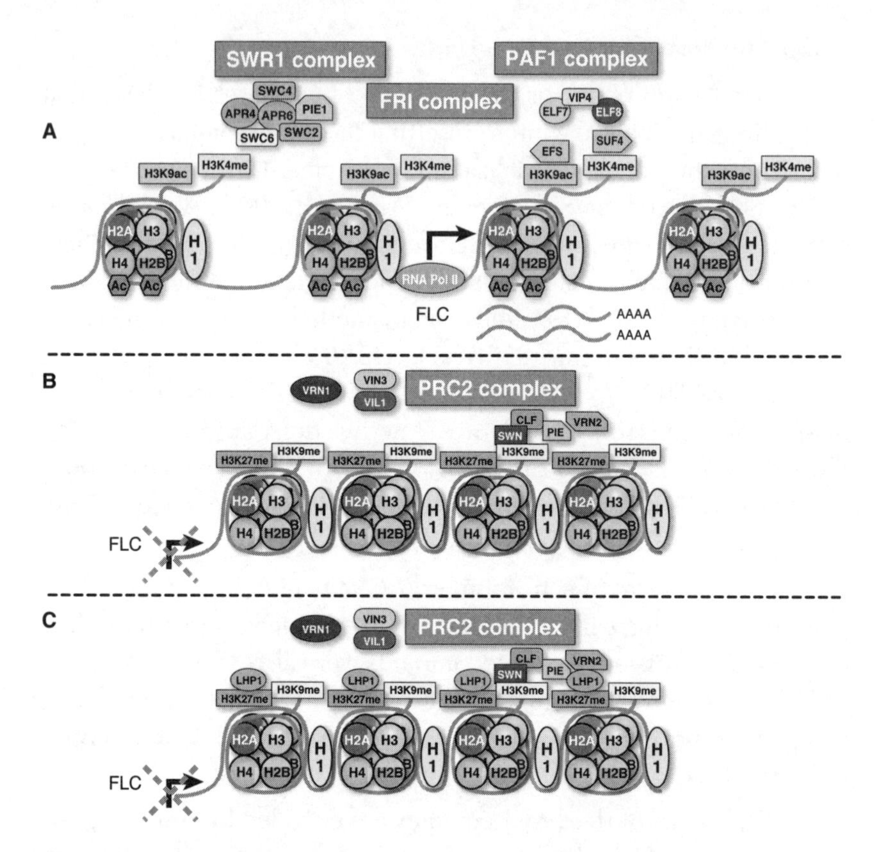

Figure 3-3 Chromatin regulation of FLC expression (modified with permission from Jarillo et al. (2009) Chromatin remodeling in plant development Int. J. Dev. Biol. 53: 1581-1596). **A.** Active chromatin state at FLC locus allows the establishment of the winter-annual habit of Arabidopsis. SWR1, FRI, and PAF complexes are required to maintain FLC chromatin activated. **B.** Exposure to cold during the winter season establishes the repressed chromatin state. **C.** Vernalization-induced stably silenced chromatin state is established through LHP1 binding to H3K27me3.

Another mutant that shows early flowering and reduced levels of *FLC* expression is SET domain-histone methyltransferase EARLY FLOWERING IN SHORT DAYS (EFS/SDG8) (Kim et al., 2005; Zhao et al., 2005). It is hypothesized that EFS works with PAF complex to regulate *FLC* expression. EFS/SDG8 H3 methyltransferase shows homology to the *Drosophila* TrxG protein Absent, Small and Homeotic Discs 1 (ASH1). EFS/SDG8 antagonizes the repressive function of PcG and thus it is suggested that this methyltransferase performs the function of TrxG proteins in plants. Indeed, high levels of H3K4me3 or H3K36me2 in the *FLC* region requires functional EFS/SDG8 (Kim et al., 2005; Zhao et al., 2005).

Putative homologues of the yeast SWR1 ATP-dependent chromatin remodeling complex EARLY IN SHORT DAYS1/SUF3/ACTIN RELATED PROTEIN 6 (ESD1/SUF3/ARP6), PHOTOPERIOD INDEPENDENT EARLY FLOWERING 1 (PIE1), ACTIN-RELATED PROTEIN 4 (AtARP4), and AtSWC6/SEF1 (SERRATED AND EARLY FLOWERING 1) are also involved in the regulation of flowering time (reviewed in Jarillo et al., 2009). In yeast, the SWR1 complex catalyzes replacement of H2A histone by the H2A.Z variant, allowing activation of the genes involved in transcriptional regulation, heterochromatic barriers and genome stability. Similarly to yeasts, plants possess H2A.Z histone variants; in fact, three homologues of H2AZ—HTA8 (At2g38810), HTA9 (At1g52740), and HTA11 (At3g54560)—are present in Arabidopsis (Yi et al., 2006). H2A.Z interacts with both PIE1 and AtSWC2. Loss of H2A.Z from *FLC* chromatin, as observed in *esd1/suf3/arp6* and *pie1* mutants, leads to reduced *FLC* expression and premature flowering (Deal et al., 2007).

These observations suggest plants, like yeasts, maintain activity of a SWR1-like complex that is targeted to different loci, including *FLC*, and that the presence of the H2A.Z histone variant in the chromatin of specific genes may enhance their transcriptional activation (Deal et al., 2007). More details about the role of histone variants in epigenetic regulation are covered in Chapter 5, "Histone Modifications and Their Role in Epigenetic Regulation."

FLC repression, the vernalized state

Long exposure to cold during the winter season results in severe repression of *FLC* expression, a state that is maintained for the entire plant life cycle, even when cold exposure ends (Sung and Amasino, 2006). Arabidopsis vernalization causes massive changes in various histone modifications at the *FLC* locus (refer to Figure 3-3B). These modifications include a decrease in permissive chromatin marks, such as histone H3 and H4 acetylation and H3K4me (Sung and Amasino, 2004; Sung et al., 2006a) as well as an increase in repressive marks such as H3K9 and H3K27 di and trimethylation (Bastow et al., 2004; Sung and Amasino, 2004; Sung et al., 2006a).

Acceleration of flowering after the long cold exposure depends on the PHD-containing protein VIN3, which is needed to deacetylate histones at the *FLC* chromatin (Sung and Amasino, 2004). The expression domain of *VIN3* overlaps with that of *FLC*, and *VIN3* mRNA levels accumulate only when plants are exposed to cold for a long time. Deacetylation of chromatin at *FLC* locus is probably the first step in vernalization-dependent changes at the locus, which might facilitate establishment of histone methylation marks. Indeed, the *vin3* mutant does not carry any of the repressive marks associated with vernalization.

Histone methylation at *FLC* loci depends on both VRN2 and VRN1. The VRN2-PRC2 complex is responsible for H3K27 methylation of the *FLC* locus (Sung and Amasino, 2004). However, the activity of this complex might not be sufficient for stable *FLC* repression after vernalization, and VRN1-dependent methylation of H3K9 might also play an important role. *vrn1* mutants do not maintain stable *FLC* repression and lack H3K9 methylation but retain H3K27me3 in the *FLC* locus. It has been suggested that VRN1 methylates H3K9 following methylation of H3K27 by VRN2-complex (Sung and Amasino, 2004).

Establishment of repressive chromatin marks requires parallel removal of permissive chromatin marks, for example H3K4me3. Although the exact details are unclear, it is thought that the PHD domain of VIN3 can bind trimethylated H3K4 and recruit chromatin-remodeling complexes to the *FLC* locus.

A VIN3-interacting protein, VIN3-LIKE1/VERNALIZATION 5 (VIL1)/VRN5, is also required to establish or maintain proper vernalization status (refer to Figure 3-3B). *vil1* plants exhibit a vernalization-insensitive phenotype and lack repressive histone modifications at the *FLC* locus. VIN3 together with VIL1/VRN5, VIL2/VEL1, VIL3/VEL2, and VIL4/VEL3 are a small family of PHD finger proteins in Arabidopsis.

VIN3, VIL1, and VIL2/VEL1 are a part of the VRN2-PRC2 complex involved in the initial steps of vernalization-mediated *FLC* repression. The number of proteins in the complex increases during cold exposure and decreases when plants are returned to warm temperatures. Association of some proteins, such as VRN2 with FLC chromatin, does not require exposure to cold, whereas recruitment of other proteins, such as VIL1, depends on cold. Recruitment of VIL1 to *FLC* chromatin also requires VIN3, suggesting that the VRN2-VIN3 complex may be necessary to direct histone deacetylase and histone methyltransferase activities to *FLC* chromatin.

The process of vernalization is a two-step process where the first step includes initial downregulation of *FLC* and the second step is the maintenance of this repressed state by a Polycomb complex. Initial repression and maintenance of *FLC* chromatin might target different regions of *FLC* locus. The deletion of a **vernalization response element (VRE)**, a portion of *FLC* intron 1 required for maintenance of the repressed state, does not prevent the initial repression of *FLC* by cold from being established. Therefore, it is not clear which *FLC* locus regions are required for the initiation of repression in response to cold.

After returning to warm temperatures, the epigenetically repressed state of *FLC* is maintained by LHP1 through conservation of increased H3K9 methylation at the VRE region of *FLC* chromatin (refer to Figure 3-3C). At the same time, LHP1 is not required for initiation of this methylation during cold exposure, suggesting there are other histone-methylating components besides VRN2-PRC2 that are responsible for vernalization.

Regulation of FLC expression in summer annuals

In summer annuals, lack of FRI function leads to *FLC* silencing. Instead, the *FLC* locus is repressed by the activity of chromatin-remodeling factors FVE (AtMSI4), FLOWERING LOCUS D (FLD), and RELATIVE OF EARLY FLOWERING 6 (REF6), the components of the autonomous floral promotion pathway, involved in flowering control and regulated by photoperiod and vernalization (reviewed in Schmitz and Amasino, 2007). Human homologues of these proteins are part of histone deacetylase (HDAC) complexes. *FVE* encodes a homologue of yeast MSI and mammalian retinoblastoma-associated proteins RbAp46 (also known as Rbbp7) and RbAp48 (Rbbp4). The human FLD homologue, KIAA0601/Lysine Demethylase 1 (LSD1), is a polyamine oxidase (PAO) that can also demethylate H3K4. REF6 is the protein belonging to jumonji-family; the yeast and mammalian homologues of this family function as histone demethylases. *fve* and *fld* mutants exhibit hyperacetylation of histones H3 and H4 and increased expression of *FLC*, confirming the possible involvement of FVE and FLD in histone deacetylation. REF6 also likely plays a role in histone deacetylation, as supported by delayed flowering in *ref6* mutants, although this late flowering is suppressed by *flc* mutations.

Other proteins regulating FLC expression

Several other components of the autonomous regulatory pathway involved in chromatin remodeling of the *FLC* locus are known. One such factor is LSD1-like (LDL1)/AtSWP1, a homologue of FLD and a member of the Arabidopsis family of PAO/LSD1 proteins. AtSWP1 interacts with the plant-specific C2H2 zinc finger-SET domain protein AtCZS, an HMTase. Together, AtCZS and AtSWP1 use histone hypoacetylation and the generation of heterochromatic histone methylation marks to repress target genes. Both *czs* and *swp1* mutants exhibit hyperacetylation of histone H4 and reduced levels of H3K9 and H3K27 methylation at the *FLC* chromatin, in conjunction with increased *FLC* expression and late flowering. Another protein, LDL1 homologue LDL2, also reduces H3K4 methylation levels in the genomic regions of *FLC* and *FWA*.

A final chromatin-remodeling factor potentially involved in regulating *FLC* chromatin status is SHK1 BINDING PROTEIN 1(SKB1), a homologue of a human arginine methylase PRMT5 responsible for symmetric dimethylation of H4R3 (H4R3sme2). SKB1 binds to the *FLC* promoter and thus loss-of-function *skb1* mutants show decreased H4R3sme2 in regulatory regions of *FLC*, resulting in upregulation of *FLC* and late flowering. This suggests that the H4R3sme2 mark might also be required for *FLC* repression and the induction of flowering. Thus, LDL1, LDL2, AtCZS, and SKB1 are additional components of plant-specific repressor complex involved in regulation of chromatin structure around *FLC* (Jarillo et al., 2009).

Regulation of FLC expression by small RNAs

Expression of *FLC* is also controlled by small RNAs. Swiezewski et al. (2007) reported identification of 24- and 30-nucleotide small RNAs complementary to the sense strand of *FLC* just 3' to the major poly(A) site. The authors characterized accumulation of these siRNAs and showed the presence of a corresponding antisense transcript whose levels were reduced in mutants defective in polymerase IV. Moreover, the corresponding genomic region of *FLC* was enriched for histone 3 lysine 9 dimethylation (H3K9me2) and this accumulation was dependent on function of DCL proteins as a triple mutant—*dcl2,3,4*—had reduced levels of H3K9me2. T-DNA insertion into this *FLC* genomic region resulted in altered expression of *FLC* and delayed flowering, suggesting its functional importance in *FLC* regulation. The authors propose that through the activity of NRPD1a, small RNAs are derived from the *FLC* 3' region, generating an antisense transcript that becomes a target for DCL3. They suggest that siRNAs of different sizes are produced, among which the 24- and 30-nt siRNAs are the most stable. It is possible that these small RNAs recruit chromatin complexes to specific *FLC* regions leading to histone methylation at nucleosome(s) positioned downstream from poly(A) site. Repressive chromatin marks result in reduced expression of *FLC*, which might be modulated through different mechanisms, including reduced transcription and improper polyadenylation. The latter might result in production of aberrant transcripts that become a target of DCL3, further reinforcing silencing at the *FLC* 3' end.

Regulation of the expression of floral activator *AGAMOUS-LIKE19*

AGAMOUS-LIKE19 (*AGL19*) is a gene that encodes MADS-box protein that acts to promote flowering in response to prolonged cold exposure, a characteristic that was recently shown to be controlled by chromatin-remodeling processes that are mediated by PcG proteins. In contrast to *FLC*-mediated repression of flowering, so far this poorly understood pathway is not known to require SOC1 to promote the floral transition. In the absence of cold, *AGL19* expression is maintained at a very low level, and its chromatin is heavily enriched in H3K27me3 repressive marks. Plant PRC2 polycomb proteins such as MSI1, CLF, and EMF2, but not VRN2, appear to be necessary for the high level of H3K27me3 as well as *AGL19* repression in the absence of cold. Following vernalization, H3K27me3 decreases in the promoter and ATG regions of *AGL19*, reducing repression via a mechanism that requires VIN3. Induction of *AGL19* results in activation of the floral meristem identity genes *LEAFY* (*LFY*) and *APETALA1* (*AP1*) and subsequent flowering.

Snf2 family of chromatin-remodeling factors

As of yet, no actual Snf2 complex has been isolated in plants, although the Arabidopsis genome contains 41 Snf2-like genes belonging to at least six subfamilies (http://www.chromdb.org).

Photoperiod-Independent Early flowering 1 (PIE1)

Photoperiod-Independent Early flowering 1 (PIE1) belongs to the Swr1-like group of the Swr1 subfamily of Snf2 proteins. Proteins belonging to this group in yeast and human, Swr1 and SRCAP, respectively, form conserved complexes consisting of more than 10 subunits. In yeast and human, the main function of these complexes is the regulation of transcription by replacing the canonical H2A histone with the H2A.Z histone variant.

PIE1 is the main catalytic component of the Arabidopsis Swr1-like complex, which also includes Serrated leaves and Early Flowering (SEF) and Actin-Related Protein 6 (ARP6) proteins as well as H2A.Z variant (Deal et al., 2007). PIE1 is the main interacting component for the H2A.Z histone; a *PIE1* null mutation (*pie1-5*) causes disruption of the normal deposition of H2A.Z at several distantly positioned loci. H2A.Z-dependent regulation of transcription might be PIE1's only activity as transcriptome of *hta9-1/hta11-2* plants deficient for two genes coding for H2A.Z proteins and transcriptome of *pie1-5* plants overlap by over 65% (Deal et al., 2007).

PIE1 is actively involved in salicylic acid (SA)-dependent pathogen response. Genes upregulated in response to the synthetic SA analogue benzothiadiazole (BTH) are dramatically enriched in *pie1-5*—nearly 40% of total number of genes with altered expression found in *pie1-5* mutant are also found to be altered by BTH application in the wild type plants. Moreover, *pie1-5* plants, and to a lesser extent *sef* and *hta9/hta11* plants, are more tolerant to a virulent *P. syringae* strain and cells of infected plants exhibit higher incidences of spontaneous cell death. It was proposed that PIE1 establishes a negative control over the SA pathway at normal growth conditions and that this activity involves SEF and/or H2A.Z (March-Diaz et al., 2008). However, the exact mechanism of this process is unknown.

In yeast, the Swr1 complex is targeted to specific DNA motifs within a given gene as well as to specific patterns of histone acetylation. DNA-binding factor Reb1 and Swr1 subunit Bdf1 might facilitate the binding to the acetylated tails of H3 and H4.

The importance of Swr1 components for proper regulation of *FLC* locus is shown by the fact that Arabidopsis plants deficient for Swr1 components have a reduced level of H2A.Z histone acetylation and H3K4 trimethylation at the *FLC* locus (Deal et al., 2007).

A model was proposed suggesting that Swr1 complex mediates negative control of SA-sensitive genes (see Figure 3-4). The components of the complex—PIE and SEF—might direct the incorporation of H2A.Z at promoters of genes encoding major repressors of the SA pathway. Binding of H2A.Z to these genes prevents DNA methylation and maintains the competence of these genes for activation or repression, a characteristic that can be passed onto daughter cells because H2A.Z remains associated with chromosomes during mitosis (Deal et al., 2007). It remains to be shown what the direct targets of Swr1 in the process of infection are.

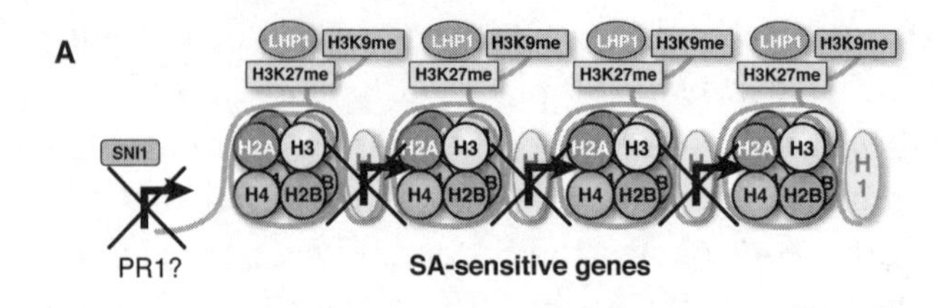

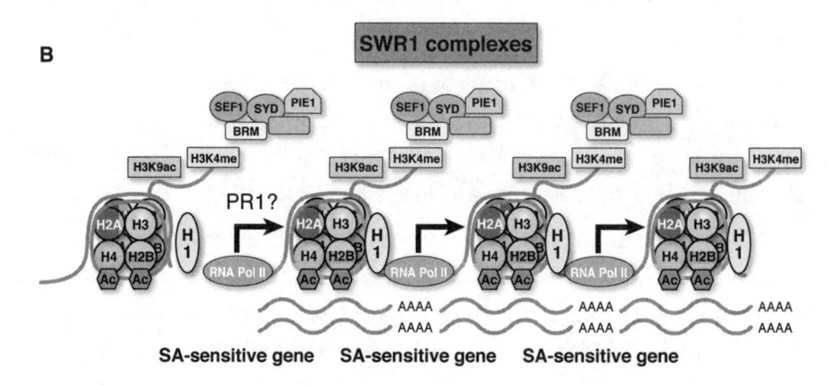

Figure 3-4 Chromatin regulation at SA-sensitive genes. **A.** Repressed states are maintained by LHP1 binding of methylated DNA. SNI1 prevents establishment of euchromatin state at *PR1* gene. **B.** Components of SWR1 complex promote the open chromatin at SA-sensitive genes. Replacement of H2A with H2AZ variant prevents binding of repressive LHP1 protein.

Splayed (SYD) and Brahma (BRM)

AtSYD and AtBRM are two more proteins belonging to the Snf2 subfamily that control vegetative and reproductive processes in plants. Both Arabidopsis *syd* and *brm* mutants are viable, although several developmental genes are misregulated. Among misregulated genes in *syd* mutant are jasmonic acid (JA)-sensitive genes *VSP2* and *PDF1.2a*. Several genes appeared to be co-regulated by AtSYD and AtBRM (Bezhani et al., 2007). One of the BRM mutants, *brm101*, had a higher spontaneous level of pathogenesis-related (*PR*) genes, suggesting that AtBRM, like PIE, might maintain basal repression of the SA pathway. As SA inhibits auxin signalling during plant defense, it was not surprising that auxin-related genes *SAUR66* and endo-xyloglucan transferase *EXGT-A1* were repressed in *brm101* plants (Bezhani et al., 2007).

Silencing *AtBRM* resulted in mutant plants that flowered earlier than wild-type plants, as well as having increased expression levels of genes promoting flowering, including *CONSTANS* (*CO*), *FT*, and *SOC1*. This data demonstrated that AtBRM represses the expression of genes that participate in the promotion of long day and flowering (Imaizumi and Kay, 2006).

PIE and BRM regulate pathogen defense pathways that are sensitive to Suppressor of NPR1, Inducible 1 (SNI1)

Suppressor of NPR1, Inducible 1 (SNI1) is a protein unique in respect to the fact that although it participates in chromatin remodeling, it lacks homology with chromatin-modifying proteins and DNA-binding motives. The protein was originally identified from a genetic screen for *npr1-1* suppressors. NPR1 is a key protein involved in the regulation of SA-dependent response to pathogens, and the mutant *npr1-1* has low pathogen resistance and does not express *PR1* when infected with pathogens. Recessive *sni1* mutation results in the return of *PR1* gene expression upon infection and allows the *npr1-1* plants to mount resistance. Further studies demonstrated that SNI1 protein represses transcription through chromatin remodeling and inhibits homologous recombination in somatic cells (Durrant et al., 2007); both of these genomic responses require histone-modifying enzymes (refer to Figure 3-4). Untreated *sni1* plants overexpress BTH-sensitive genes and have high levels of H3Ac and H3K4me2 at the *PR1* gene promoter, an effect also observed in *pie1-5* and *brm101* plants. It is suggested that SNI1 reduces euchromatic marks at the promoter of *PR1* gene, thus decreasing its expression. The defense phenotypes of *sni1* plants are suppressed by the mutation of RAD51D, a protein involved in chromatin modification and promotion of homologous recombination, providing evidence for the hypothesis that SNI1 modulates plant immunity through chromatin remodeling (Durrant et al., 2007).

Comparing the gene expression profile of three mutants—*sni1*, *brm101*, and *pie1-5*—revealed 13 commonly constitutively upregulated genes, including 11 genes inducible by BTH, such as *PR1*, *PR2*, *PR5*, and others. This finding further suggests that SA-dependent pathogen response pathway is controlled epigenetically. It remains to be shown through what exact regulatory mechanism SNI1, BRM, and PIE1 exert their repressive function.

Decrease in DNA Methylation 1 (DDM1)

DDM1 is yet another SWI2/SNF2-like protein encoding a helicase from the Lsh subfamily that regulates genomic DNA methylation. In Arabidopsis, *ddm1* mutants display more than 50% reduction in cytosine methylation, modification of histone marks in heterochromatic repeats and activation of transposable elements. Intriguingly, deficiency in the mouse homologue *Lsh* leads to similar alterations in heterochromatin structure (Dennis et al., 2001). DDM1 does not encode for enzymes that methylate or demethylate cytosine nucleotides but instead regulates DNA methylation indirectly, presumably by modulating the access of DNA methyltransferases and DNA demethylases to the genome (Dennis et al., 2001). Indeed, DDM1 remodels nucleosomes *in vitro* without modifying DNA methylation.

The *ddm1* mutant is characterized by changes in phenotype that progress with each round of self-propagation; many of the new appearing alleles are stable and are maintained even when the mutation is crossed out (Kakutani et al., 1996). One of the alleles is associated with dwarfism, curled leaves, and enhanced pathogen resistance. These so-called *bal* phenotypes arise due to changes in EDS1 and appear to be associated with overexpression of the resistance (*R*) genes from the Recognition of *Peronospora parasitica* 5 (*RPP5*) locus. One of the genes present at the *RPP5* locus is the Suppressor of *npr1-1*, Constitutive 1 (*SNC1*). Positive regulation of *SNC1* expression involves an SNC1-dependent amplification loop, whereas negative regulation occurs through the function of smRNAs generated at the locus (Yi and Richards, 2007). Although the *bal* allele appears to be epigenetic in nature, *bal* plants carry a tandem duplication of a 55-kb fragment from the *RPP5* locus and duplications in several other *R* genes, and it is not clear if such duplications are due to the lack of DDM1 activity. Mammalian DDM1 homologue Lsh interacts with *de novo* DNMTs and HDACs, generating inactive chromatin. In plants, it remains to be shown whether DDM1 functions in a similar manner by reinforcing chromatin compaction and preventing recombination between clustered *R* genes (see Table 3-1).

Table 3-1 Plant factors involved in epigenetic regulations

Name and function	Effect on chromatin	Effect of mutation and involvement in stress response	Modification/ Transcription
METHYL-CpG-BINDING DOMAIN PROTEINS (AtMBDs) / 5-methylcytosin binding proteins	Bind methylated CpG and change local chromatin structure via modification of core histone proteins; promote heterochromatin formation and repeat silencing	Late flowering and reduced fertility (*mbd11*); shoot branching and early flowering due to transcriptional repression of *FLC* (mbd9)	Local/Repression, activation
LIKE HETEROCHROMATIN PROTEIN 1(LHP1)	Chromodomain protein. Binds to histone H3K9; chromatin condensation and coating	Inability to repress expression of euchromatic genes associated with specific developmental stage	Global/Repression
DECREASED DNA METHYLATION (DDM1)/SWI2/ SNF2 DNA helicase	Control of DNA methylation, possibly through the binding methyl-CpG binding domain proteins and affecting their subnuclear localization	Decondensation of centromeric heterochromatin, redistribution of remaining DNA methylation, changes in histone methylation. Silencing of R genes and retrotransposons; DNA damage response	Global/Repression

Table 3-1 Plant factors involved in epigenetic regulations

Name and function	Effect on chromatin	Effect of mutation and involvement in stress response	Modification/ Transcription
DRD1/SWI/ SNF-like protein	Directing non-CpG DNA methylation in response to RNA signal; interacts with DNA methyltransferases and DNA glycosylases; targets promoter and LTRs in euchromatin	No significant defects in CpG methylation; loss of non-CpG methylation on previously silenced promoters and transposons; down regulation of *ROS1* and *DME*	Local, promoters/ Repression, activation
RNA POLYMERASE V (pol V) (subunits NRPD2a and NRPD1b)/RNA polymerase	Guides cytosine methylation using smRNA signals; NRPD1b possibly recruits DNA methyltransferases to asymmetric sites; works together with DRD1	Do not show significant defects in CpG methylation but exhibit loss of non-CpG methylation on previously silenced euchromatic promoters and transposons	Local, promoters/ Repression, possibly activation
MAINTENANCE OF METHYLATION 1 (MOM1)/ Similar to SWI2/SNF2	Regulation of transcription of silent heterochromatic regions; transgene silencing; preventing transcription of 180-bp satellite repeats but not of transposons	Release of transcriptional gene silencing and of 5S repeat silencing; no effect on heterochromatin organization, and DNA methylation.	Global/ Repression

Conclusion

As in animals, developmental stages in plants are promoted by antagonistic action of TrxG and PcG homologs. Plants form the germline late during the development. Thus, they require particularly flexible mechanisms that maintain a pluripotent state and are able to generate cells in which differentiation pathways are rapidly activated. Also, such developmental processes as seed germination and flowering require fine-tuned mechanism of response to environmental conditions, such as temperature and light. The central role in the process is played by the WUS protein; its activity is tightly regulated by several protein complexes. Whereas the CAF1 complex restricts expression of *WUS* to the organizing centre, SWI/SNF proteins ensure appropriate levels of *WUS* transcript in the other tissues.

Exercises and discussion topics

1. Which three groups of proteins repress the seed transcriptome?
2. What is the role of CAF1? What is one of its subunits? What happens if there is a mutation in the subunit?
3. How does organization and maintenance of RAM occur?
4. Why is chromatin remodeling important for the regulation of meristem activity?
5. What are the names of two PRC2 complexes present in plants, and what are their functions?
6. What is the role of PRC1 complex, and do plants contain one? If not, what other proteins may perform this role?
7. Name a TrxG homolog in Arabidopsis and describe its function.
8. What is the difference between epigenetic regulation of flowering in summer and winter annuals?
9. Describe the relationship between vernalization and FLC repression/expression?
10. What is the role of VIN3 in plants? What is significant about the *vin3* mutant?
11. How is FLC regulated in summer annuals?
12. What is REF5 and what possible role does it have in flowering?
13. How do small RNAs influence FLC expression?

14. What makes the AGAMOUS-LIKE19 pathway different from FLC?
15. What group and subfamily does PIE1 belong to? What pathogen response pathway is PIE1 actively involved in? What happens in a *pie1* mutant?
16. What do AtSYD and AtBRM control? What does silencing BRM do?
17. How does SNI1 protein repress transcription?
18. What is the role of DDM1 in plants?

References

Ahmad et al. (2010) Decoding the epigenetic language of plant development. *Mol Plant* 3:719-728.

Alvarez-Venegas et al. (2003) ATX-1, an Arabidopsis homolog of trithorax, activates flower homeotic genes. *Curr Biol* 13:627-637.

Bezhani et al. (2007) Unique, shared, and redundant roles for the Arabidopsis SWI/SNF chromatin remodeling ATPases BRAHMA and SPLAYED. *Plant Cell* 19:403-416.

Dennis et al. (2001) Lsh, a member of the SNF2 family, is required for genome-wide methylation. *Genes Dev* 15:2940-2944.

Desvoyes et al. (2010) Impact of nucleosome dynamics and histone modifications on cell proliferation during Arabidopsis development. *Heredity* 105:80-91.

Durrant et al. (2007) Arabidopsis SNI1 and RAD51D regulate both gene transcription and DNA recombination during the defense response. *Proc Natl Acad Sci USA* 104:4223-4227.

Imaizumi T, Kay SA. (2006) Photoperiodic control of flowering: not only by coincidence. *Trends Plant Sci* 11:550-558.

Jarillo et al. (2009) Chromatin remodeling in plant development. *Int J Dev Biol* 53:1581-1596.

Kakutani et al. (1996) Developmental abnormalities and epimutations associated with DNA hypomethylation mutations. *Proc Natl Acad Sci USA* 93:12406-12411.

March-Díaz et al. (2008) Histone H2A.Z and homologues of components of the SWR1 complex are required to control immunity in Arabidopsis. *Plant J* 53:475-487.

Pien S, Grossniklaus U. (2007) Polycomb group and trithorax group proteins in Arabidopsis. *Biochim Biophys Acta* 1769:375-382.

Ramirez-Parra E, Gutierrez C. (2007) The many faces of chromatin assembly factor 1. *Trends Plant Sci* 12:570-576.

Schmitz RJ, Amasino RM. (2007) Vernalization: a model for investigating epigenetics and eukaryotic gene regulation in plants. *Biochim Biophys Acta* 1769:269-275.

Sung S, Amasino RM. (2004) Vernalization and epigenetics: how plants remember winter. *Curr Opin Plant Biol* 7:4-10.

Swiezewski et al. (2007) Small RNA-mediated chromatin silencing directed to the 3' region of the Arabidopsis gene encoding the developmental regulator, FLC. *Proc Natl Acad Sci USA* 104:3633-3638.

Turck et al. (2007) Arabidopsis TFL2/LHP1 specifically associates with genes marked by trimethylation of histone H3 lysine 27. *PLoS Genet* 3:e86.

Wagner D, Meyerowitz EM. (2002) SPLAYED, a novel SWI/SNF ATPase homolog, controls reproductive development in Arabidopsis. *Curr Biol* 12:85-94.

Yi H, Richards EJ. (2007) A cluster of disease resistance genes in Arabidopsis is coordinately regulated by transcriptional activation and RNA silencing. *Plant Cell* 19:2929-2939.

Zhang H, Ogas J. (2009) An epigenetic perspective on developmental regulation of seed genes. *Mol Plant* 2:610-627.

4

DNA methylation as epigenetic mechanism

DNA methylation is a covalent modification of nucleotides, in which a methyl group is added to a cytosine residue at position C-5 or N-4 or to an adenine residue at position N-6. In terms of evolution, it is one of the most ancient epigenetic mechanisms for the regulation of gene expression and the timing and sequence specific targeting of its addition or removal from DNA affects the timing and targeting of events of cellular regulation. DNA methylation is found in almost all life forms, including bacteria, plants, and mammals, but it is lost in organisms such as *S. cerevisiae, S. pombe, C. elegans*, and Drosophila. However, several reports suggest that DNA methylation does occur in the fruit fly; whether the same is true for yeast and worms needs to be determined. This chapter focuses on DNA methylation as a process and explains the mechanism of its establishment, maintenance, and removal in bacteria, animals, and plants.

DNA methylation in bacteria

Methylation at the C-5 position of cytosine and at the N-6 position of adenine is found in the genomes of many fungi, bacteria, protists, and Archaea (only N-6 methylation of adenine), whereas methylation at the N-4 position of cytosine has only been found in bacteria. Methylation in bacteria occurs on both cytosine and adenine nucleotides and is imposed by **DNA methyltransferases**, also called **MTases**. These enzymes use S-adenosyl-methionine as a donor of methyl groups.

In Eubacteria and Archaea, methylation primarily serves the purpose of protection against the cell's own restriction enzymes, which are part of the so-called **restriction-modification (RM) system**. The RM system consists of two enzymes, a restriction enzyme (restriction) and a cognate DNA adenine or cytosine methyltransferase (modification), both of which recognize the same sequence. This system provides the cell with a primitive form of an immune system; it allows it to digest the DNA of an invading bacteriophage while restriction target sites in its own DNA are protected through methylation. Thus, the function of RM systems allows bacteria to differentiate between "self" and "non-self." Besides this, RM systems are also suggested to play an important role in genetic exchange and maintenance of species identity (reviewed in Wion and Cadadesus, 2006).

RM systems are broadly divided into three different types based on structural features, the position of DNA cleavage and cofactor requirements (reviewed in Wion and Cadadesus, 2006). Whereas type I and III RM systems involve multi-subunit enzymes that catalyze both restriction and modification, type II RM systems consist of two separate enzymes, a DNA-adenine/cytosine methyltransferase and a restriction enzyme that recognizes the same target sequence.

Another interesting view on RM systems is presented by Kobayashi (2001). The author suggests that RM systems represent one of the most ancient life forms, which is similar to viruses, transposons, and homing endonucleases rather than a simple protection system for bacterial genomes. According to Kobayashi, RM systems can increase their relative frequency within a cell population by three strategies. First, they defend themselves (and the host bacterium) from invading phages by digesting "non-self" DNA (see Figure 4-1A). Second, they kill cells in which RMs are eliminated through the acquisition of a new genetic element—for example, an alternative RM system (see Figure 4-1B). Cell death occurs because residual methyltransferase activity is unable to protect all the restriction sites against cleavage by the remaining endonuclease molecules. Third, RMs can move themselves between genomes (see Figure 4-1C). They are often linked with mobile genetic elements such as plasmids, prophages, viruses, transposons, and integrons. However, activity of RMs can also be deleterious to the bacterial genome. Imbalances between methyltransferase and restriction enzyme expression can result in genomic instability. Further, introduction of new RM gene complexes can induce massive genome rearrangements in some bacterial genomes *in vitro*. In summary, Kobayashi suggests that RM systems behave as selfish genetic elements, which are in host-parasite interactions with bacteria and act primarily with one purpose—to survive.

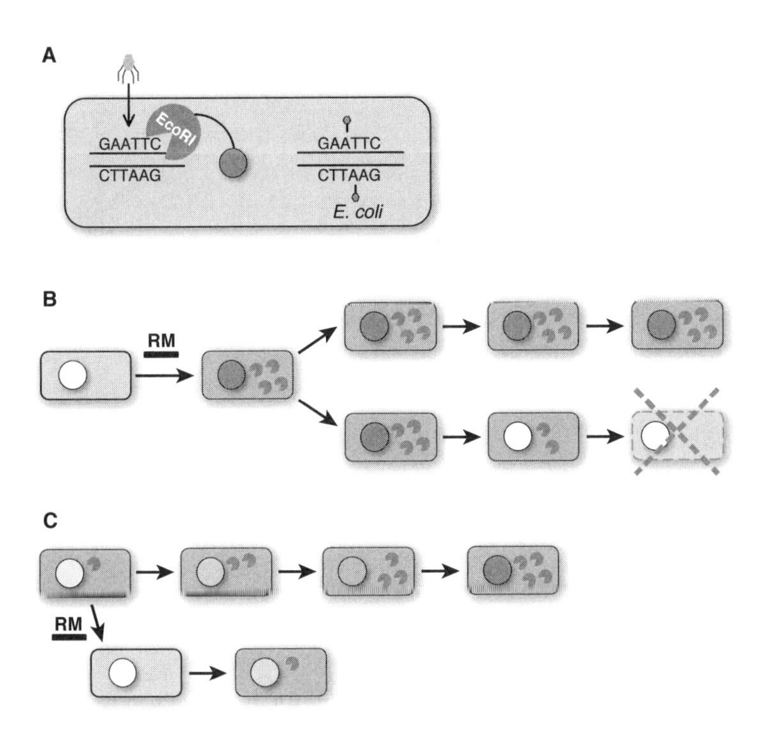

Figure 4-1 Mechanisms by which RMs increase their relative frequency in the cell population. **A.** Unmethylated phage DNA is attached by RM restrictases, whereas E. coli genomic DNA is protected. **B.** Cells that lose RMs through acquisition of other genetic elements are eliminated through post-segregational cell killing by residual restrictases. **C.** RMs self-amplify in the cells or transfer themselves to innocent cells.

Besides their role in protection against restrictases, DNA methyltransferases in bacteria also regulate the processes of DNA replication, transcription, DNA repair, and others. Many MTases in bacteria are not associated with cognate restriction enzymes. These enzymes include the *Escherichia coli* **DNA adenine methyltransferase (Dam)**, which methylates adenine nucleotides at the N-6 position within GATC sequences; the *Caulobacter crescentus* **cell cycle-regulated**

methylase (CcrM) enzyme, which methylates the N-6 position on adenines within GAnTC sequence; and the *Escherichia coli* **DNA cytosine methyltransferase (Dcm)**, which methylates the C-5 position of cytosine in CC(A/T)GG sequences. The respective recognition sequence is critical for MTase binding and a single nucleotide exchange substantially reduces or completely abolishes the efficiency of methylation.

The following sections present the non-RM MTases in more detail.

DNA adenine methyltransferase (Dam)

Based on amino acid sequence similarity, Dam belongs to a family of enzymes with homologs in various bacteria (Low et al., 2001). Although the majority of these enzymes are not involved in protection against restriction, some MTases that belong to the α group, such as the DpnII methylase, are part of the restriction-modification system.

Dam is a *de novo* MTase that methylates non-methylated and hemimethylated GATC sites with similar efficiencies. There are approximately 130 Dam molecules in each growing cell, which is a sufficient number to methylate all GATC sites during a single cycle of DNA replication. Dam-mediated methylation regulates many important cellular functions such as gene expression, DNA replication, chromosome segregation, DNA repair, and transposon activity. However, Dam-mediated methylation is not essential for *E. coli*. In contrast, Dam is an essential gene in *Yersinia pseudotuberculosis* and *Vibrio cholerae* (Mahan and Low, 2001).

Expression of the *E. coli dam* gene is controlled by a combination of five promoters. The expression from one of the main promoters, P2, is regulated by growth rate, leading to enhanced Dam protein levels in cells with high growth rates. The regulatory mechanisms that control the activity of the other promoters are unknown, but may also be dependent on various growth conditions.

An important function of Dam is the regulation of gene expression. Methylation at the N-6 position of adenine lowers the thermodynamic stability of DNA and alters the DNA curvature. These structural changes influence DNA-protein interactions because the majority of DNA-binding proteins recognize their cognate DNA-binding sites by both primary sequence and structure (reviewed in Wion and Casadesus, 2006). Decreased levels of methylation in bacteria may have either positive or negative effects on gene expression. If Dam methylation is absent at GATC in either polymerase or transcription activator binding sites, it facilitates their binding and results in gene activation. In contrast, in GATC sites that are located in areas of repressor binding, undermethylation results in more efficient binding of repressors and thus in inactivation of gene expression. For instance, it was shown that adenine methylation at GATC sites either inhibits (*trpR* and *Tn10 transposase*) or enhances transcription (*DnaA*) by altering the interaction between RNA polymerase and the consensus RNA polymerase binding sites. Dam is also involved in the methylation-dependent regulation of the expression of phage genes such as the *mom* gene in bacteriophage Mu (Hattman, 1982). The interesting fact about that is that the *mom* gene encodes a DNA modification function, which converts adenine to acetamido-adenine in a sequence-specific manner, thus protecting the phage genome against degradation by bacterial enzymes.

Another mechanism of methylation-dependent control of gene expression in bacteria is the regulation via the establishment of **DNA methylation patterns (DMPs)**. DMPs are formed in specific genomic areas in which a GATC sequence is located in close proximity to binding sites of regulatory factors or even overlaps with them. However, binding of regulatory proteins to those sites protects DNA from methylation and therefore results in undermethylated GATCs. Such binding of regulatory factors to DNA seems to be bacteria-specific and can be affected by environmental conditions such as nutrient availability or stress. It is not surprising that most non-methylated GATC sites are present in non-coding regions, as methylation of the adenine sites in these regions regulates gene expression of adjacent coding areas. One of the good examples is the regulation of expression of the glucitol utilization (gut) operon (see Figure 4-2). The expression of the operon is suppressed by the sequence-specific GutR repressor that binds upstream of the operon and blocks adenine methylation at the GATC site (van der Woude et al., 1998). When glucitol (also known as sorbitol) is present, GutR does not bind to the aforementioned site and thus no longer blocks DNA methylation, allowing expression of the operon.

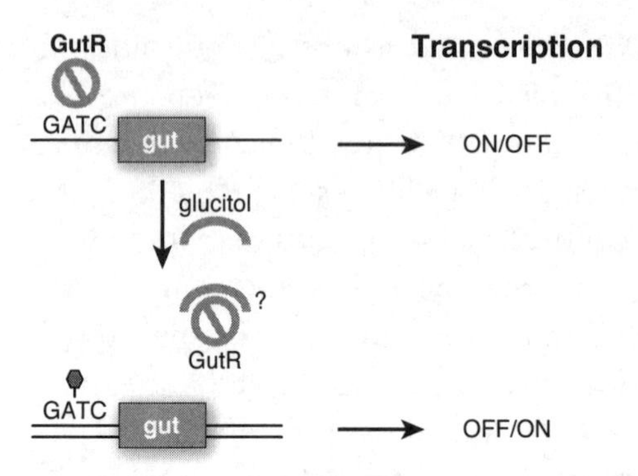

Figure 4-2 Regulation of DMPs. See text for details. Methylation of GATC sequences might result in activation or inactivation of transcription.

Dam plays an important role in DNA replication

Chromosome replication starts when the initiator protein DnaA binds to the replication origin (*oriC*) and separates the two strands of the DNA double helix. The replication origin (*oriC*) of the *E. coli* chromosome contains 11 GATC sites distributed over 254 bp. DnaA binding to the *oriC* region is only possible if the *oriC* GATCs are methylated—a hemimethylated origin is inactive. Immediately after replication, SeqA binds to hemimethylated GATC sites within *oriC*, thus preventing methylation of the newly replicated strand. This ensures that the bacterial chromosome is replicated only once per cell cycle, as the hemimethylated *oriC* cannot reinitiate replication and this hemimethylation is maintained for up to one-third of the total length of the cell cycle.

Note that *dnaA* expression itself is regulated by Dam methylation. The *dnaA* gene promoter is only active if adenines at three different GATC sites are methylated. Methylation at those sites is also affected by SeqA binding in a way analogous to *oriC* and given the *dnaA* gene and *oriC* are only 50 kb apart, they might be simultaneously regulated by SeqA (reviewed in Wion and Casadesus, 2006).

Dam methylation also plays a role in the organization of the nucleoid region (see box 4-1) immediately following replication. Hemimethylated *oriC* provides a signal for nucleoid segregation by providing binding sites for segregation-driving proteins that attach the DNA to the membrane. SeqA might also play a role in this process presumably by binding to hemimethylated GATC sites

4-1. Nucleoid Region

Unlike eukaryotic cells, prokaryotic cells do not have a cell nucleus. Most of their genetic material is nonetheless accumulated in a defined area called the nucleoid. The major component of nucleoids is DNA, but they also contain some RNA and structural, as well as regulatory, proteins.

behind replication forks and interacting with proteins involved in the attachment of DNA to the membrane (reviewed in Wion and Casadesus, 2006).

The role of Dam in DNA repair

Dam plays an important role for strand discrimination in **mismatch repair (MMR)** in *E. coli* (see Figure 4-3). It promotes long-patch excision repair that is dependent on the function of *mutH*, *mutL*, *mutS*, and *mutU (uvrD)* gene products. Post-replicative MMR is initiated by the MutS-dependent recognition and binding of mispaired nucleotides (reviewed in Fishel et al., 1998). This is followed by binding of the MutL protein that accelerates ATP-dependent translocation of the MutS–MutL complex to a hemimethylated GATC site bound by MutH. The MutS–MutL complex stimulates the endonuclease activity of MutH, which binds to hemimethylated DNA and cleaves the nonmethylated strand. Thus, the differentiation between the template and nascent strands depends on transient undermethylation of the GATC adenine nucleotides. The newly replicated DNA strand containing the mismatch is degraded by exonucleases, such as Rec J, Exo I, or Exo VII, and then resynthesized by the Pol III holoenzyme complex resulting in the completion of methylation-directed MMR (reviewed in Fishel et al., 1998). The importance of Dam in MMR is further underlined by the fact that mutants impaired in Dam function have an increased spontaneous mutation rate.

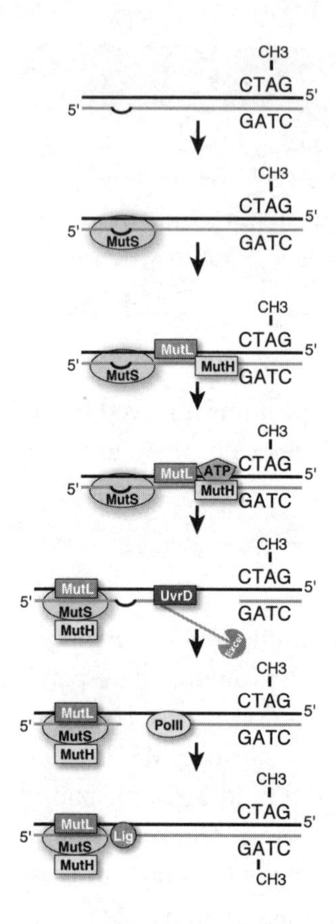

Figure 4-3 Methylation-dependent mismatch repair.

The role of Dam in transposition

In the bacterial transposons IS*10* and IS*50*, Dam-dependent methylation represses transposition by two independent mechanisms:

- Methylation of GATC sites in the promoter region (at position -10) of the IS*10* or IS*50* transposase prevents efficient binding of RNA polymerase, thus inhibiting transcription of the transposase gene. Transcription of the transposase is permitted immediately after replication when the GATC site becomes transiently hemimethylated (reviewed in Wion and Casadesus, 2006).
- Methylation of GATCs at the ends of IS*10* and IS*50* transposons inhibits the transposase activity.

Transposition that occurs right after replication might be less harmful for bacteria because the potentially deleterious effect of transposition occurs only in one cell, whereas the other cell retains an original copy of the chromosome. However, if the transposition event has positive effects for the cells, only one of the cells benefits from it. If we assume that it might also be beneficial to parasitic transposable elements, host cells will more likely survive transposition events. These might be just some of the reasons why transposons are allowed to transpose after replication and why Dam methylation plays an important role in the process.

4-2. Transposition

Transposition is a mechanism that results in "cut and paste" or "copy and paste" of a stretch of DNA—called transposon or transposable element. During "cut and paste" transposition a transposon moves from one location within the genome to another, assisted by a transposase that cuts it from its locus, DNA polymerase that fills in gaps at the cut ends and DNA ligase that ligates the transposon into its new location. During "copy and paste" transposition, the transposon sequence gets transcribed and reverse transcribed to form a DNA template that can be transposed. The latter mechanism is common to viral/parasitic and endogenous transposable elements.

The role of Dam methylation in repression of conjugation

In *Salmonella,* Dam methylation regulates the process of conjugation by preventing transfer of the Fertility factor and virulence plasmid (pSLT) (reviewed in Wion and Casadesus, 2006). In the case of pSLT, the repression of conjugation by methylation involves two steps:

1. Transcriptional activation of the *finP* gene that encodes a small RNA that inhibits mating. This activation depends on Dam methylation-dependent inhibition of binding of the nucleoid protein H-NS.

2. Transcriptional repression of the *traJ* gene that encodes a transcriptional activator of the transfer operon. This is regulated by Dam methylation-dependent inhibition of the binding of Lrp to the *traJ* promoter region. The expression of *traJ* is positively regulated by Lrp, which recognizes two Lrp-binding sites in the **upstream activating sequence** (**UAS**).

Methylation at a GATC site in one of the binding sites reduces Lrp binding and results in formation of an Lrp/UAS complex that does not allow expression from the *traJ* promoter. On contrast, both unmethylated and hemimethylated GATC are available for Lrp binding and the formation of a promoter-activating complex. Regulation of *traJ* transcription by Dam methylation and Lrp binding depends on the position of hemimethylated GATC. Methylation of GATC on the *traJ* coding strand prevents transcription, whereas methylation on the non-coding strand does not inhibit transcriptional activation. Because newly replicated plasmids exist in two different epigenetic states (one with methylated GATC on the coding strand and one with methylated GATC on the non-coding strand), transcription of *traJ* and conjugation is repressed in one case and permitted in another. Such regulation might represent an epigenetic (safety) switch that enables bacteria to conserve metabolic reactions by limiting the activation of conjugal transfer to one copy of a plasmid at a time (reviewed in Wion and Casadesus, 2006).

Dam methylation regulates bacterial virulence

One of the first lines of evidence for a possible role of Dam methylation in the control of bacterial virulence came from a study that demonstrated that the expression of adhesin-encoding genes is regulated by Dam. Later on, direct evidence for the role of Dam in bacterial virulence was provided by the observation that *dam* mutations impair virulence of pathogens such as *Salmonella enterica*, *Y. pseudotuberculosis*, and *V. cholerae* (reviewed in Low et al., 2001). Dam methylation is known to control the expression of several virulence genes and deletion of *dam* results in elimination of DMPs, thus altering the binding of regulatory proteins and their downstream effects on gene expression.

In Salmonella, it was found that Dam⁻ mutants induced an attenuated effect in infected mice when compared to infection with Salmonella with functional Dam. A weaker infection was paralleled by the following bacterial defects:

- Leakage of proteins and release of membrane vesicles due to the instability of the bacterial envelope
- Export of highly immunogenic fimbrial proteins due to ectopic expression of the Dam-repressed fimbrial operon *stdABC*
- Reduced secretion of the SipC protein translocase
- Sensitivity to bile (reviewed in Wion and Casadesus, 2006)

It is not exactly clear whether these defects have a direct or indirect effect on attenuation of the infection capacity of Salmonella Dam⁻ mutants. Although mutant bacteria persist in infected mice, a high level of immune response provides a strong protection against infection. Wion and Casadesus (2006) suggest that attenuated Dam⁻ mutants may be used as live Salmonella-based vaccines.

Transcriptomic analysis of Dam⁺ and Dam⁻ of *E coli* and *S. enteric* showed that Dam methylation regulates metabolic pathways, respiration, and motility in response to environmental cues; it also regulates genes encoding flagellar subunits and genes required for invasion of epithelial cells (reviewed in Wion and Casadesus, 2006). These data confirm that the requirement for Dam methylation in the regulation of virulence genes is not limited to Salmonella but instead represents a common regulatory pathway for controlling bacterial virulence.

Maintenance and inheritance of Dam DMPs

As previously described, post-replicative GATC sites remain hemimethylated for up to one-third of the cell cycle mediated by SeqA-mediated sequestration. Because this is temporary, the adenines at SeqA-sequestered GATC sites become fully methylated before entering the next round of cell division. More stable regulation of Dam methylation occurs through the activity of methylation-blocking proteins, such as Lrp and OxyR, which have the high affinity for unmethylated GATC sites, thus maintaining their unmethylated status throughout the entire cell cycle (reviewed in Wion and

Casadesus, 2006). For example, the *E. coli* chromosome has 50 sites that are stably hemimethylated or unmethylated, many of them located in putative regulatory regions. The distribution of undermethylated GATC sites in the *E. coli* genome depends on growth conditions, suggesting that environmental cues can alter the DMPs of the *E. coli* genome (Polaczek et al., 1997).

One example for such inheritance of undermethylation in regulatory regions is found in the promoter region of the pyelonephritis-associated pilus (*pap*) operon. In this region, Dam methylation regulates the switch between *pap* expression and nonexpression states by dictating the binding of Lrp (see Figure 4-4). At low levels of PapI, the protein that binds to Lrp, Lrp binds to the promoter proximal GATC sites ($GATC^{prox}$) protecting them from methylation and inhibiting transcription from the *papBA* promoter. The GATC site that is located 5' from $GATC^{prox}$, named the promoter distal GATC site or $GATC^{dist}$, becomes methylated and prevents the movement of the Lrp protein from $GATC^{prox}$ to $GATC^{dist}$, thus locking cells in the phase of *pap* nonexpression. This condition is inherited and persists until another round of replication generates a hemimethylated $GATC^{dist}$ site, which becomes available for binding by Lrp (reviewed in Low et al., 2001; Hernday et al., 2002) (see Figure 4-4C).

Because DNA demethylases have not been described in bacteria, it is possible that only **passive DNA replication-dependent demethylation** occurs. Undermethylation is thus maintained by binding of proteins (such as SeqA, Lrp, OxyR, GutR, and H-NS) and possibly other proteins involved in gene regulation or nucleoid organization. Sequence-specific differences in methylation levels of GATC sites within the genome can also be generated by sequence context-dependent differences in the processivity of Dam methylase or differences in accessibility of the sequences to Dam (reviewed in Low et al., 2001).

The CcrM methylase

In contrast to Dam, CcrM was originally discovered in *Caulobacter crescentus* and belongs to the β group of DNA methylases. It shares a significant homology (49% identity) with the *Hinf*I MTase of *H. influenza*. Like Dam, but unlike *Hinf*I, CcrM is not associated with a

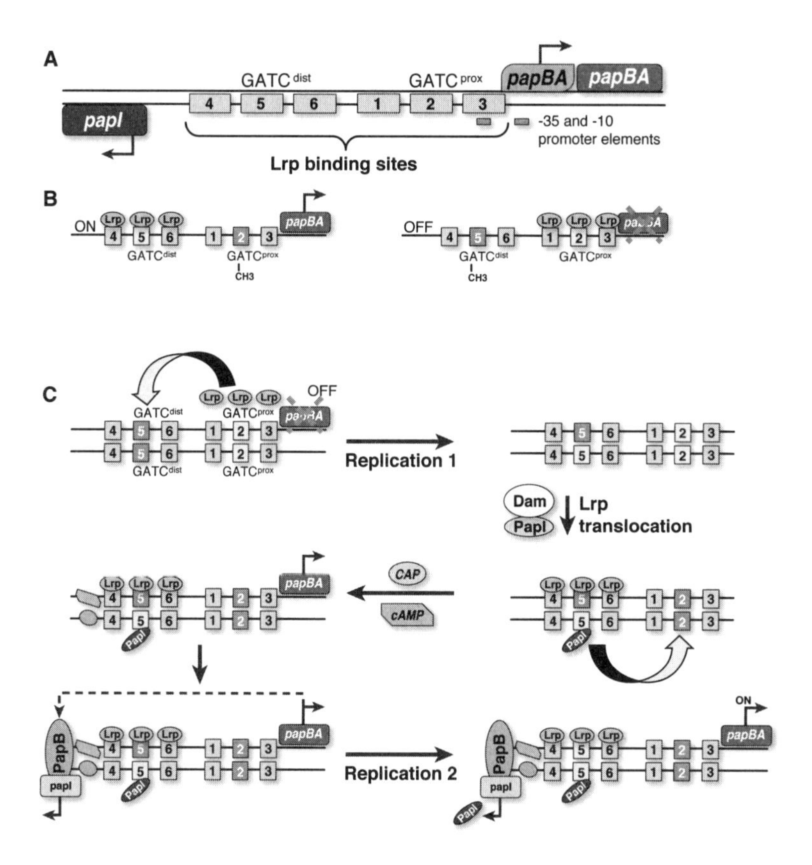

Figure 4-4 Regulation of expression from pap operon (reproduced with permission from Hernday et al. (2002) Self-perpetuating epigenetic pili switches in bacteria. Proc Natl Acad Sci USA 99:16470—16476). **A.** Structure of Pap operon. 1-3 GATC^{prox} and 4-6 GATC^{dist} are Lrp-binding sites. papBA and papI are the expressed genes with oppositely oriented open reading frames. papBAp — papBA promoter. **B.** ON and OFF states of papBA expression. **C.** During the OFF state, Lrp is tightly bound to GATCprox binding sites, protecting GATC at position 5 from methylation. Although GATC^{dist} at position 2 is methylated, the methylation process is somewhat inhibited by the presence of Lrp proteins. During replication, Lrp proteins dissociate and hemimethylated state of GATC^{dist} is generated. Lrp proteins translocate to GATC^{dist}, PapI proteins bind unmethylated GATC^{dist} at the position 2, and Dam proteins methylate unprotected GATC^{prox}. Again, the presence of Lrp proteins inhibits the methylation at the adjacent GATC^{prox}. Next, cAMP/CAP complex is recruited to the DNA and papBA expression is finally activated. PapB activates papI expression. Second round of replication removes the methyl groups from GATC^{dist}, leading to papBA ON state.

cognate restriction enzyme and therefore is not part of any known restriction modification system.

CcrM is another DNA adenine methyltransferase that specifically methylates GAnTC sequences. Similar to Dam, CcrM is a highly processive enzyme. However, unlike Dam, the activity of CcrM is higher on hemimethylated DNA than on unmethylated DNA (reviewed in Wion and Casadesus, 2006). This suggests that CcrM functions as a maintenance MTase rather than a *de novo* MTase.

Homologs of CcrM exist in several α-proteobacteria, such as *Agrobacterium tumefaciens* and *Rhizobium meliloti*. All CcrM homologs play an essential role in the regulation of the cell cycle earning them their other name: **cell cycle-regulated methyltransferases**. They are highly conserved across species, and complementation analysis showed that they were functionally interchangeable.

CcrM functions in regulation of gene expression

The first evidence that CcrM is involved in the regulation of gene expression was obtained from analysis of the expression of the *ccrM* gene itself. The promoter of the *ccrM* gene contains several GAnTC sites, and methylation of adenines at these sites inhibits transcription. In addition, single nucleotide substitutions at these sites prevent *ccrM* expression that normally occurs after replication (Stephens et al., 1995). A detailed analysis of the genome sequence of *C. crescentus* revealed that GAnTC sequences occur at a much lower frequency than randomly predicted and have a strong bias to intergenic regions. Also, global genome expression analysis showed that depletion of CcrM activity resulted in changes in the expression of nearly 100 genes (reviewed in Low et al., 2001). This and other evidence indicates a role of CcrM as a global regulator of gene expression.

CcrM functions as a cell cycle regulator

The regulation of the cell cycle in *C. crescentus* is a fascinating process. *C. crescentus* is a **dimorphic** bacterium, meaning it has two different cell types: the **stalked cell** capable of replication and the **swarmer cell** which has a flagellum and can move by chemotaxis.

The stalked cell undergoes asymmetric division and produces one swarmer cell, which contains the flagellum, and one stalked cell (Marczynski and Shapiro, 2002). Swarmer cells ultimately differentiate into stalked cells. Because CcrM is produced during the late stage of replication, the chromosome of stalked cells remains hemimethylated for the period of time between replication and production of CcrM. As in *E. coli*, the *Caulobacter* origin of replication (*Cori*) is inactive in the nonmethylated and hemimethylated states. Thus, complete methylation of *Cori* and re-activation of the replication origin is achieved fairly late when most, if not all, DNA is replicated. Synthesis of CcrM can be considered as a hallmark for cell cycle completion and cell ability to initiate another round of division.

Shortly after cell division is complete, CcrM is degraded in both stalked and swarmer cells. Chromosome replication in the swarmer cell is prevented by CtrA, which binds to methylated *Cori*. During differentiation into a stalked cell, CtrA is degraded by the Lon protease and is no longer produced due to hypermethylation of CtrA locus, which prevents the gene expression until chromosome replication renders the promoter hemimethylated. The expression of the *ccrM* gene is regulated by CtrA, which accumulates in replicating stalked cells and is able to activate *ccrM* transcription as soon as the replication fork passes through the *ccrM* genomic region, hemimethylating two GAnTC sites located in its leader sequence (reviewed in Wion and Casadesus, 2006).

The complexity of the regulation mechanism does not end here as the expression of the *ctrA* gene is also controlled by GAnTC methylation. As soon as the replication fork reaches the *ctrA* promoter bringing it to the hemimethylated state, transcription is activated. It can be hypothesized that bacteria that have *ctrA* located closer to the replication origin may have a faster turn-around of CcrM and *Cori* activities. On the other hand, if the *ctrA* gene is located downstream of *ccrM* (in comparison to the replication origin), both CcrM accumulation and *Cori* activation might be inhibited. Indeed, experimental evidence of such events exists: If the *ctrA* gene is moved to an ectopic position closer to the replication terminus, transcription from the methylated *ctrA* promoter is repressed, leading to a slow accumulation of CtrA (Reisenauer and Shapiro, 2002).

DNA methylation in eukaryotes

Methylation of cytosines is vastly predominant in eukaryotes and for long was believed to be the only type of DNA methylation. More recent reports showed, however, that adenine methylation also occurs. DNA of *Chlamydomonas reinhardtii*, *Tetrahymena pyriformis*, and of several other lower eukaryotes was found to contain m6A (reviewed in Wion and Casadesus, 2006). In protozoa, adenine methylation was found to be restricted to the DNA in the somatic nucleus (macronucleus), one of two distinctive nuclei in protozoan cell, with another one being germline nucleus (micronucleus). You can find more details about nuclear dimorphism in protozoa in Chapter 9, "Non-Coding RNAs Across the Kingdoms—Protista and Fungi." Because the process of "maturation" of macronuclear genome consists of steps of chromosome fragmentation and elimination of sequences, adenine methylation might somehow play a role in regulation of these processes. It remains to be shown how adenine methylation is established and maintained.

Methylation of the C-5 position of cytosine is a heritable epigenetic modification occurring in eukaryotes such as animals, fungi, and plants. It provides different levels of gene regulation: from a simple method for controlling gene expression to more complex processes involved in the control of chromatin architecture and mechanisms of imprinting (see Chapters 2, 3, and 5). Cytosine methylation occurs in almost all eukaryotes, except for roundworms, fruit flies, and yeast. In these organisms, regulation of gene expression, chromatin organization, and heritable changes in epigenetic regulation rely on other mechanisms of epigenetic regulation, such as histone modifications, activities of small RNAs, and differential binding of chromatin-binding proteins (this is discussed in Chapters 2, 3, and 5 through 12).

In animals and plants, methylation occurs at different frequencies and in different sequence contexts. Despite these differences, the establishment and maintenance of cytosine methylation is supported by a family of conserved enzymes known as DNA methyltransferases. These enzymes transfer methyl groups from **S-adenosylmethionine** (**SAM** or **Adomet**) to cytosine. Synthesis of SAM involves choline, methionine, and folate metabolism that interact at the point at which homocysteine is converted to methionine.

Methionine adenosyltransferases convert methionines to SAMs. For details on biochemistry of SAM synthesis and the importance of diet in the maintenance of SAM, please see the review by Niculescu and Zeisel (2002).

The importance of DNA methyltransferases is reflected by the fact that in eukaryotes (such as yeast, roundworm, and fruit fly) that lack these enzymes, DNA methylation most probably does not occur.

Also, there is a high degree of conservation between the function of DNA methyltransferases and the role DNA methylation plays in different eukaryotes. There are substantial differences between mechanisms of DNA methylation that justify their separate description in fungi, animals and plants.

Cytosine methylation in fungi

Methylation in yeasts has not yet been described, whereas methylation in *Neurospora crassa* is quite common—approximately 1.5% of cytosines are methylated in the *Neurospora* genome. Nearly all methylated cytosines are located in transposons-rich sequences, whereas protein-coding genes are void of cytosine methylation completely. Methylation at transposons in *Neurospora* is believed to be associated with self-defense against transposons movement. This genome defense is called **repeat-induced point mutation (RIP)** (reviewed in Rountree and Selker, 2010). RIP apparently protects the genome from the assault of invasive DNAs and consists of two steps, massive post-fertilization transition mutations (G:C to A:T) at duplicated sequences, followed by methylation of remaining cytosines. This process mutates up to 30% of G:C pairs in duplicated transposons and retrotransposons. Methylation in *Neurospora* is not restricted to symmetrical sequences as any cytosine in the duplicated sequences can be methylated.

Mutagenesis of *Neurospora* revealed two mutants defective in DNA methylation, *dim-2* and *dim-5*. Whereas the former one was found to be deficient in DNA methyltransferase (DMT), the latter one was deficient in histone methyltransferase. DIM-2 is required for cytosine methylation as *dim-2* mutant does not have any detectable methylated cytosines. Several other mutants were found to be partially required for DNA methylation. DNA methylation modifier 1 (DMM1) and 2 (DMM2) have altered pattern of cytosine methylation, however the protein function is still unknown (reviewed in Rountree and Selker, 2010). The role of histone modifications in RIP and DNA methylation is represented by two facts: *dim-5*, impaired in the activity of histone methyltransferase is in part repaired in cytosine methylation; inhibition of the histone deacetylase activity by Trichostatin-A also inhibits DNA methylation (Selker, 1998). More studies are necessary to unravel all components required for cytosine methylation in Neurospora. It should be stressed, however, that cytosine methylation is not essential for survival and proper growth of this fungus.

Cytosine methylation in animals

In animals, cytosine methylation occurs predominantly **symmetrically** at **CpG dinucleotide** pairs, meaning that both strands of a double-stranded DNA helix carry the methylation mark. It is estimated that in the animal genome, about 80% of all cytosines in CpGs are methylated, with the remaining 20% of unmethylated CpG dinucleotides located predominantly in promoter regions (reviewed in Law and Jacobsen, 2010). The GC content of the human genome is approximately 42%; thus, it is expected that CpG dinucleotides occur with a frequency of $0.21 \times 0.21 = 4.41\%$. Instead, CpGs are severely underrepresented in the genome, occurring at a frequency of less than 1%. However, they are frequently found in clusters termed **CpG islands** which are defined by a 50% content of GC, and the ratio of observed to expected CpG occurrence is higher than 0.6 in a stretch of 200 nt. CpG islands play a crucial role in the regulation of gene expression and transcriptional silencing (Cedar and Bergman, 2009). Only about 10% of all cytosines in the animal genome are methylated. Although it is not common in animals, low levels of non-CpG methylation occur in embryonic stem cells.

In mammals, DNA methylation patterns are established by a family of ***de novo* methyltransferases**, DNA methyltransferase 3 (DNMT3), and maintained by the **maintenance methyltransferase**, DNMT1. DNMTs have 10 conserved motifs (I–X) located in the carboxy-terminal region. In the three-dimensional protein structure, motifs I and X form the SAM-binding site, whereas motif IV contains a prolylcysteinyl dipeptide, which provides a thiolate at the active site mediating the addition of a methyl-group from SAM onto cytosine. The target recognition domain is located between motifs VIII and IX, and motif VI functions in protonating N3 positions of the target cytosine via a glutamyl residue. Refer to the review by Lan et al. (2010) for the details of protein structure and molecular and biochemical properties of the mammalian Dnmt enzymes.

In mammals, there are three large families of DNMTs: DNMT1, DNMT2, and DNMT3.

The DNMT1 family consists of DNMT1s, DNMT1o, DNMT1b, DNMT1$^{\Delta E3-6}$ and DNMT1p genes. The DNMT3 family has three members including DNMT3a, DNMT3b, and DNMT3L; DNMT3a has four isoforms (DNMT3a1 to DNMT3a4); and DNMT3b has eight isoforms (DNMT3b1 to DNMT3b8). Whereas DNMT1 and DNMT3 have members with defined functions, DNMT2 appears to be a severely N-terminally truncated protein with undetectable MTase activity and low expression levels in all analyzed tissues. Reports show that human DNMT2 has some MTase activity *in vitro* and can catalyze RNA methylation. Further, a novel DNMT2 isoform, DNMT2-gamma (DNMT2γ), was isolated from testes and preimplantation embryos, but it also appeared to be a nonfunctional protein (reviewed in Lan et al., 2010). The mammalian DNMT families and their functions are discussed in more detail in the following sections.

De novo *DNA methylation*

Both Dnmt3a and Dnmt3b are DNA methyltransferases that perform *de novo* methylation at most of the CpG dinucleotides, excluding CpG islands. Methylation at CpG islands might be maintained by cooperation of Dnmt1 and Dnmt3b enzymes, as mutation of DNMT1 alone does not result in dramatic changes in methylation at CpG islands, whereas mutation of both DNMT1 and DNMT3B results in severe loss of methylation (Robert et al., 2003).

Dnmt3a and Dnmt3b enzymes play a crucial role in embryo development. Mice with mutation of the *DNMT3* gene fail to develop to term, whereas mice with a loss of DNMT3a activity live only for a short period of time after birth (reviewed in Lan et al., 2010). The expression pattern of these two enzymes suggests that they might act at different stages of development: DNMT3b is expressed from the formation of the blastocyst until day 9.5 of mouse embryonic development (E9.5), which is the time when DNMT3a starts to get expressed.

De novo DNA methylation occurs twice during organismal development following waves of demethylation, which erase the DNA methylation imprints established in the previous generation. The first round of *de novo* DNA methylation occurs early during embryonic development, probably immediately after implantation (Watanabe et al., 2004). Two different cell lineages exist in the blastocyst, **trophoectoderm**, which later gives rise to the placenta, and the **embryoblast**, which is the inner cell mass that gives rise to the embryonic tissues. The initial *de novo* DNA methylation affects the cells of the embryoblast. This leads to an epigenetic asymmetry between the inner cell mass and trophoectoderm cells. Such differences in methylation might reflect the fact that the inner cell mass gives rise to the embryonic and adult tissues, whereas the trophoectoderm tissues are discarded at birth (Hemberger et al., 2009). The second wave of *de novo* DNA methylation occurs later during post-implantation development in **primordial germ cells (PGCs)**. Following the first round of reprogramming, in female PGCs the level of methylation drops by 70%, whereas in male PGCs it drops by 60% (Popp et al., 2010). *De novo* methylation of DNA in the male germ line occurs several days after the first round of reprogramming, between E15 and E16, whereas in the female germ line, the process takes place only postnatally during the oocyte growth phase (Sasaki and Matsui, 2008). After this second wave of *de novo* methylation, the methylation state of PGCs is similar to that of somatic cells, although substantial differences in methylation patterns can be observed in the **parental imprinting control regions (ICRs)** (see box 4-3) (Hemberger et al., 2009).

DNA methylation patterns at various imprinted loci, repetitive sequences and transposons are re-established during gametogenesis by Dnmt3a and a non-catalytic paralog, Dnmt3L (reviewed in Law and Jacobsen, 2010).

4-3. Parental imprinting control regions

Certain genes are expressed in a parent-of-origin-dependent manner (for example, *IGF2, H19*) or in case of the X-chromosome in mammals, in females only one of the two homologous chromosomes is active. The mechanism that enforces those selective expression patterns is called genomic imprinting and relies on methylation of the inactive copy, which is then called the imprinted allele. Imprinting control regions are the regulatory elements that control the imprinting of the according genes.

The finding that Dnmt3L is able to interact with Dnmt3a and unmethylated H3K4 tails led to the suggestion that Dnmt3L might direct *de novo* methylation performed by Dnmt3a to the loci with high numbers of histone 3 molecules with unmethylated lysine 4 (see Figure 4-5) (Law and Jacobsen, 2010).

As it was expected, an inverse correlation was found between levels of H3K4 methylation and allele-specific DNA methylation at several imprinted loci. Also, recent studies demonstrated that an oocyte-specific H3K4 demethylase—lysine demethylase 1B (KDM1B)—was required for the establishment of DNA methylation at several **differentially methylated regions (DMRs)** of imprinted genes. The lack of demethylation of H3K4 resulted in DNA methylation defects leading to a loss of imprinting in developing embryos (reviewed in Law and Jacobsen, 2010). Indeed, the level of H3K4 methylation inversely correlates with that of DNA methylation in many types of mammalian cells. A similar inverse correlation was found between H3K4me2/H3K4me3 and DNA methylation in plants.

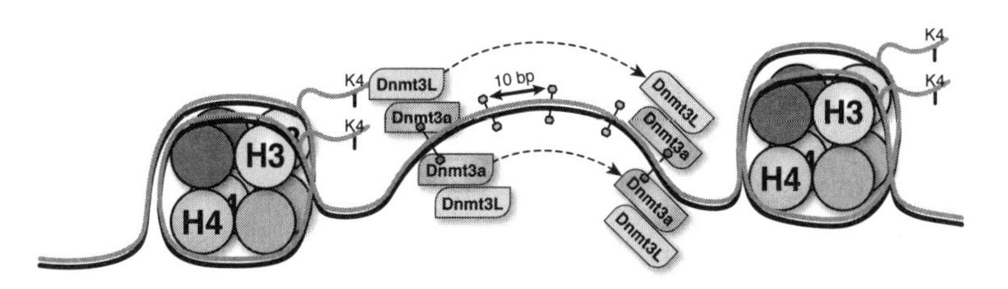

Figure 4-5 Dnmt3L-assisted *de novo* methylation of DNA by Dnmt3a. The dashed line indicates possible process of oligomerization of the Dnmt3L/Dnmt3a complex.

The interaction of Dnmt3a/Dnmt3L with H3K4 might not be sufficient for the establishment of methylation imprints. Evidence indicates that transcription of DMRs might be an initial requirement for the function of Dnmt3a and Dnmt3L. The work of Chotalia et al. (2009) shows that *de novo* methylation in growing oocytes coincides with active transcription of the imprinted *Gnas* locus. The authors suggest that either transcription or the growing mRNA chain may lead to chromatin alterations at imprinted loci that are sufficient for the recruitment of the histone-modifying enzymes and the Dnmt3a/3L complex leading to the establishment of DNA methylation imprints (refer to Figure 4-5).

Although the molecular mechanism of *de novo* methylation mediated by the Dnmt3a-Dnmt3L complex is not entirely clear yet, several possible mechanisms were recently described by Law and Jacobsen (2010). One of them involves a periodic pattern of DNA methylation by Dnmt3a/Dnmt3L tetramers. Co-crystallization of the C-terminal regions of Dnmt3a and Dnmt3L proteins showed that these two proteins interact as dimers forming a tetramer complex in which two Dnmt3a proteins are positioned together with their active sites facing each other and two Dnmt3L proteins are located on either side of the Dnmt3a dimer (Jia et al., 2007). Residues of Dnmt3L may help stabilize the conformation of the active site loop in Dnmt3a, thus explaining the stimulatory effect of Dnmt3L on the activity of Dnmt3a and Dnmt3b proteins. Further *in vitro* analysis revealed that tetramer complexes form oligomers alongside DNA. Because the two active sites of the Dnmt3a dimer appeared to be separated by one turn of DNA helix, it was suggested that each tetrameric complex is able to methylate two cytosines simultaneously, if they happen to occur within a spacer sequence of 8 to 10 nucleotides (Jia et al., 2007).

In DRM sequences, CpG dinucleotides occur with an 8 to 10 bp periodicity more frequently than randomly expected. In general, CpG methylation in the human genome occurs with a certain periodicity, and to a lesser extent in the mouse genome as well (reviewed in Law and Jacobsen, 2010). In contrast to polymerases, DNA methyltransferases are **non-processive**. Thus, oligomerization may help provide a faster and more consistent process for methylation and could explain how the observed periodic pattern of DNA methylation is established.

Still, many questions remain unanswered. It is not clear whether the formation of long Dnmt3a/Dnmt3L DNA/protein filaments occurs *in vivo*. Also, interactions of Dnmt3a/Dnmt3L tetramers with unmethylated H3K4 tails need to be demonstrated *in vivo*. In contrast, in humans, symmetrical methylation of histone 4 tails at the arginine 3 residue (H4R3me2s) recruits Dnmt3a to the β-globin locus, thus resulting in silencing of this gene. Nevertheless, according to the current knowledge, it is highly likely that Dnmt3a/Dnmt3L can directly or indirectly interact with unmethylated H3K4 tails (refer to Figure 4-5).

Other mechanisms that might function in parallel with the one described rely on interactions between Dnmt3a/b and the histone methyltransferases G9a, Suppressor of variegation 3–9 homolog 1 (SUV39H1), Enhancer of zeste homolog 2 (EZH2), and SET domain bifurcated 1 (SETB1) (reviewed in Law and Jacobsen, 2010).

De novo DNA methylation in the germline

Yet another mechanism of ***de novo* DNA methylation** might operate in germline cells. In male gametes, DNA methylation of **transposable elements (TEs)** can occur through the activity of **PIWI-interacting RNAs (piRNAs)** (see Chapters 7 and 10 for more details on piRNA biogenesis). Although initially it was unclear whether piRNAs target TEs, recent analysis demonstrated that the population of piRNAs isolated from mouse embryonic tissue is enriched with piRNAs with homology to repetitive DNA elements. Several recent studies have also demonstrated that piRNAs can regulate the activity of TEs at the level of expression by *de novo* methylation. It was shown that DNA methylation defects in *mili* mutants occurred at the stage in development when *de novo* methylation in male germ cells was observed (Kuramochi-Miyagawa et al., 2008). Further, *mili* mutants exhibit derepressed LINE1 TEs, demonstrating that MILI (and piRNA biogenesis) are important for TE suppression. Also it was demonstrated that piRNAs isolated from tissues at this developmental stage were highly enriched in transposon sequences (Aravin et al., 2008).

Although the exact process of *de novo* methylation by piRNAs is not clear, it has been suggested that PIWI-piRNA complexes either directly interact with nascent transposon transcripts and recruit *de novo* methyltransferases or recruit them indirectly through interaction with intermediate chromatin modifiers thereby targeting *de novo* methylation to transposon sequences. Because no interaction between PIWI argonautes and Dnmt3 proteins has been demonstrated, the indirect mechanism of recruitment of Dnmt3 proteins currently seems more likely (Aravin et al., 2008). The mouse PIWI homologs, Miwi, Miwi2, and Mili, were shown to interact with various Tudor domain-containing (Tdrd) proteins. *tdrd1* mutants exhibit transposon reactivation and Mili-associated piRNAs contain a lower proportion of transposon-related sequences, and decreased levels of antisense Miwi2-bound piRNAs were observed.

The maintenance of DNA methylation

There are several active and passive mechanisms for loss of DNA methylation, including the processes of DNA replication and repair. Replication results in hemimethylated DNA. In mammals, the Dnmt1 enzyme is the main maintenance DNA methyltransferase that has been found to associate with replication foci. An *in vitro* study demonstrated that Dnmt1 has a higher affinity to unmethylated and hemimethylated DNA. Up-to-date, five *DNMT1* genes have been identified including DNMT1s, DNMT1o, DNMT1b, DNMT1$^{\Delta E3-6}$, and DNMT1p, although the latter one is not translated. DNMT1s is the main, active maintenance MTase in mammals and is also known as DNMT1 or DNMT1a. It consists of an N-terminal regulatory domain and a C-terminal catalytic domain connected by a glycine–lysine (GK) repeat (reviewed in Lan et al., 2010).

The proliferating cell nuclear antigen (PCNA) is involved in methylation by recruiting Dnmt1 to replication foci. However, disruption of the interaction between PCNA and Dnmt1 only leads to minor reductions in DNA methylation. The interaction between Dnmt1 and Ubiquitin-like PHD and RING finger domain 1 (UHRF1) or between Dnmt1 and the chromatin-remodeling factor Lymphoid-specific Helicase (LSH1) might also be important for maintenance of DNA methylation. UHRF1 might be required to direct Dnmt1 to chromatin as the

UHRF1 mutant exhibits severe hypomethylation. Further, the SET and RING-associated (SRA) domain of UHRF1 protein is known to bind to hemimethylated CpG dinucleotides, which might help targeting DNMTs to hemimethylated DNA. The finding that UHRF1 interacts with Dnmt3a and Dnmt3b suggests that it might be a universal protein that helps recruiting DNA methyltransferases to unmethylated or hemimethylated DNA.

The role of LSH1 in controlling histone modifications and DNA methylation still remains unclear. Mutation of *DDM1* results in a decrease in histone 3 lysine 9 (H3K9) methylation, a modification that correlates with DNA methylation and silencing. In contrast, mutations in *LSH1* do not result in a decrease in histone 3 lysine 9 (H3K9) methylation, and thus they do not influence levels of DNA methylation and silencing (reviewed in Law and Jacobsen, 2010).

Active and passive mechanisms of DNA demethylation

In mammals, the enzymatic machinery (or machineries) involved in the active removal of methyl groups from CpG sites remained enigmatic for a long time. Initial attempts to identify orthologs of plant 5-methyl-cytosine DNA glycosylases such as ROS1 and DEMETER (DME) failed. In the search for potential active DNA demethylases, much attention has been given to two classes of enzymes that are expressed in cells undergoing active demethylation: **activation-induced cytosine deaminase (AID)** and **apolipoprotein B RNA-editing catalytic component 1 (APOBEC1)** (reviewed in Sanz et al., 2010). AID catalyzes deamination of 5-methylcytosine, resulting in T:G mismatches. In developing B cells, AID functions as a single-strand DNA deaminase required for somatic hypermutation and class switch recombination at immunoglobulin genes. The AID-mediated U-G mismatches initiate repair processes resulting in increased levels of recombination at immunoglobulin genes.

The finding that AID is expressed in PGCs led to the hypothesis that it is involved in active hypomethylation (Popp et al., 2010). The analysis of methylation in PGCs in *aid* mutants showed that they have a significantly higher methylation level than wild-type mice. Similarly, in heterokaryons obtained by fusing mouse embryonic stem cells and human fibroblasts, the removal of DNA methylation from the *NANOG* and *OCT4* genes in the fibroblast-derived genome was dependent on AID; in *aid* mutants, *NANOG* and *OCT4* promoters remain highly methylated. These two studies showed that in mammals, active demethylation might occur through the process of cytosine deamination. However, AID-dependent demethylation is not likely to be the main process in mammals as PGCs in *aid* mutants still exhibited much lower increase in methylation levels as compared to other cell types of the mutant. Similarly, the offspring of *aid* mice did not show any significant developmental defects (Popp et al., 2010). Therefore, some other processes might be involved in active demethylation. Indeed, another component was identified in the recent work of Yuki Okada and co-workers (Okada et al., 2010). In order to study DNA demethylation in the paternal pronucleus in zygotes, the authors transfected those nuclei with siRNAs against various potential targets for siRNA-mediated knock-down. Targeting of a component of the elongator complex involved in transcription elongation—elongator protein 3 (ELP3)—impaired DNA demethylation and therefore showed that this is a protein required for the removal of DNA methylation in the zygotic, paternal pronuclei (Okada et al., 2010). Authors demonstrated that the knockdown of Elp3 in zygotes prevented paternal DNA demethylation, along with two other components of the elongator complex—ELP1 and ELP4—that were equally involved. Because the elongator complex was shown to have histone acetylase activity, the potential involvement of ELP proteins in the process of DNA demethylation is a very appealing idea for further investigation.

Passive demethylation is a process in which methylation of hemimethylated and unmethylated sequences generated during DNA replication and repair does not occur or is substantially inhibited. As such, it requires the inhibition of DNA maintenance methylation by Dnmt1 and may be directed to specific loci. In mammals, exclusion of the oocyte-specific form of Dnmt1—Dnmt1o—from nuclei until just prior to blastocyst formation results in passive demethylation during preimplantation development of the embryo (reviewed in Law and

Jacobsen, 2010). In mammals, the expression of Dnmt1 is regulated by the retinoblastoma (Rb) pathway, which includes the transcription factor Rb and RbAp48, homologs of the plant RETINOBLASTOMA RELATED 1 (RBR1) and MULTICOPYSUPPRESSOR of IRA1 (MSI1) proteins. During gametogenesis, RBR1 and MSI1 repress the plant maintenance DNA methyltransferase (MET1) expression. Thus, although not much is known about the role of passive DNA demethylation in mammals, its mechanisms may be similar to those occurring in plants (refer to "Active and passive loss of methylation in plants," later in this chapter).

Cytosine methylation in plants

In plant genomes, DNA methylation is more versatile than in the genome of animals, with cytosine methylation occurring symmetrically at both **CpG** and **CpHpG** (where H = A, T, or C) sites and **asymmetrically** (methylation of cytosine on only one strand of the DNA double-helix) at **CpHpH** sites. In *Arabidopsis thaliana*, cytosine methylation in CpG, CpHpG, and CpHpH nucleotides occurs with frequencies of 24%, 6.7%, and 1.7%, respectively (Cokus et al., 2008). This frequency, however, depends on the sequence context. For example, in gene coding sequences, methylation at CpG sites occurs with a frequency of 30%, whereas at CpHpG and CpHpH sites, it is less than 1% (Widman et al., 2009). Methylation of duplicated genes occurs at 20%, 0.7%, and 0.4% at CpG, CpHpG, and CpHpH sites, respectively. In contrast, tandem and inverted repeats are methylated at CpG, CpHpG, and CpHpH sites with frequencies of more than 50%, 20%, and 5%, respectively. Unique tandem repeats are methylated at lower levels—39.4%, 12.6%, and 2.7% at CpG, CpHpG, and CpHpH sites, respectively (summarized in Widman et al., 2009). In plants, levels of overall cytosine methylation appear to be higher than those in animals, with roughly twice as many cytosines being methylated in plants as compared to animals. Unlike in animals, DNA methylation in plants occurs preferentially at repetitive DNA elements, including transposons (Zhang et al., 2006).

Type of DNA methylations and factors involved in DNA methylation in plants

DNA methylation at symmetrical CpG and CpHpG sites is inherited during replication in its hemimethylated forms, thus establishing methylation imprints on the parental DNA, which can guide the activity of methyltransferases. In contrast, methylation at asymmetrical CpHpH sites can only be reestablished *de novo* and must occur after each replication cycle because no hemimethylated sequence is available to guide the remethylation process.

Experimental evidence suggests the existence of three distinct classes of enzymes responsible for cytosine methylation in plants (see Table 4-1). In plants, *de novo* methylation occurs by DOMAINS REARRANGED METHYLTRANSFERASE 2 (DRM2), a homolog of the mammalian Dnmt3 family. The maintenance of DNA methylation relies on three different pathways depending on the sequence context. CpG methylation is maintained by the plant homolog of Dnmt1, DNA METHYLTRANSFERASE 1 (MET1), whereas CpHpG methylation is maintained by a plant specific DNA methyltransferase, CHROMOMETHYLASE (CMT3). **Asymmetric CpHpH methylation** is maintained through persistent *de novo* methylation by DRM2 (reviewed in Law and Jacobsen, 2010).

Plants defective in MET1 activity show a lack of widespread CpG methylation. Whereas *dnmt1* mutations are lethal for animals, plants deficient in MET1 activity are viable. However, when *met1* plants are inbred for several generations, they exhibit a strong progression of abnormalities. Experimental evidence exists that suggests that MET1 might not be required for establishing new methylation imprints. However, the establishment and maintenance of *de novo* CpG methylation at targeted sequences during **RNA-directed DNA methylation (RdDM)** requires the activity of MET1.

The second class of MTases, CMT3, is unique to plants (refer to Table 4-1). Loss-of-function *cmt3* mutants are characterized by a genome-wide loss of CpHpG methylation (especially at centromeric repeats and transposons) and a decrease in CpHpH methylation in several genomic regions. The difference in phenotypes between *dnmt1* and *cmt3* mutants may be explained by the fact that MET1 can substitute for CMT3 in CpG and CpHpG methylation-dependent silencing of *CACTA* transposons.

Table 4-1 The plant methyltransferases involved in DNA methylation

Name	Effects on Chromatin	Effects of Mutation	Modification/ Transcription
METHYLTRANSFERASE1 (MET1)	Methylation of symmetrical CpG sites; not required for establishing new methylation imprints; involved in the RdDM pathway	A lack of CpG methylation; a passive loss of DNA methylation throughout generations.	Global/ Repression
CHROMOMETHYLASE3 (CMT3)	The maintenance of CpHpG methylation; functionally redundant with MET1 and DRM2 in methylation of CpG and CpHpG sites, respectively; targets centromeric repeats and transposons	Loss of CpHpG methylation	Global/ Repression
DOMAIN REARRANGED METHYLTRANSFERASES (DRM1, DRM2)	*De novo* methylation of asymmetric sites; functionally redundant with CMT3 in CpHpG methylation; the maintenance of CpHpH; DRM2 mediates establishing of *de novo* CpG methylation in the RdDM pathway	Loss of *de novo* asymmetric methylation at non-CpG sites	Global/ Repression

The last class is represented by DRM1 and DRM2 (refer to Table 4-1) that direct *de novo* methylation of cytosines in different sequence contexts, including *de novo* non-CpG methylation of RdDM-targeted sequences.

De novo DNA methylation in plants

De novo DNA methylation in plants occurs in both gametic and somatic cells. An initial DNA methylation imprint can be established by DRMs at symmetric and asymmetric sites in response to environmental stimuli and is then perpetuated by MET1 and CMT3 at symmetric CpG and CpHpG sites, respectively (Cao and Jacobsen, 2002a, b). In Arabidopsis, nucleotide-resolution DNA methylation mapping revealed an element of periodicity, which is similar to the periodicity of *de novo* DNA methylation observed in mammals. Genome-wide methylation of CpHpH sites by DRM2 occurs with a periodicity of about 10 bps. Methylation of CpHpG sites by CMT3 occurs with a periodicity of 167 bp, reflecting the size of a nucleosome. RdDM is the main mechanism responsible for directing *de novo* DNA methylation to specific sites.

RdDM relies on the activity of RNA Polymerase V (Pol V) and siRNAs targeting a particular locus. SUPPRESSOR OF TY INSERTION 5-LIKE (SPT5-like) and the Pol V subunit NUCLEAR RNA POLYMERASE E1 (NRPE1) interact with AGO4. It is speculated that SPT5-like serves as an adaptor protein that binds to both AGO4 and nascent Pol V transcripts, aiding in the recruitment of AGO4/siRNA complexes to Pol V transcribed loci (reviewed in Law and Jacobsen, 2010). It is further hypothesized that this interaction recruits DRM2 to establish DNA methylation.

INVOLVED IN DE NOVO 2 (IDN2) is another downstream effector of RdDM. IDN2 is homologous to SUPPRESSOR OF GENE SILENCING 3 (SGS3), a protein involved in post-transcriptional gene silencing (PTGS) (see Chapters 11 and 15). Both SGS3 and IDN2 have an XS domain required for binding to double-stranded RNA with 5' overhangs (Ausin et al., 2009). It is hypothesized that IDN2 binds to AGO4/siRNA complexes and that this interaction serves as a signal recruiting DRM2, which is responsible for *de novo* DNA methylation (see Chapter 11, "Non-Coding RNAs Across the Kingdoms—Plants," for more details).

Silencing of some loci may also involve Pol II-dependent noncoding transcripts, as several genes were found to be transcriptionally activated in weak mutants of the Pol II subunit nuclear RNA polymerase B2 (*nrpb2*), indicating a reduction in silencing of those genes. Also, NRPB2 aids in the association of NRPE1 and NRPD1 with chromatin. Although the exact relationship between Pol II, Pol V, and Pol IV remains elusive, Pol II transcripts are suggested to act as scaffolds for the recruitment of RdDM factors including AGO4, and possibly Pol IV and Pol V. Genetic screens for RdDM factors identified a conserved protein, RNA-DIRECTED DNA METHYLATION 4 (RDM4)/DEFECTIVE IN MERISTEM SILENCING 4 (DMS4), with similarity to the yeast protein Interacts with Pol II (IWR1) (reviewed in Law and Jacobsen, 2010). Several proteins involved in Pol IV- and Pol V-dependent *de novo* methylation have been recently identified. NRPE1 association with chromatin and the accumulation of IGN transcripts require DEFECTIVE IN RNA-DIRECTED DNA METHYLATION 1 (DRD1), a chromatin-remodeling factor, and DEFECTIVE IN MERISTEM SILENCING 3 (DMS3), an RdDM component with similarity to Structural Maintenance of Chromosome (SMC) proteins (reviewed in Law and Jacobsen, 2010). The mechanism of RdDM and its possible role in siRNA-directed DNA methylation are covered in more details in Chapters 10, 11, and 15.

De novo DNA methylation in the germline

Plants do not establish germline cells early in their development and their gametes are produced by additional post-meiotic mitotic divisions, whereas in animals gametes are the direct products of meiotic cell divisions. Like animals, plant gametes undergo a substantial genome-wide loss of cytosine methylation. Due to differences in gamete formation, reprogramming of DNA methylation in plants occurs in a different way. During male gametogenesis, three cells are produced in the pollen grains—a **vegetative cell** nucleus and two **sperm cells** produced by division of the **generative cell**. The two sperm cells then fertilize the central cell and egg cells in the female gametophyte generating the embryo and endosperm, respectively. Whereas the chromatin of sperm cells is found in its condensed form, the chromatin in vegetative cells is decondensed, thus allowing for the expression of various genes. It is important to note that genetic information is passed to the progeny via sperm cells, not via vegetative cells. Slotkin et al. (2009) isolated sperm cells and analyzed their cytosine methylation status. Surprisingly, cytosines in all sequence contexts were highly methylated on both the sense and antisense strands in sperm cells; methylation levels at CpHpG and CpHpH sites were increased compared to whole pollen and leaf DNA extracts, with slightly decreased methylation levels in pollen compared to leaf. These data allowed the authors to conclude that hypomethylation primarily occurs in vegetative cells rather than sperm cells. The data on the extent of changes in DNA methylation in pollen are scarce. Huang et al. (2010) found that in David lily, methylation levels decreased from 54.8% to 33.5% during prophase of meiosis I in pollen.

Furthermore, Slotkin et al. also observed a transient re-activation of transposons in pollen, which appeared to be the result of the observed hypomethylation in the genome of vegetative cells (Slotkin et al., 2009). This is paralleled by downregulation of RdDM components, with DECREASED DNA METHYLATION 1 (DDM1), a chromatin-remodeling factor required for the maintenance of CpG methylation, being excluded from the vegetative nucleus (reviewed in Law and Jacobsen, 2010). When analyzing siRNA sequences in pollen and isolated sperm cells, Slotkin et al. (2009) noticed an accumulation of 21 nt siRNAs from transposons which are expressed only in vegetative cells. They hypothesized that the vegetative cells produce these siRNAs to reinforce silencing of transposons in the sperm cells (see Figure 4-6).

In respect to the female gametophyte, recent reports suggest that global methylation patterns in the genome of **endosperm cells** originate in demethylation of the genome in the central cell of the female gametophyte (Hsieh et al., 2009). Similarly, in *Zea mays*, a reduction in DNA methylation in the endosperm of approximately 13% reduction was found as compared to somatic tissue (Lauria et al., 2004). Despite a decrease in global DNA methylation in endosperm cells compared to embryonic tissue and somatic tissue, Hsieh et al. (2009) found increased CpHpH methylation in both the endosperm and embryo tissues relative to adult shoot tissue. The authors hypothesized that an increase in non-symmetrical cytosine methylation might have occurred due to activation of RdDM. Indeed, siRNAs derived from maternal tissues accumulate to high levels in the endosperm. This enabled them to further speculate that similar to silencing of transposons in the sperm cells by siRNAs derived from the vegetative cells, in the female gametophyte, siRNAs generated in the central cell might maintain silencing in the **egg cell** and possibly in the developing embryo (reviewed in Law and Jacobsen, 2010) (see Figure 4-6).

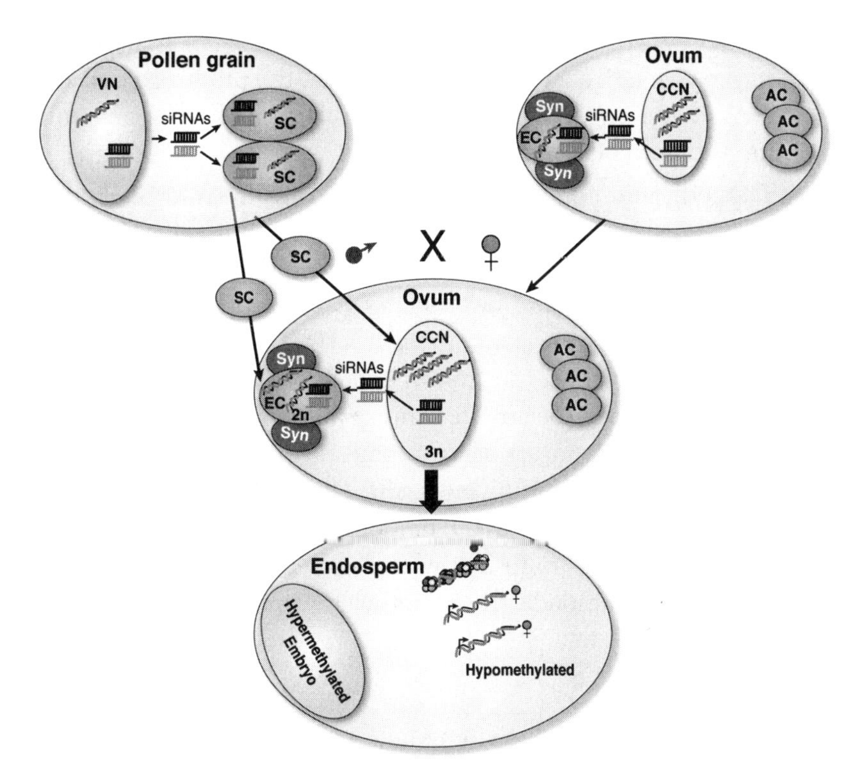

Figure 4-6 Changes in methylation in plant gametes and upon fertilization. Male gametogenesis generates tricellular pollen grain consisting of vegetative nucleus (VN) and two sperm cells. SCs are hypomethylated and produce siRNAs that are presumably transported to VN and reinforce methylation of transposons. Female gametogenesis generates a diploid (2n) central cell nucleus (CCN), a haploid egg cell (EC), three antipodal cells (ACs), and two synergid cells (Syn). Global demethylation in CCN leads to siRNA accumulation that might reinforce silencing of transposons in EC. Double fertilization events generate triploid endosperm and diploid embryo. Male genome in the endosperm is imprinted. Whereas endosperm is hypomethylated, embryo is hypermethylated.

Why do gametic tissues need to be "reminded" to silence the transposons? Transposon activity in germ line cells is mostly deleterious for the plant genome, as it induces genome rearrangements that would directly be inherited by the progeny. But if transposons are not expressed, how then does the plant cell know what to silence? The activation of transposons and thus transposons-derived siRNAs may serve a dual purpose, namely the following:

- DNA demethylation followed by the activation of transposons in the vegetative and central cells allows the production of siRNAs that can serve as constant reminders of what genomic areas should be silenced.
- Demethylation may function as a "test run" that indicates what transposons may be activated if the genome is hypomethylated, thus ensuring that these elements are efficiently silenced, as suggested by Law and Jacobsen (2010).

This makes perfect sense from the view of adaptation to a constant genetic flow that occurs in plant genomes packed with transposons as every new insertion of a transposon generates a likely candidate for jumping/multiplying again. Therefore, not surprising is the finding of Teixeira et al. (2009) demonstrating that siRNA loci but not other genomic loci are re-methylated if methylation is lost in previous generations, although the remethylation process requires several generations.

Maintenance of DNA methylation in plants

Plants appear to maintain CpG methylation in a similar way observed in mammals: All proteins involved in the maintenance DNA methylation appear to be similar in function to those in mammals. MET1, the VARIANT IN METHYLATION/ORTHRUS (VIM/ORTH) family of SRA domain proteins and DDM1 all have mammalian counterparts: Dnmt1, UHRF1, and LSH1. In plants, DNA methylation is maintained through the function of MET1 and its cofactor VIM1. VIM1 mediates the recognition of hemimethylated DNA sequences at replication foci and ensures a correct transfer of epigenetic information to the newly synthesized DNA strand. The maintenance of CpG methylation in plants might also require the activity of histone deacetylase HDA6. It still remains to be established whether plant proteins maintain DNA methylation by the same molecular mechanism as the Dnmt1/UHRF1/LSH1 group. However, it is known that in *ddm1* but not in *lsh1*, H3K9 methylation (see details in Chapter 5, "Histone Modifications and Their Role in Epigenetic Regulation"), a modification that positively correlates with DNA methylation, is decreased. Thus, there might be some specific differences in the regulation of histone modifications and maintenance of DNA methylation between plants and animals.

One of the substantial differences observed between plant and animal genomes in terms of methylation of coding regions is the fact that in Arabidopsis, nearly one-third of genes are methylated at CpG sites (reviewed in Law and Jacobsen, 2010). The analysis of gene expression and methylation in coding regions showed that genes that are moderately expressed appear to be methylated in coding regions, whereas genes of low or high expression levels tend to lack methylation in those regions. The expression of genes methylated within the coding sequences is upregulated in *met1*.

This is in striking contrast to what is observed for methylation of transposons, in which methylation at the gene body results in silencing. Although CpG methylation within gene bodies rarely occurs in mammals, it is found in invertebrates, suggesting that it might be a common feature of eukaryotic genomes (reviewed in Law and Jacobsen, 2010).

In contrast to non-CpG methylation that is substantially reduced when siRNA signals are removed, CpG methylation can be maintained for many generations. This is supported by a study showing that CpG methylation of the 35S promoter driving GFP expression was inherited by the progeny of virus infected plants. Therefore, the maintenance of CpG methylation may play a more crucial role in the stable inheritance of epigenetic marks as compared to other symmetric and asymmetric methylation marks. Indeed, plants deficient in the maintenance of CpG methylation display uncoordinated formation of aberrant epigenetic patterns over successive generations.

Maintenance of non-CpG methylation in plants

In plants, non-CpG methylation is maintained by the activity of CMT3 and DRM2. CpHpG methylation seems to require H3K9 methylation; in this process, H3K9 methylation and DNA methylation may reinforce each other. A loss of either CMT3 or SU(VAR)3—9 HOMOLOG 4 (SUVH4)/KRYPTONITE (KYP) responsible for H3K9me2 leads to a severe decrease in DNA methylation (Jackson et al., 2002). Besides its histone methyltransferase domain, KYP also has an SRA domain that allows binding to CpHpG sites. In turn, CMT3 possesses a **chromatin organization modifier domain (chromodomain)** that allows it to bind methylated histone H3 tails. So, it can be hypothesized that KYP first binds the CpHpG sites and then recruits CMT3 that methylates cytosines at these sites. Again, it remains to be established whether KYP and CMT3 actually interact *in vivo*. Two other H3K9 histone methyltransferases, SUVH5 and SUVH6, also contribute to global levels of CpHpG methylation, and these proteins are also able to interact with CMT3.

Asymmetric DNA methylation does not require maintenance methylation *per se* because DNA replication or repair removes cytosine methylation from the area of asymmetric methylation completely. Thus, CpHpH methylation is maintained through persistent *de novo* methylation by CMT3, DRM2, and the RdDM process.

Similar to **symmetrical methylation**, **non-symmetrical methylation** via RdDM also requires proteins with SRA domains, such as SUVH9 and SUVH2, that bind CpHpH or CpG, respectively, and these proteins might also recruit CMT3 or DRM2 to the targeted CpHpH sites (reviewed in Law and Jacobsen, 2010).

Active and passive loss of methylation in plants

Although the presence of DNA-methylating enzymes is well proven, the existence of mechanisms of active DNA demethylation still remains controversial. In the past, it was suggested that **nucleotide excision repair** (**NER**) could actively demethylate DNA. However, because activation of NER requires a distortion of the double-stranded DNA helix, it is unclear how methylated cytosines can recruit NER. In contrast, a passive loss of DNA methylation may occur due to inhibition of *de novo* DNA methylation or an inability to maintain the parental imprint after DNA replication and repair of DNA damage.

Glycosylases specific to methylated cytosines, such as the REPRESSOR OF SILENCING1 (ROS1) protein or DME protein, are considered to be ideal candidates for active DNA demethylation. Gong et al. (2002) isolated the *ROS1* gene that encodes a DNA glycosylase/lyase and showed that it has high affinity for methylated but not unmethylated cytosines (see Table 4-2). In *ros1* mutants, transcriptional gene silencing is greatly increased but can be released by *ddm1* mutations or treatment with DNA methylation inhibitors, supporting a role of ROS1 in active DNA methylation (Gong et al., 2002). Another protein involved in active DNA demethylation is the DNA glycosylase DME that regulates the gametophyte-specific activation of flowering time (*FWA*) gene expression (see Table 4-2). It was also shown to reverse imprinting of maternal copies of the *MEDEA* allele in the endosperm, further supporting the role of DME as an active DNA demethylase. Overall, the protein domains of DME are highly similar to those in ROS1. These proteins together with two DME-LIKE proteins, DEMETER-LIKE2 (DML2) and DML3, form the DEMETER family of DNA glycosylases.

Another protein, ROS3, that contains an RNA-binding motif and is a member of the ROS1-mediated DNA demethylation pathway, has been identified (Zheng et al., 2008). ROS3 binds to small RNAs *in vitro* and *in vivo* and colocalizes with ROS1 in discrete foci dispersed throughout the plant nucleus. It can be hypothesized that ROS1 can be targeted to specific genome loci using smRNAs bound to ROS3. This places ROS3 as an important functional link between smRNA biogenesis and DNA demethylation pathways.

Passive loss of methylation in the central cell occurs in parallel with active loss guided by DME. Recently Jullien et al. (2008) demonstrated that *MET1* expression is substantially decreased in female gametes and showed that MSI1 and RBR1 repress *MET1* transcription. The authors also showed that MSI1 and RBR1 are required for maternal expression of the imprinted *FIS2* and *FWA* genes, suggesting that active demethylation by DME and passive DNA demethylation due to the down-regulation of *MET1* expression work together for activation of imprinted genes. As summarized by Law and Jacobsen (2010), several observations make this prediction plausible. First, the activity of DME on hemimethylated DNA is higher than its activity on fully methylated DNA. Because female gametes are enriched in hemimethylated DNA due to low levels of MET1 activity, DME might be more active in female gametes. Second, hemimethylated DNA is less likely to give rise to double-strand breaks (DSBs) that typically occur upon the removal of symmetrically methylated cytosines by DME. Because DME removes methylated cytosines across from basic sites with lower efficiency, the result is lower rates of DSB in female gametes. Third, MET1 might be downregulated in order to prevent remethylation of hemimethylated CpG sites generated by DME activity on one strand of the DNA.

Table 4-2 The plant factors involved in active cytosine demethylation

Name and Function	Effects on Chromatin	Effects of Mutation	Modification/ Transcription
DEMETER (DME)/DNA glycosylase	Demethylation of previously silenced sequences, possibly in a tissue-specific manner.	Inability to activate imprinted genes; inheritance of the mutant maternal allele results in seed abortion.	Local, promoters/ Activation
REPRESSOR OF SILENCING1 (ROS1)/DNA glycosylase/lyase	Demethylation activity on methylated and not on unmethylated DNA substrates.	Hypermethylation and transcriptional silencing of specific genes; enhanced sensitivity to genotoxic agents.	Local, promoters/ Activation
DEMETER-LIKE (DML) proteins: DML2 and DML3/DNA glycosylase/lyase	Demethylation activity is primarily localized at the 5' and 3' ends of genes preventing the accumulation of methylation at or near genes. DML proteins remove aberrant 5' and 3' methylation from genes and prevent the formation of highly methylated stable epialleles.	Hypermethylation of gene sequences at either the 5' or 3' end; *dml* mutant hypermethylation has a negligible effect on gene expression.	Local, 5' and 3' ends of genes/ Mostly unaffected

Finally, passive demethylation might also rely on exclusion of other proteins from being expressed in a given gametic cell. Like in mammals, in plants, Dnmt1o protein is excluded from the nucleus just before the blastocyst is formed. Also, the expression of *DDM1* is observed in the sperm cells but not in the vegetative nucleus. Hence, plants and mammals might utilize similar strategies to passively demethylate the genomes of reproductive cells.

Conclusion

Methylation of DNA as an epigenetic modification is observed in organisms of all evolutionary levels—from prokaryotes to eukaryotes. However, there are differences in biochemistry, mediating enzymes and function. In bacteria and Archaea, the most abundant form is methylation of the N6 position of adenine, whereas in eukaryotes the prevalent modification is C5-cytosine methylation.

In bacteria and Archaea, many methyltransferases are part of restriction-modification systems and are important to protect "self"-DNA from restriction. However, there are also a few examples that lack cognate restriction enzymes, such as Dam and CcrM. Those mediate their function by regulating the expression of certain target genes or by affecting DNA curvature and thus regulatory protein binding. They play key roles in the regulation of various cellular processes ranging from cell cycle regulation to DNA repair (strand recognition) to virulence. Similar to eukaryotes, they establish and maintain specific DNA methylation patterns (DMPs).

In eukaryotes, adenine methylation is way less common; instead, eukaryotic genomes exhibit abundant cytosine methylation. However, the specific targets for methylation are different between different kingdoms of life. In mammals, CpG cytosines are the prevalent target for modification, whereas in plants also CpHpG and CpHpH sequences get methylated. Methylation patterns are established by *de novo* methyltransferases (DNMT3a, b in mammals and DRM2 and CMT3 in plants). Plants also utilize an RNA-dependent DNA methylation (RdDM) mechanism for the establishment of sequence-specific methylation patterns. As replication and DNA repair may result in loss of methylation if not maintained, maintenance methyltransferases (DNMT1 in mammals and MET1 in plants) act on hemimethylated sequences.

DNA methylation in eukaryotes generally represses gene expression; however, this also depends on the sequence context that is methylated (gene body versus promoter region). Further, DNA methylation enforces imprinting patterns. Those are established early on during development, which is why in both mammals and plants demethylation during gamete and early embryonic development is crucial for proper establishment of imprinting patterns.

Finally, both establishment and function of DNA methylation patterns also interact with other epigenetic marks, such as histone modifications. Implying that considering the entirety of epigenetic modifications in a specific sequence context might be crucial to the understanding of the outcome.

Exercises and discussion topics

1. What chemical modification(s) do we call DNA methylation and how can they affect gene expression?
2. What is the restriction modification system? Why can it be considered a parasitic mechanism?
3. Explain the regulation and function of DMPs in bacteria using an example.
4. Highlight the role of epigenetic regulation in bacterial DNA replication.
5. Predict how MMR will be affected in a Dam-impaired mutant of *E. coli*.
6. What is transposition, and why might it be beneficial to allow for transposition just after replication?
7. How does Dam affect the expression of *traJ* and what is the effect of methylation, hemimethylation, and no methylation on conjugation in this case?
8. List similarities and differences between Dam and CcrM methylases. Compare their roles in cell cycle regulation.
9. Which are the target sequences for cytosine methylation and what is the frequency of methylation in comparison between animals and plants?
10. What is the difference between *de novo* and maintenance methylation? Why are some DNMTs called *de novo* methyltransferases and some maintenance methyltransferases?

11. How does DNMT3a know which cytosine to methylate? Describe one possible molecular mechanism.
12. Compare and contrast piRNA-mediated *de novo* methylation in mammals and RNA-dependent DNA methylation in plants.
13. Why is it necessary to have mechanisms for the maintenance in addition to *de novo* methylation mechanisms? How are methylation patterns maintained during DNA replication and repair?
14. At what developmental stage does DNA demethylation occur in plants as compared to animals and how is it mediated?
15. Why do germ line tissues need to be reminded to silence transposons and how is it achieved in male as compared to female gametophytes in plants?
16. How are the different types of methylation (symmetric/asymmetric) inherited/maintained in plants?
17. Compare the mechanisms for demethylation in plants and mammals. Describe a possible molecular pathway for active demethylation in plants.

References

Aravin et al. (2008) A piRNA pathway primed by individual transposons is linked to de novo DNA methylation in mice. *Mol Cell* 31:785-799.

Ausin et al. (2009) IDN1 and IDN2 are required for de novo DNA methylation in Arabidopsis thaliana. *Nat Struct Mol Biol* 16:1325-1327.

Chotalia et al. (2009) Transcription is required for establishment of germline methylation marks at imprinted genes. *Genes Dev* 23:105-117.

Cokus et al. (2008) Shotgun bisulphite sequencing of the Arabidopsis genome reveals DNA methylation patterning. *Nature* 452:215-219.

Fishel R. (1998) Mismatch repair, molecular switches, and signal transduction. *Genes Dev* 12:2096-101.

Gong et al. (2002) ROS1, a repressor of transcriptional gene silencing in Arabidopsis, encodes a DNA glycosylase/lyase. *Cell* 111:803-814.

Hattman S. (1982) DNA methyltransferase-dependent transcription of the phage Mu mom gene. *Proc Natl Acad Sci USA* 79:5518-5521.

Hemberger et al. (2009) Epigenetic dynamics of stem cells and cell lineage commitment: digging Waddington's canal. *Nat Rev Mol Cell Biol* 10:526-537.

Hernday et al. (2002) Self-perpetuating epigenetic pili switches in bacteria. *Proc Natl Acad Sci USA* 99:16470-16476.

Hsieh et al. (2009) Genome-wide demethylation of Arabidopsis endosperm. *Science* 324:1451-144.

Huang et al. (2010) Developmental changes in DNA methylation of pollen mother cells of David lily during meiotic prophase I. *Mol Biol (Mosk)* 44:853-858.

Jackson et al. (2002) Control of CpNpG DNA methylation by the KRYPTONITE histone H3 methyltransferase. Nature 416:556-560.

Jia et al. (2007) Structure of Dnmt3a bound to Dnmt3L suggests a model for de novo DNA methylation. *Nature* 449:248-251.

Kobayashi I. (2001) Behavior of restriction-modification systems as selfish mobile elements and their impact on genome evolution. *Nucleic Acid Res* 29:3742-3756.

Kuramochi-Miyagawa et al. (2008) DNA methylation of retrotransposon genes is regulated by Piwi family members MILI and MIWI2 in murine fetal testes. *Genes Dev* 22:908-917.

Lan et al. (2010) DNA methyltransferases and methyl-binding proteins of mammals. *Acta Biochim Biophys Sin* 42:243-252.

Lauria et al. (2004) Extensive maternal DNA hypomethylation in the endosperm of Zea mays. *Plant Cell* 16:510-522.

Law JA, Jacobsen SE. (2010) Establishing, maintaining, and modifying DNA methylation patterns in plants and animals. *Nat Rev Genet* 11:204-220.

Low et al. (2001) Roles of DNA adenine methylation in regulating bacterial gene expression and virulence. *Infect Immun* 69:7197-7204.

Mahan MJ, Low DA. (2001) DNA methylation regulates bacterial gene expression and virulence. *ASM News* 67:356-361.

Marczynski GT, Shapiro L. (2002) Control of chromosome replication in *Caulobacter crescentus*. *Annu Rev Genet* 56:625-656.

Niculescu MD, Zeisel SH. (2002) Diet, methyl donors, and DNA methylation: interactions between dietary folate, methionine, and choline. *J Nutr* 132(8 Suppl):2333S-2335S.

Okada et al. (2010) A role for the elongator complex in zygotic paternal genome demethylation. *Nature* 463:554-558.

Polaczek et al. (1997) Role of architectural elements in combinatorial regulation of initiation of DNA replication in Escherichia coli. *Mol Microbiol* 26:261-275.

Popp et al. (2010) Genome-wide erasure of DNA methylation in mouse primordial germ cells is affected by AID deficiency. *Nature* 463:1101-1105.

Reisenauer A, Shapiro L. (2002) DNA methylation affects the cell cycle transcription of the CtrA global regulator in *Caulobacter*. *EMBO J* 21:4969-4977.

Robert et al. (2003) DNMT1 is required to maintain CpG methylation and aberrant gene silencing in human cancer cells. *Nat Genet* 33:61-65.

Rountree MR, Selker EU. (2010) DNA methylation and the formation of heterochromatin in Neurospora crassa. *Heredity* 105:38-44.

Sanz et al. (2010) Genome-wide DNA demethylation in mammals. *Genome Biol* 11:110.

Sasaki H, Matsui Y. (2008) Epigenetic events in mammalian germ-cell development: reprogramming and beyond. *Nat Rev Genet* 9:129-140.

Selker EU (1998). Trichostatin A causes selective loss of DNA methylation in Neurospora. *Proc Natl Acad Sci USA* 95:9430-9435.

Slotkin et al. (2009) Epigenetic reprogramming and small RNA silencing of transposable elements in pollen. *Cell* 136:461-472.

Teixeira et al. (2009) A role for RNAi in the selective correction of DNA methylation defects. *Science* 323:1600-1604.

van der Woude et al. (1998) Formation of DNA methylation patterns: nonmethylated GATC sequences in gut and pap operons. *J Bacteriol* 180:5913-5920.

Widman et al. (2009) Determining the conservation of DNA methylation in Arabidopsis. *Epigenetics* 4:119-124.

Wion D, Casadesús J. (2006) N6-methyl-adenine: an epigenetic signal for DNA-protein interactions. *Nat Rev Microbiol* 4:183-192.

Zhang et al. (2006) Genome-wide high-resolution mapping and functional analysis of DNA methylation in arabidopsis. *Cell* 126:1189-1201.

Zheng et al. (2008) ROS3 is an RNA-binding protein required for DNA demethylation in Arabidopsis. *Nature* 455:1259-1262.

5

Histone modifications and their role in epigenetic regulation

Eukaryotic DNA is organized in a DNA-protein complex called **chromatin**. A nucleosome core is the basic building block of chromatin and consists of an **octamer** of small basic proteins named histones. Two loops of DNA (approximately 147 bp) are wrapped around a histone octamer called the **histone core**. The octamer consists of two copies each of histone proteins H3, H4, H2A, and H2B, each of these is encoded by 10 to 50 intronless copies of the histone gene. The H3/H4 tetramer appears to be more stable in chromatin than the H2A/H2B dimer. A **nucleosome** array provides an approximately sevenfold DNA condensation. Another histone, H1, binds to the linker DNA region (approximately 80 bp in length) between two nucleosomes, probably stabilizing the nucleosome structure. This allows compacting of the nucleosome-based array into a 30-nm chromatin fiber, providing additional condensation (see Figure 5-1). For more details on chromatin compaction and chromatin dynamics, see Chapter 2, "Chromatin Dynamics and Chromatin Remodeling in Animals."

Histone modifications represent a complex layer of highly interactive epigenetic information influenced by developmental and environmental cues. The high complexity of information carried by histone epigenetic marks arises from a large number of possible histone modifications together with possible combinatorial effects, which might lead to the emerging of new epigenetic information. The fast reversibility of histone modifications and multiple crosstalks between histone-modifying pathways, DNA methylation, and chromatin remodeling make histone modifications an ideal choice for regulating genome transcription in changeable growth conditions. This chapter describes the role of histones in epigenetic regulation of the chromatin structure via ATP-dependent nucleosome repositioning, histone modifications, and replacement with histone variants.

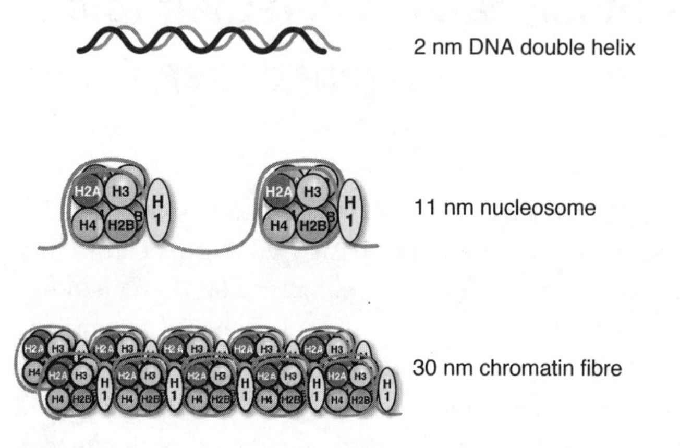

Figure 5-1 Nucleosome units and chromatin fibre.

Nucleosome is a transcriptional regulator

One of the main functions of the nucleosome is the regulation of transcription. The areas with a tighter packaging of DNA around histones are typically less transcriptionally active. Nucleosomes prevent the recognition and binding of transcription factors to promoter regions. On the other hand, the correct positioning of the nucleosomes might bring DNA loops with remote DNA sequences into proximity to active areas of transcription.

The level of nucleosome packaging depends on the affinity of positively charged histones for DNA and the interaction between histone and non-histone chromatin binding proteins. At least eight different histone modifications are known to change the ability of histones to bind DNA. The most common histone tail modifications include acetylation, methylation, phosphorylation, ubiquitination, biotinylation, and sumoylation. There are a quite substantial number of various modifications at different histone residues. For example, mass spectrometry of histone peptides in animals revealed 13 modification sites in histone H2A, 12 modification sites in histone H2B, 21 modification sites in histone H3, and 14 modification sites in histone

H4. Combinations of these modifications allow making a unique pattern of chromatin structure in a given organism, tissue, or cell.

To date, the most well-studied effects of histone modifications on gene expression are those associated with histone H3 acetylation and methylation. Histone acetylation is performed by the activity of histone acetyl transferase (HAT) enzymes, whereas the deacetylation is promoted by histone deacetylase (HDAC) enzymes. Whereas the former promotes transcription, the latter represses transcription.

Histone acetylation is normally associated with promoters and the 5' end of transcribed sequences of regulated genes. Histone acetylation acts directly by decreasing the positive histone charge and loosening the association of histone with DNA, which leads to transcriptional activation. On the contrary, histone methylation regulates binding of other **effector proteins** and their complexes and thus regulated transcription either in a positive or negative way. Finally, the type of modified amino acids (for example, lysine (K) or arginine (R)), their positioning in the histone and the degree of modification (for example, mono-, di-, or trimethylation) further specify the effect on transcription and define the chromatin localization of a given histone mark.

For example, whereas repressive H3K9me and H3K9me2 localize to heterochromatin, trimethylation at this position (H3K9me3) is associated with repression of euchromatin. Similarly, trimethylated histone H3K4me3 is associated with transcription activation, whereas H3K27me3 is associated with silencing. Trimethylation of H3K4me3 is mediated by trithorax group (trxG) protein complexes at the 5' end of actively transcribed genes. The presence of this mark allows recruiting histone acetyltransferases and chromatin-remodeling complexes, such as NURF, thus leading to chromatin decondensation at a given locus. Specific histone demethylase enzymes reverse the positive effects of H3K4me3 on transcription. Repressive histone modifications, such as trimethylation of H3K27me3, are mediated by the *Polycomb* Repressive 2 (PRC2) complex; the complex directly promotes methylation of K27 and recruits additional repressive chromatin marks, including DNA methylation. In addition, the modification of histones can result in nuclear compartmentalization of a given region of chromatin.

The assembly of mature nucleosomes requires the activity of histone **chaperones**, proteins that specialize in incorporating either histones H2A and H2B or histones H3 and H4 into nucleosomes. Nucleosome assembly protein-1 (NAP-1) and NAP-related protein (NRP) that function as nuclear-cytoplasmic shuttle chaperones bind and deposit histones H2A and H2B into nucleosomes. Several other nuclear factors, including nucleoplasmin and nucleolin, aid in the exchange of histones H2A and H2B during nucleosome remodeling, they are also involved in storage of histone H2A and H2B. In contrast, *de novo* deposition of histones H3 and H4 onto DNA during DNA replication and synthesis associated with repair is facilitated by the multi-subunit chromatin assembly factor-1 (CAF-1).

Permissive, restrictive, and bivalent states of gene promoters

Acetylation/deacetylation and methylation are the most common histone modifications. Whereas the acetylation and deacetylation of lysines directly regulates chromatin accessibility and compaction, respectively, thus influencing transcription, methylation of histone residues may have different effects on chromatin structure, depending on the position of methylated lysine amino acid or/and the number of attached methyl groups.

All gene promoters are found in three fundamental states of gene expression activity, namely, **restrictive or inactive**, **permissive or active**, or both restrictive and permissive, which is known as **bivalent state**. Gene expression states are determined by histone modification marks. Restrictive marks are mainly associated with trimethylated H3K9 and/or H3K27; permissive marks are associated with trimethylated H3K4 and acetylated H3K9; and bivalent marks are associated with trimethylated H3K27 and trimethylated H3K4.

Methylation of histone tails is performed by SET domain-containing histone methyltransferases. H3K9 methylation, for example, is performed by Clr4 (Cryptic locus regulator) in *Schizosaccharomyces pombe*, Kryptonite/SUVH4 in Arabidopsis, and SUV39H1 and SUV39H2 in humans. Genomic regions enriched with H3K9me3 serve as binding sites for Heterochromatin Protein 1 (HP1), further condensing the chromatin. In contrast, open chromatin and gene expression are associated with methylation and acetylation of histones H3K4, H3K36, and H3K79.

Non-expressed genes and some genes that are expressed at low levels carry both permissive (H3K4) and repressive (H3K27) chromatin marks. Such bivalent states of chromatin are often found in undifferentiated cells such as human embryonic stem cells. Transcription of tissue-specific regulatory genes may occur in two steps. First, the promoters of these genes are primed prior to the entry into a specific differentiation pathway and then these promoters are placed in either an active state via the recruitment of permissive marks such as H3K27me2 or a silent state through the recruitment of H3K4me2 marks. Similar bivalent states were found in the genome of Arabidopsis and rice somatic cells; in these plants, many genes carry permissive chromatin marks—such as H3K4me3, acetylated H3K9, H3K36me3, and ubiquitinated H2B—and repressive chromatin marks—such as H3K27me1 and H3K27me3 (Roudier et al., 2011).

Chromatin structure in trypanosomes

Trypanosomatids have several copies of genes encoding major histones, H1, H2A, H2B, H3, and H4, which are extremely divergent from histones found in other organisms. Histone variants such as H2AZ have also been identified in trypanosomatids. As in other organisms, nucleosomes are the basic structural unit of chromatin in trypanosomatids. Electron microscopy showed a regular pattern of nucleosomes, and digestion of chromatin with micrococcal nuclease (MNase) revealed a typical ladder consisting of approximately 200 bp monomers (reviewed in Martínez-Calvillo et al., 2010).

Digestion of *L. tarentolae* chromatin with MNase followed by hybridization with a probe against the SL RNA gene showed that the promoter and coding sequences were not organized into nucleosomes, whereas the non-transcribed intergenic region revealed the presence of nucleosomes. The analysis of nucleosome positioning in *L. major* uncovered that the rRNA promoter region had no nucleosomes, whereas the coding region was packed with nucleosomes. At the same time, tRNA and 5S rRNA genes containing internal Pol III promoters were also void of nucleosomes. In contrast, genomic regions containing protein-coding genes and their intergenic regions showed a distinct pattern of nucleosome organization (reviewed in Martínez-Calvillo et al., 2010).

All the experiments mentioned suggest that as in other eukaryotes, in trypanosomatids the positioning of nucleosomes at promoters as well as within the coding and non-coding regions of protein-coding and non-protein-coding genes and their intergenic regions controls transcription initiation by all three RNA polymerases. One of the significant differences in chromatin structure in trypanosomatids is that chromatin does not fold in the 30-nm fiber, and thus chromosomes do not condense during mitosis.

In trypanosomes, core histones undergo a substantial number of post-translational modifications found in other eukaryotes, including acetylation and methylation of lysine residues of histone H4. Several novel modifications are also found in *T. brucei*, including N-methylalanine at the N-termini of histones H2A, H2B, and H4. Sequence analysis of trypanosomatid genomes reveals the presence of several histone-modifying enzymes, acetyltransferases, histone deacetylases, and methyltransferases. A more specific analysis of *T. brucei* indicates that histone acetyltransferase 2 (HAT2) is necessary for histone H4 lysine 10 acetylation, and HAT3 is required for the acetylation of lysine 4 of histone H4. These and other histone modifications establish the **histone code** in a given region of chromatin. Special **effector complexes** read the code translating it to a specific functional outcome, such as a permissive or repressive state of the locus. Effector complexes are typically multi-subunit complexes consisting of proteins with domains that bind to histones with specific modifications. These domains include chromodomains, bromodomains, and SANT domains. It remains to be proved whether in trypanosomatids, proteins with these domains are indeed able to bind to specifically modified histones as only the Bromodomain Factor 2 (BDF2) found in *T. cruzi* has been shown to associate with acetylated histones.

The genome of trypanosomatides is organized into **polycistronic gene clusters (PGCs)**. These transcription units are frequently separated by strand-switch regions that allow for changes in the coding strand. In the *T. cruzi* genome, such regions are found to be enriched with acetylated (H3-K9/K14 and H4-K5/K8/K12/K16) and methylated (H3-K4) histones. In *Leishmania major*, PGCs and strand-switch regions between them are also associated with the presence of acetylated H3K9 and H3K14 (see Figure 5-2).

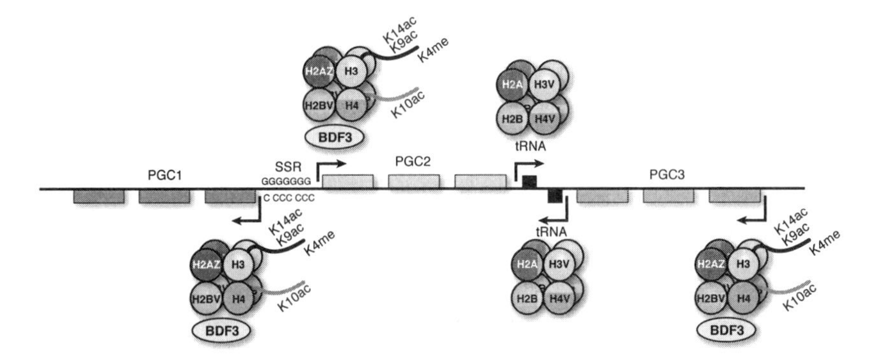

Figure 5-2 Structure of polycistronic gene clusters (adapted from Martinez-Calvillo et al., 2010 with permission). SSR is the strand-switch region. Regions of transcription initiation are associated with the presence of histone variants H2AZ and H2BV. H3 histones are acetylated at lysins 9 and 14 and trimethylated at lysine 4, whereas H4 is acetylated at K10. Transcription of PGC3 is terminated at convergent strand-switch region between PGC2 and PGC3s. Expression of tRNA genes is associated with nucleosomes containing histone variants H3V and H4V.

Some levels of acetylated histones are also observed in the vicinity of clusters of tRNA and snRNA genes. In *T. brucei*, the transcription start sites of Pol II are enriched with acetylated H4K10 and histone variants H2AZ and H2BV as well as the bromodomain factor BDF3. The authors showed that nucleosomes containing histone variants H2AZ and H2BV were less stable than **canonical nucleosomes**. Moreover, Chip-seq analysis revealed that regions with high rates of transcription are enriched with histone variants H3V and H4V that are unique to trypanosomatids. Although more research is required to understand the role of histones and various histone modifications in the regulation of transcription and genome stability, it is clear that actively transcribed regions are more often associated with active posttranscriptional modifications than random genomic regions. The role of post-translational modifications other than acetylation and methylation in the regulation of gene transcription needs to be further established.

Chromatin structure in animals

Histone modifications in animals are well studied. Table 5-1 summarizes various types of known modifications of histones in humans.

Table 5-1 Type of histone modifications in humans

Histone Type	Histone Modifications
H2A	H2AK5ac, H2AK9ac, H2AZ
H2B	H2BK120ac, H2BK12ac, H2BK20ac, H2BK5ac, H2BK5me1, UbH2B
H3	H3K14ac, H3K18ac, H3K23ac, H3K27ac, H3K27me1, H3K27me2, H3K27me3, H3K36ac, H3K36me1, H3K36me3, H3K4ac, H3K4me1, H3K4me2, H3K4me3, H3K79me1, H3K79me2, H3K79me3, H3K9ac, H3K9me1, H3K9me2, H3K9me3, H3R2me1, H3R2me2, H3ac
H4	H4K12ac, H4K16ac, H4K20me1, H4K20me3, H4K5ac, H4K8ac, H4K91ac, H4Kac, H4R3me2, H4ac

Table reproduced with permission from Yan Zhang, Jie Lv, Hongbo Liu, Jiang Zhu, Jianzhong Su, Qiong Wu, Yunfeng Qi, Fang Wang, and Xia Li. HHMD: the human histone modification database, *Nucl. Acids Res.* (2010) 38(suppl 1): D149-D154. Oxford University Press.

Histone acetylation and methylation are the most common histone modifications in animals. An open chromatin configuration is associated with high levels of histone acetylation and trimethylation at H3K4, H3K36, or H3K79. In contrast, condensed heterochromatin is associated with deacetylated histones and histones enriched in H3K9, K3K27, and H4K20 trimethylation (Kouzarides 2007).

Histone acetylation/deacetylation

In animals, HAT proteins belong to three major families: the GNAT family (Gcn5-related acetyltransferases) that preferentially acetylate histone H3, the MYST family that targets histone H4, and the CBP/p300 family that targets both H3 and H4. HATs are able to acetylate both histone and non-histone proteins. Table 5-2 lists several major human HAT proteins.

Table 5-2 Histone acetyltransferases and deacetylases

	Histone H2A	Histone H2B	Histone H3	Histone H4
Acetyltransferases				
GCN5, PCAF			K9, K14, K18	
HAT1				K5, K12
CBP, P300	K5	K12, K15	K14, K18	K5, K8

Table 5-2 Histone acetyltransferases and deacetylases

	Histone H2A	Histone H2B	Histone H3	Histone H4
TIP60/PLIP			K14	K5, K8, K12, K16
HBO1				K5, K8, K12
Deacetylases				*K5, K12*
SirT2-3				K16
RPD3/HDAC2/HDAC3			K9, K14, K16	K16

HDAC proteins are classified into three groups: Type I, Type II, and Type III, also known as Sir2-related proteins. Whereas Type I and II proteins do not require a cofactor and are similar in the mechanisms of deacetylation, Type III enzymes depend on a nicotinamide adenine dinucleotide (NAD+) catalytic cofactor. Type III enzymes frequently function as a part of large complexes, with Rpd3 HDAC being one of the main protein (see Table 5-2). Most of the acetylated residues reside in N-terminal tails of histones except for H3K56, which resides in the core domain.

Histone methylation/demethylation

Histone methyltransferases (histone KMTases) are enzymes that catalyze histone methylation. All KMTases, excluding Dot1, contain the evolutionary conserved domain (a B130 amino-acid SET (Su(var), E(z), Trithorax) (Zhang and Reinberg, 2001) (see Table 5-3).

Table 5-3 Human histone methyltransferases and dimethylases

	Methyltransferases	Demethylases
H3K4	ASH1, MLL1-5, SET1A-B, SET7/9	JARID1A-D, LSD1/BHC110
H3K9	CLL8, ESET/SETDB1, EuHM-Tase/GLP, G9a, RIZ1, SUV39H1-2	JHDM2a-b, JMJD2A/JMJD3A, JMJD2B-C, JMJD2D, LSD1/BHC110
H3K27	EZH2	JMJD3, UTX
H3K36	SET2, NSD1, SYMD2	JHDM1a-b, JMJD2A/JMJD3A, JMJD2B-C
H3K79	DOT1	
H4K20	SUVH4 20H1-2, Pr-SET7/8	

Compared to acetylation, methylation is a more complex modification. There are three different methylation states (mono-, di-, and trimethylation). Methylation can target lysine and arginine and activate or repress transcription depending on the position of modifications and the number of attached methyl groups.

In animals, the H3K27me3 repressive chromatin mark is a typical signature of the activity of the **Polycomb-group (PcG)** protein complex. There are two major types of PcG complexes: **Polycomb Repressive Complex 2 (PRC2)** and **Polycomb Repressive Complex 1 (PRC1)** (see Chapter 2). Whereas PRC2 deposits the H3K27me3 repressive chromatin marks, PRC1 recognizes and maintains it by inducing histone H2AK119 ubiquitination. Ubiquitination of histone 2 at the position of lysine 119 confers a stable gene repression (reviewed in Desvoyes et al., 2010).

Methylated H3K9 serves as a binding site for the chromodomain-containing protein HP1, which further promotes transcriptional repression and heterochromatinization (see Figure 5-3). A local heterochromatinization of the former euchromatic loci is mediated by co-repressors, among which are a retinoblastoma protein pRb and KAP1 (summarized by Kouzarides, 2007). Histone demethylases, such as the lysine-specific demethylase 1 (LSD1) that demethylates H3K4, serve to counteract the events of local heterochromatinization in euchromatic regions. In a FAD-dependent oxidative reaction, LSD1 removes one or two methyl groups from histone H3K4me2 forming H3K4me1 or H3K4me0, respectively. In a complex with the androgen hormone receptor, LSD1 demethylates H3K9me2/H3K9me1 and promotes transcription of androgen-responsive genes. JMJD2C, a newly identified Jumonji C (JMJC) domain-containing protein, is the first histone tridemethylase that regulates androgen receptor function. JMJD2C interacts with the androgen receptor and co-localizes with LSD1. The ligand-bound androgen receptor and JMJD2C assemble on androgen receptor-target genes resulting in tridemethylation of H3K9me3 and stimulation of androgen receptor-dependent transcription. Other histone demethylases, such as JHDM2A, specifically demethylate mono- and dimethyl-H3K9 and require cofactors Fe(II) and γ-ketoglutarate for proper functioning (summarized by Kouzarides, 2007). In contrast, JHDM1 demethylates H3K36me2 that converts active chromatin marks to an unmodified state. Curiously, in mammals, H3K9me3 represses gene transcription if it is associated with promoters, whereas it positively regulates transcription if it is associated with transcribed regions.

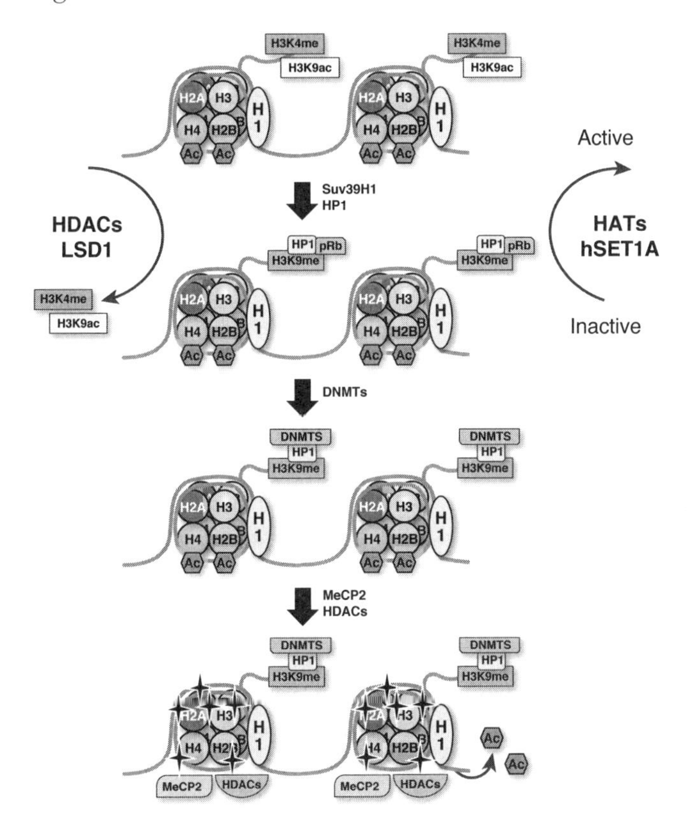

Figure 5-3 Transcriptional activation/inactivation (adapted with permission from Bártová et al., 2008). Transcriptionally active chromatin is associated with H3K4me and H3K9ac, established by K4-HMTs such as hSET1A and HATs such as GNAT and MIST, respectively. Transcriptionally inactive chromatin is achieved through the removal of H3K4me by histone demethylase LSD1, and H3K9ac by HDACs of classes I-III, followed by establishment of H3K9me by K9-HMTase Suv39H1. H3K9me is recognized by HP1 protein resulting in further spreading of inactive chromatin. Chromatin inactivation at euchromatic loci is also associated with binding of retinoblastoma pRb protein. This chromatin state recruits DNMTs that result in dramatic increase in DNA methylation. MeCP2 proteins bind methylated DNA and HDACs remove remaining acetyl groups from histones.

Table 5-4 summarizes major types of histone modifications and their influence on transcription in humans.

Table 5-4 Histone modifications and their effect on transcription in humans

Modification Type	Histone						
	H3K4	*H3K9*	*H3K14*	*H3K27*	*H3K79*	*H4K20*	*H2BK5*
me1	+	+		+	+	+	+
me2		-		-	+		
me3	+	-		-	+/-		-
ac		+	+				

+ represents activation of transcription; - represents repression of transcription. The data are collected from Barski et al. (2007).

Histone phosphorylation

Other less-common histone modifications, such as phosphorylation and ubiquitination, are also known to play a role in chromatin structure and transcription regulation. Phosphorylation of serine 10 of histone H3 is involved in two opposite processes: decondensation of chromatin fibers required for transcriptional activation and compaction of chromosomes during cell division. H3S10ph can increase the levels of H3K14ac and decrease the levels of H3K9ac and H3K9me. There are also reports showing that during mitosis, H3 is also phosphorylated at serine 28, and histones with this modification are associated with chromosome condensation during mitosis. Multiple mitosis- and meiosis-specific kinases phosphorylate H3 during mitosis and meiosis, including Msk1/2 and Rsk2 in mammals and SNF1 in yeast (reviewed in Hans and Dimitrov, 2001).

Several models propose that H3S10ph and H3S28ph might have an effect on chromatin condensation. According to the first model, phosphorylation of histone H3 results in the reduction of the positive charge of histone H3 tail from +14 to +12, thus leading to a local decondensation of the chromatin fiber. It is hypothesized that the local decondensation assists the interaction of chromosome assembly factors with chromatin allowing for a more efficient chromosome condensation. This model is not believed to be sustainable. A decrease in the charge from +14 to +12 is unlikely to decondense chromosomes substantially.

The second model proposes the existence of direct interactions between chromosome condensation factors and histone H3 initiated upon phosphorylation of the histone tail. The fact that the phospho-binding adapter protein 14-3-3 can recognize the H3S10ph modification at the promoters of inducible genes supports this model. On the other hand, the factors involved in chromatin condensation (topoisomerase II and condensins) have an equal affinity to modified and non-modified histones, which speaks against the second model.

The third model suggests that phosphorylation of histone H3 dislodges proteins that are normally bound to the unmodified H3 histone. In support of this hypothesis, mitosis-specific phosphorylation of H3S10 decreases the affinity of HP1 proteins to particular chromatin regions. Several other models suggest that histone H3 phosphorylation either may be used by cells to mark the metaphase chromosomes or may regulate other processes that take place after the metaphase (reviewed in Hans and Dimitrov, 2001).

Histone ubiquitination

Ubiquitination is histone modification that nearly doubles the size of histone proteins. Histones are usually monoubiquitinated; monoubiquitination is different from polyubiquitination associated with protein degradation via the 26S proteasome. Similar to methylation, ubiquitination results in either repressive or permissive effects on transcription. Whereas ubiquitination of H2A is associated with transcription activation, ubiquitination of H2B results in transcription repression. However, ubiquitination of both H2A and H2B histones plays a critical role in the regulation of transcription initiation and elongation, gene silencing, and DNA repair in all eukaryotes (although H2A modification is not reported for *S. cerevisiae*).

Ubiquitination is a three-step process. First, ubiquitin is activated by a ubiquitin-activating enzyme (E1); second, the activated ubiquitin is transferred to a cysteine residue in a ubiquitin conjugating enzyme (E2). Finally, with the help of a ubiquitin-protein isopeptide ligase (E3), ubiquitin is transferred from the E2 enzyme to a target lysine residue of a substrate protein. Monoubiquitination is a reversible process that is catalyzed by a class of thiol proteases known as ubiquitin-specific proteases UBPs (in yeast) or USPs (in mammals) (reviewed in Weake and Workman, 2008).

H2B monoubiquitination is catalyzed by two enzymes, Rad6 (the E2 enzyme; human homologs hHR6a and hHR6b) and Bre1 (the E3 enzyme; human homologs RNF20/RNF40). The details of H2B ubiquitination and a potential role of various human proteins have been recently reviewed (Weake and Workman, 2008). H2B ubiquitination influences other histone modifications, H3K4 methylation being the most important one. Transcription activators recruit Rad6 and Bre1 to promoters where they associate with RNA polymerase II in the process of transcription initiation. However, some additional factors may be required that are associated with an elongating form of RNA polymerase II and promote H2B ubiquitination in transcriptionally active regions. H2B monoubiquitination appears to be an essential modification for the recruitment of the Cps35 subunit of COMPASS which activates di- and trimethylation of lysine 4 of histone H3 and, possibly, H3-lysine 79 by Set1. Importantly, H2B ubiquitination specifically affects H3K4me2 and H3K4me3 but does not eliminate H3K4me. This histone crosstalk is unidirectional because mutations in the H2B ubiquitination pathway alter the level of H3K4 methylation, but not *vice versa*.

UbH2B is deubiquitinated by two ubiquitin-specific proteases in *S. cerevisiae*: Ubp8 and Ubp10. Human homologs include USP22 and possibly USP3 (refer to Table 5-4). Recent studies show that ubH2B deubiquitination is also important for transcription elongation. Ubp8 specifically recruits the H3 methyltransferase Set2 and the Ctk1 kinase that phosphorylates Ser-2 of the RNA polymerase II CTD toward the 5' end of the open reading frames. The Ubp8 mutant is defective in this process, but the *Ubp8ΔBre1Δ* double mutant gains the ability to localize Ctk1 at the 5' end, which indicates that H2B ubiquitination prevents Ctk1 recruitment to the elongating RNA polymerase II.

Weake and Workman (2008) hypothesized that H2B ubiquitination might act as a checkpoint for pausing of RNA polymerase II during early transcription elongation. The model has been proposed in which Ubp8-mediated deubiquitination of H2B recruits the Ctk1 kinase, which phosphorylates RNA polymerase II. Phosphorylated RNA polymerase II then is bound by the H3 methyltransferase Set2 that methylates lysine 36; H3K36 methylation is a necessary step for further transcription elongation.

Histone variants

Among the variety of histone variants deposited at different regions of chromatin, H2A.Z and H2AX are perhaps the most important ones. The H2A.Z histone variant is believed to primarily associate with transcriptionally active genomic regions and is mainly deposited at promoter regions. The ATP-dependent chromatin-remodeling complex SWR1 is responsible for the deposition of H2A.Z marks at the 5' end of genes in many eukaryotes, including yeast, animals, and plants. H2A.Z is mutually exclusive with DNA methylation and mainly associates with methylated and acetylated histone isoforms. Some evidence exists that H2A.Z is removed from nucleosomes during the process of transcription. The significance of the deposition and removal of H2A.Z in cells still remains unclear.

H2AX is a histone variant found in its phosphorylated form in the regions of DNA strand breaks. It has been named "the histone guardian of the genome," and its primary role might be the recruitment of the DNA damage-repair machinery to DNA damage sites (van Attikum and Gasser, 2009). The recognition of strand breaks in the genome involves phosphorylation of the histone variant H2AX in the position of Ser139 producing γH2AX. γH2AX is believed to be required for the assembly of DNA repair proteins at sites of damaged chromatin and for the activation of checkpoints proteins which trigger cell cycle arrest. Several phosphoinositide 3-kinase-related protein kinases (PIKKs), such as ATM (ataxia teleangiectasia mutated), ATR (ATM and Rad3-related), and DNA-dependent protein kinase (DNA-PK) are able to phosphorylate histone H2AX.

Strand breaks are recognized by a protein complex consisting of three proteins, MRE11, RAD50, and NBS1 (the MRN complex) that recruit ATM to the site of a DNA strand break and target ATM to initiate phosphorylation of the respective substrates (reviewed in Podhorecka et al., 2010). Along with other proteins, ARM also phosphorylates BRCA1, 53BP1 (the p53-binding protein 1), MDC1 (mediator of DNA damage checkpoint protein 1), and checkpoint proteins Chk1 and Chk2. The activation of these proteins stops the progression of the cell cycle and activates DNA repair proteins. Phosphorylation of H2AX by ATR occurs in response to single-stranded DNA breaks and upon replication fork arrest. In contrast,

DNA-PK phosphorylates H2AX upon apoptotic DNA fragmentation and when cells are under hypertonic stress. At the same time, a severe stress, such as ionizing radiation, results in the activation of all three PIKK kinases, and each of them phosphorylates H2AX (reviewed in Podhorecka et al., 2010). H2AX might function as an anchor that aids in the assembly of the multiprotein repairsome complex consisting of MRN, MDC1/NFBD1, 53BP1, and some other proteins. The assembly of this complex prevents the separation of broken DNA ends and facilitates end-joining. Chromatin compaction might further help inhibit the irreversible disassociation of broken ends and prevent the potential gross chromosomal rearrangements. The involvement of H2AX in the process might also explain the reason why homologous recombination that requires more open chromatin structures is rarely used for strand break repair.

Histone H3 has three major variants—H3.1, H3.2, and H3.3, with H3.1 and H3.2 being different in a single amino acid. Whereas more than 10 intronless genes encode H3.1 and H3.2, a single intron-containing gene encodes H3.3. H3.1 and H3.2 deposition into chromatin is replication-dependent, whereas deposition of H3.3 histones is replication-independent; moreover, H3.3 is the predominant H3 variant associated with chromatin of quiescent, G1, and G2 cells. A histone chaperone HIRA, the homolog of the yeast Hir1 protein in higher eukaryotes, interacts with the anti-silencing factor 1 (ASF1), the chromatin remodeler CHD1 and the protein complex CAF-1, thus assisting in the assembly of chromatin associated with the H3.3/H4 tetramer. However, it should be noted that H3.3 is incorporated into chromatin even if HIRA or CHD1 are absent. Although HIRA is required to deposit H3.3 into genic regions in mouse embryonic stem (ES) cells, the SWI/SNF-type chromatin remodeler ATRX deposits the histone H3.3 variant at telomeres.

The histone H3.3 variant is primarily incorporated into nucleosomes associated with promoters of transcriptionally active genes. It is frequently found to be associated with H3 acetylation and H3K4 tri-methylation, thus forming a stable permissive chromatin mark that can persist through multiple mitotic divisions. The ability to survive mitosis may allow H3.3 to be a potential mediator of epigenetic memory of active transcriptional states.

Patterns of histone acetylation and methylation in the interphase nucleus

As described in Chapter 2, within the interphase nucleus, there are regions that are enriched in chromatin, called **chromosome territories (CTs),** and chromatin-poor regions that are called the **interchromatin compartments (ICs)**. In the majority of nuclei of eukaryotic cells, the gene-rich CTs occupy the interior regions of the nucleus, whereas gene-poor CTs are located at the nuclear periphery. Hence, it can be hypothesized that specific histone modifications, such as acetylation, methylation, and ubiquitination, as well as deposition of specific histone variants, are characterized by their own specific distribution in the nucleus.

Many structural studies have been performed to analyze the distribution of methylated and acetylated histones in interphase nuclei. The analysis of patterns of the nuclear distribution methylated histones in the nuclei of normal and malignant cells showed variations in histone methylation profiles in individual cells that were dependent on the cell cycle stage (Cremer et al., 2004). In normal quiescent lymphocytes, no specific pattern of histone methylation was found, and its distribution was rather homogeneous. On the contrary, the leukemic cells were characterized by distinct clusters of methylated histones and a peripheral accumulation of methylated lysines in colon carcinoma cell lines. In breast carcinoma cells exiting the cell cycle, histone methylation was found to concentrate in several compact clusters of chromatin that were predominantly located in close proximity to nucleoli; only a few small clusters were found at the periphery of the nuclei (reviewed in Bartova et al., 2008).

In mitotically active non-malignant cells, the G1 phase is characterized by the dispersion of methylated histones throughout the nucleus, whereas the DNA synthesis phase is associated with localization of methylated lysines at the nuclear periphery and around nucleoli. In addition, sites of highly methylated lysines, possibly H3K4me2 and H3K36me2, were observed at centromeres in the S phase, and lysine methylation marks have been suggested to be associated with centromeres (Bergmann et al., 2011). Centromeres are known to contain histone-like CENP proteins that tend to replace one or both copies of the H3 protein in the H3/H4 histone tetramer. Chromatin at the centromere contains CENP-A/H4 tetramers that are periodically interrupted by nucleosomes containing H3K4me2/H4 tetramers. The specific chromatin characterized by the presence of CENP-A/H3K4me2 is distinct from flanking heterochromatin associated with di- and trimethylated H3K9, H3K27, and H4K20. It is not clear what the functional significance of local euchromatic regions found at the centromeres is, but it is believed that it is associated with the maintenance of centromere organization. Thus, centromeres appear to have local euchromatic regions with an active transcription of satellite DNA (Bergmann et al., 2011).

The three-dimensional arrangement of H3K4me3, H4K20me1, H3K9me1, H3K27me3, H4K20me3, and H3K9me3 patterns in different human cell lines was studied by Zinner et al. (2006). Again, there was found the spatial proximity of regions containing methylated lysines to centromeric regions and regions associated with new RNA synthesis. If several of the aforementioned lysine methylation patterns are overlaid in the same nucleus, methylation patterns have been found to be arranged in multiple distinct nuclear layers that only partially overlap (Zinner et al., 2006). Having calculated the coefficient of overlap, the authors reported that for active chromatin marks H4K20me1 and H3K4me3, the percentage of co-localization was 38%. Similarly, the higher percentage of co-localization (43%) was found for heterochromatin marks H3K9me3 and H3K27me3. The overlapping repressive chromatin marks were primarily found at the nuclear periphery.

The aforementioned reports confirm the existence of compartmentalization in the interphase nucleus associated with distinct nuclear layers of specifically modified histones (Zinner et al., 2006). The model that takes into consideration the existence of nuclear domains with individual types of histone modifications has been supported by the report that the chromatin mark H3K9me2 is preferentially located at the nuclear periphery, whereas the permissive chromatin mark H3K9me1 is found in the nuclear interior. Also, H3K9 dimethylation is primarily associated with genomic regions replicating during mid-S phase, whereas H3K9 trimethylation is found in the pericentric heterochromatic regions that replicate in late-S phase (reviewed in Bartova et al., 2008). Thus, it is possible that the time of replication of individual chromatin domains is regulated by the presence of specific histone modifications.

In dividing cells, chromatin is associated with high levels of H3K4, H3K9, H3K27, and H4K20 methylation as compared to G0 cells. G0 B-lymphocytes also lack heterochromatin protein 1β (HP1 β) and Ikaros proteins that are normally present at pericentric regions. If B-lymphocytes enter mitosis, histone methylation and binding of associated HP1 β and Ikaros proteins to chromatin are restored. Similarly, the nuclei of mouse embryonic erythrocytes possess high levels of HP1 proteins and H3K9me3 accompanied by a parallel loss of H3K27me3 (reviewed in Bartova et al., 2008). However, this pattern is not preserved in all cells. For example, in maturing chicken erythroid cells, relatively low levels of HP1 proteins, H3K9me3, and H3K27me3 can be observed. No HP1 isoforms are detected in terminally differentiated human granulocytes (Lukášová et al., 2005). It can be hypothesized that patterns of binding of HP1 to chromatin change upon cell differentiation. Malignant processes may alter these patterns, as it has been observed in differentiated myeloid leukemia cells in which residual levels of HP1 protein and H3K9 methylation were actually detected (Lukášová et al., 2005). Therefore, Bartova et al. (2008) suggest that different patterns of histone modification may be used by the same cell lineage to form heterochromatin.

Histone modifications in plants

Repressive and permissive histone modifications are largely similar in both plants and animals.

Histone acetylation and deacetylation

Based on homology to other eukaryotic HATs, plant histone acetyltransferases are classified into four main families: HAG, HAM, HAC, and HAF. The HAG family contains GCN5 and GCN5-related N-acetyltransferases; GCN5 preferentially acetylates histone H3 at lysine 14. The HAM subfamily members are HAT enzymes with a MYST (MOZ-YBF2/SAS3- SAS2-TIP60) domain; the members of the HAM subfamily, *Arabidopsis* HAM1 and HAM2, acetylate lysine 5 of histone H4. It is not clear whether HAM enzymes do play an active role in regulating gene expression. Nevertheless, their role in the regulation of replication is well established in animals; a human homolog of HAM1 and HAM2 proteins can bind proteins of pre-replication complexes such as a large subunit of the origin recognition complex1.

The HAF subfamily members are related to the TATA-binding protein-associated factor 1. The members of the HAC subfamily are involved in transcriptional regulation and hormonal perception, acetylating histone and non-histone proteins alike.

Transcriptional repression in plants is accomplished, in part, through the activity of three families of histone deacetylases (HDAC). The first family consists of homologs of the yeast RPD3 (reduced potassium deficiency 3) and HDA1 proteins, whereas the second family contains NAD-dependent HDACs, homologs of the yeast silent information regulator 2 (Sir2). Finally, the third family appears to be plant-specific and contains HD-tuins proteins (HDT1, HDT2, and HDT3).

Histone methylation

The *Arabidopsis thaliana* genome contains 39 different SET domain-containing proteins; based on their substrate specificities, these proteins belong to six different families. In addition, other residues in histones H3 (K9 and K36) and H4 (K20) can be methylated by SET domain-containing histone KMTases. The protein Dot1 specific to yeasts and animals that belongs to a separate class of non-SET domain KMTases and is responsible for H3K79 methylation has not been identified in Arabidopsis.

H3K4me

Methylation of histone H3 lysine 4 (H3K4me) is one of the most common histone methylations. It is mainly present in euchromatic regions and is absent in heterochromatic regions. Whereas H3K4me3 and, to a lesser degree, H3K4me2 are found in actively transcribed genes, H3K4me1 is primarily associated with transcriptionally silent genes. Two-thirds of Arabidopsis genes are associated with H3K4me modifications, with H3K4me2 and H3K4me3 being localized in the promoter regions or/and at the 5' end of coding regions. In contrast, H3K4me1 is associated with the transcribed regions only.

H3K4me3 is established by the SET domain-containing ARABIDOPSIS HOMOLOG OF TRITHORAX-1 (ATX1). Together with ATXR7, another Set1-related H3K4 methyltransferase, ATX1 regulates developmental processes such as flowering time. H3K4me3 and H3K4me2 are recognized by effector proteins containing plant homeodomain (PHD) such as ING and Alfin1-like proteins (reviewed in Desvoyes et al., 2010). These proteins further promote permissive chromatin structures.

H3K27me

Similar to animals, in *Arabidopsis*, H3K27me3 is also associated with a repressed chromatin state and is promoted by PRC2 complexes. In contrast to animals where H3K27me3 is spread across large genomic regions, in *Arabidopsis*, H3K27me3-enriched domains are mostly restricted to transcribed regions of genes, suggesting that the establishment and spreading of H3K27me3 marks may occur differently in plants and animals. As little as 18% of *Arabidopsis* genes contain H3K27me3 in their promoters (Zhang et al., 2007), which means that these marks establish tissue-specific gene expression patterns.

Several observations indicate that the LIKE HETEROCHROMATIN PROTEIN (LHP1) may also aid in methylation of H3K27 in plants. LHP1 associates with H3K27me3 *in vitro* and is required for the maintenance of the stable H3K27me3-mediated gene repression in euchromatin. Moreover, disruption of LHP1 chromodomain prevents binding to H3K27me3 and results in a release of silencing of PcG target genes, although LHP1 is not a histone methylase per se. In addition, LHP1 has a similar pattern of binding to euchromatic loci that colocalizes with H3K27me3 marks (Zhang et al., 2007). These features of LHP1 allowed to hypothesize that LHP1 may represent part of PRC1, a complex that is not known to exist in plants.

Another two proteins that may methylate lysine 27 of histone 3 are ATXR5 and ATXR6 involved in heterochromatin formation and gene silencing. They both methylate H3K27 *in vitro* and *in vivo*. Because ATXR5 and ATXR6 are also involved in the regulation of cell cycle and replication, there might be a connection between heterochromatin formation and cell cycle regulation.

H3K9me

Mono- and dimethylation of lysine 9 of histone H3 is performed by proteins belonging to the SUVH family, including SUVH4/ KRYPTONITE (KYP), SUVH2, SUVH5, SUVH6, and several other proteins. In contrast to animals, in *Arabidopsis*, H3K9me3 is associated with euchromatin and active promoters, whereas H3K9me2 is found in repressed genes, transposons, pseudogenes, and pericentromeric regions. The importance of the SUVH proteins for methylation of other lysines and even other histones is proven by alterations in the levels of H3K9me1, H3K9me2, H3K27me1, H3K27me2, and H4K20me1 in *suvh2* mutants.

H3K9me1 and H3K9me2 are associated with DNA methylation; SUVH4, 5, and 6 have been also shown to be required for the establishment of CMT3-dependent CNG DNA methylation. Methylated H3K9 recruits other proteins, such as HP1, that help spread heterochromatin to neighboring chromosomal regions.

H3K36me

In *Arabidopsis*, Set2 homolog ASHH1/SDG8 is a likely candidate for H3K36 methylation because the *sdg8* mutant plants have a reduced level of H3K36 methylation at the *FLC* gene.

Histone variants in plants

Histone variants are non-canonical forms of histones that replace canonical histones, mostly H2A and H3. They are involved in the regulation of many biological processes: the prevention of heterochromatin spreading to euchromatin regions, control of gene expression, the progression of cell cycle, the maintenance of genome stability, the suppression of antisense RNAs, the stabilization of condensin association with mitotic chromosomes, and the ability of plants to percept changes in temperature. There are two variants of histone H2A that are known in plants.

The H2A.Z is found throughout the plant genome but mainly in nucleosomes associated with transcriptional start sites. The H2A.Z variant is encoded by the *HTA8* (*At2g38810*), *HTA9* (*At1g52740*), and *HTA11* (*At3g54560*) genes. It assists in transcriptional regulation and the formation of heterochromatin boundaries that prevent DNA methylation from spreading into euchromatin regions. Therefore, H2A.Z is normally excluded from heavily methylated regions and is preferentially deposited by the SWR1/SRCAP ATPase complex that is associated with transcriptionally active genes.

Plants also contain the H2AX histone variant. Although little is known about the function and regulation of this variant, it is hypothesized that in plants, as in animals, phosphorylation of this variant plays an important role in response to DNA damage.

In plants, there are also two variants of histone H3: H3.3 and CenH3. The H3.3 variant is incorporated by a histone chaperon, HIRA (histone gene repressor A) into chromatin regions of active nucleosome remodeling. Specifically, H3.3 is deposited into promoter regions, gene regulatory elements, and transcribed genomic regions. Another H3 variant, CenH3, is incorporated at centromeres where it is involved in chromosome segregation.

Histone deposition during cell cycle

Cell cycle progression is characterized by extensive chromatin remodeling and histone modifications. The expression of E2F targets, such as HDACs HDT1-4 or the HAM1-2 histone acetyl transferases, is cell-cycle regulated.

In mammalian cells, the Rb protein, a regulator of cell cycle progression, interacts with HDAC, thus possibly assisting in the recruitment of HDAC to E2F-bound promoters at early G1 phase. This Rb/HDAC-dependent mechanism allows maintaining the repressed state of promoters responding to E2F. A similar mechanism exists in plants and has been experimentally demonstrated in maize and tomato.

During S-phase, new unmodified or modified histones are deposited into DNA of daughter cells. Histone modifications may be either similar to modifications found in parental cells or substantially different, thus possibly reflecting various experiences encountered by parental cells. Because both active replication and transcription occur during S-phase, a higher ratio of acetylated histones associated with DNA can be observed. Indeed, Arabidopsis cells have high levels of H3K18ac and H4K16ac during S-phase.

During S-phase, CAF-1 deposits dimers of acetylated histones H3 and H4 into the growing DNA chain. This is followed by the incorporation of H2A/H2B dimers by NAP-1. CAF-1 is a heterotrimeric complex formed in *Arabidopsis* by proteins encoded by the *FASCIATA1* (*FAS1*), *FASCIATA2* (*FAS2*), and *MULTICOPY SUPPRESSOR OF IRA1* (*MSI1*) genes. At the same time, CAF-1 is required to maintain heterochromatic patterns because the *fas1* and *msi1* mutants exhibit reduced levels of heterochromatin. It also represses genes involved in the G2 checkpoint. Cells deficient of one of the components of CAF-1 switch to the endoreplication program. A similar effect is observed in *nap1* mutants. Normal cell cycle progression also requires histone monoubiquitination; mutations in the *HUB1* (*HISTONE MONO-UBIQUITINATION 1*) gene that encodes a RING E3 ligase that establishes ubiquitination of histone H2B lead to a delay in cell cycle progression and a premature switch to the endoreplication program.

The onset of mitosis is characterized by chromosome condensation associated with several histone modifications. Like animal mitosis, plant mitosis is characterized by increased levels of phosphorylation of histone H3 at serine 10 (H3S10ph). In addition, in plants, phosphorylation of T3, T11, and S28 residues of H3 occurs during both mitosis and meiosis. Plants possess several kinases with the histone H3 phosphorylation activity. The *Arabidopsis* homologs of AURORA, kinases—AUR1, AUR2, and AUR3—are highly expressed in mitosis and can phosphorilate H3. Another kinase that phosphorylates H3, the TOUSLED (TSL) kinase, is characterized by a constitutative expression during the entire cell cycle with the pick of activity occurring between late G2/M and the next G1 phase.

The correlation between histone modifications and DNA methylation

Several experimental evidences suggest that DNA methylation and histone methylation are interdependent in plants. In the *met1* mutant, loss of CpG methylation leads to the loss of H3K9 methylation. On the other hand, KYP mutations that are responsible for H3K9 methylation loss results in the loss of H3K9 methylation but do not alter the level of CpG methylation. It can be hypothesized that CpG methylation acts upstream or even may be a prerequisite for methylation of H3K9 and further heterochromatin reinforcement. It should be noted that CpNpG methylation is partially dependent on the activity of KYP as methylation at the *Ta2* and *Ta3* retrotransposons is lost in the *kyp* mutant. It has been also shown that the maintenance of CpNpG methylation at the sequences that are targets for RdDM-induced transcriptional silencing requires the activity of KYP (Jackson et al., 2002). Likewise, histone deacetylase HDA6 is also required for maintaining RdDM-induced CpG methylation; methylated transgene reporters are reactivated in the *hda6* mutant. Methyl-CpG-binding domain proteins (MBDs) recognize methylated CpGs, bind them and recruit enzymes that modify core histone proteins, thus further changing local chromatin structure (reviewed in Zemach and Grafi, 2007).

Conclusion

Histone proteins are the core of the chromatin organization. Although DNA methylation plays a significant role in regulating of chromatin structure and controlling gene expression, regulation via histone binding appears to provide a more drastic result and is by far more flexible. Different modifications of the histone tails and different degree of these modifications allow organisms to control their development and respond to various internal and external stressors in very efficient manner. Regulation of gene expression via histone modifications is largely similar in animals and plants. Histone acetylation and trimethylation of histone H3 at lysine 4 (H3K4me3) serve as permissive chromatin marks and are associated with euchromatin, whereas histone deacetylation together with H3K9me3 and H3K27me3 are repressive chromatin marks associated with heterochromatin. Also, histone variants H2A.Z and H2AX are present in both animals and plants and seem to play similar role in both organisms; whereas H2A.Z is associated with transcription start in active genes, the phosphorylated form of H2AX is essential for response to DNA damage and namely, strand break.

Exercises and discussion topics

1. How do histone modifications influence the chromatin structure?
2. What are the histone chaperones? What is their function?
3. Describe three fundamental states of gene expression.
4. Describe the two-step model in which transcription of tissue-specific regulatory genes occurs.
5. What are the similarities and differences in nucleosome positioning in trypanosomatids as compared to animals?
6. What are the effector complexes? What domains do the effector proteins normally contain?
7. What are the canonical and non-canonical nucleosomes?
8. Describe the organization of polycistronic gene clusters (PGCs) in trypanosomatids.
9. Describe steps of transcriptional activation/inactivation with H3K4me and H3K9ac modifications in animals.
10. What are the roles of LSD1 and Jumonji C (JMJC) domain-containing proteins in chromatin structuring?
11. Describe ubiquitination steps and explain the role of histone ubiquitination in transcription.
12. Describe the role of H2A.Z histone variant.
13. Describe the role of H2AX histone variant.
14. Describe different families of histone acetyltransferases in plants and explain their role.
15. Describe the role of H3K4, H3K9, and H3K27 methylation for transcription regulation.
16. Describe the role of histone modifications in cell cycle regulation.
17. Explain interrelationship between histone modifications and DNA methylation.

References

Barski A, Cuddapah S, Cui K, Roh TY, Schones DE, Wang Z, Wei G, Chepelev I, Zhao K (May 2007). "High-resolution profiling of histone methylations in the human genome." *Cell* 129:823-837.

Bártová et al. (2008) Histone modifications and nuclear architecture: a review. *J Histochem Cytochem* 56:711-721.

Bergmann et al. (2011) Epigenetic engineering shows H3K4me2 is required for HJURP targeting and CENP-A assembly on a synthetic human kinetochore. *EMBO J* 30:328-40.

Desvoyes et al. (2010) Impact of nucleosome dynamics and histone modifications on cell proliferation during Arabidopsis development. *Heredity* 105:80-91.

Hans F, Dimitrov S. (2001) Histone H3 phosphorylation and cell division. *Oncogene* 20:3021-3027.

Kouzarides T. (2007) Chromatin modifications and their function. *Cell* 128:693-705.

Lukášová et al. (2005) Methylation of histones in myeloid leukemias as a potential marker of granulocyte abnormalities. *J Leukoc Biol* 77:100-111.

Martínez-Calvillo et al. (2010) Gene Expression in Trypanosomatid Parasites. *J Biomed Biotechnol* doi:10.1155/2010/525241.

Podhorecka et al. (2010) H2AX Phosphorylation: Its Role in DNA Damage Response and Cancer Therapy. *J Nucleic Acids* doi:10.4061/2010/920161.

Roudier et al. (2011) Integrative epigenomic mapping defines four main chromatin states in Arabidopsis. *EMBO J* 30:1928-1938.

van Attikum H, Gasser SM. (2009) Crosstalk between histone modifications during the DNA damage response. *Trends Cell Biol* 19:207-217.

Weake VM, Workman JL. (2008) Histone ubiquitination: triggering gene activity. *Mol Cell* 29:653-63.

Zemach A, Grafi G. (2007) Methyl-CpG-binding domain proteins in plants: interpreters of DNA methylation. *Trends Plant Sci* 12:80-85.

Zhang et al. (2007). Whole-genome analysis of histone H3 lysine 27 trimethylation in Arabidopsis. *PLoS Biol* 5:e129.

Zhang et al. (2010) HHMD: the human histone modification database *Nucl Acids Res* 38(suppl 1):D149-D154.

Zhang Y, Reinberg D. (2001) Transcription regulation by histone methylation: interplay between different covalent modifications of the core histone tails. *Genes Dev* 15:2343-2360.

Zinner et al. (2006) Histone lysine methylation patterns in human cell types are arranged in distinct three-dimensional nuclear zones. *Histochem Cell Biol* 125:3-19.

6

Realm of non-coding RNAs: From bacteria to human

The central dogma of molecular biology has been "DNA makes RNA makes protein" for the past 50 or 60 years. The development of this postulate was largely influenced by the publication of an article by George Beadle and Edward Tatum (1941) describing genetic mutations in the mold *Neurospora crassa*. This work allowed Norman Horowitz to develop the "one gene-one enzyme hypothesis." Although the assumption that each gene makes a protein might partially apply to unicellular prokaryotes, it is definitely far from being true for multicellular eukaryotes. For a long time, we used to believe that RNA molecules are intermediate messengers between a gene and the protein it encodes. Again, it was thought that primary regulatory functions in the cell are carried out by proteins, and most of the non-protein-coding sequences present in eukaryote genomes have no specific function. More recently, an assumption has been made that non-coding sequences in the genome exist as "protective cushions" against DNA-damaging agents. When the idea of the existence of chromatin was established, another suggestion was made that additional DNA sequences function as structural and organizational motifs. Although this is true, it is only a small portion of the actual function of non-coding DNA sequences. As the majority of DNA in the human genome is transcribed but not translated, current knowledge enables us to speculate that these non-coding RNA (ncRNA) molecules work alone and in concert with proteins to create a complex interactive regulatory network controlling enzymatic, metabolic, and heritable processes in the cell.

Both the type and the number of discovered ncRNAs in different species have grown exponentially. Although there are some serious gaps in understanding the role of various ncRNAs, many of them have been assigned a specific function. Some of the ncRNAs, such as tRNA and rRNA, have structural or enzymatic functions and are highly conserved across all domains of life. Other ncRNAs involved in the regulation of transcriptional activity, mRNA stability, and mRNA translation are less conserved and have unique features among genera. In nearly all domains of life, there are some means of ncRNA-dependent regulation of gene expression. Therefore, the focus of this chapter is an overview of non-coding RNAs across the species, including some examples of ncRNAs that are involved in epigenetic regulation.

History of RNA world

The question, "How do organisms pass on their acquired traits to the offspring?" has been around for perhaps 200 years. In 1866, the first substantial discovery was made by Gregor Johann Mendel who suggested that certain characteristics that determined the color of pea flowers, the shape of pea seeds, and other easily observable factors were passed from parental plants to their progeny. It took nearly 80 years until a factor responsible for a certain trait/phenotype was identified. In 1944, Avery, Macleod, and McCarty showed that transformation of the avirulent strain of Pneumococcus with DNA from the virulent strain made the former strain virulent, thus suggesting that DNA was a "causative" hereditary factor (Avery et al., 1944). On the other hand, in 1939, Caspersson and Schultz demonstrated by cytochemical methods that ribonucleic acid was involved in protein synthesis.

In 1953, two breakthrough publications further supported the role of DNA as the genetic material. Sanger reported the amino acid sequence of a part of the insulin protein (Sanger and Thompson, 1953), and Watson and Crick (1953) proposed a model of DNA structure and its replication. The work of Watson and Crick was preceded by the initial discovery of DNA by Friedrich Miescher in 1869; later, it was followed by the discovery of the composition of nucleic acids, sugar-phosphate-base made by Phoebus Levene (1869–1940) who also coined the term nucleotide and described the structure of ribose and deoxyribose in RNA and DNA. These works—as well as the work by Erwin Chargaff on the ratio between certain nucleotides in any given species and that of Rosalind Franklin on an X-ray diffraction image of the DNA structure—enabled Watson and Crick to deduce the structure of DNA.

Thus, the aforementioned works and other studies (not mentioned here) enabled Francis Crick in 1970 to put forward the central dogma of molecular biology: DNA carries inheritable determinants, called genes, and is able to replicate itself (Crick, 1970). The main function attributed to RNA was that of a mere messenger molecule communicating the information from DNA to direct protein synthesis.

The first RNAs that are not translated, thereafter referred to as **non-coding RNAs** (**ncRNAs**), were discovered in 1956 by Hoagland et al. who described **transfer RNA** (**tRNA**) as a carrier of amino acids. Another group of ncRNAs, **ribosomal RNAs** (**rRNAs**), was described by Scherrer et al. in 1963. Since then, the field of ncRNAs has expanded dramatically, and the current notion is that essentially the whole human genome is transcribed into RNA, with the majority of this being ncRNAs (Birney et al., 2007).

A detailed analysis of an essential role of ncRNAs in a variety of functions of multicellular organisms has been recently done by John Mattick (2010). He compared the complexity of function of two multicellular organisms with a relatively similar number of genes—humans and nematodes. Whereas an anatomically complex and cognitively advanced organism, a human contains about 10^{14} cells, with approximately 10^{10} of them being neurons with an estimated 10^{14} synaptic connections in the neocortex alone (Andersen et al., 2003), fairly simple organized nematodes have a mere 1,000 somatic cells. Mattick suggested that the difference in organization and function is majorly defined by the structural organization of the genome and, what is more important, by the complexity of regulatory information that might mainly be in the form of ncRNAs. Indeed, in humans, the extent of non-protein-coding DNA reaches 98.8%, and with the notion that virtually all DNA in humans is transcribed, one can suggest that 98.8% of all RNAs are ncRNAs that are potentially involved in the regulation of various cellular processes.

Embracing the complexity of ncRNAs

The classification of ncRNAs is a difficult task because they have been found in different organisms and have been often assigned different functions. Most of the newly described ncRNAs seemed to be unique in their structure and function and thus formed new classes, with more than 100 different classes being known to date. Because such a non-systemic approach to classification was impractical, many attempts were made to establish a new, simpler system to classify and catalog ncRNAs. Table 6-1 presents some of the challenges facing scientists in their efforts to classify ncRNAs.

Table 6-1 Criteria for classification of ncRNAs

	Basis	Problem
Size	Long ncRNAs (> 200 nt in length to longer than 100 kb) Small ncRNAs (19–30 nt)	The biological meaning is unclear. Long ncRNAs still include a big variety of sizes and functions.
Secondary structure	Conservation in secondary structure is defined by conservation of sequences.	snoRNAs are structurally conserved, whereas telomerase RNAs are structurally diverse.
Biogenesis	ncRNAs such as miRNAs and siRNAs have a specific biogenesis pathway.	There is a large group of ncRNAs that are transcribed by RNA Pol II and are processed like mRNA, but not much is known about their purpose.
Subcellular Localization	Some of the ncRNAs localize to specific cellular compartments, such as nucleolus (snoRNA) or cytoplasm (tRNA).	The classification is useless for prokaryotes that have no nucleus.
Biological Function	The majority of ncRNAs have a defined function in a certain biological process (for example, tRNA, snRNA, rRNA).	The biological function of many of the recently discovered ncRNAs is not known. Some ncRNAs, such as tRNAs or siRNAs, have several functions.

The most recent approach to ncRNA (NONCODE classification) is based on a biological function as a criterion. He et al. (2008) integrated 212,527 public sequences from 861 different organisms into a publicly accessible database (www.noncode.org) and came up with 26 functional classes, ranging from **DNA imprinting**, DNA repair, the regulation of replication and transcription, RNA processing, and regulation of translation to protein transport. Figure 6-1 is a simplified overview of the main functional groups of ncRNAs.

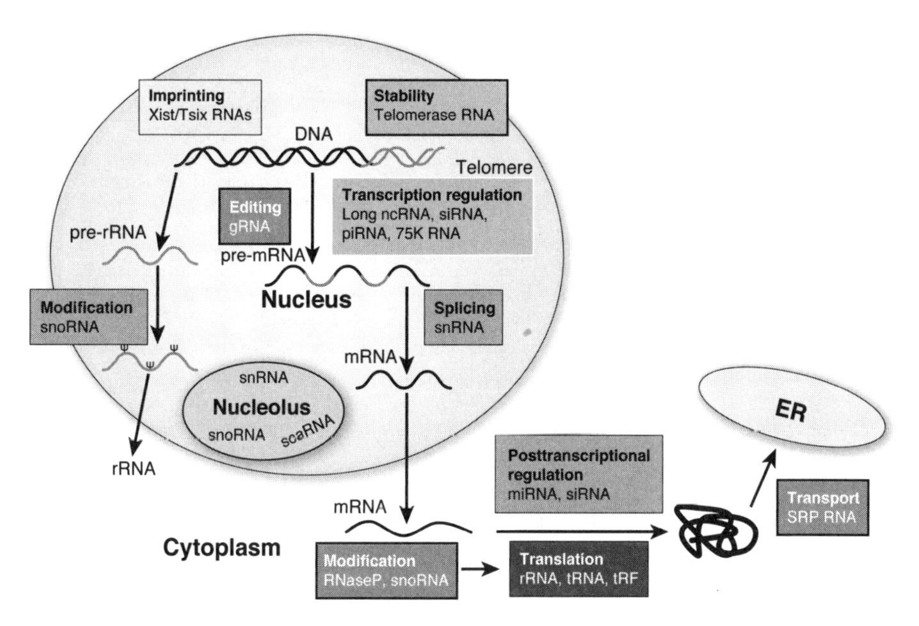

Figure 6-1 The main groups of ncRNAs and their possible roles in a human cell. The boxes represent different functional groups of ncRNAs, each of them combines several groups from NONCODE classification. snoRNA and RNaseP are ncRNAs that modify other ncRNAs.

Some of the examples of ncRNAs from different functional groups are summarized in Table 6-2. This table does not represent a complete list of existing ncRNAs; it gives an example of the main groups according to their function.

Table 6-2 Groups of ncRNAs

Functional group	Bacteria	Archaea	Eukaryotes
DNA stability	-	-	Telomerase RNA
Imprinting	-	-	Diverse in animals
Transcriptional regulation	copA RNA 6S RNA	?	SRA RNA 7SK RNA
Splicing	-	snRNA	snRNA
Translation	tRNA, rRNA, tmRNA	tRNA, rRNA	tRNA, rRNA

Table 6-2 Groups of ncRNAs

Functional group	Bacteria	Archaea	Eukaryotes
Post-transcriptional regulation miRNA siRNA	DicF RNA Spot 42 RNA GcvB RNA OxyS RNA	?	Animals: miRNA, siRNA, piRNA Plants: miRNA, siRNA, rasiRNA Fungi: siRNA Protozoa: siRNA, scan-RNA
Transport	SRP RNA	SRP RNA	SRP RNA

The examples presented in the table were taken from the NONCODE-database and two reviews on the topic (Storz, 2002; Muljo et al., 2009). No clear examples of ncRNAs regulating transcription and posttranscriptional modifications are known for Archaea, as indicated by question marks.

Evolutionarily, ncRNAs involved in basic cellular processes, such as translation (tRNA and rRNA) or splicing (snRNA), are highly conserved. However, when it comes to ncRNAs with regulatory roles, post-transcriptional regulators, such as miRNA, are abundant in animals and plants, but it is not completely clear yet whether they exist in bacteria and Archaea.

New classes of small RNAs are continuously discovered. The group of John Mattick has recently described the following new ncRNA classes: tiny RNAs associated with transcription initiation sites (**tiRNAs**) that are probably related to nucleosome positioning; similarly sized RNAs associated with splice junctions (**spliRNAs**); a range of small RNAs derived from snoRNAs (**sdRNAs**), some of which appear to function as miRNAs (reviewed in Mattick, 2010).

The next section discusses several specific examples of different functional groups across various kingdoms.

tRNAs—their known and largely unknown roles

tRNA is one of the best known ncRNA groups. tRNAs interpret the genetic code by physically connecting nucleic acid codons to the sequence of amino acids. The human nuclear genome codes for more than 500 tRNAs. Nuclear tRNAs are transcribed as longer **precursor**

tRNAs (pre-tRNAs) by RNA polymerase III (Pol III), trimmed at each end, and spliced to remove introns. Mature tRNAs are 73 to 93 nucleotides in length and have similar tertiary structures. The process of maturation of tRNAs involves extensive modification in both the nucleus and cytoplasm. You can find a more detailed description of tRNA function elsewhere (Phizicky and Hopper, 2010); this book describes only a few "unusual" functions.

Another group of ncRNAs derived from tRNA, so-called **tRNA fragments (tRFs)** were recently recognized as a major RNA species in human and Drosophila cells. tRNAs undergo endonucleolytic cleavage in the cells exposed to oxidative stress. The importance of tRFs on cell physiology and the mechanism and regulation of this process remain largely unproven. It is hypothesized that nicked but otherwise fully folded "cleaved tRNAs" may function by regulating protein synthesis via competitive inhibition by being involved in translation instead of normal tRNAs.

One of the surprising roles of tRNA in the regulation of apoptosis has been recently suggested by Mei et al. (2010). Treatment of mammalian S100 extracts with RNase strongly increased cytochrome *c*-induced caspase-9 activation, whereas the addition of RNA to the extracts impaired caspase-9 activation. This enabled the authors to suggest that some RNA species inhibit a factor required for caspase-9 activation. Further analysis showed that several cytosolic and mitochondrial tRNAs specifically associate with cytochrome *c*. The involvement of tRNAs in the process of apoptosis was further confirmed via several *in vitro* experiments. The microinjection of tRNA inhibited the ability of cytochrome *c* to induce apoptosis, whereas degradation of tRNA by a tRNA-specific RNase—Onconase—resulted in an increase in the apoptotic activity. Cytochrome *c* was shown to bind both mitochondrial and cytosolic tRNAs with higher affinity to the mitochondrial form. More details on the possible function of tRNAs in apoptosis can be found in Mei et al. (2010).

snRNA and snoRNA function

Small nuclear ribonucleic acid (snRNA) is a class of ncRNAs molecules that are primarily found in the nucleus of eukaryotic cells. Their main function includes RNA splicing, regulation of the activity of transcription factors and RNA polymerase II (Pol II) as well as the maintenance of the telomere length. snRNA are frequently associated with specific proteins, making the complexes referred to as **small nuclear ribonucleoproteins** (**snRNPs**). Recent publications also suggest that snRNAs might function not only as structural molecules but also as regulatory molecules. Some of the proposed functions include chromosomal DNA replication, mRNA transcription, and global RNA Pol III transcription. You can find more information on the details of snRNA function in the review published by Jawdekar and Henry (2008).

Small nucleolar RNAs (**snoRNAs**) represent a subclass of snRNAs involved in chemical modifications of other RNAs, including ribosomal, transfer, and small nuclear RNAs. Two main modifications include 2'O-ribose-methylation and pseudouridylation (the isomerisation of uridine to pseudouridine). Their function in ribosome maturation includes a key cleavage step to generate individual rRNAs and a guided site-specific modification of rRNA. The biosynthesis of a single ribosome in the budding yeast *Saccharomyces cerevisiae* includes nearly 100 snoRNA-guided modifications, and in humans, their number is approximately 200. 2'O-ribose-methylations are carried out by C/D-box small nucleolar ribonucleoproteins (snoRNPs) which consist of a guide snoRNA acting in concert with various proteins, including Nop1p, the RNA methylase (see Figure 6-2). Pseudouridylations are guided by H/ACA-box snoRNPs, with the Cbf5p protein performing the isomerisation reaction (see Figure 6-2). Although most snoRNAs fall into two classes, C/D-box and H/ACA-box snoRNAs, there are several composite snoRNAs that, in contrast to all other snoRNAs accumulating in the nucleolus, have been found to accumulate in another nuclear structure called the **Cajal body** (formerly known as the coiled body). These snoRNAs are frequently referred to as small **Cajal body-specific RNAs** (**scaRNAs**). Nevertheless, not all snoRNAs found in Cajal bodies are composite. scaRNAs are apparently involved in guided modifications of RNA Pol II-transcribed spliceosomal RNAs U1, U2, U4, U5, and U12.

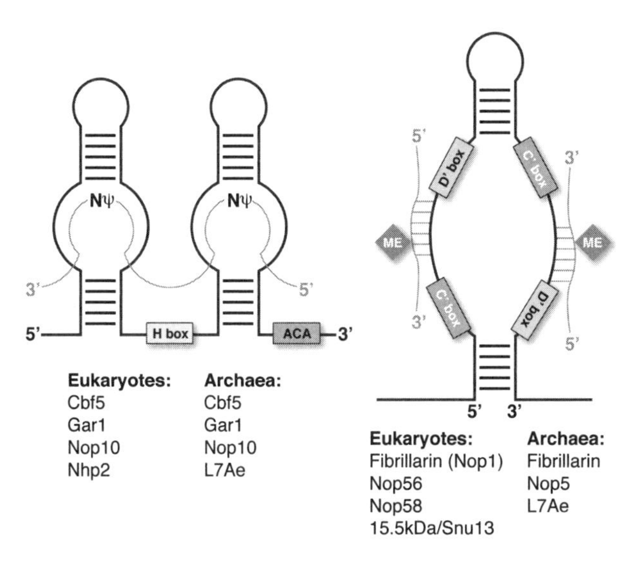

Figure 6-2 Structure of H/ACA-box (A) and C/D box (B) snoRNAs (reproduced with permission from Gardner PP, Bateman A, and Poole AM (2010) SnoPatrol: how many snoRNA genes are there? Gardner et al. J Biology 9:4). The RNA target is in gray (red in ebook version). Proteins aiding snoRNAs are listed. See more details in the text.

rRNA function

rRNAs are an essential structural part of every ribosome. The simplest ribosomal complex, the bacterial 30S subunit, contains an approximately 1530 nt 16S rRNA and more than 20 proteins. Crystal structures of large and small ribosomal subunits revealed that the rRNA forms the core of the ribosome.

In *E. coli* cells growing with a doubling time of 24 min, 73% of total RNA synthesis is the synthesis of rRNAs. A combination of multiple copies of the rRNA genes per genome and high transcription rates per each gene allows achieving high rRNA transcription rates. In the *E. coli* genome alone, there are seven copies of the rRNA genes. Their location close to the origin of replication allows a fast increase in the gene copy number per cell. Also, the transcription rate per gene, approximately 70 transcripts/min, is much higher than transcription rates for protein-encoding genes.

The genes coding for rRNAs are the most conserved and the most utilized genes in eukaryotes. In eukaryotes, rRNA genes are repeated head to tail at chromosomal loci known as **nucleolus organizer regions** (**NORs**). In the plant *Arabidopsis thaliana*,

NORs are composed of approximately 375 rRNA genes that border the telomeres on the 5' ends of chromosomes 2 and 4. Each rRNA gene repeat is transcribed by RNA polymerase I (Pol I) to produce a 45S pre-rRNA primary transcript. The primary transcript is then processed into the 18S, 5.8S, and 25S rRNAs that partner with 5S rRNA, transcribed by Pol III, and approximately 80 proteins, encoded by mRNAs transcribed by Pol II to form the 40S and 60S ribosomal subunits. The locus encoding a single rRNA of the small ribosomal subunit (18S RNA) and two of the rRNAs of the large ribosomal subunit (5.8S and 28S RNA) thus represents one transcription unit (see Figure 6-3A)(Eickbush and Eickbush, 2007). Each rRNA gene is separated from the adjacent rRNA gene by an **intergenic spacer** (**IGS**) that contains multiple regulatory elements. In *Arabidopsis thaliana*, the IGS includes the gene promoter, spacer promoters, and repetitive elements known as Sal repeats (for the presence of *Sal*I restriction sites) (see Figure 6-3B).

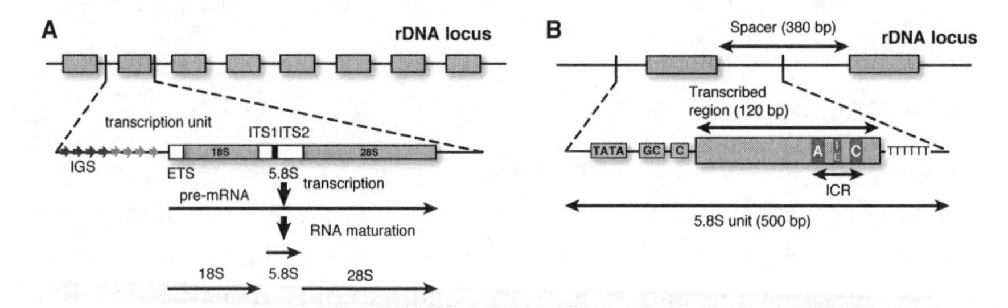

Figure 6-3 Organization of the rRNA genes in eukaryotes (reproduced with permission from Eickbush TH, Eickbush DG. (2007) Finely orchestrated movements: evolution of the ribosomal RNA genes. Genetics 175:477-85). **A.** The rDNA loci are organized into tandemly repeated units (top). The structure of rDNA unit is shown at the bottom. Open boxes show the regions that are processed from the primary transcript: ETS (external transcribed spacer) and ITS (internal transcribed spacer). The IGS are located between the transcription units. **B.** Structure of *Arabidopsis thaliana* 5.8S rDNA units. Internal control region (ICR) contains the internal promoter consisting of box A (A), intermediate element (IE), and box C (C). The motifs required for transcription, TATA, GC, and C nt are positioned at -28, -13, and -1, respectively.

In eukaryotes, there are hundreds copies of rRNA genes organized as large tandem arrays in the genome. The number of rRNA gene repeats varies greatly among organisms, ranging from less than 100 to more than 10,000, and transcription of these genes is also characterized by high rates. In mammals, rRNA transcription is regulated at the transcript elongation stage: stimulation by growth factors increases the elongation speed by nearly fivefold, allowing increasing the transcription rate by fivefold.

Because the main function of rRNA is to help the cell produce a sufficient number of proteins, it would be logical to assume that the number of rRNA gene copies per genome of an organism should be proportional to the organism's growth rate. Unfortunately, there is no good correlation between the cellular growth rate and the number of rRNA genes. The fact that only a fraction of the existing rRNA genes (rDNA) is used for transcription suggests that there is a considerable redundancy and plasticity in the number of their structures. In metabolically active cells of humans and mice, approximately half of the ~400 rRNA gene copies are transcribed, whereas the other half are silent.

Curiously, the relative ratio of active to silent genes is similar in both growing and resting cells and during interphase and metaphase. This suggests that the chromatin structure at the silent rRNA loci is rather stable, and it does not change upon cell division. Indeed, chromatin configuration of active rRNA loci is different from that of silent loci. Active genes have low levels of CpG methylation and high levels of histone acetylation and other histone modifications associated with euchromatin, whereas silent rRNA genes exhibit heterochromatic features, including high levels of CpG methylation, H3K9 and H4K20 methylation, and association with the heterochromatin protein 1 (HP1). The heterochromatin at the silent rRNA loci is in part maintained by the action of transcripts originating from IGS that separate rDNA (reviewed in McStay and Grummt, 2008). Such chromatin remodeling apparently occurs through the function of the NoRC complex, which consists of 150 to 300 nucleotide RNAs that are complementary in sequence to the rDNA promoter. Mutation in the large subunit of the NoRC complex—TIP5—prevents the association of NoRC with rDNA and results in failure to establish H3K9 and H4K20 methylation and HP1 recruitment at the rDNA promoter (McStay and Grummt, 2008).

Analysis of 67 independent Arabidopsis promoters showed that a large subset of sequences was heavily methylated, whereas a small subset remained almost completely unmethylated. A good plant model for the analysis of silencing of rDNA loci is A. *suecica*, the allotetraploid hybrid of A. *thaliana* and A. *arenosa*, in which rRNA genes are transcriptionally silenced due to the effect of nucleolar dominance. Analysis of the methylation and chromatin structure showed that the silent A. *thaliana* rRNA genes were associated exclusively with H3K9me2, and their promoters were hypermethylated. The application of 5-aza-2' deoxycytosine, a DNA methylation inhibitor or trichostatin, a histone deacetylation inhibitor, allowed reversing the gene silencing. Repressive chromatin marks, H3K9me2 and DNA hypermethylation of promoters were replaced with permissive marks, H3K4me3 and the loss of promoter methylation. **RNAi**-mediated suppression of specific histone deacetylases showed that the repression of silent rDNA loci in the hybrid was primarily carried on by HDT1, a nucleolar-localized member of a plant-specific HDAC protein family. Thus, the mechanism of silencing and derepression of rRNA genes might function as an "on/off" mechanism associated with specific epigenetic changes (see the model in Figure 6-4; Lawrence and Pikaard, 2004).

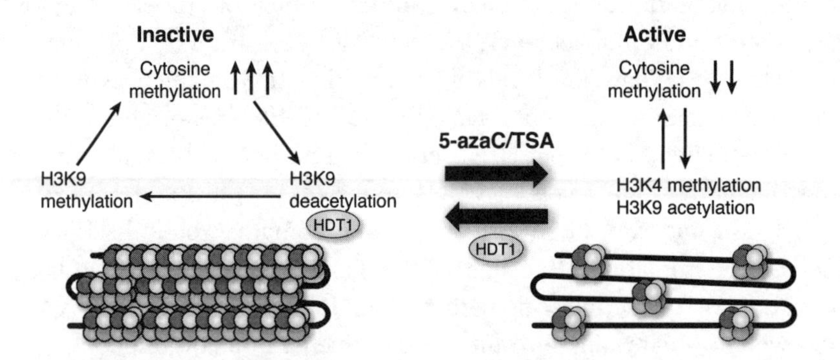

Figure 6-4 Transcriptional gene silencing at rRNA gene loci (adapted from Lawrence and Pikaard, 2004, with permission). Transcription repression operates through self-reinforcing loop. Methylation of H3K9 regulates cytosine methylation, which recruits the histone deacetylase HDT1 to deacetylate H3K9, in turn allowing it to be methylated. Transcription activation may be achieved through inhibition of DNA methylation with 5-azaC or through inhibition of HDT1 via TSA. Establishment and/or maintenance phases of silencing might require the activity of HDT1.

The recent work by Earley et al. (2010) enables us to suggest that histone deacetylation via the action of HDA6 is critical for a correct regulation of rDNA expression. In the *hda6* mutant, symmetric cytosine methylation at CG and CHG motifs is reduced, and false Pol II transcription occurs throughout the intergenic spacers. Because the spacers consist of inverted repeats, the resulting sense and antisense spacer transcripts trigger the overproduction of siRNAs that, in turn, direct *de novo* cytosine methylation of corresponding gene sequences. Surprisingly, the resulting *de novo* DNA methylation fails to suppress Pol I or Pol II transcription in the absence of HDA6 activity. The authors propose that spurious Pol II transcription throughout the intergenic spacers in *hda6* mutants, combined with losses of histone deacetylase activity and/or maintenance DNA methylation, eliminates repressive chromatin modifications needed for developmental rRNA gene dosage control.

Although this model is attractive and might apply to some rDNA loci, there apparently exists a much more complex system of control of the epigenetic status of various rRNA genes. For example, *mom1* mutant plants exhibit a release of silencing of **transcriptionally silent information** (**TSI**) repeats at chromocenters without altering their DNA methylation status. The effect of *mom1* on 5S gene silencing was tested and confirmed by the higher proportion of both minor and 5S-210 transcripts in *mom1* than in wild type plants, although the dense cytosine methylation of 5S genes remained unaffected. This suggests that methylation is not the only mechanism that regulates rRNA gene silencing. The complexity of regulation of repression at rDNA loci is reflected in the long list of mutants exhibiting changes in 5S siRNA accumulation and 5S rDNA gene silencing (Douet and Tourmente, 2007).

RNAi is the pathway that is proposed to be involved in the silencing of 5S rRNA genes in *A. thaliana*. A particular pathway, **transcriptional gene silencing** (**TGS**), represses repetitive sequences through DNA methylation, histone methylation, and changes in chromatin structure. siRNAs of 21 to 24 nt, processed from **double-stranded RNAs** (**dsRNAs**) derived from the endogenous transcripts, direct the sequence-specific methylation changes in repetitive DNA. A key role in the process is played by **RNA-directed DNA methylation** (**RdDM**) accomplished by the *de novo* cytosine methyltransferase domains rearranged methylase 2 (DRM2) and defective in RNA-directed methylation 1 (DRD1).

In *A. thaliana*, silencing at endogenous repeat loci involves histone H3K9 methylation and RdDM. The fact that repetitive 5S rDNA is controlled by the function of RdDM is supported by two observations: 5S siRNAs are detected in *A. thaliana* plants and 5S rDNA is highly methylated at CNN, a hallmark of RdDM.

Recent progress in understanding the mechanisms of control over the expression of repetitive DNA, such as AtSN1 (small retroelements), 45S and 5S rDNA, suggests the following model: DRM-mediated *de novo* methylated sequences stall or slow Pol I/III and the produced transcripts are recognized as aberrant, thus becoming templates for RNA polymerase IV (Pol IV) (Vaucheret, 2005) (see Figure 6-5). Precursor RNA generated from the endogenous repeats moves to the nucleolus where the antisense strands are generated by RNA-dependent RNA polymerase 2 (RDR2) transcription. The resulting double-stranded RNA is then diced by **Dicer-like 3** (DCL3) and loaded into an argonaute 4 (AGO4)-containing effector complex or RNA-induced silencing complex (RISC) within the siRNA processing centers in the nucleolus. Then a subunit of the Pol V would associate with AGO4-RISC. It is not clear how this complex directs specific changes in methylation, but the resulting compacted chromatin is methylated at DNA, and methylated and deacetylated at histones.

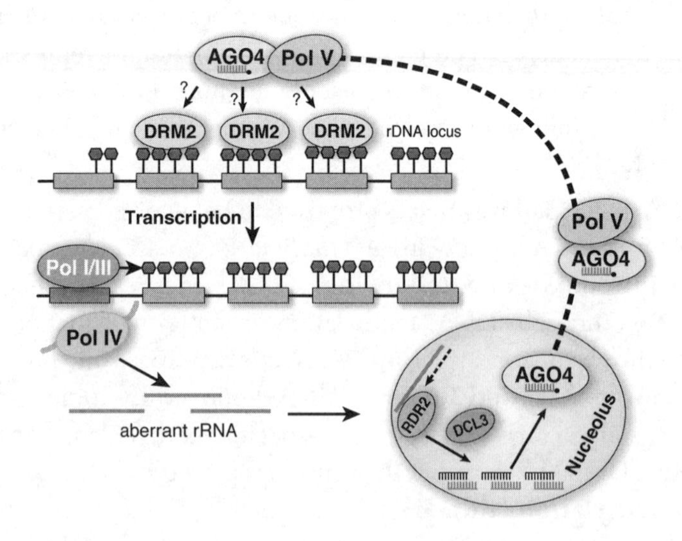

Figure 6-5 Mechanism of control over the expression of repetitive elements and rDNA.

Recently yet another player in silencing of rDNA loci has been identified. Henderson et al. (2010) identified catalytically mutated DRM2 paralogs, termed DOMAINS REARRANGED METHYLTRANSFERASE3 (DRM3). Despite being catalytically mutated, *DRM3* was shown to be required for the normal maintenance of non-CG DNA methylation, the establishment of RNA-directed DNA methylation triggered by repeat sequences and for the accumulation of repeat-associated small RNAs.

Guide RNAs (gRNAs)—non-coding RNAs involved in editing

Guide RNAs (gRNAs) are small non-coding RNAs (presumably processed by Pol II) that are involved in the process of RNA editing—the guided post-transcriptional modification of mRNA upon which nucleotides are inserted, deleted, or replaced in the targeted mRNA sequences, frequently resulting in new translational products. gRNAs serve as guides that encode the editing information in the form of complementary sequences. Editing consists of several major types, including the insertion of multiple uridines (U) in flagellates, C-to-U and A-to-I substitution editing in protists, plants, and animals as well as some cases of insertion/deletion of C or G nucleotides in viruses.

The mechanisms of C-to-U and A-to-I editing both include deamination processes. Whereas deamination of cytosine generates uracil, deamination of adenine produces inosine, which is decoded as guanine in the process of translation. All the aforementioned processes of editing create diversity and flexibility in the regulation of gene expression required for either different tissues or different environmental conditions. The following sections describe the most common types of editing occurring in protists, plants, animals, and viruses.

Editing in flagellated protists

Although any organism containing whip-like organelles called flagella might be referred to as "flagellate," we use the term only for protists. *Trypanosoma brucei* is one of the examples of Kinetoplastid protists in which the mitochondrial genome undergoes extreme cases of editing—post-transcriptional modifications of RNAs that involve either specific insertion/deletion or modifications of nucleotides. Two different types of RNA-editing events have been described in the kinetoplast-mitochondrion of trypanosomatid protists (Simpson et al., 2000). The first type involves the precise insertion and deletion of U residues in the coding regions of maxicircle-encoded mRNAs to produce open reading frames. This editing is mediated by short overlapping complementary gRNAs encoded in both the **maxicircle** and the **minicircle** molecules and involves a series of enzymatic cleavage-ligation steps. To date, it has only been described for protists. The second type of editing system is a C34 to U34 modification in the anticodon of the imported tRNATrp, resulting in the decoding of the UGA stop codon as tryptophan.

Insertion/deletion of uridines

The transcripts of several maxicircle protein-coding genes (approximately 12, but the number varies with the species) are edited post-transcriptionally by the insertion and occasional deletion of uridine (U) residues. This occurs mostly within coding regions and results in the correction of frameshifts and production of translatable mRNAs. The minicircles encode gRNAs, which are complementary to edited sequences; in addition to canonical base pairs, there also exists a G:U pairing. Editing is catalyzed by a ribonucleoprotein complex, termed the editosome. Exact sequence information that is necessary for mRNA editing is provided by gRNAs, which base-pair to pre-edited mRNA downstream to an editing site. gRNAs possess three functional domains: an anchor domain, which anneals to the pre-edited mRNA; a guiding domain, which directs U insertion or deletion; and a 3'-oligo(U) tail, which is added post-transcriptionally. The 3'-oligo(U) tail presumably tethers the purine-rich 5'-cleavage fragment of the pre-mRNA intermediate to the editosome complex (Madej et al., 2007). The tail might also be used for stabilizing the initial interaction of the gRNA and the mRNA by either RNA-RNA or RNA-protein interactions.

The mechanism of U-insertion/deletion editing involves a series of enzymatic cleavage-ligation steps. The precise cleavages are determined by base pairing with the cognate gRNAs. In this process, the entire mRNA region can be edited by multiple overlapping gRNAs, each being able to insert/delete up to 10 uridines. gRNAs bind to the complementary sequences and promote downstream editing, thus establishing the overall 3' to 5' polarity of editing site selection. pre-mRNA is targeted for endonuclease cleavage immediately upstream of the anchor gRNA/pre-mRNA duplex. The generated 5'-cleavage fragment is then the substrate for either a 3' terminal uridylyl transferase (TUTase) in case of U insertion or a 3' to 5' exonuclease in case of U deletion (Madej et al., 2007).

Although the editing process is fairly specific, a certain level of "misediting" has been observed at the junction regions between fully edited and unedited sequences, presumably as a result of either correct guiding by an incorrect gRNA or stochastic errors in the editing process. Misedited sequences are typically re-edited correctly (Simpson et al., 2000).

C-to-U modification editing

Curiously, the UGA stop codon is used to encode tryptophan in the mitochondrial genome of all kinetoplastid species including Diplonema, but not in the euglenoids. C-to-U editing is found in many phylogenetically diverse organisms, both in organellar and nuclear genomes, suggesting that this site-specific modification represents an ancient evolutionary activity. It is interesting that up to 7% of the UGA tryptophan codons in *L. tarentolae* mitochondria are created by U-insertion editing. More than 40% of the mitochondrial $tRNA^{Trp}$ is edited at the position of C34. The exact mechanism of C-to-U editing in protists is not exactly clear, but it might include the function of specific cytidine deaminase, as in the case of similar editing in mRNAs encoded by human nuclear genes.

C-to-U editing in humans

In humans, the best characterized example of C-to-U RNA editing represents the nuclear transcript encoding intestinal apolipoprotein B (apoB). Such editing changes a CAA to a UAA stop codon, generating a truncated protein, apoB48 (Blanc and Davidson, 2003). C-to-U editing of apoB RNA occurs on a single-strand template with well-defined characteristics in the immediate vicinity of the edited base. The process involves a multi-protein editosome complex with a minimal core consisting of two cofactors: apobec-1 and an apobec-1 complementation factor (ACF). Apobec-1 is the catalytic deaminase, whereas ACF functions as an adaptor protein that is able to bind both the deaminase and mRNA. ApoB RNA editing occurs in the nucleus and is selective for spliced mature mRNA transcripts rather than pre-mRNA (see Blanc and Davidson, 2003). Moreover, the assembly of spliceosome or targeting RNA to the splicing pathway inhibits C-to-U editing. It is assumed that the inhibition is due to some protein-RNA or protein-protein interactions between the spliceosome and editosome complexes. C-to-U editing requires several *cis*- and *trans*-acting factors. Among the *cis*-acting factors are the sequence specificity and the secondary structure of target mRNA. RNase mapping and folding algorithms predict that the apoB template folds into a stem-loop structure with the targeted cytidine located within an exposed loop of RNA. Editing requires an 11-nucleotide mooring sequence located 4 to 6 nucleotides downstream of the edited base. There is also a requirement of a consensus binding site (UUUN(A/U)U) for apobec-1 located three nucleotides downstream of the editing site, partially overlapping the mooring sequence at the apex of a stem-loop (see Blanc and Davidson, 2003).

As mentioned before, *trans*-acting regulators consist of multiple factors, with the most critical being the cytidine deaminase apobec-1 and ACF. *In vitro* editing of synthetic apoB mRNA can be performed just by using these two proteins. Like most cytidine deaminases, apobec-1 functions as a dimer (Blanc and Davidson, 2003). Active sites of the dimer allow asymmetric contributions from each monomer that permits both substrate binding and deamination. ACF is a novel 65-kDa protein containing three non-identical RNA recognition motifs at the amino terminus, with a putative double-stranded RNA binding domain and 6 RG repeats within the carboxyl terminus. ACF is able to specifically bind a 12-nucleotide sequence in apoB RNA that surrounds the editing site and partially overlaps the proximal end of the mooring sequence. Mutagenesis experiments have identified the functional domains involved in apoB RNA binding, apobec-1 interaction, and apobec-1 complementation of C-to-U editing.

One of the levels of regulation of C-to-U editing of apoB RNA is the subcellular localization of the components of the editosome. It appears that whereas ACF has the nuclear localization signal (NLS) and is primarily found in the nucleus, apobec-1 lacks NLS and is primarily located in the cytoplasm. It is hypothesized that editing requires partial assembly of the editosome in the cytoplasm upon which ACF brings both enzymes into the nucleus. Although it is not clear what factors facilitate this process.

A second example of C-to-U RNA editing in mammals involves site-specific deamination of a CGA to UGA codon in the neurofibromatosis type 1 (NF1) mRNA. The generated stop codon presumably truncates the protein, but a direct proof of this does not exist. C-to-U RNA editing is also common for plant mitochondria.

A-to-I modification/editing in humans and plants

A-to-I editing is a nucleotide substitution editing in which adenosine is replaced by inosine; it is fairly common in plants and animals (Farajollahi and Maas, 2010). During translation, inosine is interpreted as guanosine, thus A-to-I editing can potentially result in changes in protein sequence. Because the editing might occur only to a certain degree or at certain conditions, the cells in which such editing occurs typically carry a mixture of two versions that differ by one amino acid (Farajollahi and Maas, 2010). Often, this single amino acid difference plays an important role in regulating protein function. The best examples in human are the ion-permeability, kinetic properties, and trafficking of glutamate receptor channels in the brain; the signaling properties of the serotonin 5-HT_{2C} receptor; and the conduction of K_v potassium channels (Maas, 2010).

Other essential targets of A-to-I RNA editing are repetitive sequence elements located in 5' and 3' UTRs of mRNAs and in introns. Additionally, miRNAs precursors as well as viral RNA sequences are also among targets of A-to-I RNA editing (for reviews see Farajollahi and Maas, 2010).

The site-selective and RNA folding-dependent modification of adenosine to inosine is mediated by a small family of enzymes termed **ADARs** (adenosine deaminases acting on RNA). In humans, two

ADARs (ADAR1 and ADAR2) are responsible for all currently known A-to-I editing activities and modifications of RNAs with distinct but overlapping specificities (Nishikura, 2010). Most of the eukaryotes have both ADARs present in their genome. ADARs are expressed ubiquitously in all tissues, and the proteins share a highly conserved catalytic deaminase domain in their C-terminal half and a number of **double-stranded RNA-binding domains** (**dsRBDs**) in their N-terminal half. The proteins function as homodimers. They have NLSs and appear to be able to shuttle in and out of the nucleus presumably facilitated by importin and exportin molecules. In the nucleus, primary localization of ADARs is in the nucleolus, where ADARs might facilitate the editing of miRNA precursors.

Surprisingly, the ADAR proteins display little intrinsic substrate specificity when perfectly base-paired RNA is involved. In contrast, the editing of partially base-paired structures, including bulges and loops, is highly site-specific. Thus, it appears that the overall three-dimensional structure of the substrate rather than a sequence of target RNA determines which nucleotide is modified (Nishikura, 2010). The highly dynamic nature of RNA folding makes it very difficult to predict the potential sites of editing.

mRNAs are a perfect target for ADARs because they form the double-stranded structure required for ADAR editing because their introns contain a sequence, termed the **editing-site-complementary sequence (ECS)**, able to base pair with the exonic regions (Wulff and Nishikura, 2010). The folded pre-mRNA creates stretches of imperfectly base-paired dsRNA suitable for specific editing by ADAR (see Figure 6-6). Thus, A-to-I editing takes place in the nucleus co-transcriptionally, before splicing has happened. The fact that ECSs tend to occur downstream of their base pairing partner exons makes it possible to allow for base pairing to take place even before all of the intron has been transcribed. As a result, the editing might occur before the splicing machinery has a chance to act upon pre-mRNA. You can find more details on the mechanism of A-to-I editing in several reviews (Farajollahi and Maas, 2010; Nishikura, 2010; Wulff and Nishikura, 2010).

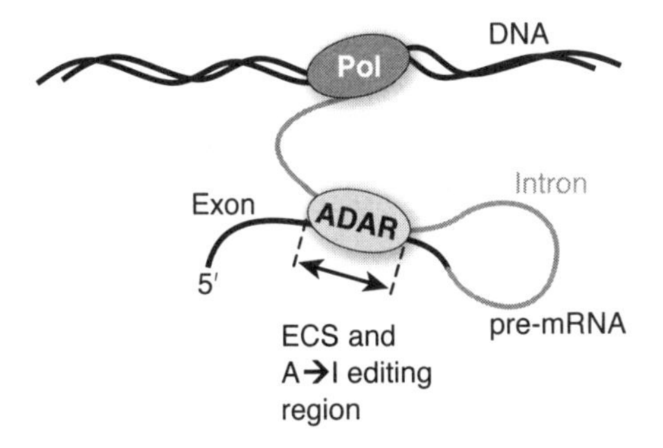

Figure 6-6 A-to-I editing occurs cotranscriptionally.

Insertion/deletion of C or G nucleotides

In several rare cases, the editing events include insertions and deletions of C or G nucleotides. In late 1980s and early 1990s, an insertion of several G nucleotides into the P-gene mRNA transcript of the paramyxoviruses simian virus 5 (SV5), measles virus, Sendai virus, and mumps virus has been reported. An insertion of G residues at this sequence would permit an access to a reading frame encoding a cysteine-rich domain in each case homologous to the C-terminal domain in the V protein of SV5. The paramyxovirus P protein together with the L protein is thought to form the paramyxovirus transcriptase, therefore being an essential protein. An interesting feature of paramyxovirus mRNA editing is that each particular virus inserts G residues in a pattern matched to the organization of its P gene ORFs. There is increasing evidence to suggest that the G insertion event is a virus-encoded co-transcriptional mechanism and that this form of RNA editing may be a common feature in paramyxovirus P-gene-derived mRNAs.

Two alternative models for G insertions in these viruses have been proposed. One is called a **stuttering model**. All negative-strand virus RNA polymerases (RNAPs) that polyadenylate their mRNAs are thought to do so by stuttering on a short run of template U residues (4 to 7 nt long), and it was this observation that first suggested that the G insertions would similarly occur by pseudo-templated transcription. Paramyxovirus mRNA editing is thought to take place as follows (see Hausmann et al., 1999 for details):

1. The viral polymerase pauses before the end of the template C run (nt 1051 to 1053).
2. The nascent chain, whose 3' end is base paired to the template, slips backward by one (SeV and morbilliviruses) or two (rubulaviruses) template positions.
3. As a result, one or two of the template C residues are copied a second time when transcription resumes processively.

In the realignment of nascent mRNA and template, U:G (but not A:C) pairs are permitted, and in analogy to ribosomal frameshifting, the region where alternate base pairing occurs after realignment is called the slippery sequence (A_6G_3 in paramyxoviruses) (Hausmann et al., 1999). For most viruses, this cycle of slippage and pseudo-templated synthesis occurs only once, but for bPIV3 it is postulated to occur repeatedly, generating a range of multiple G insertions. The presence of a counting mechanism has therefore been invoked to explain these different patterns of G insertions.

In a competitive **kinetic model**, the elongation and stuttering processes are characterized by specific overall rate constants ($k_{forward}$ and $k_{stutter}$) at each template position. The magnitude of these rate constants is expected to depend on template and nascent chain sequences and in particular on the base-pairing possibilities of the realigned upstream sequences. More details on this model can be found in Hausmann et al. (1999).

Reasons for RNA editing and its significance

What are the reasons for RNA editing? One of the main possibilities is to increase genetic variability without fixing it to the genomic sequence. More complex organisms seem to employ more A-to-I editing thereby increasing their genetic diversity. Apparently, A-to-I editing is tissue-specific; editing in the central nervous system occurs at a particularly high frequency. Being the most complex organ at molecular, cellular, and functional levels, the nervous system might benefit from variations in transcripts. Natural selection may favor systems with increasing RNA editing potential because the editing might be considered as one of the epigenetic mechanisms generating variations in gene expression, thus enhancing the adaptability of the organism (Gommans et al., 2009). Stefan Maas (2010) suggested that novel RNA editing sites could be tested out in different mRNA transcripts and favorable editing events became more prevalent through the selection of genetic variants that stabilized those RNA folds that supported editing. He drew a parallel between the emergence of novel, low-level alternative splicing events that create additional protein variants and the generation of edited transcripts that ensures the continuous production of the established, conventional gene product whereas the new variant can be tested out by the organism at a small cost for fitness. In this case, RNA editing might represent an epigenetic mechanism that facilitates and/or accelerates the evolution of more complex biological systems (Gommans et al., 2009).

Indeed, as it has been recently reported by Paz-Yaacov et al. (2010), the editing level in humans is significantly higher compared with non-primates, chimpanzees, and rhesus monkeys, due to unique editing within the primate-specific Alu sequences. They found that new editable species-specific Alu insertions, subsequent to the human-chimpanzee split, are significantly enriched in genes related to neuronal functions and neurological diseases. The authors further suggest that the enhanced editing level in the human brain and the association with neuronal functions both hint at the possible contribution of A to I editing to the development of higher brain function.

Other classes of ncRNAs

A limited number of small RNAs carry out specific housekeeping functions, among them are the *4.5S* RNA component (also known as 7SL or 6S) of the **signal recognition particle** (**SRP**), the *Ribonuclease P (RNase P)* RNA responsible for processing of tRNAs and other RNAs, and **tmRNA**, which acts as both a tRNA and mRNA to tag incompletely translated proteins for degradation and to release stalled ribosomes.

The SRP ribonucleoprotein complex consists of a single 4.5S RNA (7S in eukaryotes) and several proteins, and it directs the trafficking of the proteins to be secreted to cell compartments or to the cell's exterior. This occurs through structure-specific binding between SRP RNA and the short signal peptide at the N-terminus of the protein.

RNase P is an essential endoribonuclease responsible for removal of the 5' leader sequence from the pre-tRNAs. Bacterial RNase P also processes precursor 4.5S RNA, 30S preribosomal RNA, tmRNA, and some protein-coding RNAs. Bacterial RNase P is the simplest form of the holoenzyme, with one large RNA subunit and a single small protein subunit. Eukaryotic nuclear RNase P is a far more complex enzyme than its prokaryotic counterpart, employing multiple essential protein subunits in addition to the catalytic RNA subunit.

tmRNA (also known as 10Sa RNA or SsrA) is employed in a remarkable **trans-translation** process. In this process, tmRNA enters ribosomes in the same way as tRNA but then uses its mRNA-like sequence for translation. This results in the addition of the tmRNA-encoded peptide tag onto the C terminus of the nascent polypeptide, thus rescuing the stalled ribosome. Then different proteases recognize specific sequences in the tmRNA-encoded peptide tag and degrade the released protein. The mistranslated mRNA is also rapidly degraded. The process of trans-translation plays an essential role in the development, pathogenesis and response to stress in bacteria. You can find more information on this fascinating process in recent reviews (Dulebohn et al., 2007).

Other classes of bacterial ncRNAs capable of binding proteins, RNA, or DNA molecules, and thus regulating gene expression, are described in the chapter on ncRNAs involved in epigenetic regulation (Chapter 7, "Non-Coding RNAs Involved in Epigenetic Processes: A General Overview").

Groups of cis- and trans-acting ncRNAs functioning in epigenetic processes

Finally, a large group of non-coding RNAs—such as miRNA, siRNA, piRNA, and so on, whose primary function is to regulate epigenetic processes—are described in Chapters 7 through 12.

Future directions and perspectives

The completion of the human genome sequence in 2003 has allowed substantial progress in identification and classification of ncRNA molecules. Soon after that, the National Human Genome

Research Institute (NHGRI) launched another public research consortium named the Encyclopedia Of DNA Elements (ENCODE). The goal of this project is to establish a catalog of the functional elements in the human genome. To pursue this purpose, participating research groups use a combination of high-throughput methods and bioinformatic analysis for studying the primary transcriptome, for analyzing the chromatin structure using DNaseI hypersensitivity assays, and for identifying the regions of binding regulatory factors using ChIP-chip experiments and many other applications. The integration of newly generated data with those available from sequencing allows the formation of complex models that are able to predict which regions in the human genome are transcribed or replicated and even which regions are evolved and at what rate. One of pilot studies utilizing only approximately 1% of the genome sequence enabled researchers to confirm the previously established models that determined the role of histone modifications in transcription regulation and to uncover some of the so far unexplained phenomena such as overlapping transcripts. As the pilot results suggest, only approximately 5% of the bases in the human genome are evolutionarily conserved and appear to be in some sort of evolutionary constraint. The sequences of many functional elements do not fall into this 5% category. Because the majority of the genome is actively transcribed but not translated, it is a matter of time to identify novel groups of ncRNAs and discover new mechanisms of action.

Conclusion

From the time when the first discovery of the role of RNA as a messenger between the genetic information in the form of DNA sequence and the protein synthesis machinery was made, a great variety of ncRNAs have been discovered that affect every single step in transcription, post-transcriptional modifications and translation. Although some ncRNA classes with either structural or enzymatic functions, such as tRNAs, rRNAs, and snRNAs, are evolutionarily conserved and well understood, many regulatory ncRNAs are less conserved and are still not well understood. The finding that virtually the entire genome can be transcribed, and the discovery of abundant long non-coding RNAs that have not been characterized, yet make it possible to predict that in the future new classes of ncRNAs will be identified, and their mechanisms of action in the cell will be elucidated.

As it is put by John Mattick, "The emerging evidence suggests that, rather than oases of protein-coding sequences in a desert of junk, the genomes of humans and other complex organisms should be viewed as islands of protein-coding sequences in a sea of regulation, most of which is transacted by RNA." Thus, rather than being mere messenger molecules, RNAs may be the computational engine of the cell and the functional mechanism of interaction with the environment (Mattick, 2010). The study of what had for a long time been dismissed as junk might hold the key to understanding species evolution, intricate processes of development and cognition, complexities of human behavior and emotions, as well as individual differences in susceptibilities to complex diseases (Mattick, 2009).

Exercises and discussion topics

1. List the facts that challenge the central dogma of molecular biology.
2. Which ncRNAs are highly conserved among all domains of life? Explain the reasons why they are highly conserved.
3. How would you classify ncRNAs? Use Table 6.1 to justify your point of few.
4. What are the "unusual roles" of tRNAs?
5. Compare and contrast snRNAs and snoRNAs.
6. Why do snRNAs and snoRNAs modify other RNAs?
7. Explain why there is no correlation between the cellular growth rate and the number of rRNA genes.
8. Describe the mechanism of control over the expression of repetitive elements.
9. List RNA polymerases involved in synthesis of tRNA, snRNA, snoRNAs, rRNA, and gRNAs.
10. Describe structure and function of apobec-1 and ACF.
11. What is the common feature of RNA editing in viruses? Is the editing beneficial for viruses?

12. Describe pseudo-template transcription and the slippery sequences in viruses.
13. Describe function of tmRNA.
14. Describe importance of RNA editing in evolution.

References

Andersen et al. (2003) Aging of the human cerebellum: a stereological study. *J Comp Neurol* 466:356-365.

Avery et al. (1944) Studies on the Chemical Nature of the Substance Inducing Transformation of Pneumococcal Types : Induction of Transformation by a Desoxyribonucleic Acid Fraction Isolated from Pneumococcus Type Iii. *J Exp Med* 79:137-158.

Birney et al. (2007) Identification and analysis of functional elements in 1% of the human genome by the ENCODE pilot project. *Nature* 447:799-816.

Blanc V, Davidson NO. (2003) C-to-U RNA editing: mechanisms leading to genetic diversity. *J Biol Chem.* 278:1395-1398.

Caspersson T, Schultz J. (1939) Pentose Nucleotides in the Cytoplasm of Growing Tissues. *Nature* 143:2.

Crick F. (1970) Central dogma of molecular biology. *Nature* 227:561-563.

Douet J, Tourmente S. (2007) Transcription of the 5S rRNA heterochromatic genes is epigenetically controlled in Arabidopsis thaliana and Xenopus laevis. *Heredity* 99:5-13.

Dulebohn et al. (2007) Trans-translation: the tmRNA-mediated surveillance mechanism for ribosome rescue, directed protein degradation, and nonstop mRNA decay. *Biochemistry* 46:4681-4693.

Earley et al. (2010) Mechanisms of HDA6-mediated rRNA gene silencing: suppression of intergenic Pol II transcription and differential effects on maintenance versus siRNA-directed cytosine methylation. *Genes Dev* 24:1119-1132.

Eickbush TH, Eickbush DG. (2007) Finely orchestrated movements: evolution of the ribosomal RNA genes. *Genetics* 175:477-485.

Farajollahi S, Maas S. (2010) Molecular diversity through RNA editing: a balancing act. *Trends Genet* 26:221-230.

Gardner et al. (2010) SnoPatrol: how many snoRNA genes are there? Gardner *et al. Journal of Biology* 9:4.

Gommans et al. (2009) RNA editing: a driving force for adaptive evolution? *Bioessays* 31:1137-1145.

Hausmann et al. (1999) Two nucleotides immediately upstream of the essential A6G3 slippery sequence modulate the pattern of G insertions during Sendai virus mRNA editing. *J Virol* 73:343-351.

He et al. (2008) NONCODE v2.0: decoding the non-coding. *Nucleic Acids Res*. 36 (database issue).

Henderson et al. (2010) The de novo cytosine methyltransferase DRM2 requires intact UBA domains and a catalytically mutated paralog DRM3 during RNA-directed DNA methylation in Arabidopsis thaliana. *PLoS Genet* 6:e1001182.

Hoagland et al. (1956) Enzymatic carboxyl activation of amino acids. *J Biol Chem* 218:345-358.

Jawdekar GW, Henry RW. (2008) Transcriptional regulation of human small nuclear RNA genes. *Biochim Biophys Acta* 1779:295-305.

Lawrence RJ, Pikaard CS. (2004) Chromatin turn ons and turn offs of ribosomal RNA genes. *Cell Cycle* 3:880-883.

Maas S. (2010) Gene regulation through RNA editing. *Discov Med*. 10:379-86.

Madej et al. (2007) Small ncRNA transcriptome analysis from kinetoplast mitochondria of Leishmania tarentolae. *Nucleic Acids Res* 35:1544-1554.

Mattick JS. (2010) The central role of RNA in the genetic programming of complex organisms. *An Acad Bras Ciênc* 82:4.

Mattick JS. (2009) Deconstructing the dogma: a new view of the evolution and genetic programming of complex organisms. *Ann NY Acad Sci* 1178:29-46.

Mattick JS. (2010) RNA as the substrate for epigenome-environment interactions: RNA guidance of epigenetic processes and the expansion of RNA editing in animals underpins development, phenotypic plasticity, learning, and cognition. *Bioessays* 32:548-552.

McStay B, Grummt I. (2008) The epigenetics of rRNA genes: from molecular to chromosome biology. *Annu Rev Cell Dev Biol* 24:131-157.

Mei et al. (2010) tRNA binds to cytochrome c and inhibits caspase activation. *Mol Cell* 37:v668-678.

Muljo et al. (2009) MicroRNA targeting in mammalian genomes: genes and mechanisms. *Wiley Interdiscip Rev Syst Biol Med* 2:148-161.

Nishikura K. (2010) Functions and regulation of RNA editing by ADAR deaminases. *Annu Rev Biochem* 79:321-349.

Paz-Yaacov et al. (2010) Adenosine-to-inosine RNA editing shapes transcriptome diversity in primates. *Proc Natl Acad Sci USA* 107:12174-12179.

Phizicky EM, Hopper AK. (2010) tRNA biology charges to the front. *Genes Dev* 24:1832-1860.

Sanger F, Thompson, EO. (1953) The amino-acid sequence in the glycyl chain of insulin. I. The identification of lower peptides from partial hydrolysates. *Biochem J* 53:353-366.

Scherrer et al. (1963) Demonstration of an unstable RNA and of a precursor to ribosomal RNA in HeLa cells. *Proc Natl Acad Sci USA* 49:240-248.

Simpson et al. (2000) Evolution of RNA editing in trypanosome mitochondria. *Proc Natl Acad Sci USA* 97: 6986-6993.

Storz G. (2002) An expanding universe of noncoding RNAs. *Science* 296:1260-1263.

The ENCODE (ENCyclopedia Of DNA Elements) Project (2004). *Science* 306:636-640.

Vaucheret H. (2005) RNA polymerase IV and transcriptional silencing. *Nat Genet* 37:659-660.

Watson JD, Crick FH. (1953) The structure of DNA. *Cold Spring Harb Symp Quant Biol* 18:123-131.

Wulff BE, Nishikura K. (2010) Substitutional A-to-I RNA editing. *WIREs RNA*. 1:90-101.

7

Non-coding RNAs involved in epigenetic processes: A general overview

Chapter 6, "Realm of Non-Coding RNAs: From Bacteria to Human," introduced various **non-coding RNAs** (**ncRNAs**), attempted to classify them, and described some of their basic roles in the cell. This chapter specifically focuses on the role of ncRNAs in epigenetic processes, such as transcriptional and post-transcriptional gene silencing, imprinting, and regulating heritable responses to changing environmental conditions.

Since their discovery, small regulatory RNA species have been revealed to play major roles in gene regulation of most eukaryotic organisms (Vaucheret, 2006). They have been shown to have essential functions from embryonic development through sexual maturity in regulating such processes as cell patterning, differentiation, cell cycle progression, genome stability, and apoptosis. Indeed, deregulation of these key processes have been implicated in a number of human diseases, including fragile X syndrome, DiGeorge syndrome, and, most notably, cancer. Further, in plants, loss of small RNA production can lead to enhanced disease susceptibility.

Small regulatory RNAs are a group of single-stranded ncRNAs usually ranging in size from 20 to 24 nucleotides. They are derived from complementary or semi-complementary **double-stranded RNA** (**dsRNA**) molecules and have functions in transcriptional and post-transcriptional gene silencing. **Post-transcriptional gene silencing** (**PTGS**) refers to downregulation of gene expression after transcription and mRNA maturation. In animals and plants, PTGS can refer to the sequence-specific binding of ncRNAs to mRNA molecules in the cytoplasm resulting in the inhibition of translation. This sequence-specific binding, which occurs more commonly in plants and rarely in animals, can result in the degradation of mRNA transcripts.

The biological activity of small RNAs has been harnessed with the advent of RNA interference (RNAi) as a tool used in molecular biology to knock down genes using PTGS. Transient or stable expression of synthetic small RNA species can efficiently target particular genes of interest, allowing the study of phenotypes in the absence of target transcripts. As it can be imagined, RNAi has already become a hot topic in disease therapy being a therapeutic agent with sequences designed to control the expression of genes deregulated in disease (Martin and Caplen, 2007).

Small RNA function and mechanism is somewhat conserved between various kingdoms; however, substantial changes in their biogenesis and function do exist. This chapter describes the biogenesis and mechanism of action of the main groups of ncRNAs involved in epigenetic regulation, whereas the following chapters include a more detailed analysis of ncRNAs in various kingdoms of life.

Biogenesis and function of ncRNAs in epigenetics

ncRNAs involved in epigenetic processes are artificially classified as **long ncRNA (lncRNA)** ranging in size from 200 to more than 100,000 nucleotides and as **small ncRNAs** with sizes ranging from 20 to 30 nucleotides (see Figure 7.1). The ncRNAs that are in the size range of 30 to 200 nucleotides are less frequently found. The classification of ncRNAs by size is proposed for mere convenience and reflects the separation of RNA molecules in common protocols of RNA preparation.

lncRNAs are perhaps the most diverse but the least understood group of ncRNAs. With 90% to 100% of the human genome (and possibly genomes of other organisms) being transcribed, there are tens of thousands of potentially produced lncRNAs. They are less conserved in sequence and structure as compared to small ncRNAs. This is a possible indication of lack of functionality; however, lncRNAs such as Xist have a defined function. It should be noted that some regions of lncRNAs are more conserved than the others, which possibly indicates important sequence or structure recognition areas.

Many lncRNAs should have biological functions as their transcription is regulated by developmental processes and differentiation. Many lncRNAs are localized to specific cellular compartments, suggesting their importance in specific cellular processes. Changes in the expression of many lncRNAs were shown to associate with various human diseases (Wilusz et al., 2009). The evidence for selection of certain lncRNAs in the process of evolution also suggests their importance.

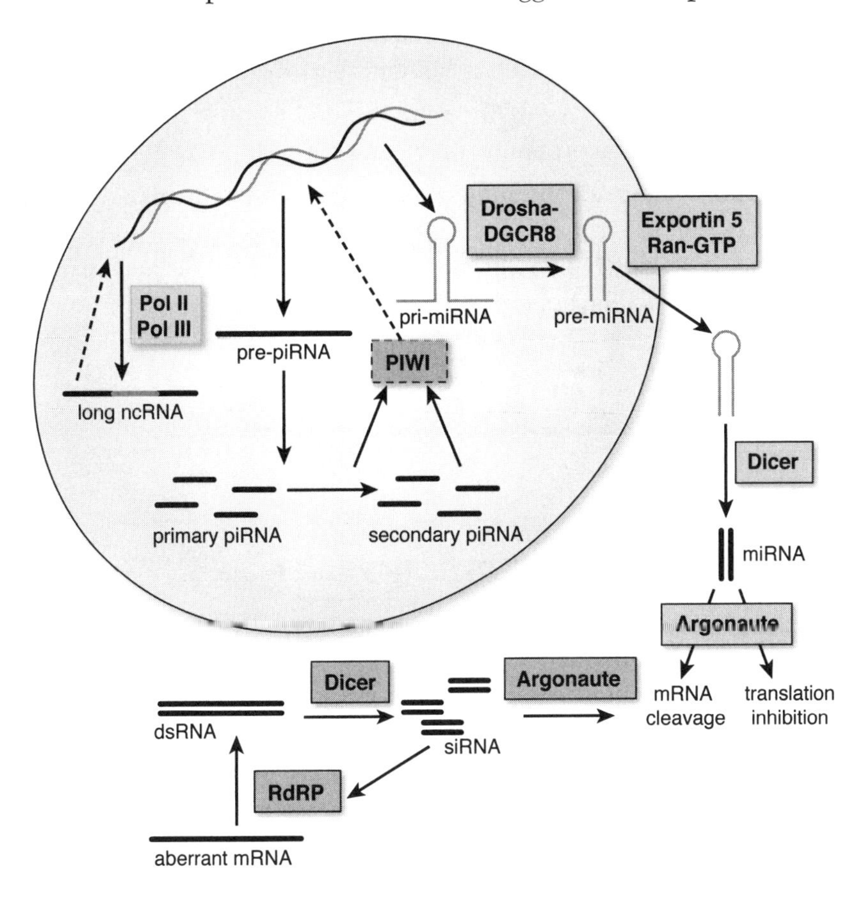

Figure 7-1 An overview of the biogenesis of long and small ncRNAs in mammalian cells. Key players in the biogenesis of different ncRNAs are highlighted in different color shades. The light gray box (blue in color ebook) represents miRNA biogenesis; the dark gray box (purple) represents siRNA biogenesis; the box outlined with a black dashed line indicates piRNA biogenesis; and the box outlined with a black solid line represents ncRNA biogenesis. The solid arrows indicate the sequence of events, whereas the dashed arrows indicate silencing. For the sake of simplicity, the function of siRNA in mediating transcriptional silencing is not represented in this figure.

Depending on their genome localization, lncRNAs can be attributed to one of the following five categories (A-E in Figure 7-2):

- Sense ncRNAs that overlap with exons of the downstream gene on the same strand (A)
- Antisense ncRNAs that overlap with exons of the neighboring gene on the opposite strand (B)
- Bidirectional ncRNAs, if the expression of lncRNA-coding gene and the neighboring gene on the opposite strand occurs in close genomic proximity (but does not overlap) (C)
- Intronic ncRNAs, if they are located entirely in the intron sequence of a protein-coding gene (D)
- Intergenic ncRNAs, if their sequence does not overlap with any coding or intronic sequence on either sense or antisense strand and the lncRNA coding gene is at a substantial distance from a protein-coding gene (E) (Ponting et al., 2009)

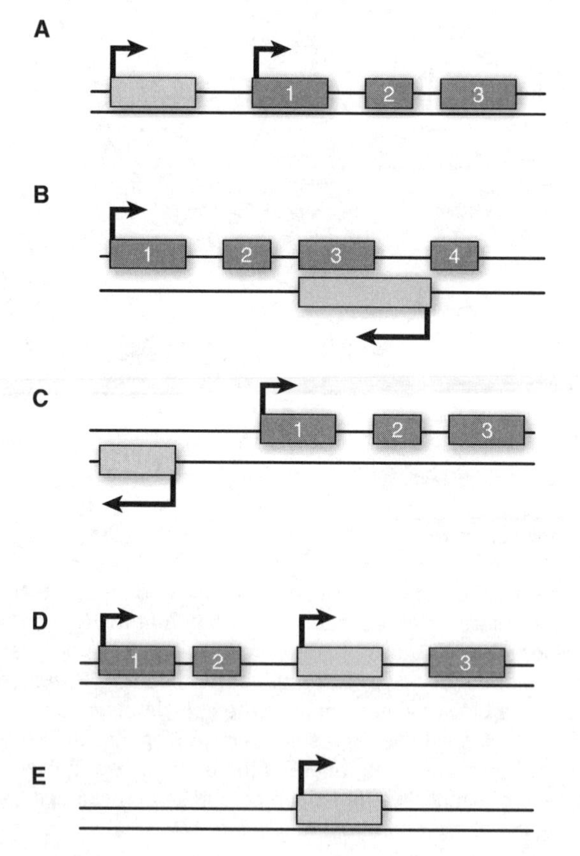

Figure 7-2 Various sources of lncRNAs production (see the text for description). Dark gray (dark blue in ebook) boxes are exons of mRNA-coding genes, whereas light gray (light blue) boxes are ncRNA-coding genes.

The category of antisense lncRNAs is referred to as ***cis* natural antisense transcripts** (***cis*-NATs**). They are common in plants and mammals and may exist in about 10% of all gene transcripts.

Like mRNAs, most lncRNAs are transcribed by Pol II, though examples of ncRNAs that are transcribed by Pol III (Alu ncRNA) or by Pol IV (repetitive methylated sequences) in plants have been also reported. In the majority of cases, the lncRNA transcripts are not processed, and they do not undergo capping at the 5' end, polyA, or splicing. As a result, they escape nuclear export and, by doing this, they stay in close proximity to their targets (refer to Figure 7-1).

Some of the lncRNAs might have evolved from mRNAs, with the **Xist RNA** being the most well-known example. The promoter and part of the transcribed sequence of Xist RNA are derived from the *Lnx3* gene, and during the course of evolution, they were modified through mutations.

The main cellular function of lncRNAs is transcriptional repression (see Figure 7-3A), although for the majority of the transcripts in this group, the biological and mechanistic function is largely unknown. A prominent exception is Xist and **Tsix** RNA, whose biological function in animals is imprinting. Some of other known functions include the regulation of stress response, such as stress-induced silencing mediated by INDUCED BY PHOSPHATE STARVATION1 (IPS1), which negatively regulates miR399 to allow phosphate accumulation in Arabidopsis.

The epigenetic mechanisms by which lncRNAs mediate their functions are very heterogeneous:

- Xist RNA coats the X chromosome and recruits histone-modifying complexes, such as PRC2, which mediates H3K27 trimethylation and results in the formation of heterochromatin at one of the X chromosomes.
- In Arabidopsis, intergenic ncRNAs interact with siRNAs to induce heterochromatin formation.
- **Alu ncRNAs** are induced upon heat shock in human cells, bind and inhibit the function of Pol II; the same mechanism, apparently, operates in bacteria.
- IPS1 RNA binds to and inhibits the function of miR399 in Arabidopsis.

Overall, the mode of action of long ncRNAs can be divided into several major categories (see Figure 7-3):

- Regulation of expression of neighboring genes
- Blocking of splicing of protein-coding genes by using antisense transcripts
- Interaction with proteins making them more or less capable of fulfilling specific functions
- Serving as precursors for smaller ncRNAs

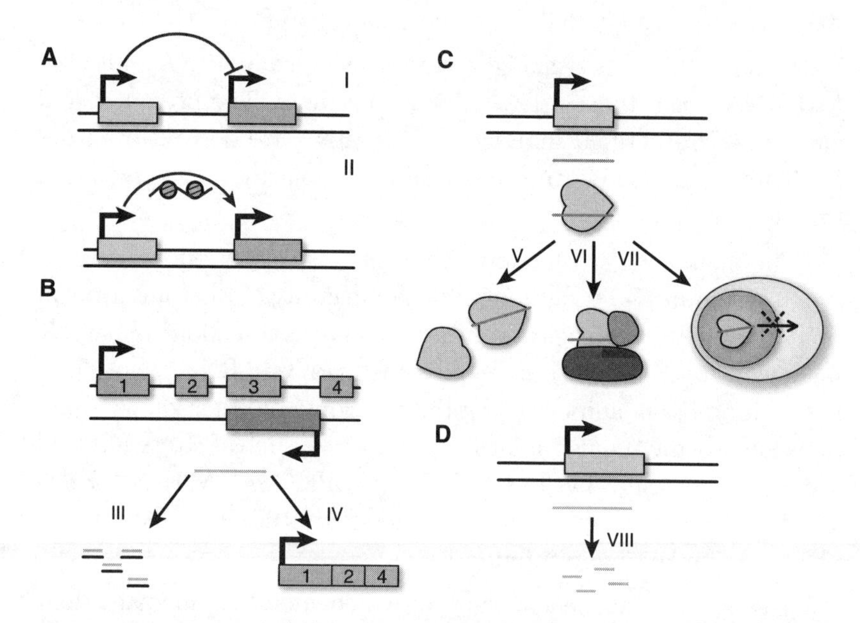

Figure 7-3 The major modes of action of lncRNAs. **A.**) Transcription from an lncRNA gene regulates the transcription of neighboring genes. I. Transcription from an upstream noncoding promoter (light gray/light blue) can regulate the expression of a downstream gene in a negative manner by blocking the Pol II access. II. Transcription from an upstream noncoding promoter (light gray/light blue) can regulate the expression of a downstream gene (gray/blue) in a positive manner by inducing chromatin remodeling. **B.**) Transcripts from antisense lncRNAs. III. The production of endogenous siRNAs from sense/antisense hybrids. IV. A long antisense transcript of ncRNA (light gray/light blue) can bind the overlapping sense transcript blocking splicing and leading to the production of alternative transcripts. **C.**) lncRNAs bind various proteins. V. lncRNA transcripts bind to specific proteins modifying their activity. VI. Protein/lncRNA serves as a structural component that allows a larger RNA—protein complex to be generated. VII. Binding to the protein changes its localization. **D.**) lncRNAs as small ncRNA precursors. VIII. lncRNA transcripts serve as precursors for the production of small ncRNAs.

There are, of course, more mechanisms of action and examples of function of lncRNAs that demonstrate the diversity of their involvement in various cellular processes. With more knowledge and better characterization of lncRNAs, the mechanistic classification of them will be more possible. For more details as to function, structure, and evolutionary significance of lncRNAs, see the review by Ponting et al. (2009).

As mentioned earlier, small ncRNAs are ncRNAs that are 20 to 30 nt long. Depending on their origin, mode, and place of function, ncRNAs involved in the epigenetic regulation can be classified into major categories: **microRNA** (**miRNA**), **small interfering RNA** (**siRNA**), and **PIWI (P-element-induced wimpy testis)-interacting RNA** (**piRNA**). The biogenesis of small RNAs broadly involves the following major steps:

1. Transcription of the genomic locus by one of three polymerases—Pol II, Pol III, or Pol IV enzymes—resulting in the formation of dsRNA.
2. RNase III type endonucleases (Drosha and Dicer) process these long precursor dsRNAs into duplex RNAs of 19 to 28 nt in length. Argonautes process duplex RNAs into single-stranded, mature small RNAs.
3. These mature small RNAs guide molecules to the multi-protein complex (the **RNA-Induced Silencing Complex** (**RISC**) or the **RNA-Induced Transcriptional Silencing Complex** (**RITS**)) depending on the downstream effect.
4. The complexes regulate gene expression by degrading target mRNAs, inhibiting the translation, or inhibiting transcription processes via ncRNA-induced chromatin modifications.

This chapter provides an overview of the biogenesis of specific types of ncRNA, but you can find more details as to the organism-specific biogenesis and function of specific ncRNAs in the epigenetic process in Chapters 8 through 12.

Micro-RNAs

Micro-RNAs (miRNAs) are small 22-nt lncRNAs initially discovered in *Caenorhabditis elegans* (*C. elegans*), which by now have been reported to exist in all metazoans. miRNAs are expressed either from individual miRNA genes or as a part of other mRNA-coding genes. Genes coding for miRNAs are present in the genome either as single genes or clustered genes. Frequently, such genes have their individual promoters and regulatory elements, but many miRNA genes are also present within the introns of other protein-coding genes. Thus, similar to the genomic location/position of lncRNA genes, miRNA genes, depending on their genomic position, can be broadly classified into intergenic and intragenic; the latter ones can be subdivided into intronic and exonic, which overlap with introns and exons and those that are found in the 3'UTR of protein-coding genes.

miRNAs are mainly transcribed by Pol II as long primary transcripts, although examples of miRNAs transcribed by Pol III, such as the human C19MC cluster, also exist (Borchert et al., 2006). The initial RNA product synthesized by Pol II is called a **primary transcript (pri-miRNA)** (refer to Figure 7-1). The miRNA precursors are predicted to form irregular hairpin structures containing various mismatches, internal loops, and bulges.

In humans, these molecules are processed into **precursor-miRNAs (pre-miRNAs)** by the Drosha-DGCR8 complex, in which Drosha is an RNaseIII-like endonuclease, whereas DGCR8 is a double-stranded RNA-binding protein (dsRBP) encoded by DiGeorge syndrome Critical Region gene 8 (see Figure 7-4).

First, the pri-miRNA is processed in the nucleus into a 60 to 70 nt–long pre-miRNA by the Drosha-DGCR8 complex in mammals and by Drosha-Pasha in flies. The resulting pre-miRNA has a hairpin structure: a loop flanked by base-paired arms that form a stem. Pre-miRNAs have a two-nucleotide overhang at their 3' ends and a 5' phosphate group, which reflect their production by an RNase III. Sometimes pre-miRNAs are edited by the adenosine deaminase acting on RNA (ADAR) enzyme at specific positions (mostly +4 and +44), thus bringing additional diversity to the precursor sequence.

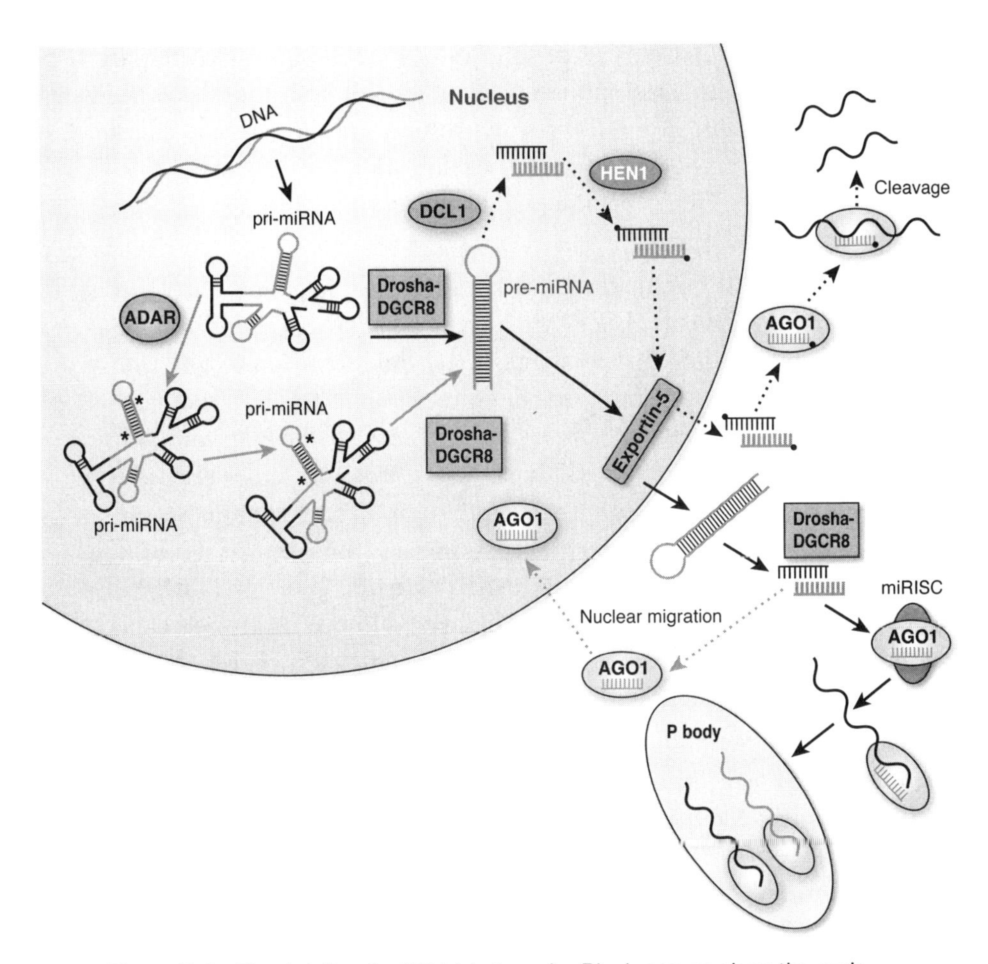

Figure 7-4 The details of miRNA biogenesis. Black arrows show the main pathway, whereas gray (red in ebook) arrows are secondary steps. Black dashed arrow shows the miRNA biogenesis in plants. In animals, transcribed pri-miRNAs are processed into pre-miRNAs by Dorsha/DGCR8, and in plants they are processed by Dcl1. Some pri-miRNAs are edited by ADARs and returned to the pri-miRNA pool. Pre-miRNAs are either directly exported into the cytoplasm (in animals) or are further processed into miRNA/miRNA* duplex (in plants). Asterisk indicates an "antisense" miRNA strand. In plants and Drosophila, these duplexes are methylated by the HEN enzyme and then are exported to the cytoplasm. In animals, pre-miRNAs are processed by Dicer into miRNA/miRNA* duplexes, and then they join the miRNA-RISC complexes including Argonaute and other proteins. In animals, these complexes inhibit mRNA translation sending the inhibited mRNA into P bodies. In plants, miRNAs target mRNAs for degradation by promoting sequence-specific cleavage. Some miRNAs are transferred back to the nucleus where they are apparently involved in the formation of heterochromatin, splicing, and other processes.

In plants, DCL1 cleaves pre-miRNAs generating miRNA duplex in the nucleus. In Drosophila and Arabidopsis, these duplexes are further methylated at the 2'-hydroxyl residues of the terminal ribose sugars. It is hypothesized that this modification protects methylated duplexes either from further modifications by uridylation or from degradation.

Pre-miRNAs (or mature miRNA duplexes in plants) are then exported into the cytoplasm by exportin-5 and Ran-GTP. The functions of this binding are to stabilize the duplex, to protect it from degradation, and to transport the pre-miRNA to the cytoplasm. Exportin-5 and RanGTP target the duplexes to the nuclear pores where the hydrolysis of RanGTP to RanGDP results in the release of the pre-miRNA to the cytoplasm. In the cytoplasm (in animals), pre-miRNAs are further processed by another endonuclease of the RNaseIII superfamily: Dicer and its dsRBD partner proteins (TRBP in mammals or Loqs in Drosophila). The main difference between Dicer and Drosha is that Dicer contains a PAZ domain that probably allows it to bind a two-nucleotide overhang at the 3' end left by Drosha.

The recognition and binding of Dicer stimulates the cleavage of pre-miRNA at ~22-nucleotides from the 3'-OH and produces a mature miRNA duplex with two-nucleotide 3' overhangs. The cleavage by Dicer results in a double-stranded (duplex) molecule consisting of miRNA and miRNA* corresponding to the siRNA guide and passenger strands. The identification of whether the strand becomes miRNA or miRNA* depends on that whose 5'-end is less tightly paired to its complement. Typically, the one whose 5'-end is less tightly paired is selected to enter into the RISC, whereas the opposite strand is degraded. It should be noted, however, that miRNAs can arise from either arm of the pre-miRNA stem. Some pre-miRNAs produce mature miRNAs from both arms, whereas others predominantly produce them from only one arm, with an opposite arm being completely degraded.

In addition to Dicer, RISC contains Argonaute proteins and several other proteins. The Dicer protein then dissociates from the RISC complex, leaving the active machinery for silencing termed the **miRNA/AGO ribonucleoprotein** (**miRNP**). The fate of miRNAs in RISCs depends on many factors, and usually it is defined by sequence specificity. In animals, some miRNAs can travel back into the nucleus and can be involved in diverse processes such as heterochromatin formation. Those cytoplasmic miRNAs regulate the level of target transcripts. The mode of action of the miRNA-containing RISC primarily depends on the degree of complementarity between sense-strand miRNAs and target mRNAs. Low levels of complementarity are associated with inhibition, or sometimes activation, of translation, and high levels of complementarity are related to endonucleolytic cleavage of miRNA-mRNA (refer to Figure 7-4).

In animals, mature miRNAs within the miRNP complex direct gene silencing in a sequence-specific manner using miRNAs as guides to bind to target mRNAs at **miRNA recognition elements** (**MREs**). These MREs are usually found in the 3' untranslated region (UTR) of the mRNA; however, recent evidence for mRNA 5' UTR binding has also been presented (Lytle et al., 2007).

The capacity for translation inhibition is dictated by small sites with perfect complementarity, called **seed regions**, between the miRNA:mRNA duplex. In Drosophila, it was thought that nucleotides 2 through 8 from the 5' end were involved in targeting recognition and some miRNAs required varying degrees of pairing at the 3' end for greater stability (Brennecke et al., 2005). More recent work on mammals has shown the importance of the seed site, nucleotides 2 through 7 from the 5' end; several other important observations of pairing that boost the miRNA:target binding affinity have also been outlined. These sites include an additional pairing at the eighth nucleotide and/or the presence of adenosine at the first position of the 5' end. Further, as in Drosophila, the binding sites at the 3' region, especially at nucleotides 13 to 16 from the 5' end, can greatly enhance binding affinity (Grimson et al., 2007).

Binding affinity affects the degree of translational inhibition, but other factors also seem to have an effect on the efficacy of inhibition. For example, MREs that are flanked with AU-rich sequences and/or MREs that are located at the ends of the 3' UTR at least 15 nucleotides away from the stop codon have greater translational repression ability. Logically, the strongest inhibition of translation was observed when multiple binding sites for the same or different miRNAs were clustered in the 3'UTR (Grimson et al., 2007).

In animals, the repressed target mRNAs are transported into vesicles called **Processing (P) bodies** and are potentially degraded. The process is initiated via the interaction between AGO1, a member of the RISC, and P-body proteins of the GW182 family such as GW1, AIN1, AIN2, and others. P-bodies contain RNA-protein complexes linked to cytoplasmic RNA decay pathways including mRNA decapping, nonsense-mediated decay and small RNA-mediated decay. Translationally repressed mRNAs stored in P-bodies are either degraded or released, if necessary. The translational activity of such mRNAs is usually not altered. Plants also contain P-bodies; however, homologs of GW182 family proteins are yet to be identified.

A single miRNA can target multiple mRNAs, thus regulating the expression of numerous genes. Also, several miRNAs with imperfect complementarity to the target mRNA can be involved in translation inhibition achieved by binding to the 3' UTR. miRNAs exhibit high levels of tissue specificity and are often expressed at a particular time during development. This allowed scientists to hypothesize that the main role miRNAs play in the organism is to regulate the developmental processes.

How are miRNAs regulated? The hypothesized feedback loop mechanism suggests that low levels of target mRNAs are sensed by the presence of secondary proteins that communicate with transcription factors or nucleases. miRNAs can thus be regulated at the level of transcription by blocking the expression of miRNA-coding (or miRNA-containing) genes. They also can be controlled by exonucleolytic degradation. Editing the seed region by ADARs might also play a significant role in reducing (as well as in increasing) the specificity of miRNAs to their target mRNAs. Besides editing, binding of miRNAs to their targets in animals can be apparently prohibited by the presence of highly conserved flanking sequences in mRNA targets. It is hypothesized that specific proteins such as HuR and Dnd1 (Dead End 1) bind to the AU-rich elements (ARE) in the 3'-UTR and prevent the repressive activity of miRNAs (Bhattacharyya et al., 2006). Indeed, Dnd1 was shown to negate the miR122-mediated translation repression of p27 in human cells.

The details of biogenesis and function of miRNAs in different organisms have been extensively reviewed (see, for example, Davis and Hata, 2009) and are covered in more detail in Chapters 8 through 12.

Mirtrons as a novel concept of miRNA biogenesis

The recent progress in understanding miRNA biogenesis in animals shows that miRNAs situated in introns may be produced through a different pathway. These miRNAs were termed **mirtrons**. Several mirtrons from Drosophila and *C. elegans* have been identified. It is expected that many miRNAs might actually derive from introns. Such miRNAs can be processed through a canonical biogenesis pathway that depends on RNA folding (see Figure 7-5). Hairpins generated upon folding are then processed by the enzyme Drosha. Mirtron biogenesis occurs through a different mechanism. Splicing of introns generates the characteristic 2'-5' phosphodiester bonds resulting in loops or loop-like structures called **lariats**. These introns are then either degraded (the experiments show that the steady-state level of introns is 10 times lower than that of exons) or undergo debranching by the lariat debranching enzyme (in humans, it is hDBR1). The released RNAs have a potential to refold, forming hairpin structures that can be exported by Exportin-5 into the cytoplasm, picked up by Dicer, and then further processed into miRNAs involved in RISC (see Figure 7-5). The analysis of Drosha and Pasha mutants revealed that mirtrons are still processed into functional miRNAs, thus escaping the canonical biogenesis route involving Drosha/DGCR8 proteins (Martin et al., 2009). Furthermore, knockdown of Dcr-2 and RDR-2 (proteins that are responsible for siRNA biogenesis) does not inhibit the production of mirtrons, suggesting that the siRNA biogenesis pathway is also not involved in generating mirtrons (reviewed in Naqvi et al., 2009). The production of both canonical miRNAs and mirtrons is dependent on Dicer and Argonaute proteins, suggesting that biogenesis in cytoplasm follows the same route.

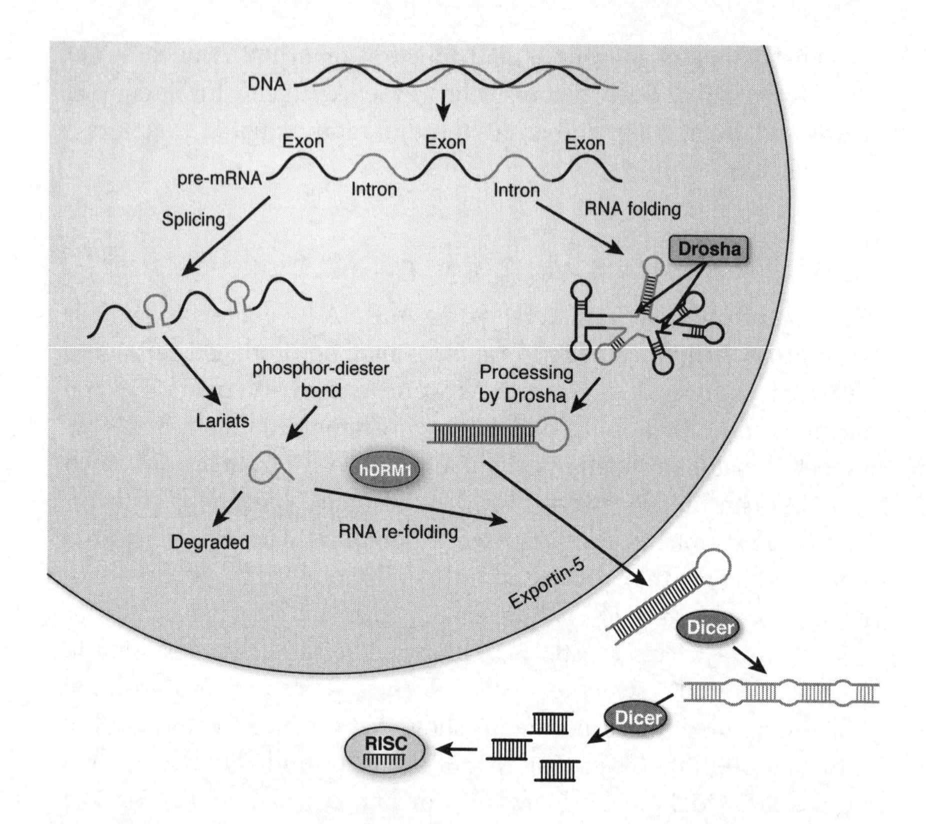

Figure 7-5 Biogenesis of miRNAs form mirtrons. Pre-mRNAs containing miRNAs can fold into secondary structures containing pri-miRNAs. The canonical miRNA biogenesis pathway includes processing of these pri-miRNAs into pre-miRNAs by Drosha. Alternatively, splicing of pre-mRNA releases the intron structure termed a lariat. The debranching enzymes, such as hDBR1, cleave phosphodiester bonds and allow RNA introns to re-fold. The refolded RNAs with a 2 nt 3' overhang is picked up by Exportin-5 and is exported to the cytoplasm. These pre-mirtrons merge into the mainstream pre-miR pool where they are processed by Dicers and incorporated into RISC.

Small interfering RNAs

Small interfering RNAs (siRNAs) are 20 to 24 nt–long molecules produced mainly through processing of double-stranded RNA (dsRNA) precursors. These regulatory molecules protect cells against the intrusion of exogenous nucleic acid such as viruses and are involved in the maintenance of genome integrity by preventing transcription from retrotransposons and other loci coding for repetitive sequences.

The discovery of dsRNA (a substrate for Dicer-mediated cleavage and production of small RNAs) is credited to the recent Nobel Prize laureates Andrew Fire and Craig Mello (2006). They demonstrated that exogenously introduced dsRNAs induce silencing in *C. elegans* more efficiently than sense or antisense small single-stranded RNA molecules (Fire et al., 1998).

There are many other endogenous and exogenous substrates for siRNA production, including viral dsRNA genomes, intermediates of reverse transcription, and inverted repeats. Many organisms, including plants, fungi, and worms (*C. elegans*), possess a specific enzyme, **RNA-dependent RNA Polymerase** (**RdRP**), which is able to target any aberrant transcript or even any high-level transcript to yield dsRNA. dsRNA is cleaved into double-stranded siRNAs by Dicer. siRNA duplexes produced by Dicer comprise two ~21 nt strands, each bearing a 5' phosphate and a 3' hydroxyl group, thus forming two-nucleotide overhangs at the 3'-end.

The dsRNA precursors are cleaved by Dicer and then are integrated into either RISC or RITS. Both of them contain Argonaute proteins to mediate the function of siRNA (refer to Figure 7-1).

The major difference between these two complexes is the mode of action of siRNA; in the RISC, siRNAs function very much like miRNAs, whereas in the RITS, siRNAs target the homologous regions of the genome for methylation and heterochromatization. Although the mechanism of RITS-mediated function is unclear, it is believed that siRNAs might interact with DNA methyltransferases and histone-modifying enzymes (reviewed in Bernstein and Allis, 2005). Furthermore, studies in yeast showed that the RITS complex interacts with an RNA-dependent RNA polymerase, thus resulting in the amplification of a siRNA signal and enforcement of silenced states (Motamedi et al., 2009).

The strand that directs silencing is called the guide strand, whereas the degraded strand is called the passenger strand. Similarly to miRNAs, the relative thermodynamic stability of the 5'-ends of siRNA strands in the duplex defines which strand is used as a guide and which one is degraded. Argonaute protein is activated by the siRNA-RISC complex; the RNase activity of Argonaute protein allows a specific cleavage of the target sequence at the position complementary to 10 and 11 nt counting from the 5' end of the siRNA. At the same time, the seed region covering nucleotides 2 through 7 (from the 5' end) confers siRNAs their target specificity and the ability to cleave/repress a specific target. The complementarity of other regions of siRNA to mRNA is also critical for cleavage.

siRNAs are involved in almost all possible nucleic acid regulatory pathways: mRNA degradation, translation inhibition, transcriptional gene silencing, and even DNA elimination (see Chapter 9, "Non-Coding RNAs Across the Kingdoms: Protista and Fungi"). The choice between mRNA cleavage and translation inhibition depends on the complementarity between siRNAs and mRNA targets. Whereas perfect complementarity leads to mRNA target cleavage, lesser complementarity typically leads to translational inhibition. Target cleavage and degradation seem to be predominant mechanisms in plants, whereas translational inhibition is the mechanism that is more common in animals.

Read Chapters 10 through 12 for more details on the function of various siRNAs in cellular processes, including trans-acting siRNAs (tasiRNAs), repeat-associated siRNAs (rasiRNAs), scan RNAs (scnRNAs), and long siRNAs (lsiRNAs).

PIWI-associated RNAs

piRNAs (PIWI-associated RNAs) have traditionally been studied in germline cells of animals and were thought to be restricted to them and their neighboring cells (Aravin et al., 2007). However, a recent publication described the discovery of piRNAs in HeLa cells by employing a high-throughput sequencing strategy (Lu et al., 2010). piRNA loci exist as clusters of 10 to 4500 piRNA genes and are transcribed from bidirectional promoters as long single-stranded RNA precursors. The majority of piRNA sequences are mapped to the genomic regions that transcribe only non-coding RNAs; however, some piRNAs exist in intergenic, exonic, and intronic regions of coding RNAs. The exact mechanism of how small RNAs are generated from the precursor is still under debate.

Several proteins are known to interact with piRNAs. Among those is the protein PIWI (**P element-induced wimpy testis**), a member of the Argonaute protein family. Another one is ATP-dependent helicase Rec Q1. This protein/RNA complex is termed PIWI-interacting RNA complex (piRC). The current model suggests that the precursors are cleaved into primary piRNAs of two distinct size categories (24 to 28 nt and 29 to 31 nt) (see Figure 7-6). In contrast to miRNAs and siRNAs that are always antisense to their targets, piRNAs appear to have a strand bias as they can be sense or antisense to the transcript they are targeting. In mice, the expression pattern of two proteins, MILI and MIWI (PIWI orthologs), resembles the temporal pattern of production of two classes of piRNAs, 24 to 28 nt and 29 to 31 nt, respectively. This suggests that MILI and MIWI may be involved in the production of some specific classes of these piRNA species (reviewed in Naqvi et al., 2009).

Single-stranded molecules of these two classes of piRNAs are transported into cytoplasm where they are methylated at the 2'-O of 3' ends by the Drosophila homolog of Arabidopsis HEN1 methyltransferase, piRNA methylase (Pimet). Mature piRNAs then bind to PIWI proteins and AUBERGINE (Aub) or Argonaute (AGO3) proteins that initiate piRNA amplification via a ping-pong mechanism. Whereas sense piRNAs associate with AGO3, antisense piRNAs associate with Piwi/Aub (see Figure 7-6) (Naqvi et al., 2009). These two complexes promote binding of sense and antisense piRNAs that are complementary at the first 10 bases; the 5' end of antisense piRNAs typically contains a uridine base, and the tenth nucleotide of sense piRNAs—an adenine base.

Many questions as to how piRNAs are produced still remain unanswered. It is not clear why AGO3 strictly binds sense piRNAs derived from transposon mRNAs. It is also unknown how the 3' ends of piRNAs are made. Ghildiyal and Zamore (2009) propose to test the hypothesis that different forms of Pol II transcribe primary piRNA transcripts and transposon mRNAs and that the specialized form of Pol II that transcribes the primary piRNA precursor recruits Piwi and Aub, but not Ago3. This idea still remains to be tested.

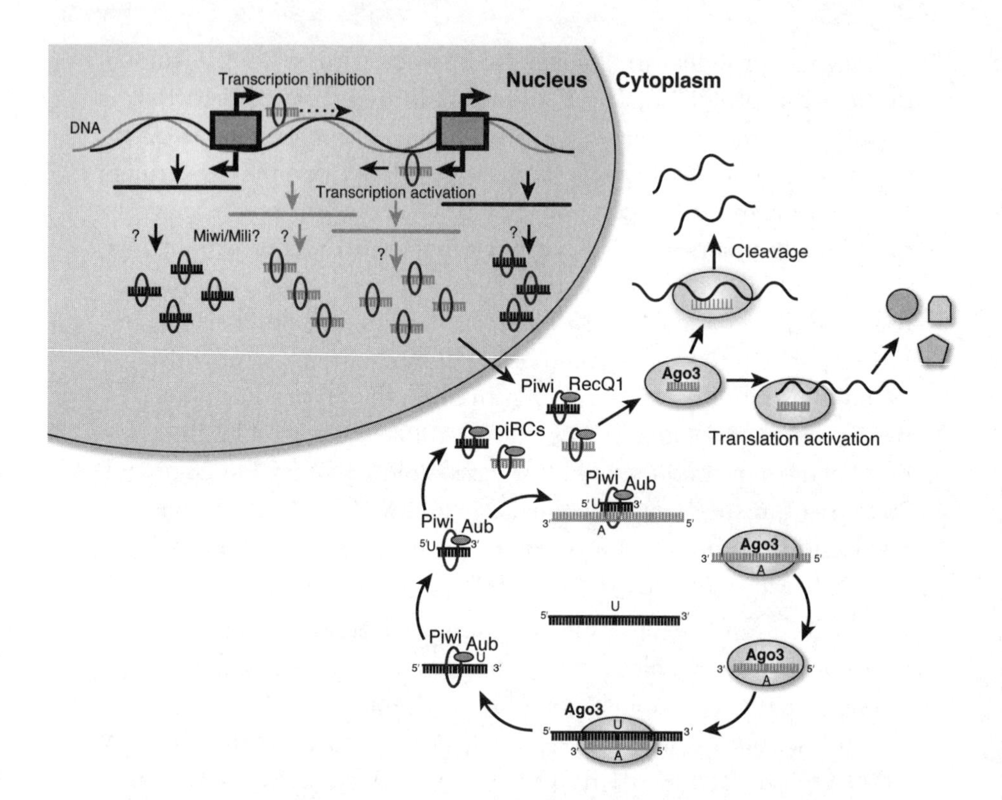

Figure 7-6 The mechanism of biogenesis and function of piRNAs. piRNAs are transcribed from bidirectional promoters as long precursors. Their processing into functional piRNAs is not clear, but newly generated piRNAs can function in the nucleus by either inhibiting or activating transcription. A fraction of piRNAs is transported into cytoplasm where they are joined by Piwi/RecQ1 and, perhaps, other proteins forming piRCs. piRNAs from these complexes may regulate translation either negatively (through cleavage) or positively. piRNAs apparently have a self-amplifying circle known as a ping-pong mechanism. First, antisense piRNAs (short black) are recognized by a pair of the proteins. Piwi and Aub bind the mRNA target (long light gray—red in ebook), promoting cleavage between the tenth and eleventh nucleotides of the piRNA. The cleaved mRNA is picked up by AGO3 that trims it further preparing for binding to the piRNA cluster transcript (long black). This results in cleavage of the piRNA transcript, which is then picked up by the pair of Piwi/Aub proteins that further trim the transcript, thus either preparing it to enter the cycle again or directing it to the process of regulation of translation. (Adapted from original figure from Naqvi AR, et al. The Fascinating World of RNA Interference. Int. J. Biol. Sci. 2009; 5:97-117. Available from http://www.biolsci.org/v05p0097.htm.)

The ping-pong mechanism can quickly generate a substantial number of piRNAs to inactivate any harmful retroelement. Both the level of target transcripts and the level of piRNA cluster transcripts apparently work as regulatory mechanisms of piRNA production. The mechanism of piRNA action can be quite diverse. It includes transcriptional gene silencing that primarily represses the loci that generate piRNAs and other regions with homologous transposon sequences; positive regulation of gene expression via transcription stabilization; post-transcriptional regulation via transcript cleavage; and positive regulation of translation via translation stabilization. The main function of piRNAs appears to be the protection of germline cells against genomic instability, but it might also include other processes involved in germ cell development. A more detailed analysis of piRNA function is presented in Chapter 10, "Non-Coding RNAs Across the Kingdoms—Animals."

Structure and function of RNase III–type endonucleases

Dicer is one of the main RNaseIII–type endonucleases involved in miRNA and siRNA generation. Whereas animals typically use a single type of Dicer to generate different classes of small RNAs, Drosophila and *C. elegans* encode two dicers; one of them, Dicer-1, is involved in miRNA processing; the second one, Dicer-2, is involved in siRNA processing. In contrast, plants have several **Dicer-like** (DCL) proteins, with Arabidopsis possessing 4 DCLs and rice - 6 DCLs.

The Dicer protein has six domains including DExH Helicase, DUF283, PAZ (Piwi, Argonaute, and Zwille), RNase IIIa, RNase IIIb, and RNA Binding Domain (RBD) (see Figure 7-7). Based on crystal structures of several individual domains, their function is the following: RBD recognizes the duplex RNA structure, whereas the PAZ domain binds to the 3' 2nt overhangs of the cleaved RNA substrate (reviewed in Naqvi et al., 2009). The RNase IIIa and IIIb domains form an intra-molecular dimer upon binding to dsRNA. The RBD and the PAZ domains might assist the RNase III domains in forming such dimer. The RNase III domains position themselves around dsRNA precursors in such a manner that the ~21 nt duplex RNA molecules are cleaved from these precursors.

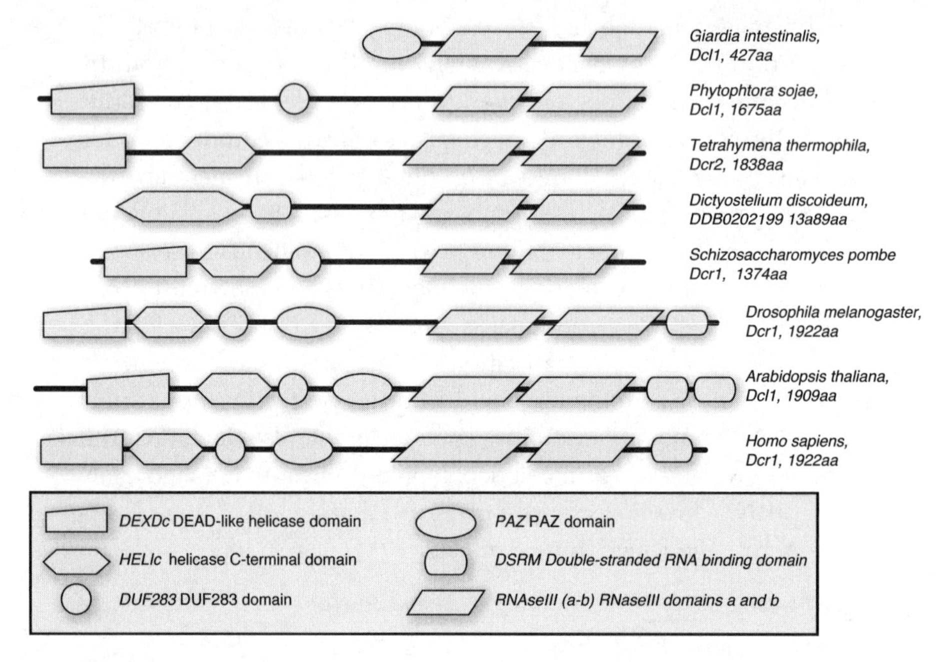

Figure 7-7 Structure of different Dicer and Dicer-like proteins.

Argonaute is a helicase protein recruited by Dicer that further unwinds duplex ncRNAs. The protein contains three main domains: PAZ, MID, and PIWI. Similarly to Dicer proteins, the PAZ domain in Argonaute also recognizes the 3' end and binds to two protruding nucleotides of the cleaved siRNA substrate. The MID domain of Argonaute protein accepts the 5' end into its basic pocket. The PIWI domain apparently has a dual role interacting with other proteins such as Dicer and participating in the actual cleavage of the target mRNA. Argonautes are part of the RNAi machinery in both animals and plants. Table 7-1 summarizes the conservation of key players in small RNA biogenesis across kingdoms.

Table 7-1 Evolution of proteins involved in sRNA biogenesis. Bacteria and Archaea lack a clear Dicer ortholog, but they have proteins of the same family.

sRNA biogenesis	Bacteria	Archaea	Protozoa	Fungi	Plants	Animals
RdRP	–	–	+	+	+	+ (not in mammals and Drosophila)
Dicer	RNAseIII	RNAseIII	+	+	+	+
Argonaute	*Aquifex*	+	+	+	+	+
PIWI protein	–	–	–	–	–	+

gRNAs—editing of miRNA and siRNA

Guide RNAs (gRNAs) and the process of editing are introduced in Chapter 6. During the process of A-to-I editing, the ADAR proteins promote deamination of adenine in the target RNAs. We mentioned that any RNA can be a potential target of editing, including pre-mRNAs, viral RNAs, and miRNAs and siRNAs (as it has recently been shown).

miRNAs are transcribed from the genome as pri-miRNAs that fold up into a number of double-stranded stem regions. pri-miRNAs are then processed into individual hairpins of approximately 70 nucleotides in length by the enzyme Drosha with help of the dsRBD-containing DGCR8 protein. It appears that both components of the miRNA biogenesis pathway and ADARs make use of dsRBDs, and the consequent overlap of substrate pools provides the ground for much crosstalk between the two pathways (see Figure 7-8). Editing of pri-miRNAs prevents processing by either the Drosha-DGCR8 or Dicer-TRBP complexes. It might prevent loading of an edited miRNA onto the RISC or affect which strand is chosen for loading. Editing also results in a RISC-loaded edited miRNA that can target a different set of transcripts compared to its unedited counterpart (Wulff and Nishikura, 2010).

In animals, each miRNA molecule can target a large set of mRNAs, thus resulting in either translation inhibition or mRNA degradation. Because editing potentially changes the variability of miRNA targets, controlling the extent of editing of a single miRNA can be used by the cell to control large-scale expression patterns. Because the selection of which strand is loaded onto the RISC depends on the relative stability of the two duplex ends, editing may modify it, thus allowing a different strand to be used for targeting mRNA. Most of the reports suggest that editing primarily occurs at the level of pri-miRNA, although *in vitro* evidence suggests the possibility of editing at the pre-miRNA stage (Kawahara et al., 2007). It is not clear whether editing at this stage does occur *in vivo* and what the degree of editing is.

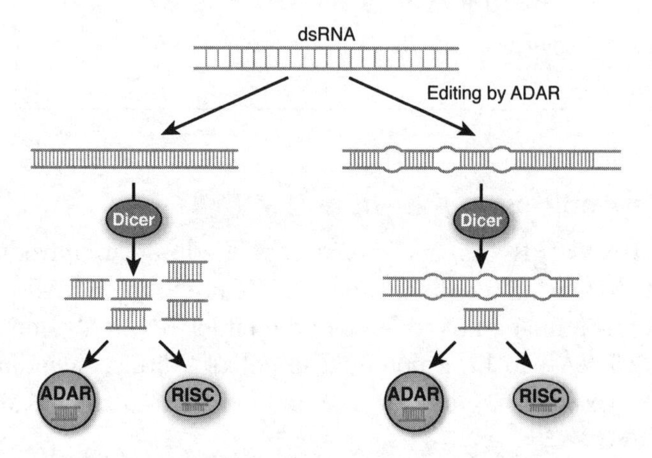

Figure 7-8 Competition between miRNA biogenesis and ADAR-based editing. The double-stranded RNA is a substrate for both Dicer-dependent cleavage and ADAR-dependent editing. ADAR editing produces mispairing that prevents efficient processing of miRNAs by Dicer. As a result, a smaller amount of small RNAs is produced, and large mispaired dsRNAs accumulate. Moreover, ADARs might interfere with small RNA loading onto RISC because ADARs bind to dsRNA intermediates and make them unavailable for further processing.

siRNAs are generated in the cytoplasm as long dsRNAs, which are cut into duplexes of approximately 23 nt per strand by the Dicer enzyme with the help of the TRBP protein. One strand of each duplex is then loaded onto the RISC. Similarly to miRNAs, siRNAs are also targets for the editing pathway. Cytoplasmic ADAR1L has the ability to compete with Dicer for the long dsRNA substrate. Moreover, *in vitro* editing of the long dsRNA has been shown to severely decrease the efficacy of cleavage by Dicer, thus resulting in reduced production of mature siRNA molecules and the RNA interference phenomenon. siRNA molecules that are still generated in the process may have a reduced efficiency either in targeting the original dsRNAs or in promoting DNA methylation due to a decrease in homology with the target template. ADAR1L can also bind tightly to any mature siRNA generated, thus competing with the RISC (Wulff and Nishikura, 2010).

Targeting Alu elements is another possible influence of editing on epigenetic regulation. An Alu element is a 280 bp-long sequence representing a short interspersed element (SINE), which is unique to primates. There are more than one million copies interspersed throughout various locations of the human genome, including the introns and UTRs of 75% of all known genes (Wulff and Nishikura, 2010). SINEs represent reverse transcribed RNA elements that are able to insert in the genome. Because they lack their own transposase, they rely on transposases encoded by other retrotransposons. Transcripts containing two adjacent, oppositely oriented Alu copies can base pair to form a hairpin structure that can be targeted by ADAR proteins. The abundance of Alu elements in the primate genome has been linked to the fact that the frequency of A-to-I editing in humans is at least an order of magnitude higher than in the mouse, rat, chicken, or fly genomes. The number of SINEs is similar in primates and rodents; however, the genome of primates has the abundance of a single type of SINE. Such SINEs (if in opposite orientation) would form the double-stranded substrate for ADAR proteins.

It is not clear why the editing level of Alu elements is so high. It is possible that editing of Alu repeats allows the control over transposable RNA elements by altering their sequences. Thus, editing of SINE RNAs might represent another type of epigenetic control of the genome stability.

Conclusion

Small RNAs play many different roles in cell biogenesis. They are able to interact with other RNAs, including small RNAs and mRNA; they can bind proteins; and they can direct reversible (methylation) and irreversible (editing) changes to DNA and RNA sequences. Most organisms contain small non-coding RNAs involved in epigenetic regulation of cellular functions, and their role is not yet completely understood. What becomes apparent is that this "dark matter" of the genome carries myriad functions that deserves more attention and will definitely be the object of many future studies.

Exercises and discussion topics

1. Describe the biogenesis of long and small ncRNAs in mammalian cells.
2. List types of lncRNAs.
3. Describe the modes of action of lncRNAs.
4. How are miRNA genes organized in the genome?
5. Describe miRNA biogenesis in animals and plants.
6. Explain the fate of translationally inhibited mRNAs.
7. Which strand of miRNA/miRNA* duplex is selected to enter into the RISC? How does the selection occur?
8. What are MREs? Why are they important?
9. Why are the repressed mRNAs sent to P-bodies? When would they be released?
10. What are ADARs? Describe their roles.
11. What are mirtrons? Describe mirtron biogenesis.
12. Describe siRNA biogenesis.
13. What are the differences between miRNAs and siRNAs in terms of their biogenesis and function?
14. Describe biogenesis and function of piRNAs.
15. Describe structure of Dicer and Argonaute.
16. What are the important roles of ADAR1L?
17. What is the importance of editing of SINE RNAs?

References

Aravin et al. (2007) The Piwi-piRNA pathway provides an adaptive defense in the transposon arms race. *Science* 318:761-764.

Bhattacharyya et al. (2006) Relief of microRNA-mediated translational repression in human cells subjected to stress. *Cell* 125:1111-1124.

Brennecke et al. (2005) Principles of microRNA-target recognition. *PLoS Biol* 3:e85.

Davis BN, Hata A. (2009) Regulation of MicroRNA Biogenesis: A miRiad of mechanisms. *Cell Commun Signal* 7:18.

Fire A. et al. (1998) Potent and specific genetic interference by double-stranded RNA in Caenorhabditis elegans. *Nature* 391:806-811.

Ghildiyal M, Zamore PD. (2009) Small silencing RNAs: an expanding universe. *Nat Rev Genet* 10:94-108.

Grimson et al. (2007) MicroRNA targeting specificity in mammals: determinants beyond seed pairing. *Mol Cell* 27:91-105.

Kawahara. RNA editing of the microRNA-151 precursor blocks cleavage by the Dicer-TRBP complex. *EMBO Rep* 2007;8: 763-769.

Lu Y. et al. (2010) Identification of piRNAs in Hela cells by massive parallel sequencing. *BMB Rep* 43:635-641.

Lytle et al. (2007) Target mRNAs are repressed as efficiently by microRNA-binding sites in the 5' UTR as in the 3' UTR. *Proc Natl Acad Sci USA* 104:9667-9672.

Martin et al. (2009) A Drosophila pasha mutant distinguishes the canonical microRNA and mirtron pathways. *Mol Cell Biol* 29:861-70.

Martin SE, Caplen NJ. (2007) Applications of RNA interference in mammalian systems. *Annu Rev Genomics Hum Genet* 8:81-108.

Motamedi et al. (2004) Two RNAi complexes, RITS and RDRC, physically interact and localize to noncoding centromeric RNAs. *Cell* 119:789-802.

Naqvi et al. (2009) The Fascinating World of RNA Interference. *Int J Biol Sci* 5:97-117.

Ponting et al. (2009) Evolution and functions of long noncoding RNAs. *Cell* 136:629-641.

Vaucheret H. (2006) Post-transcriptional small RNA pathways in plants: mechanisms and regulations. *Genes Dev* 20:759-771.

Wilusz J. (2009) RNA stability: is it the end o' the world as we know it? *Nat Struct Mol Biol* 16:9-10.

Wulff BE, Nishikura K. (2010) Substitutional A-to-I RNA editing. *WIREs RNA*. 1:90-101.

8

Non-coding RNAs across the kingdoms—bacteria and Archaea

Even though pathways, mechanisms, and functions of biogenesis have been known for some groups of non-coding RNAs (ncRNAs) (especially in eukaryotes), there are still numerous pitfalls and challenges on the way to understanding the whole complexity and evolution of ncRNAs. The following pages highlight examples of ncRNAs in Prokaryota that can be attributed to one of the previously discussed major groups (see Chapter 7, "Non-Code RNAs Involved in Epigenetic Processes: A General Overview") and act alike or represent different modes of biosynthesis and functions.

Bacteria

Bacteria contain many types of regulatory RNAs that are often called **small RNAs (sRNAs)** or ncRNAs because they predominantly are not translated. The regulatory RNAs can be broadly classified into those that act in ***cis*** and are part of the mRNAs that they regulate, and those that act in ***trans*** and have their own dedicated promoters and terminators.

Regulatory RNAs can influence gene expression at different levels, including chromatin formation, transcription, translation, and mRNA stability. Gene expression, in broader terms, may describe the entire process including all the steps from regulation of transcription until expression of a phenotype. Regulatory RNAs in bacteria were known for years before miRNAs and siRNAs were discovered in eukaryotes. Among the first examples in 1981 was the discovery that the ~ 108 nucleotide RNA I was found to block ColE1 plasmid replication by targeting the RNA that is cleaved to produce the replication primer (Tomizawa et al., 1981). Two years later, another ~180 nucleotide sRNA transcribed from the pOUT promoter of the Tn10 transposon was shown to repress transposition by preventing translation of the transposase mRNA (Simons and Kleckner, 1983). In 1984, yet another sRNA, **mRNA-interfering complementary RNA** (**micRNA**), which was capable of inhibiting translation of the mRNA encoding the major outer membrane porin OmpF through interaction with OmpF mRNA, was described in *Escherichia coli* (Mizuno et al., 1984).

In bacteria, non-coding RNAs that affect gene expression largely fall into two different categories:

- ncRNAs that bind to proteins and modify their function
- ncRNAs that bind to mRNA and either inhibit translation or induce degradation

Another group of the regulatory RNAs that technically do not belong to ncRNAs include **riboswitches**. The following sections describe these three groups in greater detail.

Riboswitches

Riboswitches represent the simplest class of bacterial regulatory elements and consist of sequences at the 5' end of mRNAs that fold into defined secondary structures. Upon binding of small molecules, these structures can undergo conformational changes—switching from one conformation to another. Those structural changes often depend on the presence of ligand molecules (such as various metabolites, nutrients, and so on), stalled ribosomes, or shifts in temperature. Riboswitches thus sense and respond to changes in the cell environment. In general, riboswitches consist of an **aptamer region**, which binds the ligand, and a so-called **expression platform**, which regulates gene expression through alternative RNA structures that affect transcription or translation (reviewed in (Montange and Batey, 2008). Binding of ligands results in the change of riboswitch conformation. These alternative RNA structures form or disrupt transcriptional terminators or antiterminators (see box 8-1) as well as block or expose ribosome-binding sites (see Figure 8-1).

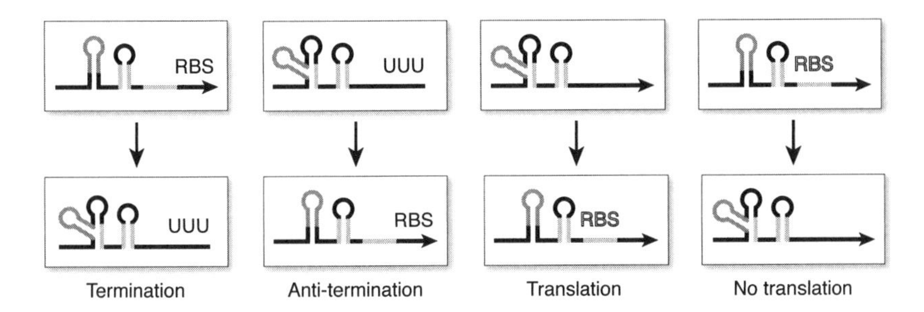

Figure 8-1 Function of bacterial riboswitches. Aptamer loops are in dark gray (red in ebook version). Expression platforms are in light gray. Poly-U stratch is associated with transcriptional termination. RBS stands for ribosome binding sites. See text for more details. (Reproduced with permission from Waters LS, Storz G. (2009) Regulatory RNAs in bacteria. Cell 136:615-28.)

These elements were first described decades ago in studies characterizing transcription attenuation. In this process, stalled ribosomes lead to changes in mRNA structure, affecting transcription elongation through the formation of terminator/antiterminator structures in the mRNA. The effect of the environment on riboswitch activity is demonstrated by the fact that temperature influences the folding of some of the identified leader sequences within riboswitches, known as "RNA thermometers." Another family of riboswitches, so called T-boxes, represent sequences in transcripts of tRNA synthetases. Binding of the T-boxes to the corresponding uncharged tRNAs results in folding into the antiterminator structure. In all these cases, the alteration of the RNA structure leads to changes in the expression of the downstream gene.

8-1. Terminator/antiterminator structures

Transcriptional antitermination is a mechanism that prevents transcriptional termination. There are different ways of achieving this: modifying polymerase binding or modifying the secondary structure of the mRNA region that induces transcriptional termination. Thus, terminator structures are secondary structures in the mRNA that attenuate and eventually terminate transcription, whereas antiterminator structures generally affect the same region of the mRNA but do not induce transcriptional termination.

The finding that the conformation of leader sequences can change upon binding of small molecules is more recent (reviewed in Mandal and Breaker, 2004). Those leader sequences that are sensitive to the presence of certain metabolites received the name **riboswitches** because their function is to directly regulate the expression of genes involved in the uptake and use of the metabolite by switching expression "on" or "off" depending on nutrient availability. The presence of specific riboswitches in a given mRNA may point to a specific physiological role of the product of the gene in which the leader sequence was found. The example of *Bacillus subtilis* illustrates how the mechanism of riboswitch-dependent gene regulation works: The expression of 2% of all genes of this bacterium is regulated via riboswitches that are sensitive to metabolites such as flavin mononucleotide (FMN), thiamin pyrophosphate, S-adenosylmethionine, lysine, guanine, and many others.

In the vast majority of cases, riboswitches inhibit rather than activate transcription or translation. It should be noted, however, that the modular nature of riboswitches allows the same aptamer to function as an activator or an inhibitor depending on the developmental stage or nutritional condition (reviewed in Nudler and Mironov, 2004). For example, the cobalamin riboswitch regulated by the coenzyme form of vitamin B12 modulates translation initiation of the *cob* operons of Gram-negative bacteria and acts as a transcription terminator in the *btuB* genes in Gram-positive bacteria. Moreover, some transcripts carry tandem riboswitches that respond to different physiological stimuli.

Further, the function of bacterial riboswitches is not necessarily limited to transcription, translation, or RNA processing. It can also be associated with any process that involves RNA, including RNA modification, localization, or splicing. An example for more complex regulation is that the glmS leader sequence acts as a ribozyme (RNA tertiary structure capable of catalyzing chemical reactions) to catalyze self-cleavage. Self-cleavage of glmS is promoted by binding of glucosamine-6-phosphate. This process inactivates the mRNA encoding the enzyme that generates glucosamine-6-phosphate.

Protein-binding ncRNAs

Protein-binding sRNAs (for example, 6S, CsrB, and GlmY) regulate the activity of their cognate proteins in bacteria by mimicking the structures of those proteins' target nucleic acids (see Figure 8-2). This is illustrated by the examples in the following paragraphs.

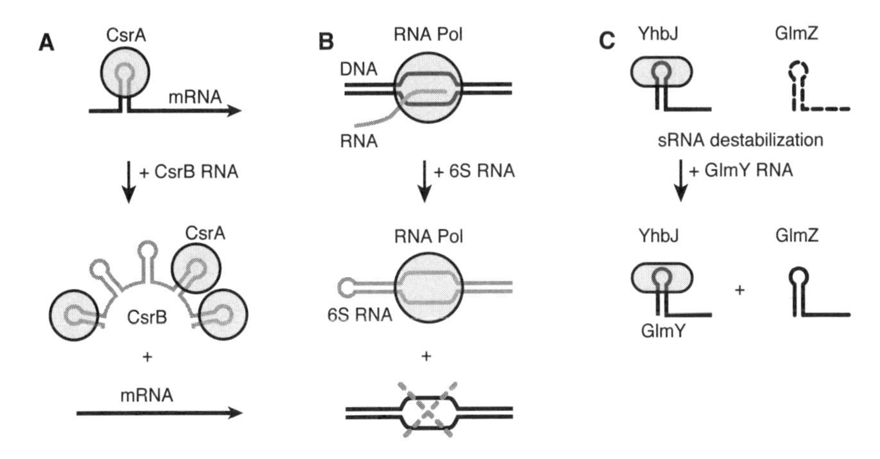

Figure 8-2 sRNAs bind regulatory proteins and sequester them from the reaction. (Reproduced with permission from Waters LS, Storz G. (2009) Regulatory RNAs in bacteria. Cell 136:615-28.) **A**. CsrB RNA mimics the structure of mRNA and sequesters CsrA protein from being able to regulate mRNA. **B**. 6S RNA mimics open DNA structure and attracts RNA Pol, preventing it from transcribing the gene. **C**. GlmY mimics the structure of sRNA, sequestering YhbJ and preventing it from destabilizing GlmZ sRNA.

CsrB and CsrC RNAs

The **CsrB** and **CsrC RNAs** of *E. coli* modulate the activity of CsrA, an RNA-binding protein that regulates bacterial motility and carbon usage under nutrient-poor conditions during entry into stationary phase. CsrA dimers bind to GGA motifs in the 5' UTR of their target mRNAs, thereby affecting their stability and/or translation. The CsrB and CsrC RNAs each contain multiple GGA sites. Thus, they are able to mimic the target mRNAs of CsrA and effectively sequester it away from its target mRNA leaders. The transcription of the *csrB* and *csrC* genes is induced by the two-component regulator BarA-UvrB under nutrient-poor growth conditions. Degradation of CsrB and CsrC RNAs is mediated by RNase E, which is recruited by the CsrD protein.

6S RNA

6S RNA was first described in *E. coli*, and since then it has been discovered in a variety of bacteria (Barrick et al., 2005). The mode of action of this RNA is tightly associated with the function of bacterial RNA polymerase, which consists of two α-, one β-, and one β'-subunits and is bound to a **sigma (σ) factor** that defines promoter specificity. The gene encoding 6S RNA is under control of two different promoters leading to two different size transcripts, which were shown to be under control of two different σ factors. The long 6S RNA precursor is generated when σ^{70} and σ^{S} are present, whereas the short precursor is produced in the presence of σ^{70} only. The biogenesis of the long precursor includes cleavage with RNase E, whereas the processing of the shorter precursor requires RNase E or RNase G. These different ways for the biogenesis of 6S RNA possibly modulate their levels under given conditions.

Structurally, 6S RNA is 184 nt in length and resembles a melted promoter. During stationary phase, the expression of 6S RNA precursor—which is able to form a stable complex with much of the σ^{70}-bound housekeeping form of RNA polymerase but is not associated with the σ^{S}-bound stationary-phase form of RNA polymerase—is induced (Trotochaud and Wassarman, 2005). The interaction between 6S RNA and σ^{70} holoenzyme results in the specific down-regulation of gene expression driven by the σ^{70}-specific promoters but not by the σ^{S}-specific promoters. This repression state is reversed through negative regulation of 6S RNA; RNA polymerase uses 6S RNA as a substrate to produce **RNA products (pRNAs)** 14 to 20 nucleotides in length that direct degradation of 6S RNA. Similarly functioning ncRNAs are present in all domains of life, including Alu mRNA in humans and FC RNA aptamers in fungi.

Another protein-binding sRNA, **GlmY**, has recently been proposed to competitively bind and inhibit the protein that processes GlmZ RNA (reviewed in Gorke and Vogel, 2008) (refer to Figure 8-2). GlmZ and GlmY RNAs are homologous in both sequence and structure, and they both promote accumulation of the glucosamine-6-phosphate synthase (GlmS). The GlmZ RNA basepairs with the glmS mRNA and activates its translation. GlmY lacks the region that is complementary to GlmS mRNA and does not directly activate GlmS translation. Instead, GlmY expression inhibits a GlmZ-processing event that would otherwise render GlmZ unable to activate glmS translation. It is possible that GlmY stabilizes GlmZ RNA by competing with GlmZ for binding to the YhbJ protein that targets GlmZ for the inactivating processing event. Thus, the GlmY RNA functions as a safeguard of the homologous GlmZ RNA by offering itself for processing to the enzyme that would otherwise deplete the pool of GlmZ RNA.

Thus, protein-binding ncRNAs in bacteria can function in a variety of ways: CsrB RNA resembles an mRNA element, 6S imitates a DNA structure, and GlmY mimics another sRNA. There is no doubt that other yet-to-be-discovered mechanisms of function of ncRNAs involved in protein-binding also exist.

Cis- and trans-encoded small RNAs

Two groups of sRNAs that bind to mRNA consist of *cis*- and *trans*-encoded small RNAs. The first group of *cis*-encoded sRNAs represents small RNAs located on the strand opposite the target RNA, which has perfect or near-perfect complementarity. The majority of these sRNAs are located on plasmids, on bacteriophages, and in transposons and are believed to regulate the copy number of these elements in bacterial cells. The second group consists of *trans*-encoded sRNAs that are similar in function to eukaryotic microRNAs (miRNAs) (Chapters 10 and 11). These sRNAs exhibit partial complementarity to the target mRNA and function either through inhibition of translation or through cleavage of their target mRNAs.

Cis-encoded ncRNAs

The name for this group of ncRNAs can be misleading. Although they are ***cis*-encoded**, meaning that they are expressed from the strand opposite to the strand encoding the target mRNA (on the same molecule of DNA), they actually function ***in trans*** (on a different molecule of RNA or DNA) as diffusible molecules, which find and bind to the target mRNA. Thus, it is more appropriate to call these ncRNAs "*cis*-encoded" rather than "*cis*-acting."

Cis-encoded antisense sRNAs expressed from phage DNA, plasmids or transposons regulate the copy number of those mobile elements through a variety of mechanisms, including inhibition of replication primer formation and transposase translation, all preventing TE transposition (reviewed in Brantl, 2007).

The action of *cis*-encoded ncRNAs follows two major mechanisms (see Figure 8-3A). The first one includes an sRNA encoded opposite to the 5' UTR of its target mRNA. Basepairing between ncRNA and the mRNA target results in the inhibition of ribosome binding and often mRNA degradation. The second group of *cis*-encoded antisense sRNAs modulates the expression of genes located in an operon (see Figure 8-3B). In this case, the ncRNA encoded opposite to the sequence separating two genes in the operon can basepair with the target mRNA and recruit RNases to that region causing mRNA cleavage. For example, in *E. coli*, the stationary phase-induced *GadY* antisense sRNA base pairs with the complementary region found in the operon of *gadXW* mRNAs, which leads to cleavage of the duplex between the *gadX* and *gadW* genes. As a result, the levels of a *gadX* transcript increase. A similar effect is observed for the iron-transport biosynthesis (ITB) operon of the virulence plasmid pJM1 of *Vibrio anguillarum*. Binding of the RNAβ antisense ncRNA to the region between *fatDCBA* and *angRT* mRNAs leads to transcription termination after the *fatA* gene, resulting in reduced expression of the downstream *angRT* genes.

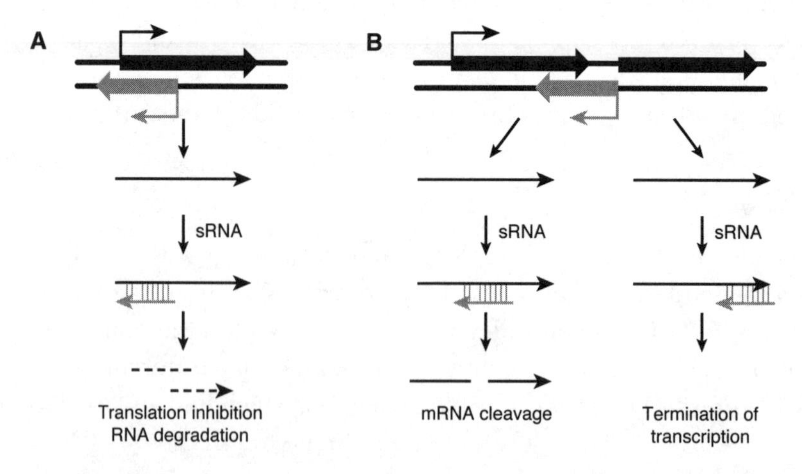

Figure 8-3 Regulatory function of cis-encoded sRNAs. (Reproduced with permission from Waters LS, Storz G. (2009) Regulatory RNAs in bacteria. Cell 136:615-28.)

A subset of *cis*-encoded ncRNAs functions as antitoxins (see box 8-2) promoting degradation and/or repression of translation of toxin mRNAs, which encode proteins that are toxic at high levels (reviewed in Fozo et al., 2008). Two sRNAs, *OhsC* and *IstR,* in *E. coli* lie directly adjacent to genes encoding potentially toxic proteins. Regardless of them not being true antisense RNAs, they contain extended regions of perfect complementarity with the toxin mRNAs and act as type I antitoxins (see box 8-2).

Some of the chromosomal antitoxin sRNAs might have been acquired through horizontal gene transfer from plasmids/phages. For example, the Hok/Sok loci present in the *E. coli* chromosome are homologous to plasmid antitoxin sRNAs, whereas the *RatA* RNA of *B. subtilis* is located in genomic regions acquired from a remnant of a cryptic prophage. However, some *cis*-encoded antisense antitoxin sRNAs do not have known homologs on mobile elements. What is the possible function of such antitoxin ncRNAs? Whereas high levels of toxins kill cells, low levels may only slow their growth. Thus, it can be suggested that chromosomal toxin-antitoxin pairs induce slow growth or stasis under conditions of stress to allow cells time to repair damage or otherwise adjust to their environment (Unoson and Wagner, 2008). Alternatively, antitoxin ncRNAs might provide protection against plasmids/phages that carry similar toxin genes and provide a primitive form of immune response.

8-2. Toxin-antitoxin system

Toxin-antitoxin systems consists of at least two genes—one of them encoding a toxin, the product of which can kill a cell, and the other one encoding an antitoxin, which can prevent the function of the toxin. Such systems are considered as selfish DNA elements, as the loss of the antitoxin gene in this case would lead to cell death. There are several types of toxin-antitoxin systems, classified depending on their mechanism of function of the antitoxin. Type I antitoxins are generally sRNAs that prevent translation of the toxin mRNA. Type II antitoxin proteins inhibit toxins through protein-protein interaction. In type III systems, the RNA antitoxin binds to the protein toxin and thereby inhibits its function.

Trans-encoded ncRNAs

In contrast to *cis*-encoded sRNAs, the genomic location of genes coding for *trans*-encoded sRNAs typically does not correlate with the genomic position of genes encoding their target mRNAs.

Also *trans*-encoded and *cis*-encoded sRNAs require different levels of complementarity to their targets. Whereas *cis*-encoded sRNAs exhibit nearly perfect homology to their target (because they are produced from the antisense strand), *trans*-encoded sRNAs have a much lower extent of homology. Although the region of potential basepairing between *trans*-encoded ncRNAs and its targets stretches between 10 and 25 nucleotides, only a core of the nucleotides seems to be important for regulation. For example, the *SgrS* sRNA forms 23 base pairs with the *ptsG* mRNA across a stretch of 32 nucleotides, but only 4 nucleotides seem to be critical for the ability of *SgrS* to downregulate *ptsG*.

Such imperfect basepairing allows *trans*-encoded sRNAs to bind to and regulate multiple mRNA targets. Due to the ability of *trans*-encoded sRNA to regulate multiple targets, changes in expression of even a single sRNA can globally modulate a particular physiological response. Well-characterized examples of changes induced by single sRNAs in an entire physiological pathway include the downregulation of Fe-Su cluster-containing enzymes under conditions of low iron (*E. coli* RyhB), repression of quorum sensing at low cell density (*Vibrio* Qrr), and repression of outer membrane porin proteins under conditions of membrane stress (*E. coli* MicA and RybB).

Interestingly, every major transcription factor in *E. coli* may control the expression of one or several sRNA regulators. There are also multiple negative feedback loop mechanisms exemplified by the fact that genomic locations for a number of sRNAs are found adjacent to the gene encoding their transcription regulator, including *E. coli* pairs such as OxyR-OxyS, GcvA-GcvB, and SgrR-SgrS. Some feedback mechanisms involved in down-regulation of sRNAs that are induced by physiological conditions that were introduced earlier are discussed in the following paragraph: In *E. coli*, *micA* and *rybB* are repressed when membrane stress is relieved upon their downregulation of outer membrane porins, and *ryhB* is repressed when iron is released after RyhB downregulates iron-sulfur enzymes (reviewed in Waters and Storz, 2009). Moreover, in *Vibrio cholerae*, the Qrr sRNAs can directly inhibit the expression of mRNAs encoding their own transcription factors. The process of regulation of gene expression by *trans*-encoded sRNAs mimics or perhaps even backs up the function of a transcription factor at a post-transcriptional level (reviewed in Bejerano-Sagie and Xavier, 2007). The major mode of action of *trans*-encoded sRNAs is translation inhibition or destabilization of target mRNAs. As a result, the outcome for gene expression in most interactions with *trans*-encoded sRNAs is negative (reviewed in Aiba, 2007) (see Figure 8-4).

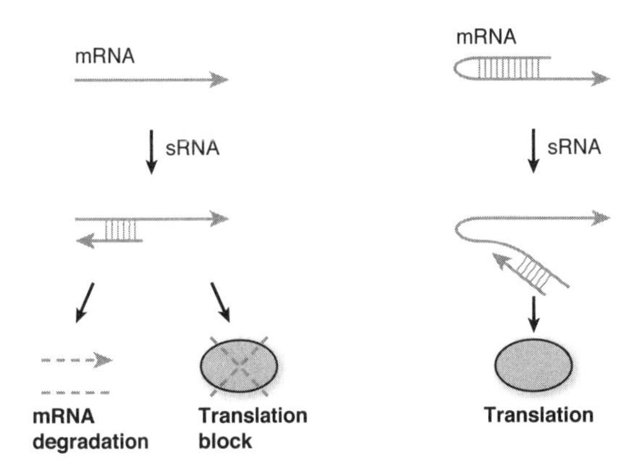

Figure 8-4 Function of trans-encoded sRNAs. (Reproduced with permission from Waters LS, Storz G. (2009) Regulatory RNAs in bacteria. Cell 136:615-28.)

Trans-encoded bacterial sRNAs primarily bind to the 5' UTR of mRNAs and mostly block the ribosome-binding site (see Figure 8-4, left panel). At the same time, examples of translation inhibition by basepairing upstream of the AUG of the repressed gene also exist (for example, in the case of GcvB and RyhB ncRNAs) (Vecerek et al., 2007). sRNA-mRNA duplexes can also be targeted by RNase E. The degradation of the sRNA-mRNA duplex makes this kind of regulation an irreversible process. Therefore, the major targets of this process might be translationally inactive mRNAs that could be depleted through this mechanism.

There are also some examples of positive effects on gene expression imposed by *trans*-encoded ncRNAs. Activation of target mRNA expression can occur through an anti-antisense mechanism that involves basepairing of the sRNA and an inhibitory secondary structure, which normally blocks the ribosome-binding site; the basepairing changes the confirmation of the mRNA and sequesters the ribosome-binding site (reviewed in Gottesman, 2005) (refer to Figure 8-4, right panel). Other theoretical mechanisms include a mode of action that is similar to *cis*-encoded sRNAs: Binding of a *trans*-encoded sRNA and its target can promote transcription termination or antitermination and alter mRNA stability through changes in polyadenylation.

OxyS RNA is a typical representative of *trans*-encoded sRNAs. It is induced under oxidative stress in *E. coli*. It affects the expression of as many as 40 genes, among which are the *fhlA* gene, which encodes a transcription factor, and a sigma factor *RpoS*, which is induced under stress conditions. The OxyS RNA represses *fhlA* expression by binding to a sequence stretch that overlaps with the Shine-Dalgarno sequence and thereby prevents the ribosome from binding (Altuvia et al., 1998). The activity of *OxyS* depends on Hfq, which is an RNA-binding protein.

Hfq has a frequent binding motif (5'-AAYAAYAA-3'), which is preferentially found in the antisense strand of protein-coding sequences. Also, Hfq regulates translation of several transcripts by binding to the 5' UTR and inhibiting ribosome binding. Ribosome binding usually prevents transcript degradation by RNAse E. Therefore by inhibiting it, Hfq may also target transcripts for degradation. In terms of sRNA-mediated functions, the RNA chaperone Hfq might be required to facilitate RNA-RNA interactions due to limited complementarity between the *trans*-encoded sRNA and the target mRNA (reviewed in Aiba, 2007). The structure of the Hfq protein consists of a hexameric Hfq ring, which is similar to that of Sm and Sm-like proteins involved in splicing and mRNA decay in eukaryotes. Thus, if Hfq also shares functional similarity with the eukaryotic proteins, it might actively remodel RNAs to melt inhibitory secondary structures. Another possibility is that Hfq might function as a platform to allow sRNAs and mRNAs to sample potential complementarity. In fact, all *trans*-encoded base pairing sRNAs examined to date coimmunoprecipitate with Hfq, and *hfq E. coli* mutants are not capable of supporting *trans*-encoded sRNA-mediated regulation of mRNA levels. At the same time, when the *E. coli SgrS* RNA is pre-annealed with the *ptsG* mRNA *in vitro*, the Hfq protein is no longer required.

Further, the stability of most *trans*-encoded sRNAs is dependent on Hfq (reviewed in Aiba, 2007). The mechanism might involve protection of sRNAs from degradation. Many of the known sRNA-mRNA pairs are subject to degradation by RNase E. This could be either due to the passive lack of protection from the released Hfq or actually due to active recruitment of the RNA degradation machinery by Hfq. The need for the Hfq protein might be species-specific, as all characterized *E. coli trans*-encoded sRNAs require Hfq for regulation of their targets, whereas in *Vibrio cholera*, mutation in Hfq does not prevent *VrrA* sRNA repression of *OmpA* mRNA expression. It is also possible that the RNA chaperone is mostly required when long stretches of base pairing between *trans*-encoded sRNAs and mRNAs cannot be provided. Indeed, the function of *cis*-encoded sRNAs that form long stretches of homology to their mRNA targets does not require Hfq.

There are many other differences between *trans*- and *cis*-encoded sRNAs, in terms of their origin and dependence of their mode of action on cofactor proteins (see Table 8-1). One additional difference is that in contrast to *cis*-encoded sRNAs, which are mostly constitutively expressed, *trans*-encoded sRNAs are synthesized under defined physiological conditions. In *E. coli*, the expression of *trans*-encoded regulatory RNAs is induced by oxidative stress (OxyR-activated OxyS), elevated glycine levels (GcvA-induced GcvB), changes in glucose concentration (CRP-repressed Spot42 and CRP-activated CyaR), low iron (Fur-repressed RyhB), outer membrane stress (sE-induced MicA and RybB), and elevated glucose-phosphate levels (SgrR-activated SgrS) (reviewed in Waters and Storz, 2009).

Table 8-1 Comparison of cis- and trans-encoded sRNAs

	***Cis*-encoded sRNAs**	***Trans*-encoded sRNAs**
Origin	Expressed from plasmid, bacteriophages, transposons, and rarely from genomic DNA (antitoxins)	Expressed from genomic DNA
Position as to the target position	In antisense to the target	No correlation
Homology	Long stretches of perfect homology	Short (10 to 25 nt) stretches of imperfect homology
Number of targets	Single	Multiple
Hfq dependence	Independent	In most cases, dependent
Mechanism	Translation inhibition, mRNA degradation	Translation inhibition, mRNA degradation
Regulation of gene expression	Mostly negative, but many cases of positive regulation are known	Predominantly negative, rarely positive
Targeted areas	5' UTRs (ribosome binding sites) and operons; rarely translation start codon	5' UTRs (ribosome binding sites)
Expression of sRNAs	Mostly constitutive	Predominantly regulated by growth conditions

Clustered regularly interspaced short palindromic repeats (CRISPRs)

Clustered regularly interspaced short palindromic repeats (CRISPRs) represent an independent group of small RNAs that mediate adaptation in prokaryotes. CRISPR loci are composed of up to 250 repeats, typically 24 to 48 bp, that are separated by unique spacer sequences and are commonly flanked by an approximately 300 to 500 bp leader sequence at one end of the locus. Transcription and integration of new spacer sequences seem to depend on the location of this leader sequence (Mojica et al., 2009). Further, comparison of CRISPR-positive with CRISPR-negative genome sequences revealed a group of **CRISPR-associated genes (Cas)**. Computational analysis of those *Cas* genes showed similarities of some of their members to Dicer, Slicer, and RNA-dependent RNA Polymerase; however, they were not orthologs.

Spacer sequences exhibit homology to extrachromosomal and phage sequences, and a higher number of repeats correlates with bacterial tolerance to phage infection. Therefore, the **CRISPR-Cas system** (**CASS**) is believed to act as an immune system in bacteria. The current model of CASS suggests that the CRISPR loci are transcribed to pre-crRNAs, which are then processed by a complex of Cas proteins to yield a mature crRNA. The mature crRNA then interacts with Cas proteins to inhibit viral proliferation (van der Oost et al., 2009).

CRISPR-mediated immunity shows similarities to the eukaryotic miRNA and siRNA systems in the way it functions and some analogy to the piRNA system as the loci are clustered, giving rise to large single-stranded ncRNA precursors. You can find a more detailed view on the structure, function, and possible evolution of CRISPR loci in Chapter 14, "Bacterial Adaptive Immunity: Clustered Regularly Interspaced Short Palindromic Repeats (CRISPR)."

Archaea

The field of non-coding RNAs in Archaea is fairly young; the first report of stably expressed antisense RNAs in an archaeal species was published only in 2005 (Tang et al., 2005). Tang et al. identified 57 ncRNAs and found that the most prominent group of antisense RNAs was encoded on the opposite strand of transposase coding sequences, possibly regulating transposase activity in a negative way and preventing transposon mobility. This group of antisense RNAs belongs to the group of *cis*-encoded ncRNAs.

The *Sulfolobus solfataricus* P2 transcriptome sequencing project, a common archaeal model system, brought further insight into the structure and possible function of ncRNAs in Archaea (Wurtzel et al., 2010). This study identified 310 ncRNAs. The majority of those ncRNAs (185 cases, 60%) corresponded to *cis*-encoded antisense RNA transcripts, which, by the extent of the phenomenon, is more similar to eukaryotes than to bacteria. All of the identified transcripts reside on the antisense strand opposite the coding region on the sense strand and are thus fully complementary with their targets (Chen and Crosa, 1996).

The analysis of the molecular functions of potential targets of these *cis*-encoded ncRNAs showed that genes involved in ion transport and metabolism were significantly overrepresented, suggesting that the regulation by *cis*-encoded ncRNAs might be a common regulatory mechanism for such genes in *S. solfataricus*. Several other dominant groups of ncRNAs in *S. solfataricus* were also found by transposon-associated ncRNAs (28 out of 310); putative regulators of transposition; CRISPR-associated small RNAs (18); ncRNAs involved in protection against phage infection; and C/D-box RNAs, which direct methylation of specific sites in rRNAs or tRNAs (13 cases).

In bacteria, the regulation by ncRNAs is mostly mediated by basepairing of the sRNAs with the 5' UTR of target mRNAs (Waters and Storz, 2009). According to studies of Wurtzel et al. (2010), archaeal mRNAs apparently lack 5' UTRs, and thus the mechanism of ncRNA/mRNA interaction likely differs between bacteria and Archaea. The fact that archaeal mRNAs have 3' UTRs of significant sizes suggests that ncRNAs regulate mRNA expression by binding the 3' UTR as in eukaryotes. The existence of long regions of complementarity found between ncRNAs and 3' UTRs of protein-coding mRNAs in *S. solfataricus* supports this hypothesis (Tang et al., 2005).

As many as 8% of *S. solfataricus* operons may be targeted by *cis*-encoded antisense transcripts. A handful of chromosomally encoded *cis*-antisense transcripts were documented in bacteria and were shown to regulate translation, mRNA stability, and mRNA degradation (Brantl 2007). Still, *cis*-antisense transcription has never been observed at the level of approximately 8% for any bacteria or Archaea to date but has been documented in many eukaryotes including Arabidopsis, Drosophila, and humans. In humans, it is estimated that 5% to 25% of transcripts have a *cis*-antisense counterpart. It is possible that in other eukaryotes there are also abundant *cis*-encoded ncRNAs, but such detailed studies in other organisms, as those available in humans, do not exist.

The fact that bacteria have only a handful of *cis*-antisense transcripts when compared to Archaea might suggest a very different mode of regulation of gene expression in these two domains of life (see Table 8-2). It should be noted, however, that the CRISPR-Cas regulatory system has also been described in Archaea. Therefore, bacteria and Archaea share some RNA-mediated regulatory mechanisms. Clearly, further studies are needed to point out the overlapping of the eukaryotic and bacterial regulatory systems as well as the uniqueness of the archaeal system. High-throughput approaches in combination with bioinformatic analysis and functional verification will likely fill some of the gaps in this field in the future.

Table 8-2 Comparison of ncRNAs in bacteria and Archaea

	Bacteria	**Archaea**
Type	Both *cis*- and *trans*-encoded	Mostly *cis*-encoded
Position	In antisense to protein-coding genes, as well as at random genomic positions	Predominantly in antisense to protein-coding genes
Target	Predominantly 5' UTR via translation inhibition	Predominantly 3' UTR: via translation inhibition and cleavage (?)
mRNA degradation	mRNAs are degraded via RNase E	mRNAs are cleaved at internal TNRNR|NTDR sites
CRISPR	Yes	Yes

The analysis of the transcripts in *S. solfataricus* also revealed several conserved internal sites in mRNA; therefore, the authors suggested that these were the recognition sites for endoribonucleases that promoted degradation and turnover of mRNAs. By calculating the base distribution surrounding 1240 dominant internal sites, the authors detected the position-specific motif TNRNR|NTDR (where | marks the cleavage position, R = A/G, D = A/G/T, N=any nucleotide), which was tenfold over-represented compared to any random position within the same genes. Thus, it can be suggested that RNA degradation in *S. solfataricus* predominantly includes cleavage at specific positions within the transcript.

The mechanism of mRNA turnover is different from the one observed in bacteria. In gram-negative bacteria, RNaseE is recruited to the ncRNA-mRNA complex and cleaves mRNA in internal AU-rich sites, followed by the 3'-5' degradation by the exosome. To date, in Archaea, no endoribonuclease was identified to be responsible for internal mRNA cleavage, and RNA degradation is presumably done by the exosome from the 3' end of mRNA. It should be noted, however, that studies in Archaea are not as advanced as in bacteria, and there is a possibility that RNase-E-like enzymes do exist in Archaea. Indeed, some *S. solfataricus* proteins were shown to have a partial endoribonucleolytic activity *in vitro* (Evguenieva-Hackenberg, 2010).

Conclusion

Despite the fact that studies of ncRNAs in Archaea are in their infancy, it becomes apparent that there are substantial differences from bacteria as to the type of ncRNAs produced, their location in the genome, their preferential binding sites and even in the ways mRNAs are degraded when targeted by ncRNAs (Table 8-2). More studies will be required to further explore possible differences and similarities of ncRNA biogenesis in these two kingdoms of life.

Exercises and discussion topics

1. Why can riboswitches be classified as regulatory ncRNAs? Do they follow in *cis* or in *trans* mechanisms? How do they mediate their function?
2. What are the components of a riboswitch?
3. Why did riboswitches get their name? What happens if they are bound by a nutrient they are responsive to? Predict possible outcomes of such an interaction.
4. What is the unifying mechanism of action of protein-binding ncRNAs?
5. How does 6S RNA switch between gene expression programs?
6. In regulation of GlmS translation, which sRNA mediates its function by binding to a protein? Which category would the other sRNA involved in the process fit (riboswitch, protein binding, in *cis* acting, or in *trans* acting ncRNA)?

7. Define "*cis*-encoded," "*cis*-acting," "*trans*-encoded," and "*trans*-acting." Give examples from the text as far as possible.
8. How do *cis*-encoded ncRNAs mediate their function?
9. Why are type III antitoxins not classical *cis*-encoded ncRNAs? What makes them similar to *cis*-encoded ncRNAs?
10. What are consequences of imperfect basepairing for the function of *trans*-encoded ncRNAs?
11. How is Hfq involved in sRNA-mediated function?
12. Why can crRNAs be considered *trans*-encoded ncRNAs?
13. Compare *cis*-encoded sRNAs in Archaea and bacteria in terms of location in the genome and mode of function.
14. What are two different mechanisms for RNA degradation in Archaea? Describe them.

References

Aiba H. (2007) Mechanism of RNA silencing by Hfq-binding small RNAs. *Curr Opin Microbiol* 10:134-139.

Altuvia et al. (1998) The Escherichia coli OxyS regulatory RNA represses fhlA translation by blocking ribosome binding. *EMBO J* 17:6069-6075.

Babitzke P, Romeo T. (2007) CsrB sRNA family: sequestration of RNA-binding regulatory proteins. *Curr Opin Microbiol* 10:156-163.

Barrick et al. (2005) 6S RNA is a widespread regulator of eubacterial RNA polymerase that resembles an open promoter. *RNA* 11:774-784.

Bejerano-Sagie M, Xavier KB. (2007) The role of small RNAs in quorum sensing. *Curr Opin Microbiol* 10:189-198.

Brantl S. (2007) Regulatory mechanisms employed by cis-encoded antisense RNAs. *Curr Opin Microbiol* 10:102-109.

Chen Q, Crosa JH. (1996) Antisense RNA, fur, iron, and the regulation of iron transport genes in Vibrio anguillarum. *J Biol Chem* 271:18885-18891.

Evguenieva-Hackenberg E. (2010) The archaeal exosome. *Adv Exp Med Biol* 702:29-38.

Fozo et al. (2008) Small toxic proteins and the antisense RNAs that repress them. *Microbiol Mol Biol Rev* 72:579-589.

Görke B, Vogel J. (2008) Noncoding RNA control of the making and breaking of sugars. *Genes Dev* 22:2914-2925.

Gottesman S. (2005) Micros for microbes: non-coding regulatory RNAs in bacteria. *Trends Genet* 21:399-404.

Mandal M, Breaker RR. (2004) Gene regulation by riboswitches. *Nat Rev Mol Cell Biol* 5:451-463.

Mizuno et al. (1984) A unique mechanism regulating gene expression: translational inhibition by a complementary RNA transcript (micRNA). *Proc Natl Acad Sci USA* 81:1966-1970.

Mojica et al. (2009) Short motif sequences determine the targets of the prokaryotic CRISPR defence system. *Microbiology* 155:733-740.

Montange RK, Batey RT. (2008) Riboswitches: emerging themes in RNA structure and function. *Annu Rev Biophys* 37:117-133.

Nudler E, Mironov AS. (2004) The riboswitch control of bacterial metabolism. Trends Biochem Sci. 29:11-17.

Simons RW, Kleckner N. (1983) Translational control of IS10 transposition. *Cell* 34:683-691.

Tang et al. (2005) Identification of novel non-coding RNAs as potential antisense regulators in the archaeon Sulfolobus solfataricus. *Mol Microbiol* 55:469-481.

Tomizawa et al. (1981) Inhibition of ColE1 RNA primer formation by a plasmid-specified small RNA. *Proc Natl Acad Sci USA* 78:1421-1425.

Trotochaud AE, Wassarman KM. (2005) A highly conserved 6S RNA structure is required for regulation of transcription. *Nat Struct Mol Biol* 12:313-319.

Unoson C, Wagner EG. (2007) Dealing with stable structures at ribosome binding sites: bacterial translation and ribosome standby. *RNA Biol* 4:113-117.

van der Oost et al. (2009) CRISPR-based adaptive and heritable immunity in prokaryotes. *Trends Biochem Sci* 34:401-407.

Vecerek et al. (2007) Control of Fur synthesis by the non-coding RNA RyhB and iron-responsive decoding. *EMBO J* 26:965-975.

Waters LS, Storz G. (2009) Regulatory RNAs in bacteria. *Cell* 136:615-628.

Wurtzel et al. (2010) A single-base resolution map of an archaeal transcriptome. *Genome Res* 20:133-141.

9

Non-coding RNAs across the kingdoms—protista and fungi

This chapter describes non-coding RNAs found in unicellular eukaryotes from **Protozoa** as well as ncRNAs from uni- and multicellular eukaryotes from **Fungi** kingdom.

Protozoa

Protozoa are unicellular heterotrophic eukaryotic organisms belonging to the kingdom **Protista**, although according to the Integrated Taxonomic Information System 2009 classification, they form their own kingdom. Classification of Protista and Protozoa organisms is not a simple task. Depending on their type of locomotion, the protozoans are divided into three phyla that use cilia, flagella, or pseudopodia. Although various miRNAs and siRNAs have been identified in multicellular animals, most unicellular animals seem to lack miRNAs, but might have siRNAs and another siRNA-like acting ncRNA, scanning RNAs, or scnRNAs. This chapter presents an overview of the knowledge of ncRNAs in some representative organisms belonging to the three phyla of protozoa. Recent discoveries suggest that unicellular chloroplast-containing protists algae do have functional miRNAs, so this section also introduces them.

Ciliates

Ciliates are unique in that they possess two genomes, each in its own nucleus. The smaller nucleus, the micronucleus, contains the **germline genome**, which is actively transcribed only during sexual reproduction. The larger nucleus or macronucleus contains the

somatic genome, which is transcribed during the entire somatic phase of the organism's growth. The fate of each nucleus is quite different. Diploid micronuclei (there are two in *P. aurelia* and one in *Tetrahymena thermophila*) undergo meiosis to produce gametic nuclei, whereas the macronucleus undergoes massive DNA rearrangements that includes the loss of a major part of genomic DNA and extreme polyploidization (up to 800 nt in *P. tetraurelia*). The macronucleus originates from a micronucleus and in the process of conjugation major genome rearrangements occur, including targeted elimination of **internally eliminated sequences** (**IES**), which account for about 15% of the genome (reviewed in Mochizuki and Gorovsky, 2004). You can read more details about biology of life cycle and development of ciliates in Pearson and Winey (2009).

The somatic genomes of ciliates contain disproportionally large numbers of genes for their genome size. This efficiency is brought about by massive DNA rearrangements, including DNA deletions and chromosome fragmentation. Up to 95% of entire genomes, primarily including repetitive elements and transposons, can be eliminated in the macronucleus. Many of the rearrangement events are homology-dependent and thus can be epigenetically regulated. Homology-dependent silencing takes a turn for the extreme in ciliates as it results not only in silencing of undesired genomic locations but also in their complete elimination. These events consist of precise and imprecise deletions of IES located between two direct repeats. Precise deletions include those occurring at the same nucleotide positions in similar copies of macronuclear chromosomes. Deleted sequences are normally encountered in coding sequences, but can also be found in intronic and intergenic regions. Imprecise deletions eliminate intergenic regions containing transposons and minisatellites.

In Paramecium, silencing was shown to be associated with presence of homologous ~ 23 nt short RNAs. These short RNAs are responsible for homology-dependent mRNA degradation and therefore they might be considered short interfering RNAs (siRNAs). Experiments showed that silencing could be triggered by cytoplasmic injection of dsRNAs (double-stranded RNA) or feeding heat-killed *E. coli* expressing dsRNAs to Paramecium cells. It should be noted, however, that ciliates might lack the RNA-dependent RNA polymerase-based amplification of short RNAs or stable transcriptional siRNA-dependent gene silencing, as silencing is quickly reversed when the source of dsRNA is no longer present.

Another ciliated protist, *Tetrahymena termophila*, is commonly used as a model organism in experimental biology. The genomic rearrangements observed in Tetrahymena require the function of the Tetrahymena ortholog of PIWI—Twi1p (Mochizuki et al., 2002). An increase 28 nucleotides long RNAs was detected simultaneously with conjugation, but was greatly reduced in cells with disrupted Twi1p function. Based on their findings, Mochizuki put forward the **scan RNA model** (Mochizuki and Gorovsky, 2004) (see Figure 9-1) in which IES are transcribed in a bi-directional fashion during conjugation and give rise to dsRNA. These dsRNAs become substrate to RNAi-like machinery, resulting in small RNAs. The authors call these small RNAs **scan RNA** (**scnRNA**), because they are thought to scan the maternal and developing macronuclei for homologous sequences. When such homologous sequences in the maternal nucleus are found, the scnRNAs are eliminated. The remaining scnRNAs reach the developing macronucleus, and, if homology is detected, the homologous sequences are eliminated. The presence of these scnRNAs during conjugation leads to local DNA deletions within the targeted region, and degradation of homologous maternal transcripts, which subsequently are unable to sequester the germline scan RNAs, resulting in the deletion of corresponding DNA in the developing nucleus of Paramecium and Tetrahymena (Mochizuki and Gorovsky, 2004). Targeted sequence elimination requires H3K9 methylation, and the loss of Twi1p function results in the loss of H3K9me. Disruption of *DCL1*, which codes for Dicer ribonucleases, prevents production of scnRNAs and causes accumulation of germline transcripts and the failure of IES excision. This suggests that scnRNA mediates accumulation of specific chromatin modifications that then result in elimination of the sequence stretch (see Figure 9-1).

This mechanism is siRNA-like because it uses dsRNA as a substrate for the production of small RNAs, however the size of the small RNAs is atypical for siRNAs. Furthermore, scnRNAs deviate from the siRNA path by targeting histones for modification in a sequence-dependent way. Thus, the model suggests that small RNAs that are produced and operated by an RNAi-related mechanism regulate DNA elimination in Paramecium and Tetrahymena. It was further

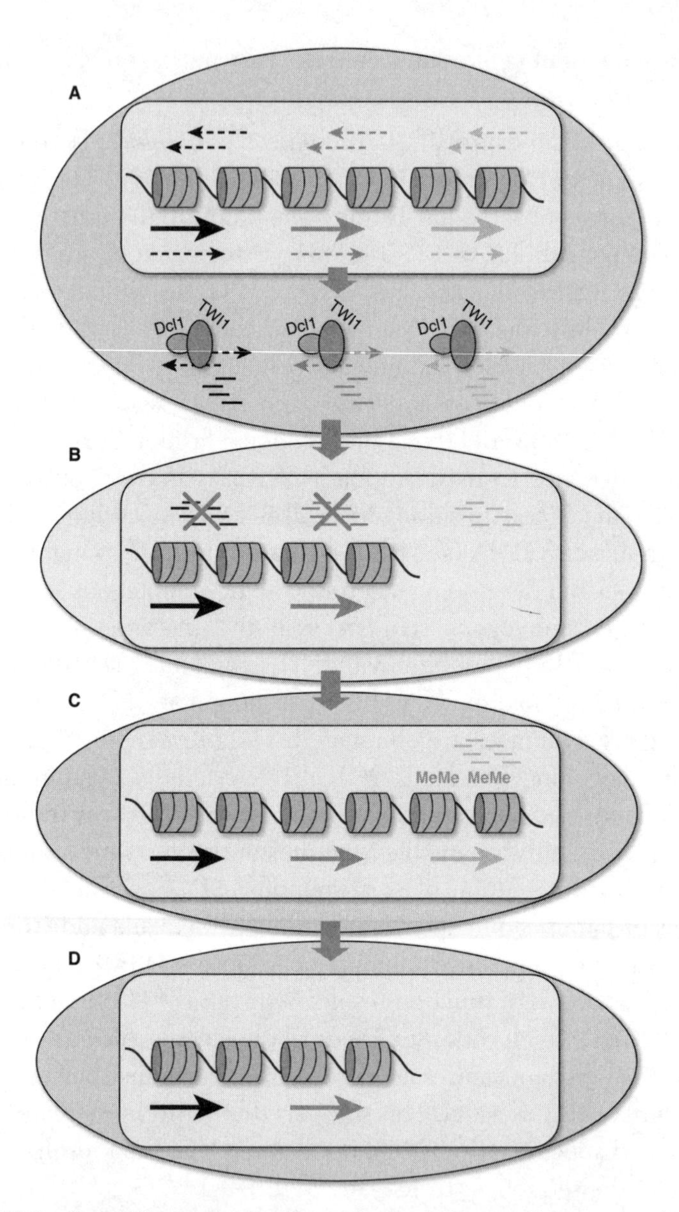

Figure 9-1 Scan RNA model. **A.** The majority of the germline genome is transcribed early in development. Dcl1 dices the formed RNA duplexes into 28 nt scnRNAs that are then incorporated into RISC-like complexes containing Twi1. **B.** scnRNAs are transported into the maternal macronucleus where they are matched against homologous sequences. scnRNAs that contain homology are eliminated. **C.** The remaining scnRNAs are transported into developing the macronucleus where they direct the deposition of H3K9me into homologous sequences. **D.** The genomic regions associated with H3K9me are eliminated from the genome.

proposed by Mochizuki (2010) that the small RNAs from the micronuclear genome are used to identify eliminated DNAs by whole-genome comparison of the parental macronucleus and the micronucleus.

scnRNA-mediated elimination of genomic sequences is epigenetically heritable. Experimental macronuclear deletions in sexual progeny also occur with the same efficiency in future generations, an inheritance that is maternally controlled. Curiously, the dsRNA that induced the deletion initially is no longer needed to induce the same deletions in the progeny. The gene deleted from the macronucleus was always presented in the micronucleus, further supporting the epigenetic mode of inheritance. The targeted genomic region, whether induced by homology-carrying transgene or dsRNA, is treated as a transposon or IES sequence and is eliminated during the development of the macronucleus. The deletions that were observed in further sexual generations were not triggered by the presence of scnRNAs or any special imprints in the micronuclear genomes, suggesting that the gene was deleted from the genome of the developing macronucleus simply because it was absent from the maternal macronucleus. Thus, the scan RNA model might suggest that although scnRNA are required for induction of deletion of homologous sequences from the macronuclear genome, the recurrent deletions occurring in the sexual progeny do not require any "marking," such as scnRNA or changes in chromatin, but instead occur as a result of comparison to the already-rearranged maternal genome (refer to Figure 9-1).

DNA unscrambling during sexual reproduction of another ciliate, Oxytricha, is guided by long ncRNAs that are produced from the somatic parental macronuclear genome. These RNAs act as templates for the direct unscrambling events and can regulate chromosome copy number in the developing macronucleus. It was recently shown that experimentally induced changes in RNA abundance can both increase and decrease levels of corresponding DNA molecules in progeny, suggesting epigenetic inheritance of chromosome copy number (Nowacki et al., 2010). Disruption of specific RNA templates was also shown to disable rearrangement of the corresponding genes. This data supports a model of template RNA-mediated epigenetic inheritance of DNA copy number in the Oxytricha macronucleus (see Figure 9-2) (Nowacki et al., 2010).

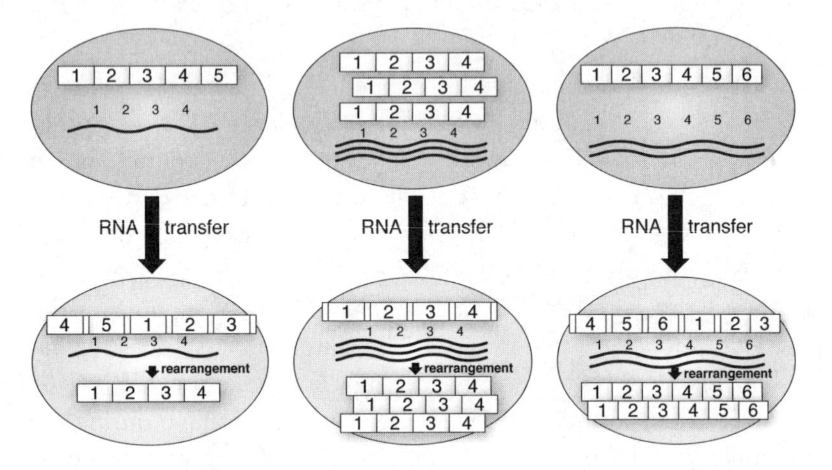

Figure 9-2 RNA-guided inheritance of DNA copy number in Oxytricha (modified with permission from Nowacki et al. (2010) RNA-mediated epigenetic regulation of DNA copy number. Proc Natl Acad Sci USA 107:22140-4). During sexual conjugation, transcripts produced by maternal somatic nuclei (darker shaded/colored) are transferred into developing nuclei (lighter shaded/colored). The transcript amounts are roughly proportional to the chromosome copy number. Maternal transcripts guide DNA rearrangements in such a manner that the number of generated mature chromosomes is proportional to the number of transcripts transferred. Alternative hypothesis suggests that RNAs simply protect from degradation those chromosomes that have a "correct" pattern of rearrangements. Boxes with numbers show chromosomal regions that are involved in rearrangements, whereas small white boxes show deleted regions.

If the transcription rate of all ncRNA-coding genes involved in control of DNA rearrangements is more or less similar, the ncRNA abundance should correlate with the DNA concentration.

The level of correctly processed chromosomes might also correlate with the concentration of available template RNA, suggesting a quantitative relationship from maternal DNA to maternal RNA to progeny DNA. This establishes the chromosome copy number very early during development. Furthermore, the effect of RNA injection on the chromosome copy number is not dependent on the type of DNA rearrangement during chromosomal maturation, influencing both scrambled and nonscrambled genes. Although these findings are consistent with a scnRNA model proposed for Paramecium and

Tetrahymena, it is not clear whether maternal RNA in these and other organisms can influence DNA copy number, as shown for Oxytricha. The results of Nowacki et al. (2010) thus suggest that maternal RNA, in addition to controlling gene expression or DNA processing, can also program DNA amplification levels.

Flagellates (protista)

Although any organism containing whip-like organelle called flagella may be referred to as "flagellate," we describe ncRNAs in protozoans only. *Trypanosoma brucei* is a representative flagellate species. The information on non-coding RNAs in protists is substantial; however, details of ncRNAs involved in epigenetic regulation in flagellates are rather scarce. *T. brucei* represents one of the most ancient eukaryotes in which RNAi has been experimentally verified (Ngo et al., 1998).

There is no evidence that miRNA regulatory pathway exists in flagellated protozoans, but initial work by Djikeng et al. (2003) showed that in trypanosomes, a portion of siRNAs generated from three different dsRNAs are associated with polyribosomes in the form of a 70-kD ribonucleoprotein particle. The authors suggest this is indirect proof for the interaction between siRNAs and translation machinery.

Trypanosoma brucei contains functional Dicer-like enzyme (TbDcl1) further supporting the existence of siRNA-dependent mechanism of regulation of gene expression. Shi et al. (2006) provided evidence that TbDcl, despite an unusual structure of its RNase III domains, is required for the generation of siRNA-size molecules and for RNAi. Split localization of RNase IIIa and IIIb domains appears to be specific to trypanosomatids. Another difference in the RNAi pathway might include the fact that, whereas many metazoans have multiple Argonautes, single-celled organisms, such as the fission yeast *Schizosaccharomyces pombe* and the protozoa *Trypanosoma brucei*, encode only one Argonaute.

Recently Patrick et al. (2009) showed the existence of another Dicer-like enzyme in *T. brucei*: *Tb*DCL2. Biochemical and genetic analyses indicated that DCL2 and DCL1 are the major and, most likely, the only enzymes of this class that drive the RNAi pathway in *T. brucei*. DCL1 and DCL2 might have both overlapping and distinct functions. Whereas both DCL1 and DCL2 are required for the generation of siRNAs from retroposon transcripts, production of CIR147 repeat-derived siRNAs might only require DCL2 activity. Generated siRNAs in *T. brucei* are then loaded into AGO1. The authors found that most retroposon- and repeat-derived siRNAs in *T. brucei* have a 5'-monophosphate, which have been generated as a result of amplification of secondary siRNAs with the help of RNA-directed RNA polymerase (RdRP), although bioinformatics analysis did not reveal a canonical RdRP in the genome of *T. brucei* (Patrick et al., 2009). The analysis by Patrick et al. (2009) also showed that siRNAs have a blocked 3' end, which is suggestive of a 2-O-methyl modification. It is thus possible that 3' modification of siRNAs is a rather ancient process.

Analysis of the phenotypes of *dcl1*$^{-/-}$ and *dcl2*$^{-/-}$ cells suggests compartmentalization of RNAi processes. First, transcriptional activity of retrotransposons and repetitive elements is upregulated in *dcl2*$^{-/-}$ but not in *dcl1*$^{-/-}$ cells. Second, dsRNAs electroporated into the cytoplasm of *dcl1*$^{-/-}$ cells are not processed, suggesting that DCL2 is not able to access/use the dsRNAs located in the cytoplasm. On the other hand, *dcl2*$^{-/-}$ cells are significantly more responsive to dsRNA or siRNA transfections than *dcl1*$^{-/-}$ or wild-type cells. The authors hypothesized that in the absence of DCL2, and consequently the absence of many endogenous siRNAs, there could be less competition by endogenous siRNAs for loading of synthesized AGO1. Thus, *dcl2*$^{-/-}$ cells might be primed to accept exogenous RNAi triggers—that is, siRNAs that are transfected directly or are derived from transfected dsRNAs (Patrick et al., 2009).

Thus, the RNAi pathway in the parasitic protozoan *T. brucei* relies on a single member of the Argonaute family of proteins—AGO1—and is initiated by two distinct Dicer-like enzymes, namely *Tb*DCL1, mostly found in the cytoplasm, and *Tb*DCL2, which is primarily localized in the nucleus (Patrick et al., 2009).

Pseudopodia-containing protists

Social amoeba *Dictyostelium discoideum* is a protist that uses pseudopodia for locomotion and is capable of forming large colonies. It is one of the model organisms used for research, such as cell differentiation and host-pathogen interactions. This unicellular organism is interesting for several reasons. First, although at normal conditions it functions as a unicellular organism, upon starvation it clumps into large colonies of approximately 100,000 cells that are possibly able to more efficiently communicate with each other and share resources. Together, the cells go through development as a multicellular organism, ultimately forming a ball of spores on top of a stalk (Hinas et al., 2007). A second interesting characteristic is that *D. discoideum* occupies an intermediate evolutionary position, branching after plants but before the division between animals and fungi.

The sequenced and annotated genome of *D. discoideum* allows the discovery of various ncRNAs. RNA interference (RNAi) via transgene delivery is a standard method to knock down gene expression in *D. discoideum*. Indeed, it is predicted that the organism carries most of the homologs necessary for proper operation of RNAi machinery. Among those are two Dicer-like proteins—*drnA* and *drnB*—and three RdRPs—*rrpA*, *rrpB*, and *rrpC*—although only *rrpA* seems to be required for the function of transgenic RNAi (Hinas et al., 2007). There are also five potential homologs of PIWI-like proteins with potentially redundant functions.

Hinas et al. (2007) isolated and characterized a large fraction of small RNAs (18 to 26 nt) from the unicellular amoeba *D. discoideum*, finding several developmentally regulated miRNAs that are potentially processed by a Dicer homolog, *drnB*. The results predicted several targets for the miRNA candidates, with near-perfect complementarity to ORFs and imperfect base pairing within 3' UTRs. Due to the intermediate evolutionary position of *D. discoideum*, it is unclear if *D. discoideum* can use its miRNA for mRNA degradation, like plants, or for translation inhibition, like animals. It is curious whether there is a differential mode of action of these miRNAs in normal conditions when the organism functions as unicellular or upon starvation when it operates as multicellular. Furthermore, if the functionality of these miRNAs were confirmed, this would be the first report of miRNAs isolated from protists.

A large number of RNAs, 21 nt in length, originated from centromeric areas constituted of clusters DIRS-1 retrotransposons. Both, DIRS-1 small RNAs and DIRS-1 mRNA are up-regulated during development. It is likely that RNAi machinery prevents mobilization of the transposon, a result known to occur in many other eukaryotes. Furthermore, because it is hypothesized that clusters of DIRS-1 constitute centromeres, DIRS-1 small RNAs might function in a similar manner as the analogous small RNAs derived from centromeric repeats in *S. pombe* and *A. thaliana* (Hinas et al., 2007). These small RNAs might appear to function as siRNAs because these molecules have been shown to direct DNA methylation of both transposons and centromeric repeats. Thus, previously reported DNA methyltransferase-dependent methylation of DIRS-1 might have been the result of RNAi activity; this, however, still needs to be demonstrated in *D. discoideum*.

Small RNAs from another retrotransposon, Skipper, were found to be less abundant then DIRS-1 small RNAs and no correlation between expression of these small RNAs and mRNA of the retrotransposon itself has been found (Hinas et al., 2007). At the same time, these small RNAs shared nearly perfect complementarity to a set of mRNA that did not derive from retrotransposons, although whether this represents any function remains to be shown. Because some small RNAs derived from repetitive elements have been implicated in miRNA pathways in mammals and plants, it is possible that Skipper-derived small RNAs might do the same.

Skipper small RNAs were significantly upregulated in amoeba strains depleted of *drnA*, and a putative RNA-dependent RNA polymerase, *rrpC*. In contrast, the expression of DIRS-1 small RNAs was not altered in any of the analyzed strains. These findings suggest the presence of multiple RNAi pathways in *D. discoideum* (Hinas et al., 2007). Similarly, multiple RNAi pathways have been demonstrated in the other species such as *C. elegans*, *A. thaliana*, and *T. thermophila*.

Hinas et al. (2007) also isolated several developmentally regulated small RNAs with antisense complementarity to mRNAs. Similar ncRNAs have been reported in the other organisms, including *C. elegans*, *A. thaliana*, and *T. thermophila*. In Arabidopsis, they are called natural **antisense** transcript-derived **RNAs**, or **nat-siRNAs**, and they are activated by stress, such as salt exposure. The genes encoding these small RNAs in *D. discoideum* express longer antisense RNAs from which the small RNAs might originate. For one small RNA, the longer antisense RNA was found to be complementary to the spliced, but not the unspliced, pre-mRNA, indicating possible involvement of RNA-dependent RNA polymerase.

The two Dicer-like proteins in *D. discoideum* do not contain the N-terminal helicase domain, which is a conserved part of Dicers in most organisms, including animals, fungi, and plants (Hinas et al., 2007). However, the motif is present on the RdRPs in *D. discoideum*, suggesting domain swapping between the Dicers and the polymerases. Nevertheless, Dicer B is required for the biogenesis of at least one of the *D. discoideum* miRNA candidates. In other protists, such as *Tetrahymena thermophila*, *Trypanosoma brucei*, and *Giardia lamblia*, dicer-like sequences also lack the helicase domain but are nevertheless able to function as a part RNAi-related mechanism, including *in vivo* and/or *in vitro* cleavage of dsRNAs. However, it should be noted that these processes do not include miRNA-biogenesis.

Algae-chloroplasts containing protists

Algae are protists capable of photosynthesis and among them *Chlamydomonas reinhardtii*, a unicellular biflagellate green alga, is used for variety of molecular genetic studies. RNAi-based gene knock out is efficient in this organism, inferring that this alga has functional RNAi machinery. Zhao et al. (2007) characterized small RNAs in *C. reinhardtii*. Using bioinformatics, they showed that among 4,182 unique small RNA sequences identified, 200 species were derived from genomic sequences with the potential to form hairpin structures resembling miRNA precursors. These ncRNAs share several common features with miRNAs, including having a 5' phosphate and a preference for uracil at the 5' end. The authors confirmed the presence of 5' phosphate because they were able to attach adapter molecules to the 5' of the miRNA by T4 RNA ligase, an enzyme that requires 5' phosphate for ligation to occur. In addition, several of the identified miRNAs have a sequenced pairing miRNA* located in the opposite arm of the precursor. As a result, a duplex bearing the 2-nt 3' overhang can be formed, which is the hallmark of Dicer cleavage (Siomi and Siomi, 2007). These findings show that *C. reinhardtii* encodes miRNAs, which is one of the first evidences of miRNAs in a unicellular organism, similar to *D. discoideum* predicted miRNAs.

Curiously, computational analysis revealed no orthologs of *C. reinhardtii* miRNAs in other green algae, plants, or animals, possibly suggesting that *C. reinhardtii* miRNA genes might have appeared after splitting from a common ancestor. Similar to plants, many *C. reinhardtii* miRNAs are located in relatively long hairpins, many of which have the potential to produce multiple small RNAs. This suggests that the evolution of such miRNAs *de novo* through inverted duplication of their future target genes might be similar to those in plants, despite the fact that *C. reinhardtii* miRNAs might have appeared independently. Siomi and Siomi (2007) suggested that these evolved recently, and therefore nonconserved miRNAs might still contain long segments with high similarity to the target gene sequence. Indeed, such miRNA loci are present in *Arabidopsis thaliana*. It remains to be shown whether recently evolved *C. reinhardtii* miRNAs also carry large stretches of similarity to target genes. Such "young" miRNAs might represent hallmarks of the response to environment and relatively recent adaptation events and thus might be particularly abundant in organisms grown in stressful environments. A physiological peculiarity of *C. reinhardtii* is the fact that in response to many stressful conditions *Chlamydomonas* haploid cells of opposite mating types become gametes and conjugate to form diploid zygotes that shed their flagella and form resting spores. At normal conditions, the spores undergo meiosis, producing four haploid cells that regenerate flagella. Interestingly, computational studies have predicted that *C. reinhardtii* miRNAs might target many genes encoding flagella-associated proteins and thus might potentially regulate the *C. reinhardtii* life cycle. Expression analysis of eight randomly chosen miRNAs in vegetative cells and gametes showed that five out of eight exhibited development-specific regulation, with some miRNA candidates being upregulated and some downregulated in gametes.

Although Zhao et al. (2007) were able to show that computationally predicted *C. reinhardtii* miRNAs are expressed, there does not yet exist any direct proof of their functionality. Fractionation of total extracts from *C. reinhardtii* cultures analysis by Zhao et al. (2007) showed that the *C. reinhardtii* miRNAs exist in complexes with a molecular size of approximately 150 kDa, a size roughly corresponding to the minimal size of RISC complex in human cells. Computational analysis showed that *C. reinhardtii* encodes three Dicer-like proteins, two AGO proteins (transcriptionally active), and one protein containing a PIWI domain that might be a part of the RISC complex in this organism. It remains to be determined whether the two AGO proteins reside in *C. reinhardtii* miRNP complexes and which cellular processes they are involved in.

The authors attempted to identify a potential mode of action of miRNAs in this alga—that is—translation repression or mRNA cleavage, and found that the predominant mode of action was mRNA cleavage, with the majority of the cleavage sites located between the target nucleotides that pair with the tenth and eleventh nucleotides of the miRNA (Zhao et al., 2007).

Thus, it appears that *C. reinhadtii* expresses a number of miRNAs, and they regulate the gene expression by mRNA cleavage. It remains to be shown whether the processes such as specification of flagellar assembly, mating, conjugation, and others are regulated by miRNAs in this unicellular alga. As some newly identified *C. reinhardtii* miRNAs appear to be differentially expressed in a vegetative state and in gametes, they might play roles in gametic differentiation.

In addition to miRNAs, Zhao et al. (2007) identified a large number of endogenous siRNAs with a strong preference for uracils at the 5' end. siRNA locations are random and include protein-coding genes and intergenic regions. Many *C. reinhardtii* siRNAs are enriched in several genomic loci and phased relative to each other, reminiscent of plant transacting siRNAs (tasiRNAs).

Curiously, repetitive areas of the genome are underrepresented: there were very few siRNAs from repetitive elements in which transposons were found. Despite the fact that in many cases endogenous siRNAs are the products of the activity of RNA-dependent RNA Polymerases (RdRPs) in plants and *C. elegans*, no RdRP homologue has been found so far in Chlamydomonas (Schroda, 2006). It remains to be shown what the preferential mode of action for siRNAs in *C. reinhardtii* is. It might be similar to that of plant tasiRNAs that silence messages from loci that are not related to those from which the tasiRNAs derive. Alternatively, siRNAs in *C. reinhardtii* might function as piRNAs in mammals by silencing the areas of the genome from which they derive.

There are many questions remaining. First of all, it is not clear whether *C. reinhardtii* has any other yet undiscovered type of small RNAs. Secondly, it is unknown whether other unicellular organisms, including other algae, also have small RNAs, or whether it is a unique feature of *C. reinhardtii*. It would also be interesting to analyze if miRNAs/siRNAs in this alga are indeed regulating transitions from vegetative to gametic stages and whether they play any role in establishing heritable changes such as silencing.

Fungi

Post-transcriptional gene silencing, analogous to co-supression described in transgenic plants, also exists in fungi. It was first discovered in *Neurospora crassa* shortly after discovery of **co-suppression** in plants and is termed **quelling** (Romano and Macino, 1992). Romano and Macino attempted to transform *N. crassa* with two exogenous genes *albino3* (*al-3*) and *albino1 (al-1)* involved in carotenoids biogenesis pathway genes and observed repression of the expression of endogenous genes. Silencing of both exogenous and endogenous *al-1* or *al-3* genes resulted a color change from orange to albino/pale yellow. It was shown that some of the phenotypes were reversible, presumably due to the reduction in the number of the exogenous gene copies. A significant correlation between the number of the exogenous gene repeats and the efficiency as well as stability of quelling was also documented.

A series of experiments showed the following picture of quelling. First of all, it was shown that the minimal sequence necessary to trigger suppression is about 130 nt and this sequence does not have to include the promoter sequence. Although mutants of *al* genes are recessive in nature, most of the *al*-quelled transformants were heterokaryons and were dominant over wild-type strain, suggesting that quelling is not nucleus-limited and can act *in-trans* by diffusible molecules. The presence of transgene-specific sense RNAs in quelled strains and not in reverted strains implied that transcription of the transgene is required for quelling. This further led to the hypothesis that if transgene-derived RNAs have aberrant nature (**aberrant RNA** or **aRNA**), post-transcriptional gene silencing is initiated.

A large mutagenesis experiment using quelled *al-1* strains identified several mutants, called **quelling deficient (qde)** mutants, belonging to three major groups: *qde-1*, *qde-2*, and *qde-3*. QDE-1 appeared to be an RdRP. The first gene from an RNAi pathway ever cloned, QDE-1, was later shown to mediate amplification of the siRNA signal, which is the rate-limiting step in quelling. From this finding, a model has been proposed suggesting that aRNAs are used as templates to produce dsRNA with the help of RdRP enzyme QDE-1. The second large group of mutants, *qde-2*, encoded an Argonaute protein homologous to *rde-1* gene from C. *elegans*. Because *rde-1* is an essential protein for production of dsRNA, it was suggested RNAi and transgene-induced PTGS are similar processes and that they might have evolved from the same ancestral mechanism (Li et al., 2010). Thus, it appears that both RdRP and Argonaute are required for RNAi and quelling, suggesting that the two processes are mechanistically linked gene silencing phenomena.

The *qde-3* group of mutants appeared to be unique to fungi. The *qde-3* gene encodes a RecQ DNA helicase homologous to the human Werner/Bloom's syndrome proteins. QDE-3 might act upstream of the dsRNA formation and could be important for the generation of aRNA/dsRNA from the transgenic loci. The enzymes involved in DNA repair are QDE-3 and another RecQ DNA helicase RecQ-2. Although the exact role of QDE-3 in quelling is still largely unknown, the involvement of helicases in silencing suggests that DNA structure and DNA repair process might be important for the process. There is no direct proof but the QDE-3 homologue in rice, OsRecQ1, seems to be required for RNA silencing induced by the introduction of inverted-repeat DNA, but not for dsRNA induced RNA silencing. Moreover, a homologue of QDE-3 in rats, rRecQ-1, was reported to be associated with piRNA-binding complex (reviewed in Li et al., 2010). Small RNAs of 25 nt produced during quelling were shown to be associated with QDE-2; however, their biogenesis was independent of *qde-2* but required *qde-1* and *qde-3*.

What is the mechanism of quelling? How do the fungi differentiate between transgenes and endogenous sequences? How are endogenous dsRNAs made? None of these questions can be answered in full. It is hypothesized that the synthesis of aberrant RNAs and their recognition by RdRP enzyme is critical for quelling initiation. The previous model suggested that the transcription of a transgene by RNA polymerase II produces aRNA, which is used as the substrate by QDE-1 to generate dsRNA. However, recent experiments show that QDE-1 is not only an RNA-dependent but also a DNA-dependent RNA polymerase and might be involved in making aRNA with the help QDE-3 helicase. It is interesting that overexpression of QDE-1 increases the rate of quelling dramatically from the normally observed rate. This finding supports the hypothesis that the amount of produced dsRNAs is the limiting factor in quelling.

Another DNA repair protein possibly required for quelling is RPA1. RPA1 is a *N. crassa* homologue of Replication Protein A, a single-stranded DNA-binding protein involved in the processes of replication, repair and recombination. QDE-1 also appears to interact with RPA1. Chromatin immunoprecipitation assays showed that QDE-1 is enriched at the transgenic *al-1* locus, possibly suggesting QDE-1 is recruited to the transgenic locus. Accumulation of siRNAs upon quelling is DNA-synthesis dependent (hydroxyurea treatment inhibited DNA replication and abolished siRNAs accumulation); therefore, it was suggested that, during replication, QDE-1 together with RPA-1 differentiate repetitive transgenes from endogenous genes and then target them for silencing (reviewed in Li et al., 2010). Because QDE-1 appears to possess functions of both an RdRP and a **DNA-dependent RNA Polymerases** (**DdRP**), it could be that QDE-3, the DNA helicase required for aRNA production, and RPA might facilitate the binding of QDE-1 to ssDNA at the transgenic region. ssDNA might be formed either in the process of replication with the involvement of RPA1 or through the activity of QDE-3 helicase that resolves the complex DNA structures at the transgenic locus. At this stage, QDE-1 acts as a DdRP to generate aRNA, which will be further converted into dsRNA by the RdRP activity of QDE-1 (reviewed by Li et al., 2010). It should be noted, however, that this is just a model and requires further proofs.

More evidence for the role of QDE-1 aRNA formation in initiation of quelling comes from studies that demonstrate that both QDE-1 and QDE-3 are not required for dsRNA-induced gene silencing. When an inverted repeat-containing transgene, which can result in the production of dsRNA, is expressed in *N. crassa*, the quelling occurs in both *qde-1* and *qde-3* mutants. Furthermore, in contrast to other organisms in which RdRP is involved in amplification of secondary siRNAs, QDE-1 in *N. crassa* is not involved in the amplification and production of secondary small RNAs. These results support the hypothesis that dsRNAs are the intermediates in quelling and that QDE-1 and QDE-3 function upstream of the pathway responsible for production of dsRNAs. As soon as dsRNAs are produced, QDE-1 and QDE-3 are no longer essential. On the other hand, QDE-2 is essential for gene silencing induced by dsRNA.

The release of the *N. crassa* genome sequence allowed identification of two genes coding for partially redundant Dicer proteins DCL-1 (Dicer-like-1) and DCL-2 (Dicer-like-2) (reviewed in Li et al., 2010). Mutation of one of the Dicers does not disrupt quelling, whereas knocking out both Dicers does, a fact that provides evidence for their redundancy. Due to the requirement of two independent mutations in these two genes to turn quelling off, these genes did not come up in the comprehensive mutant screening approach that identified *qde-1*, *qde-2*, and *qde-3*. Despite the functional redundancy of DCL-1 and DCL-2, there appears to be a significant depletion of the amount of siRNAs in *dcl-2* mutant, suggesting that DCL-2 might be the main processing enzyme.

As in plants and humans, formation of RISC in fungi requires the activity of Argonaute protein QDE-2. The siRNA duplex associated with QDE-2 and the RISC is inactive, as the antisense "passenger" strand has to be removed. Several biochemical studies attempting to analyze the possible role of QDE-2 in the process of nicking and removal of the passenger strand from siRNA duplex led to the identification of a novel QDE-2 interacting protein (QIP). QIP is a putative exonuclease and its function requires QDE-2. Wild type QDE-2 activity is required for the production of single-stranded siRNA, as the mutant *qde-2* is impaired in RNAi and namely in accumulation of single-stranded siRNA forms. However, QDE-2 cleavage of the passenger strand alone was shown to be insufficient for single-stranded siRNA production and RISC activation (reviewed in Li et al., 2010). Disruption of QIP led to the impairment of quelling as most of the siRNAs were accumulated in nicked-duplex form. QIP appears to function as an exonuclease to remove the cleaved passenger strand in a QDE-2–dependent manner. Thus, the cleavage of the passenger strand by QDE-2 and its subsequent removal by QIP are critical biochemical steps in the Neurospora RNAi pathway.

Based on these studies, Maiti et al. (2007) proposed a model for the Neurospora RNAi pathway (see Figure 9-3). According to the model, silencing of repetitive transgenes in *N. crassa* consists of the following steps:

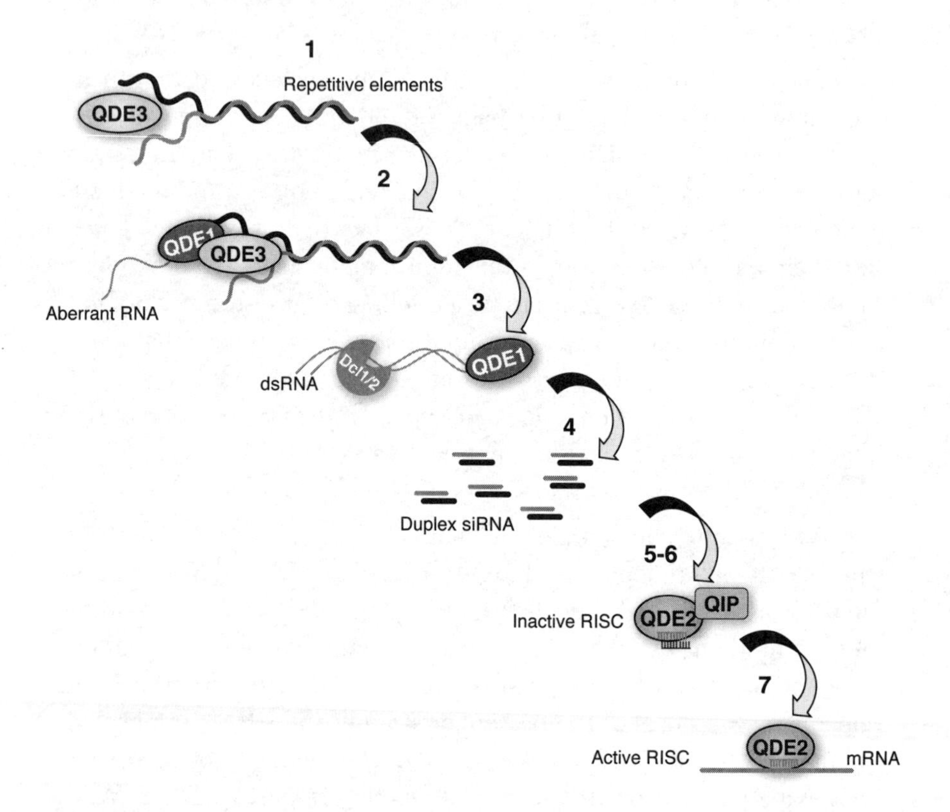

Figure 9-3 Model for quelling.

1. Replication process depends on the activity of QDE-3 helicase and RPA-1.
2. QDE-3 or/and RPA-1 recruit QDE-1, which uses its DdRP activity to synthesize aRNA.
3. aRNAs are converted to dsRNAs via QDE-1 RdRP activity.
4. DCL-1 and DCL-2 (predominantly?) cleave the dsRNAs into siRNAs.
5. siRNAs are then loaded onto the RISC consisting of QDE-2 and QIP.

6. Maturation of siRNAs includes two steps—cleavage of the passenger strands of the siRNA duplexes by Argonaute protein QDE-2 and the removal of the nicked passenger strands by exonuclease QIP.
7. Remaining mature guide siRNA together with the RISC protein cleaves homologous mRNAs, resulting in gene silencing.

The following questions still remain: Does the replication of nonrepetitive elements require QDE-3? If not, how is QDE-3 recruited to the transgenes upon replication? Can QDE-1 synthesize normal RNA or its role to synthesize only aRNA? If it can only synthesize aRNA, how does it do it? What are the main features of these RNAs that make them recognizable as aberrant? Does QDE-1 dissociate from aRNA after replication or does it stay attached and then initiate the synthesis of the passenger strand using its RdRP activity?

What is the main role of quelling in *N. crassa*? As quelling is induced by transgenes carrying sequences homologous to the endogenous DNA, it can be hypothesized that the process of quelling is the mechanism that prevents the expression of repetitive elements, such as transposons and retrotransposons. Indeed, siRNAs with the homology to transposons were detected in *N. crassa*. The *qde-2* mutant and *dcl-1 dcl-2* double mutants impaired in quelling siRNAs also had substantially higher transcript levels and copy number of the LINE1-like transposons.

Despite obvious similarities between the mechanisms of RNAi in *N. crassa* and other organisms such as yeast, plants, and animals, there are still substantial differences. The function of RNAi in formation of heterochromation and directing methylation of cytosines does not seem to require the known RNAi components in *Neurospora crassa*, including three RdRPs (QDE-1, SAD-1, and RRP-3), two Argonaute proteins (QDE-2 and SMS-2), two dicer-like proteins (DCL-1 and DCL-2), and two RecQ helicases (QDE-3 and RecQ-2) (reviewed in Li et al., 2010). It can thus be suggested that RNAi pathway in *N. crassa* is not involved in the process of transcription gene silencing. Alongside this notion, it was shown that siRNAs produced during quelling are not required for H3K9 methylation. Although siRNAs might not be necessary for H3K9 methylation, H3K9 methylation itself is important for proper quelling indicated by the fact that mutation in the methyltransferase gene *dim-5* responsible for methylation of lysine 9 at the histone H3 caused a rapid loss of the integrated transgenic copies and, as a result, a low quelling efficiency and frequent reversion of the quelled transformants.

qiRNAs

Lee et al. (2009) recently discovered another group of siRNAs, called **qiRNA** because they interact with QDE-2. The authors found that the level of this novel class of small RNAs which are approximately 21-nt in length was very low in fungi grown under normal conditions but was significantly induced upon exposure to DNA-damaging agents. qiRNAs are primarily produced from the highly repetitive rDNA locus and have a strong 5' uridine and 3' adenine preference. The experiments showed that qiRNAs production depends on QDE-1, QDE-3, and the Dicers but not QDE-2, despite the fact that QDE-2 expression is induced by DNA damage. qiRNAs produced match the transcribed rRNA regions and the untranscribed intergenic spacer regions suggesting that qiRNAs originate from aRNAs. Also, produced qiRNAs correspond to both sense and antisense rDNA strands and their long precursor RNAs accumulated to high levels in the *dcl-1 dcl-2* double mutant, indicating that qiRNAs are processed from long dsRNAs. All these facts about qiRNAs suggest that the mechanism of their production is similar to the mechanism of production for quelling siRNAs, with the main difference being the induction of the process with damage to DNA.

The origin of DNA loci giving rise to aRNAs is another similarity between the processes of quelling and production of qiRNA. In wild type *N. crassa*, the rDNA locus is the only highly repetitive DNA locus. In the transgenic *N. crassa*, the transgene copies from the second potential repetitive locus. Thus, the common trigger for generation of aRNAs and small RNAs is the repetitive nature of the rDNA and the transgene loci. Replication stress caused by the repetitive nature of the transgene and rDNA is also a possible trigger for generation of aRNA and siRNA.

Are qiRNAs important for the DNA damage response? The QDE-3 helicase was previously shown to play a role in DNA damage response and several RNAi mutants lacking qiRNAs were shown to have increased sensitivity to DNA damage. These results suggest that qiRNAs might contribute to the DNA damage response by inhibiting protein translation. qiRNAs might also play an important role in stabilization of rDNA loci; it was reported that the rDNA gene copy numbers in the earlier identified quelling mutants *qde-1*, *qde-2*, and *qde-3* are reduced as compared to the wild-type strain.

Meiotic silencing by unpaired DNA

Another RNAi-related mechanism discovered in *N. crassa* is **meiotic silencing by unpaired DNA** (**MSUD**) (Shiu et al., 2001). The mechanism of occurrence is similar to quelling but it only occurs during meiosis. *N. crassa* is a haploid organism that exists in a diploid stage (ascus cell) only transiently when the two cell nuclei fuse for mating. Ascus cells undergo two rounds of meiosis followed by one round of mitosis, resulting in eight haploid ascospores. MSUD appears to function in the first meiotic prophase by silencing all unpaired gene copies upon the pairing of homologous chromosomes.

Metzenberg and colleagues first discovered this phenomenon when they observed the dominant effect of the deletion in ascospore maturation 1 gene (*asm-1*). Strains with the deletion of this gene had difficulty with maturation of ascospores. It was initially believed that the single remaining copy of the gene was not sufficient for proper function, but two experiments proved otherwise. First, it was shown that ectopic insertion of a second functional copy did not restore the phenotype. Second, the presence of a pairing partner allele that had nearly identical sequence, but was not able to make a proper protein product, allowed the ascospore maturation. It was then suggested that the absence of pairing during meiosis triggers silencing, and the phenomenon received the name **meiotic transvection** or **meiotic transsensing**. Further experiments showed that the cross resulting in the presence of three copies of *asm-1* in ascus cells still causes silencing, preventing proper maturation of ascospores. Thus, the occurrence of MSUD and the proper maturation of ascospores require the absence of unpaired (undamaged) copies of the *asm-1*, rather than the absence of paired copies.

Further experiments aiming to uncover the mechanism of MSUD included the UVC mutagenesis of *N. crassa* strain with subsequent crosses to MSUD proficient (with heterozygous *asm-1* deletion) strain. One of the identified mutants, suppressor of ascus dominance-1 (*Sad-1*) appeared to be a paralogue of *qde-1* and thus was similar to RdRP enzymes. Because the process of MSUD involves silencing of both unpaired and paired alleles it can be hypothesized that MSUD is an RNAi-related phenomenon. The requirement of SAD-1 for MSUD further suggests that the synthesis of dsRNA is required for MSUD (reviewed in Li et al., 2010). Another loss-of-function mutant—suppressor of meiotic silencing-2 (*sms-2*) shown to be deficient for MSUD—lacks the gene homologous to Argonautes.

The protein DCL-1 is also required for MSUD to function, with mutation of DCL-1 but not DCL-2 eliminating MSUD. Curiously, DCL-2 is the protein that has a main role in production of siRNAs in quelling. It can thus be hypothesized that whereas DCL-1 plays a predominant role during meiosis (when MSUD occurs), DCL-2 plays a more important role during mitosis (when quelling occurs). Thus, the requirement of the RdRP SAD-1, the Argonaute protein SMS-2 and the Dicer DCL-1 indicates that dsRNA and small RNAs are involved in the MSUD.

The work of Lee et al. (2004) showed that MSUD occurrence resulting in effective silencing is directly dependent on the presence of the homology between the two unpaired regions. In addition, the efficiency of silencing showed a positive correlation with the size and homology between the two unpaired regions. Loss of reporter transcript directly correlated with the induction of meiotic silencing suggesting that MSUD functions post-transcriptionally.

Based on the aforementioned observations, Li et al. (2010) proposed a model of MSUD:

1. During meiosis, an unpaired DNA works as a signal for initiation of the transcription of aRNAs from the unpaired DNA region.
2. SAD-2 recruits SAD-1 to aRNA.
3. The RdRP SAD-1 converts aRNA into dsRNA.
4. dsRNA is processed by DCL-1 into small RNAs.

5. sRNAs are processed by RISC complex containing SMS-2.
6. sRNAs target homologous genes post-transcriptionally.

What are the other possible functions of MSUD? Rearrangement at any locus resulting in either deletion of duplication of chromosomal regions can potentially be targeted by MSUD silencing. Similarly, MSUD might also function as a mechanism that prevents transposon expansion during meiosis because transposon replication ultimately results in mispairing at a given genomic position.

milRNAs

Until very recently, Dicer-dependent miRNAs and Dicer-independent PIWI-interacting RNAs that associate with Argonaute family proteins had not been found in fungi. Then, in 2010, Lee et al. (2010) analyzed small RNAs associated with the Neurospora Argonaute protein QDE-2 and found that **miRNA-like small RNAs** (**milRNAs**) are produced in this filamentous fungus.

The authors found at least 25 loci with potential milRNAs and showed that many of these milRNAs are similar to conventional miRNAs from animals and plants. The similarities include that they both originate from specific stem-loop RNA precursors; most of the milRNAs seem to require Dicer for the biogenesis; and milRNAs are imperfectly complementary to their targets, suggesting that they function in translation inhibition like animal miRNA. For a given milRNA locus, most of the small RNA sequences matched to a particular arm of the hairpin (sense milRNA) and much lower numbers of sRNAs matched the complementary arm (antisense milRNA or milRNA*). Similarly to miRNAs in other eukaryotes, nearly all milRNAs originating from the same locus share the same 5' U position, but differ substantially at the 3' termini.

One major difference between miRNAs in animals and plants and milRNAs is the mechanism of their biogenesis. Whereas a Dicer-dependent pathway produces miRNAs from plants and animals, milRNAs are apparently produced in four different but possibly overlapping pathways (Li et al. 2010). The authors (Lee et al., 2010) analyzed the biogenesis of four milRNAs and found that unlike qiRNAs, production of milRNAs is independent of QDE-1 and QDE-3 proteins. The biogenesis pathway closest to the one in plants is responsible for generating milR-3. The generation of pre-milRNA and processing of milRNA requires only Dicer (1 and 2) protein. In contrast, biogenesis of milR-4 is only partially dependent on Dicer, indicating the existence of a novel nuclease. Lee at al. (2010) demonstrated that this novel nuclease might be a Neurospora homolog of the yeast mitochondrial ribosomal protein MRPL3. The protein contains a dsRNA recognition motif and a putative RNAse III domain, although it has little sequence similarity to those of Dicers and Drosha. The mutant *mrpl3* has a dramatic decrease in the production of milR-1 and milR-4 but not milR-2 and milR-3, suggesting the essential role MRPL3 plays in milR-1 and milR-4 biogenesis.

The biogenesis of milR-1 is quite complex and includes Dicers for the production of pre-milRNA and mature milRNAs production, as well as QDE-2 and the exonuclease activity of QIP for the biogenesis of mature milRNAs. According to Li et al. (2010), the steps of milR-1 might include the following:

1. Dicer processes the pri-milRNA into double-stranded pre-milRNAs.
2. QDE-2 binds to pre-milRNA and recruits the exonuclease QIP.
3. Together QDE-2 and QIP process the pre-milRNAs into mature milRNAs.

This appears to be rather a novel and a distinct mechanism from the one functioning in plants/animals, as the Argonaute protein appears to function as an adaptor by binding to pre-milRNA and recruiting other factor(s) to mediate milRNA maturation.

Biogenesis of milR-2 is also rather unique to *N. crassa*, as this milRNA is produced independently from function of Dicer and mostly relies on QDE-2 and its catalytic activity. The steps of milR-2 biogenesis include

1. milR-2 pri-milRNA exists as a hairpin structure with a very small loop.
2. QDE-2 alone or with the help of yet unknown nucleases process pri-milR-2 into long pre-milR2 molecules.

3. QDE-2 binds to a long pre-milRNA and cleaves the milRNA* strand of the pre-milRNAs.
4. Another yet unknown nuclease, possibly together with QDE-2, further processes pre-milR2 into mature milR-2 forms.

The mechanism of milR-2 biogenesis is truly remarkable, as it appears to be rather different from canonical Dicer-dependent miRNA biogenesis pathway functioning in animals and plants.

The importance of this observation is phenomenal as just the possibility of producing miRNAs without Dicers is previously unheard of. The report by Lee et al. (2010) was immediately followed by two reports from animal researchers indicating the existence of Dicer-independent pathway of miRNA biogenesis by describing the biogenesis of the mouse miR-451 (Cheloufi et al., 2010; Cifuentes et al., 2010). Remarkably, the mechanism was very similar to the one reported for *N. crassa* milR-2. More exciting perhaps is the fact that the organisms such as eubacteria and archaea that do not have functional Dicer proteins might produce miRNA-like RNAs through Argonaute-like proteins.

Dicer-independent small interfering RNAs (disiRNAs)

Dicer-independent small interfering RNAs (disiRNAs) are another novel class of sRNAs produced in Neurospora (Lee et al., 2010). Their name is based on the fact that their production is not dependent on normal RNAi biogenesis pathway, including components such as QDE-1, QDE-2, or QDE-3, but rather because the level of these molecules does not substantially change in *dcl-1/dcl-2* double mutant, it is suspected they are produced in a Dicer-independent manner. disiRNAs are 22 nt–long sRNAs with a strong 5' U preference derived from approximately 50 loci with non-repetitive DNA that carry no apparent homology to each other. disiRNAs are produced more or less symmetrically from both strands of DNA. Lee et al. (2010) showed that up to 80% of the disiRNA loci have overlapping sense and antisense transcripts, suggesting that disiRNAs are made from naturally occurring complementary sense and antisense transcripts. In this respect, disiRNAs are similar to nat-siRNAs from plants.

The finding that disiRNAs are produced without involvement of Dicers suggests that their biogenesis might be similar to animal piRNAs, which are also produced in a Dicer-independent manner. However, it was shown that mutations of Argonaute genes or other known RNAi genes does not influence the level of disiRNAs. This indicates that disiRNAs might be a completely distinct class of sRNAs, as far as biogenesis is concerned. Despite the lack of requirement of QDE-2 for their biogenesis, disiRNAs do interact with this protein, suggesting that they might still function in RNAi pathway.

Other interesting small RNA-related epigenetic phenomena in N. crassa

B2 and aptamer FC RNA can bind to RNA Polymerase II and prevent transcription (Kettenberger et al., 2006). Thus, they exhibit analogous function to the previously described bacterial 6S RNA. Kettenberger et al. (2006) solved the crystal structure of the aptamer FC RNA-Pol II complex and found that the RNA forms a stem loop structure that resides in the active cleft of Pol II. They suggest that this structure resembles the form of a melted promoter and therefore is similar to bacterial 6S RNA. Also, aptamer FC RNA employs molecular mimicry to inhibit promoter binding of Pol II (Lee et al., 2009).

A dsRNA-induced transcriptional program important for RNAi

The work of Choudhary et al. (2007) suggests that *N. crassa* might possess a mechanism similar to mammalian dsRNA-induced antiviral interferon response. The expression of dsRNA in Neurospora significantly induced the expression of *qde-2* and *dcl-2* genes. This transcriptional response was regulated by dsRNA rather than siRNA, as the transcriptional activation of **dsRNA-activated genes** (**DRAGs**) was maintained in the *dcl* double mutant, as well as in *qde-1* and *qde-2* mutants in which siRNA production was completely abolished (reviewed in Li et al., 2010).

Injection of dsRNAs into *N. crassa* activated the expression of approximately 60 genes, including *qde-2*, *dcl-2*, and several other RNAi components as well as homologs of antiviral and interferon-stimulated genes. The authors hypothesized that the induction of the antiviral and interferon-stimulated genes by dsRNA suggests that such response is a part of a conserved ancient host defense response to counter viral infection and transposons (Choudhary et al., 2007). Indeed, viral infection and expression of hairpin RNA in the chestnut blight fungus *Cryphonectria paracitica* significantly increases the expression of both, *dcl-2* and *agl2* (coding for Argonaute-like protein), indicating that the dsRNA-induced transcriptional response might be conserved in filamentous fungi.

Yeasts—Saccharomyces cerevisiae and Schizosaccharomyces pombe

Yeasts are unicellular eukaryotic organisms and are ideal for various genetic and epigenetic studies. Specific details of production of various small RNAs and their function are introduced using two different yeast models, *Saccharomyces cerevisiae* and *Schizosaccharomyces pombe*.

Saccharomyces cerevisiae (S. cerevisiae)

S. cerevisiae, a budding yeast, is a unicellular organism commonly used for beer production and for baking. The organism exists in both haploid and diploid state. For a long time, the existence of ncRNAs potentially involved in epigenetic regulation in *S. cerevisiae* was uncertain. However, it was recently shown that *S. cerevisiae* cells produce a number of non-coding transcripts from 20 to several thousand nucleotides. Because *S. cerevisiae* does not have the components of RNAi, it is an ideal model organism for the analysis of the role of RNAi-independent mechanisms of ncRNA regulation of gene expression. Major components of RISC and RITS, such as Dicer-like RNases, Argonautes, RdRPs, or PIWI-like proteins, are absent in budding yeasts.

The presence of dsRNA in *S. cerevisiae* does not trigger the repression of homologous genes or mRNAs as in other eukaryotes. Loss of such a powerful system of protection against invading molecules is expected to have evolutionary costs. Although there are obvious downsides to the loss of RNAi such as inability to regulate the expression of genes through silencing and inability to target foreign sequences for degradation, there might be selective advantages.

Some of the possible advantages include novel mechanisms of regulation of gene expression using RNA-RNA interactions and generation of more diverse pool of translatable RNAs (Harrison et al., 2009).

Did *S. cerevisiae* cells develop a mechanism that replaced RNAi? Unlikely, although several types of ncRNAs—including those produced from bidirectional transcription and nucleosome-free regions, short sense ncRNAs initiating upstream of mRNA start sites, as well as ncRNAs originating from cryptic sites within coding regions—are produced in *S. cerevisiae*.

ncRNAs produced from bidirectional transcription and nucleosome-free regions

The ncRNAs produced from bidirectional transcription are short-lived RNAs often referred to as **cryptic unstable transcripts (CUTs)** because of the absence of obvious promoter and rapid degradation. When the exosome subunit Rrp6 is absent, vast populations of ncRNAs are enriched in *S. cerevisiae*.

Recently, Hainer et al. (2011) identified a regulatory system in *S. cerevisiae* in which transcription of intergenic ncDNA (SRG1) represses transcription of an adjacent protein-coding gene (SER3) through transcription interference. They showed that SRG1 transcription causes repression of SER3 by directing a high level of nucleosomes at the SRG1 chromatin, which overlaps the SER3 promoter.

Promoters with bidirectional transcription appear to be common in yeasts as they are in humans. Recent analyses show that bidirectional transcription typically produces two classes of RNA: mRNAs and ncRNAs. Short transcript clusters of approximately 250 nt with the distribution around the transcription start of mRNAs were identified. Nearly two-thirds of these clusters are generated from the intergenic regions between tandemly arranged genes, whereas almost one-third is from intergenic regions found between divergent genes (Neil et al., 2009 and reviewed in Hainer et al., 2011). In addition, a small fraction of these CUTs stem from intergenic regions between convergent genes. Short transcripts initiated from the intergenic regions between tandemly arranged genes were found in antisense orientation to the upstream gene. One possible explanation is that these ncRNAs are produced because of bidirectionality of the downstream promoter (Hainer et al., 2011) (see Figure 9-4A). The extent of such bidirectional transcription is influenced by the convergence of the 3' end of the upstream gene with the 5' end of the downstream gene (Neil et al., 2009). As many of these CUTs were found to stem from promoter and 3' UTR regions that are typically free of nucleosomes, Neil et al. (2009) suggested that these ncRNAs do primarily come from nucleosome-free regions. Indeed, using high-density tiling arrays, Xu et al. (2009) compared nucleosome density with abundance of ncRNA in S. *cerevisiae* cells deficient in the exosome subunit Rrp6 and found a significant correlation between Rrp6-sensitive RNA abundance and nucleosome-free regions. Based on this observation, Hainer et al. (2011) suggested that ncRNAs might be passive byproducts of nucleosome-free regions. Alternatively, it is possible that transcription of ncRNAs might promote or maintain nucleosome depletion. Indeed, the majority of coding and noncoding divergent transcript pairs share an expended 5' nucleosome-free region. Moreover, all classes of mRNA and ncRNAs have nucleosome-free promoter regions (Xu et al., 2009).

Short sense ncRNAs initiating upstream of mRNA start sites

Short unstable ncRNAs are also produced from the promoter areas of coding genes. Mapping the sequenced transcripts to the genomic areas showed that transcription of these ncRNAs is initiated upstream of the mRNA transcription site, in sense orientation (Neil et al., 2009) (see Figure 9-4B). In wild type S. *cerevisiae* strain, the transcription of these ncRNAs is terminated by cooperative function of Nrd1 with Nab3 and Sen1 proteins. Nrd1 binds "aberrant" transcripts in a sequence-specific manner. It can be hypothesized that the sequence specificity of ncRNAs produced from the promoter regions is a sufficient trigger for Nrd1 to bind. The Nrd1/Nab3/Sen1 recruits the 3' to 5' exonuclease, known as the exosome, and the exosome-activating complex TRAMP. Processing and turnover of ncRNAs also occurs with help of yet another exosome component, the RNase D-type exoribonuclease Rrp6, one of the main components for ncRNA degradation.

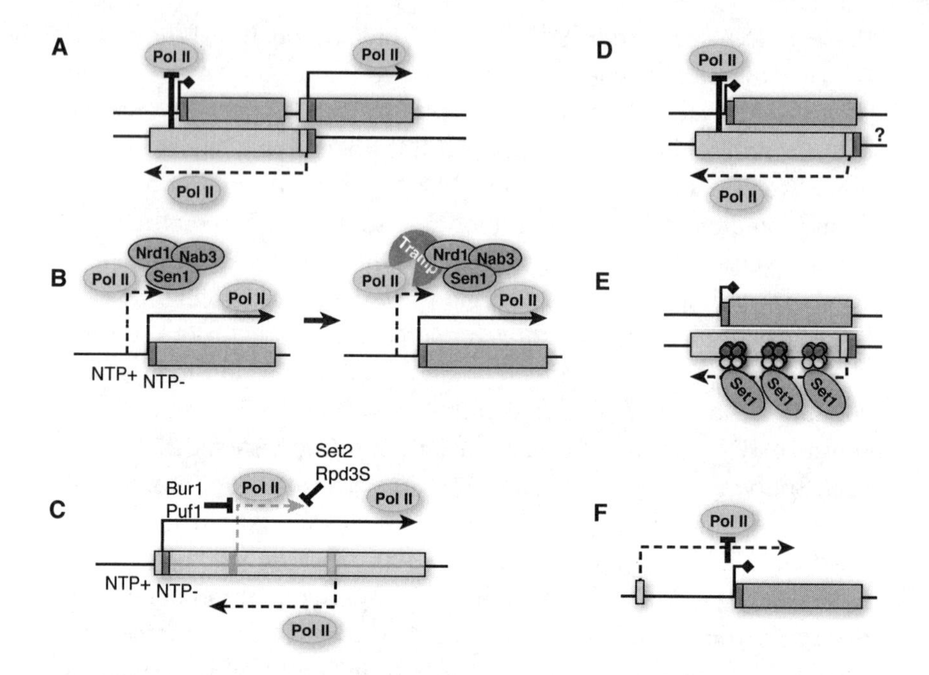

Figure 9-4 Mechanisms of biogenesis and function of ncRNAs in *S. cerevisiae*. **A.** Bidirectional promoter activity results in production of majority of ncRNAs from tandemly arranged genes. Antisense transcripts interfere with transcription initiation from the sense strand. **B.** Short sense ncRNAs can be generated from the promoter activity immediately upstream of coding gene promoter. The activity of these promoters is associated with NTP presence. The activity from this promoter suppresses transcription from the downstream promoter. **C.** Slow or abnormal reassembly of the chromatin following Pol II may result in the transient use of internal cryptic promoters for generation of ncRNAs. **D.** Synthesis of antisense ncRNAs from cryptic promoters results in suppression of transcription. **E.** Expression of antisense ncRNAs results in deposition of repressive chromatin marks such as H3Kme and H3Kme2 established by Set1 methyltransferase. **F.** Transcription of sense ncRNA upstream from the coding gene transcripts interferes with its expression. Solid arrows indicate coding transcripts, whereas dashed arrows represent ncRNA transcripts.

The ncRNAs are also targeted by the RNA-binding proteins Mpp6 and Rrp47. Whereas Mpp6 preferentially targets poly(U) sequences, sparing poly(A) sequences and structured RNAs, Rrp47 preferentially interacts with structured RNAs. It is thus hypothesized that Mpp6 processes pre-rRNA and pre-mRNA and degrades ncRNAs.

The activities of the proteins mentioned earlier keep the number of ncRNAs produced in this manner low. The relative abundance of these ncRNAs as compared to all other ncRNAs found in *S. cerevisiae* suggests that the mechanism of transcription of ncRNAs from the promoter of mRNA is the most common mechanism. A similar mechanism operates in biogenesis of snoRNAs in *S. cerevisiae*; exosome is involved in trimming of the 3' end of snoRNAs.

ncRNAs originating from cryptic sites within coding regions

The coding regions of many genes contain cryptic promoters and putative TATA boxes. One of the functions of Pol II is repression of transcription from these promoters. Activity of cryptic promoters is also repressed by transcription elongation factors Spt6 and Spt16, chromatin-remodeling protein Isw2, histone modifiers Rpd3S, Set2, and histone chaperone Rtt106. Li et al. (2007) suggested that the mechanism controlling ncRNA transcription from internal sites is regulated by quick reassembly of the chromatin following transcription elongation step by Pol II.

Two independent complexes inhibiting the transcription from cryptic promoters at the 5' and the 3' regions of the coding gene exist. Whereas the elongation factors Bur1-Paf1 inhibit the expression from 5' regions, the histone modifiers Set2-Rpd3S regulate the expression from the 3' end (see Figure 9-4C). Interestingly, nutrient availability leads to differential regulation of the activity in these complexes. Switching yeast growing on rich media to nutrient poor media results in differential expression of more than 20% of such ncRNAs, suggesting that repressive complexes are regulated by environmental factors (reviewed in Harrison et al., 2009).

The reason for the existence of transcription from cryptic promoters in *S. cerevisiae* is unclear; however, it can be suggested that this is one of the ways of producing alternative transcripts (proteins), similar to alternative splicing, internal ribosome entry, and so on. It is also possible that transcription from cryptic promoters regulates the expression of "normal" coding genes, by either inhibiting the transcription directly or through the chromatin restructuring or perhaps post-transcriptionally through the function of ncRNAs. All of these explanations still have to be demonstrated in *S. cerevisiae*.

ncRNA-mediated mechanisms of regulation of gene expression

The mechanisms of ncRNA-mediated regulation of gene expression in *S. cerevisiae* are poorly studied. These yeast cells do not have RNAi so the small ncRNAs of 20 to 24 nt in length are not readily produced. There are several examples of inverse correlation between mRNA level and the level of antisense ncRNA produced from the same locus. Moreover, there appears to be a correlation between the level of antisense ncRNAs and the mRNAs from different loci, suggesting that there might be a post-transcriptional mechanism of regulation of gene expression (reviewed by Harrison et al., 2009).

The following possible mechanisms of ncRNA-mediated regulation of gene expression might function in *S. cerevisiae*:

1. Repression of mRNA transcription via antisense ncRNA transcription or by ncRNA themselves.
2. Transcription priming or attenuation by ncRNA produced from the promoter shared with mRNA.
3. Sense ncRNAs transcription interferes with downstream mRNA transcription.
4. Antisense ncRNAs influence chromatin structure thus influencing mRNA transcription.
5. Transacting ncRNA—transcript degradation or translation inhibition.

1) The inverse correlation between mRNAs and their antisense ncRNA transcripts suggests direct regulatory effects of these ncRNAs. The expression of *IME4* mRNA is regulated by antisense transcription initiated at the 3' UTR of the gene. Curiously, the expression of the ncRNA itself is regulated by binding of a1/α2 heterodimer to the conserved region located at the 3' UTR of the *IME4* gene. It is thus speculated that repression of ncRNA transcription through the a1/α2 heterodimer alleviates transcriptional interference (see Figure 9-4D) (reviewed in Harrison et al., 2009).

Apparently not all ncRNA antisense transcripts result in negative regulation of mRNA expression. It was shown that expression of *PHO5* gene, which is repressed upon phosphate starvation, is activated by the transcription of the antisense ncRNA initiated from the 3' of the *PHO5* gene. Transcription of ncRNA is positively regulated by phosphate. The mechanism of positive regulation of ncRNA transcript is believed to involve "histone eviction." Antisense transcription through the region of *PHO5* promoter displaces the repressive histones, thus allowing transcription to occur. Indeed, when insertion of a spacer into *PHO5* locus prevented ncRNA transcription reaching the *PHO5* mRNA promoter, no histone eviction has occurred, and no *PHO5* mRNA transcript accumulated.

2) Regulation of gene expression via sense ncRNAs stemming from the same promoter is not a rare mechanism in *S. cerevisiae* (see Figure 9-4B). Autoregulation of genes involved in uracil and guanidine synthesis, namely *URA2* and *IMD2*, occur through regulation of Pol II start-site selection (reviewed in Hainer et al., 2011). In the case of *URA2*, transcription of coding and noncoding RNAs occurs from the same promoter but from different initiation sites. The choice of initiation site depends on the availability of nucleotide pool (NTPs). When nucleotides are abundant, Pol II preferentially used the initiation site that results in synthesis of ncRNAs, whereas depletion of nucleotides leads to transcription of *URA2* mRNA. Regulation of transcription of *IMD2* occurs through transcription attenuation. A GTP-rich environment results in transcription from the so-called "G-site" as well as production of attenuated transcript. According to the attenuation model, GTP-dependent regulation requires the presence of guanine nucleotide at the first or second position of the upstream noncoding transcript. Depletion of GTP results in transcription from "A-site" and production of normal *IMD2* mRNA.

3) Transcription of ncRNAs might interfere with transcription of mRNA from a different downstream promoter by occluding the Pol II complex at the mRNA promoter (see Figure 9-4F). Transcription of the *SER3* gene involved in serine biosynthesis is inhibited by the transcription of the upstream sense *SRG1* ncRNA. Similarly, the expression of *ADH1*, Zn-dependent alcohol dehydrogenase, is regulated by ncRNA *ZRR1*. Expression of *ZRR1* is itself regulated by binding of Zap1, a zinc-sensitive transcriptional activator. Bird et al. (2006) demonstrated that *ADH1* transcription was only inhibited when transcription of *ZRR1* was extended over the region that covers the binding site for Rap1—the transcriptional activator of *ADH1*. Thus, *ZRR1* transcription results in a transient displacement of Rap1 from *ADH1* promoter.

4) Regulation of transcription through chromatin modifications has been suggested for two loci: *GAL1-GAL10* and *PHO84*. In the case of the *GAL1-GAL10* locus, two different antisense transcripts are produced in the absence of galactose, a condition that is known to repress *GAL10*. Antisense transcripts are initiated at the 3' end of *GAL10* and span the entire gene region, including the promoter. It was found that antisense transcription requires a Reb1 activator that apparently changes the chromatin around the start of ncRNA transcription. Activation of ncRNA transcription recruits Set1 methyltransferase that increases the level of H3K4me2 and H3K4me3 within the *GAL10* coding region. Consensus sequences for Reb1 binding have also been found at the 3' of *PHO84* coding sequence, from which the antisense ncRNAs is also produced (reviewed in Hainer et al., 2011). These findings thus suggest that one of the mechanisms of inhibiting mRNA transcription relies on antisense ncRNA-dependent changes in chromatin (see Figure 9-4E).

5) **Trans-acting ncRNA** might be extremely rare in *S. cerevisiae*. The absence of RNAi suggests that RNA-RNA and RNA-DNA interactions common for this silencing mechanism are not common for budding yeasts. Indeed, so far, there is only one known case of transacting ncRNA in *S. cerevisiae*. In 2008, Berretta et al. identified an antisense ncRNA produced from *Ty1* elements. The expression of reporter genes integrated in the genomic locus carrying Ty1 elements and the initiation of transcription of Ty1 itself were repressed by antisense ncRNAs expressed from transiently delivered plasmids (Berretta et al., 2008). This finding suggests that the antisense ncRNAs can regulate the expression of Ty1 *in trans*. The specificity of suppression suggests that these ncRNAs possibly regulate the expression of *Ty1* mRNA through one of several mechanisms involving RNA-RNA or RNA-DNA interactions. Moreover, because Pol II is depleted at the chromatin of *Ty1* elements of the RNA-decay mutants, the antisense ncRNAs may also regulate *Ty1* transcription. Despite the fact that no mRNA-ncRNA interaction (except for tRNAs and rRNAs) has been reported yet for *S. cerevisiae*, such interactions might exist for *Ty1* ncRNAs. *Ty1* is a retrotransposon and therefore the intermediate step of integration is the generation of cDNA. Suppression of *Ty1* transcription as well as possible degradation of produced *Ty1* mRNA results in dramatic depletion of *Ty1* cDNA. It is possible that ncRNAs produced

from the *Ty1* region might also inhibit the reverse transcription of *Ty1* mRNA. Alternatively, through RNA-DNA interaction, ncRNAs might promote degradation of *Ty1* cDNA (reviewed in Harrison et al., 2009).

Possible mechanisms of regulation of expression from *Ty1* loci by ncRNAs, including regulation of transcription, RNA degradation, inhibition of translation, inhibition of cDNA synthesis, as well cDNA degradation, are described in Harrison et al. (2009). All these possible mechanisms have yet to be confirmed for *S. cerevisiae*.

Meiotic unannotated transcripts (MUTs)—meiotic ncRNAs

Lardenois et al. (2011) showed that ncRNAs not detected during mitosis in wild-type *S. cerevisiae* cells accumulate to high levels during meiosis and spore formation. The authors named these ncRNAs **meiotic unannotated transcripts** or **MUTs**. The activity of one of the main components of the exosome, Rrp6, is maintained during mitosis and diminishes when cells switch to respiration and sporulation. It can be hypothesized that Rrp6 directly targets various ncRNAs during vegetative growth and depletion of Rrp6 at the meiosis onset allows many of these ncRNAs to exert their functions.

Lardenois et al. (2011) suggest that some of the noncoding RNAs expressed in meiosis may regulate protein-coding genes by inhibiting the activity of their promoters, which is done by pausing RNA polymerase during elongation to prevent activator binding. An inverse correlation between the concentration of CLN2 mRNA, a repressor of IME1 (the inducer of meiosis) and the level of IME1 mRNA declines during sporulation. Because the 5'-regulatory region of CLN2 is also the region of production of MUT1465, it can be hypothesized that initiation of transcription of MUT1465 inhibits production of CLN2 transcript, thus de-repressing the meiosis. Another such mRNA/MUT pair is found at the locus of Cell Division Cycle 6 (CDC6), a cell cycle regulator and repressor of DNA replication. The level CDC6 mRNA decreases when a cell enters meiosis, whereas the level of ncRNA SUT200 that spans an entire region upstream of CDC6 promoter dramatically increases with the meiosis onset.

Transcription interference via MUTs might also play a role in regulation of autonomously replicating sequence (ARS) elements during sporulation. Lardenois et al. (2011) found that several MUTs cover

ARS220 and ARS607. Also, the mitotically active ARS605 showed inhibition during meiotic prophase. The authors speculated that ncRNAs might directly contribute to the regulation of ARS activity, thereby influencing the efficiency of DNA replication at different stages of growth and development.

Schizosaccharomyces pombe (S. pombe)

Schizosaccharomyces pombe (*S. pombe*) is a fission yeast representing an haploid unicellular organism frequently used for fermentation. In contrast to *S. cerevisiae*, *S. pombe* has the orthologs of the many RNAi components including Argonaute, Dicer, and RdRP genes, and as such it has an active functional RNAi system, including both RISC and RITS complexes.

Non-coding RNAs in the form of siRNAs, with the average size of approximately 21 nt, are readily produced in *S. pombe* in a Dicer-dependent manner. Produced siRNAs are incorporated into RISC and RITS complexes and require RdRP and Argonaute for propagation and proper function. The main activity of these siRNAs in RITS is the establishment of heterochromatin at the centromeric repeats.

In the *S. pombe* RITS complex, Argonaute interacts with two additional proteins: Tas3 and Chp1. Tas3, a glycine and tryptophan motif-containing protein, links Ago1 to Chp1, a chromodomain-containing protein specifically interacting with nucleosomes enriched with H3K9 me2 and H3K9me3 (Halic and Moazed, 2010). Thus, RITS in *S. pombe* associates with chromatin through base-pairing between siRNAs and nascent transcripts and through interaction of Chp1 with H3K9 methylated nucleosomes.

In fission yeast as well as in nematodes and plants, the silencing signal in the form of siRNA is amplified by the activity of RdRP (Halic and Moazed, 2010). The *S. pombe* homologue of RdRP—Rdp1—interacts with two conserved proteins—Hrr1 and Cid12—in a functional **RNA-dependent RNA polymerase complex (RDRC)**, both shown to be required for RNAi in other eukaryotes.

The RNAi machinery in the form of RITS and RDRC complexes localizes to chromatin-bound transcripts and apparently mediates their processing into siRNAs. This results in the generation of additional siRNAs and further promotes assembly of heterochromatin.

But where do original siRNAs come from? How is heterochromatin assembly initiated? There are several possible models (Halic and Mozed, 2010). The *first model* suggests that centromeric siRNAs are produced from the processing of dsRNA, formed either by RDRC-dependent activity on specific centromeric RNAs or by base pairing of sense and antisense centromeric transcripts (see Figure 9-5A). The *second model* suggests that RDRC recognizes specific features of centromeric RNAs and synthesizes the initial dsRNA substrate for Dicer to act upon (see Figure 9-5B). An alternative *third model* proposes that low levels of histone H3K9 methylation recruit the RITS and RDRC complexes to centromeric repeats to initiate siRNA generation and the amplification of H3K9 methylation (see Figure 9-5C). Clr4-Rik1-Cul4 (CLRC) methyltransferase/ubiquitin ligase complex (CLRC) plays an essential role for extra cycles of H3K9 methylation.

These models are yet to be supported experimentally.

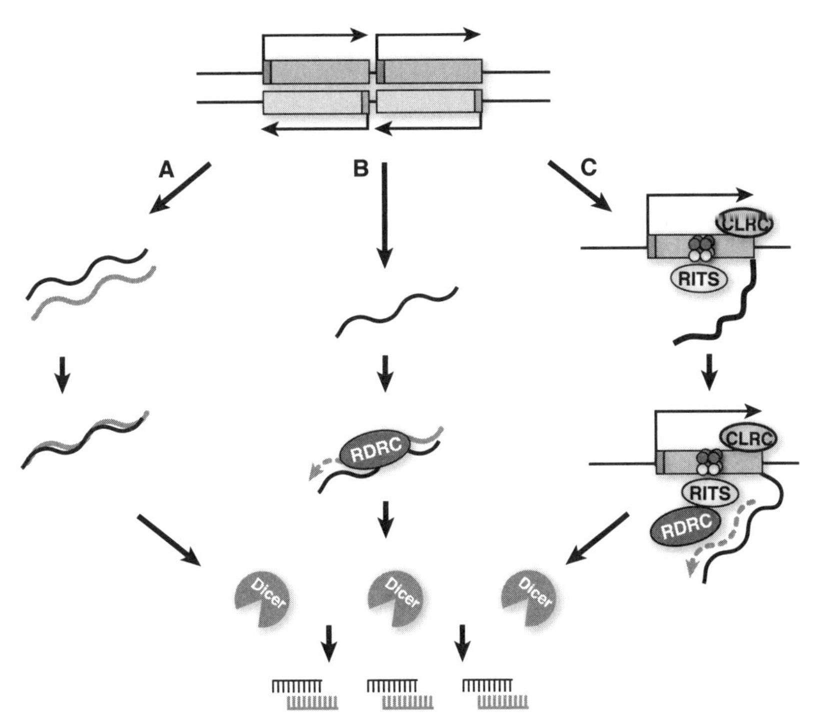

Figure 9-5 Mechanisms of generation of siRNAs from heterochromatic regions.

Most recently, Halic and Moazed (2010) demonstrated the existence of two Ago1-dependent pathways that mediate the generation of different levels of small RNAs from centromeric repeat sequences (see Figure 9-6). Both pathways might depend on the accumulation of degradation products stemming from bidirectional transcription from centromeric repeats. In the first pathway, these degradation products might either form dsRNAs that are further processed by Dicer or/and might be picked up by RDRC (see Figure 9-6A). Indeed, the analysis of small RNA profiles in heterochromatin mutants showed that the amplification of siRNAs can occur independently of H3K9 methylation involving RDRC and Dicer activity on specific ncRNAs. Because this process required Argonaute protein activity, the authors suggested that the Ago1-associated small RNAs target RDRC to centromeric transcripts (Halic and Moazed, 2010).

In the second pathway, the degradation products might associate with Ago1 due to its ability to bind all degradation RNA products in an abundance-dependent manner (see Figure 9-6). The small RNAs associated with Ago1 are trimmed to 22 to 23 nt **primal small RNAs** (**priRNAs**) representing a distinct ncRNA class. However, it is not clear why only siRNAs associated with heterochromatin and pericentromeric regions are amplified. The authors proposed that such specificity is due in part to the increased occurrence of bidirectional transcription from centromeric repeats (Halic and Moazed, 2010). It is possible that amplification depends on the presence of sense and antisense transcripts or dsRNA. priRNAs do not target genomic regions containing protein coding genes and/or non-repetitive sequences, but effectively target repetitive regions that are bidirectionally transcribed, presumably because priRNAs have antisense targets within these regions. The authors propose that priRNAs produced from the second pathway might trigger RDRC/Dicer-dependent siRNA amplification from antisense centromeric RNAs, which occur via the first pathway.

Another interesting result of the work by Halic and Moazed (2010) is the demonstration that heterochromatin-independent amplification of siRNAs occurs only on transcripts from *dg* centromeric repeats. Amplification of dg siRNAs requires the slicer activity of Ago1. Ago1, guided by the small RNA, must target the dg RNA before dsRNA is synthesized by RDRC (see Figure 9-6B). The authors show that an

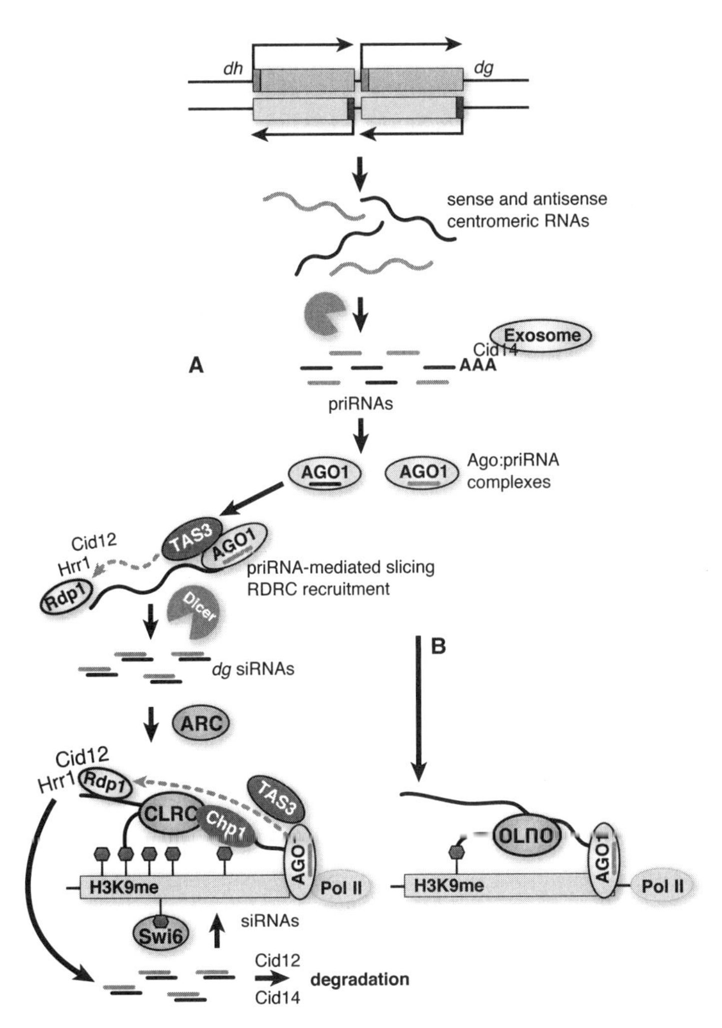

Figure 9-6 Model summarizing heterochromatin-independent and heterochromatin-dependent generation and amplification of priRNAs and siRNAs (reproduced from Halic M, Moazed D. (2010) Dicer-independent primal RNAs trigger RNAi and heterochromatin formation. Cell 140:504-16). Degradation of sense and antisense centromeric transcripts results in priRNAs that are loaded into Ago1. A.) Heterochromatin-independent generation of siRNAs occurs through targeting of centromeric transcripts by priRNAs. Ago1 and RDRC/Dicer are required for this process. The ARC complex is required for targeting the nascent transcripts, perhaps by recruiting RITS, RDRC, and CLRC complexes that maintain certain level of H3K9 methylation and induce further amplification of siRNAs. Modification of the 3' ends of siRNAs by Cid12 and Cid14 enzymes trigger degradation of small RNAs. B.) priRNAs might bound nascent centromeric transcripts without RDRC/Dicer-dependent siRNA amplification. priRNAs binding recruit the CLRC complex and induce low levels of H3K9 methylation.

Ago1/Tas3 subcomplex targets RDRC to dg RNAs. These heterochromatin-independent dg siRNAs in the form of the RITS complex target centromeric transcripts to induce both H3K9 methylation and further siRNA amplification at the dg repeat regions (Halic and Moazed, 2010).

The authors suggest that in *S. pombe*, in contrast to RNAi in other organisms, a Dicer and RDRC-independent Ago1 surveillance mechanism, rather than the processing of dsRNA formed from centromeric sense and antisense transcripts, mediates RNAi-dependent nucleosome positioning and heterochromatin assembly at the pericentromeric repeats (Halic and Moazed, 2010) (see Figure 9-6B).

Conclusion

As can be seen from this chapter, evolution of prokaryotes into simple unicellular and multicellular eukaryotes brought tremendous complexity to RNA world and to sophisticated systems of self-defense and self-regulation. Regulation of gene expression became more complex; ncRNA-mediated cleavage or repression of translation acquired more components, including Dicer and Argonaut proteins. Self-defense and self-regulation became more elaborate, evolving to RNAi and RNAi-like processes such as quelling in *N. crassa*, targeted elimination of IES in ciliates, and meiotic silencing by unpaired DNA (MSUD or meiotic transvection). Dicer- and Argonaut-dependent production of ncRNAs in fungi branched even further as several types of Dicer-independent but QDE-2-dependent small RNAs (milRNAs) are produced in Neurospora. An interesting turn in evolution of self-defense using ncRNAs has occurred in yeasts: Whereas budding yeasts *S. cerevisiae* do not support RNAi, fission yeasts *S. pombe* have active RNAi and RdRP components. It still remains to be shown whether the presence of RNAi in *S. pombe* gives this organism substantial advantage over *S. cerevisiae*.

Exercises and discussion topics

1. What is a precise and imprecise deletion? What is one example of when they occur?
2. Describe Mochizuki's scan RNA model. Make sure to mention why they are called scan RNAs and how they are formed.

3. What makes the RNAi pathway for Trypanosomatids and other unicellular organisms unique?
4. Is there any evidence for miRNAs in protists? If so, give an example.
5. Which proteins are thought to be a part of the RISC complex in *C. reinhardtii*?
6. What is quelling and what is it analogous to in plants? How was it discovered and how does it work?
7. How were DCL-1 and DCL-2 discovered in *N. crassa* and why weren't they discovered earlier? Which gene appears to be more important in quelling?
8. What is the model for the Neurospora RNAi pathway? Make sure to include all seven steps and relevant proteins.
9. What is the main role of quelling in *N. crassa*?
10. What are the substantial differences between the mechanisms of RNAi in *N. crassa* and other organisms such as yeast, plants, and animals?
11. Describe the characteristics of qiRNA as well as the mechanism of their production.
12. Are qiRNAs important for the DNA damage response? Why or why not?
13. Is a single copy of *asm-1* sufficient for proper function of MSUD? Which experiments by Metzenberg and colleagues support this?
14. Why would mutation of DCL-1 but not DCL-2 eliminate MSUD?
15. Explain the MSUD model, as proposed by Li et al. (2010).
16. How are milRNAs in fungi similar to conventional miRNAs from animals and plants? What is the major difference?
17. What are the steps of milR-2 biogenesis?
18. What may be the reason that antiviral and interferon-stimulated genes are induced by dsRNAs? Provide an example supporting this reason for the response.
19. What are the possible mechanisms for ncRNA-mediated regulation of gene expression in *S. cerevisiae*?
20. What is "histone eviction"?
21. Are transacting ncRNA common in *S. cerevisiae*? Describe an example of antisense ncRNAs in *S. cerevisiae* and how it works.

22. What are meiotic unannotated transcripts (MUTs), and what is one of their possible purposes?

23. When it comes to RNAi, what makes *S. pombe* so different from *S. cerevisiae*?

24. What are the possible models for the production of siRNAs and the initiation of heterochromatin assembly? Have these been supported experimentally?

25. What is the function of QDE-1, QDE-2, and QDE-3?

26. Describe the function of RPA1 and its suggested role in quelling.

References

Berretta et al. (2008) A cryptic unstable transcript mediates transcriptional trans-silencing of the Ty1 retrotransposon in S. cerevisiae. *Genes Dev* 22:615-626.

Bird et al. (2006) Repression of ADH1 and ADH3 during zinc deficiency by Zap1-induced intergenic RNA transcripts. *EMBO J* 25:5726-5734.

Cheloufi et al. (2010) A dicer-independent miRNA biogenesis pathway that requires Ago catalysis. *Nature* 465:584-589.

Choudhary et al. (2007) A double-stranded-RNA response program important for RNA interference efficiency. *Mol Cell Biol* 27:3995-4005.

Cifuentes et al. (2010) A novel miRNA processing pathway independent of Dicer requires Argonaute2 catalytic activity. *Science* 328:1694-1698.

Djikeng et al. (2003) An siRNA ribonucleoprotein is found associated with polyribosomes in Trypanosoma brucei. *RNA* 9:802-808.

Hainer et al. (2011) Intergenic transcription causes repression by directing nucleosome assembly. *Genes Dev* 25:29-40.

Halic M, Moazed D. (2010) Dicer-independent primal RNAs trigger RNAi and heterochromatin formation. *Cell* 140:504-516.

Harrison et al. (2009) Life without RNAi: noncoding RNAs and their functions in Saccharomyces cerevisiae. *Biochem Cell Biol* 87:767-779.

Hinas et al. (2007) The small RNA repertoire of Dictyostelium discoideum and its regulation by components of the RNAi pathway. *Nucleic Acids Res* 35:6714-6726.

Kettenberger et al. (2006) Structure of an RNA polymerase II-RNA inhibitor complex elucidates transcription regulation by noncoding RNAs. *Nat Struct Mol Biol* 13:44-48.

Lardenois et al. (2011) Execution of the meiotic noncoding RNA expression program and the onset of gametogenesis in yeast require the conserved exosome subunit Rrp6. *Proc Natl Acad Sci USA* 108:1058-1063.

Lee et al. (2004) Properties of unpaired DNA required for efficient silencing in Neurospora crassa. *Genetics* 167:131-150.

Lee et al. (2009) qiRNA is a new type of small interfering RNA induced by DNA damage. *Nature* 459:274-277.

Lee et al. (2010) Diverse pathways generate MicroRNA-like RNAs and dicer-independent small interfering RNAs in fungi. *Mol Cell* 38:803-814.

Li et al. (2010) RNA interference pathways in filamentous fungi. *Cell Mol Life Sci* 67:3849-3863.

Maiti et al. (2007) QIP, a putative exonuclease, interacts with the Neurospora Argonaute protein and facilitates conversion of duplex siRNA into single strands. *Genes Dev* 21:590-600.

Mochizuki K. (2010) RNA-directed epigenetic regulation of DNA rearrangements. *Essays Biochem* 48:89-100.

Mochizuki K, Gorovsky MA. (2004) Small RNAs in genome rearrangement in Tetrahymena. *Curr Opin Genet Dev* 14:181-187.

Mochizuki et al. (2002) Analysis of a piwi-related gene implicates small RNAs in genome rearrangement in tetrahymena. *Cell* 110:689-699.

Neil et al. (2009) Widespread bidirectional promoters are the major source of cryptic transcripts in yeast. *Nature* 457:1038-1042.

Ngo et al. (1998) Double stranded RNA induces mRNA degradation in Trypanosoma brucei. *Proc Natl Acad Sci USA* 95:14687-14692.

Nowacki et al. (2010) RNA-mediated epigenetic regulation of DNA copy number. *Proc Natl Acad Sci USA* 107:22140-22144.

Patrick et al. (2009) Distinct and overlapping roles for two Dicer-like proteins in the RNA interference pathways of the ancient eukaryote Trypanosoma brucei. *Proc Natl Acad Sci USA* 106:17933-17938.

Pearson CG, Winey M. (2009) Basal body assembly in ciliates: the power of numbers. *Traffic* 10:461-471.

Romano N, Macino G. (1992) Quelling: transient inactivation of gene expression in Neurospora crassa by transformation with homologous sequences. *Mol Microbiol* 6:3343-3353.

Schroda M. (2006) RNA silencing in Chlamydomonas: Mechanisms and tools. *Curr Genet* 49:69-84.

Shi et al. (2006) An unusual Dicer-like1 protein fuels the RNA interference pathway in Trypanosoma brucei. *RNA* 12:2063-2072.

Shiu et al. (2001) Meiotic silencing by unpaired DNA. *Cell* 107:905-916.

Siomi H, Siomi MC. (2007) Expanding RNA physiology: microRNAs in a unicellular organism. *Genes Dev* 21:1153-1156.

Xu et al. (2009) Bidirectional promoters generate pervasive transcription in yeast. *Nature* 457:1033-1037.

Zhao et al. (2007) A complex system of small RNAs in the unicellular green alga *Chlamydomonas reinhardtii*. *Genes Dev* 21:1190-1203.

10

Non-coding RNAs across the kingdoms—animals

Animals express all the groups of non-coding RNAs described in Chapter 9, "Non-Coding RNAs Across the Kingdoms—Protista and Fungi." Among several classes of small RNAs that have been discovered, three main categories prevail: **microRNAs (miRNAs)**, **short interfering RNAs (siRNAs)**, and **PIWI-interacting RNAs (piRNAs)**. These RNAs are primarily found eukaryotes, although the Argonaute proteins and RNAs associated with them can also be found in some bacteria and archaea. Among the three major groups, siRNAs and miRNAs that derive from double-stranded RNA precursor are most common. They are found in most phylogenetic groups of animals and are shown to be involved in a variety of physiological functions. In contrast, piRNAs that derive from single stranded RNA precursor are less common and are mainly found to operate in animal germline. This chapter presents details of the mechanism of biogenesis of these groups of small non-coding RNAs in *Caenorhabditis elegans* (*C. elegans*), Drosophila, and human.

ncRNAs in *C. elegans*

C. elegans is the model organism used for genetic and epigenetic studies of developmental processes. The cells of the organism produce a variety of ncRNAs, including miRNAs, siRNAs, and piRNAs. Specific characteristics of ncRNA biogenesis in *C. elegans* include the presence of only a single Dicer protein for processing miRNAs and siRNAs, the presence of a specific class of piRNAs, which are exactly 21 nt long, called **21U-RNA**, and the presence of the secondary siRNAs generated through RdRP (RNA-dependent RNA Polymerase) activity. Specialization in the processing of miRNA or siRNA in *C. elegans* may be achieved using a single Dicer as well as an additional protein, such as the double-stranded RNA-binding protein, RDE-4, as the gatekeeper for entry into the **RNA interference (RNAi) pathway** (Ghildiyal and Zamore, 2009).

C. elegans miRNAs

C. elegans was the first organism for which miRNA was described (Lee et al., 1993). The miRNA lin-4 was identified as an endogenous regulator of genes that control developmental timing in *C. elegans*. It appeared to negatively regulate the function of LIN-14 protein. The authors found two small lin-4 transcripts of approximately 22 and 61 nt with sequences complementary to a repeated sequence element in the 3' untranslated region (UTR) of lin-14 mRNA. The authors suggested that lin-4 regulates lin-14 translation via an antisense RNA-RNA interaction.

Subsequently, Fire, Mello, and colleagues reported that, also in *C. elegans*, exogenous double-stranded RNAs (dsRNAs) silence genes through a newly described mechanism of RNAi (Fire et al., 1998). You can find more details on RNAi in Chapter 15, "Gene Silencing: Ancient Immune Response and a Versatile Mechanism of Control over the Fate of Foreign Nucleic Acids."

Expression of many miRNAs in *C. elegans* is tightly regulated during development. For example, the expression of two most famous miRNAs, *lin-4* and *let-7*, are specifically upregulated at the second larval (L2) and the fourth larval (L4) stages, respectively, and are necessary for the normal transition from the first to the second larval stage and from the fourth larval stage to the adult, respectively. There also appears to be the difference in miRNA pool in male and hermaphrodite worms. Kato et al. (2009) found that about 12% of known miRNAs exhibited major differences in expression in hermaphrodites and in males.

According to the most recent annotation of miRBase, *C. elegans* has 154 identified miRNA genes that encode sequences with features of typical miRNA precursor—single-strand RNA that can be folded into hairpin structure with mismatched base-pairing and that can be processed into ~22 nt mature miRNA. Although more *C. elegans* miRNA genes may be discovered by various biochemical approaches or through deep sequencing of transcriptome under specific growth conditions, the current list is likely to be quite comprehensive for this animal (reviewed in Ibáñez-Ventoso and Driscoll, 2009).

In general, approximately 60% of *C. elegans* miRNAs are redundant and share substantial sequence similarities with other miRNAs in the nematode genome; however, it remains to be shown whether sequence redundancy is transformed into functional redundancy. It is possible that redundancy in miRNA function is avoided by spatial and temporal separation of expression of potentially redundant miRNAs; related miRNAs might not necessarily be expressed in the same cell types at the same time (Ibáñez-Ventoso and Driscoll, 2009).

The existence of redundancy in miRNA genes together with the absence of functionality of such miRNAs is demonstrated by the analysis of knockouts of individual miRNA genes. Genetic knockout of each of 92 *C. elegans* miRNA genes showed that most deletions do not confer apparent phenotypes under normal growth conditions (Miska et al., 2007). Actual miRNA mutant phenotypes were only observed for the mir-35-41 cluster, which exhibited temperature-sensitive lethality and for the mir-240/mir-786, which resulted in a prolonged defecation cycle. The role of miRNAs in developmental processes and control of aging is well described in the review by Ibáñez-Ventoso and Driscoll (2009).

There is substantial conservation of miRNAs across animal species: approximately 62% and 55% of *C. elegans* miRNAs are related to those in Drosophila and human miRNAs, respectively. Such similarity is not accidental and suggests that similar miRNAs might have common functions across phyla. The most famous example is the *let-7* gene family. In *C. elegans*, *let-7* miRNA regulates the timing of fate differentiation of multipotent seam cells during worm development. *let-7* miRNA and homologous miR-84 negatively regulate *C. elegans* RAS gene *let-60*. In humans, *let-7* has been shown to downregulate expression of RAS gene and over-expression of *let-7 in vitro* inhibits the growth of lung cancer cells. Moreover, cancer cells often have poor expression of *let-7* family of miRNA genes.

It has been mentioned that miRNAs in *C. elegans* are under the tight developmental control. Some cases of posttranscriptional regulation of miRNA biogenesis in *C. elegans* involve a double-negative feedback loop. The *let-7* miRNA inhibits translation of Lin28, and at the same time is inhibited by high amounts of Lin28 protein. The mechanism of such inhibition includes binding of Lin28 protein to specific sequences of the terminal loop of either *pri-let-7* or *pre-let-7* RNAs. Because Lin28 is primarily localized in the cytoplasm, it probably plays a greater role in regulation of cytoplasmic form of *let7*. Indeed, association of Lin28 with *pre-let-7* was also shown. The mechanisms of inhibition might include the ability of Lin28 to promote polyuridylation of the 3' terminus of *pre-let-7* and to inhibit Drosha-mediated and Dicer-mediated cleavages. Accumulation of polyuridylated *pre-let-7* suggests that Dicer does not process such pre-miRNA forms efficiently. It cannot be excluded, however, that Dicers are not able to cleave *pre-let-7* when it is associated with Lin28, and polyuridylation occurs at uncleaved *pre-let-7* as a secondary, possibly unrelated mechanism.

Primary and secondary siRNAs involved in RNAi

Primary siRNAs processed by dicing dsRNA precursors are quite rare in *C. elegans*. As per Mello and Conte (2004), RNAi in *C. elegans* is a very potent mechanism, with only several dsRNA molecules able to induce strong response. This response is maintained by secondary siRNAs from the 3' ends of freshly cleaved target mRNA via the action of **RNA-dependent RNA polymerase (RdRP)** enzymes. This generation of **secondary siRNAs** has the potential to amplify and maintain the response on relatively high level, leading to systemic silencing in the entire organism. Secondary siRNAs are then loaded into specialized **secondary Argonautes (SAGOs)**. An RdRP-dependent amplification mechanism is known to function in *C. elegans*, but is apparently absent in insects, such as Drosophila, and vertebrates, such as mice and humans. Secondary siRNAs allow targeting of mRNA and genomic areas that were not initially targeted by primary RNAi. Known as transitive RNAi, this mechanism can lead to the silencing of multiple transcripts that share a high degree of homology.

Comprehensive analysis of sequences of secondary siRNAs (Pak and Fire, 2007) has shown that the majority of them are not generated through Dicer-dependent processing. A large majority of secondary siRNAs appeared to carry homology to the antisense strand of the mRNA targeted by the primary dsRNA trigger. The original (primary) siRNAs generated through dsRNA intermediate are processed by Dicer and contain a mixture of sense and antisense guide strands, according to thermodynamic siRNA asymmetry. Also, the 5' termini of secondary siRNAs carry di- or triphosphate groups, indicating that they are not processed by Dicer but rather represent unprimed RdRP products. It can therefore be assumed that secondary siRNAs bypass the Dicer-dependent RISC assembly used by primary siRNAs and load directly into Ago proteins.

21 U-RNAs—C. elegans piRNAs

A new class of small non-coding RNAs was identified by Ruby et al. (2006) upon analysis of the small RNA profile in *C. elegans*. The main features of these ncRNAs are their size of 21 nucleotides and the presence of uridine at its 5' end, thus their name: 21U-RNAs. The authors found that the majority of the 21U-RNAs stem from two major regions on chromosome IV. Sequences flanking 21U-RNAs include two upstream elements: large motif of 34 nt and small motif of 4 nt separated by spacer of 19-27 nt long (see Figure 10-1). Large motif contains core consensus region sequence CTGTTTCA, whereas small motif has YRNT as the core sequence. High conservation of these motifs suggests their requirement for transcription of 21U-RNAs. However, each 21U-RNA is autonomously transcribed and they do not carry substantial homology to each other, suggesting that each represents an independent gene. In this case, the 5' flanking region might serve as a site of recognition for the start of transcription.

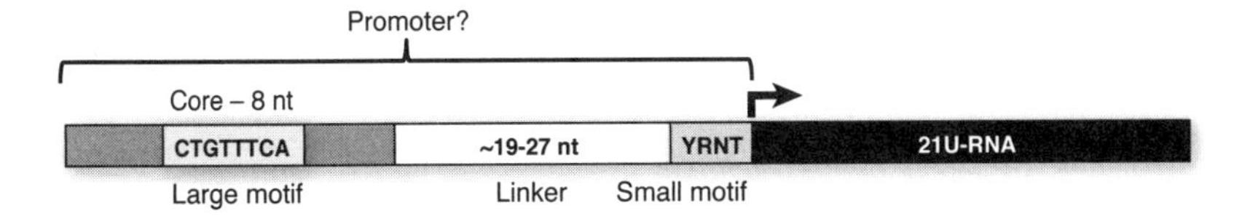

Figure 10-1 Structure of 21U-RNA locus.

The exact biogenesis of these RNAs is not known. Several reports have suggested that these ncRNAs are actually piRNAs (Wang et al., 2008). It is suspected they bind PIWI-Related Gene-1 (PRG-1), a *C. elegans* homolog of PIWI protein, and are required for maintenance of the germ line stability and organism's fertility (see Figure 10-2). In worms, these piRNAs resemble pachytene piRNAs in mammals (see later in this chapter). Similar to small RNAs in plants and rasiRNAs in flies, 21U-RNAs/piRNAs are modified at either 2' or 3' oxygen, suggesting their role in the maintenance of genome stability. To date no transcript with the homology to 21U-RNAs has been found, so it is assumed that the role of these piRNAs is primarily in the nucleus. piRNAs in mammals act as genome stabilizers, although they also function in the cytoplasm. The two known groups of piRNA are larger in size, typically larger than 24 nt, so it is not clear whether 21U-RNAs are indeed involved in genome stabilization. Because 21U-RNAs biogenesis does not include a cytoplasmic stage, they might also be involved in regulation of splicing.

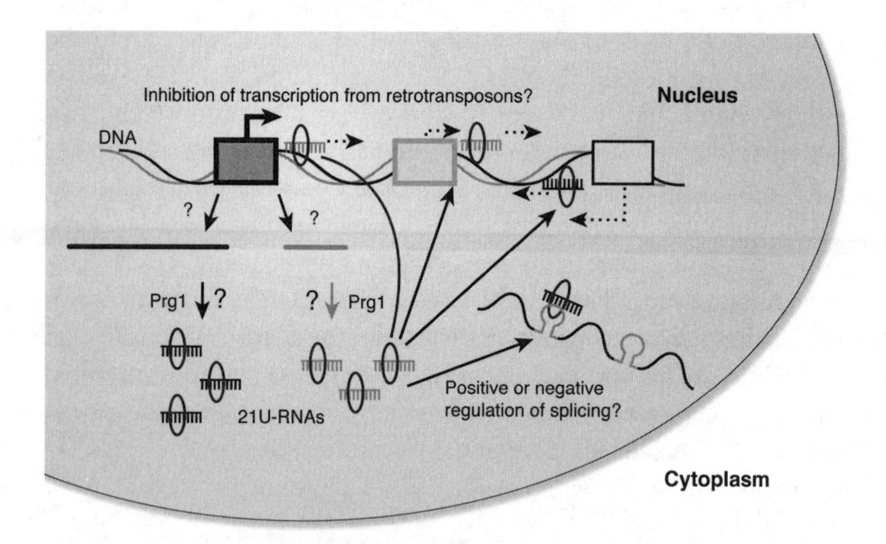

Figure 10-2 Possible function of 21U-RNAs. Prg1 proteins are involved in biogenesis of 21U-RNAs. These RNAs are either used for inhibition of transcription from their loci of origin and/or transcription from retrotransposons as well as for regulation of splicing.

So far, no precursor piRNAs in *C. elegans* has been identified, so it is unclear how piRNAs with a length of exactly 21 nt are produced.

The presence of 21U-RNAs during L1 and dauer (stasis stage) larvae stages suggests that they may play an essential role in worm development and possibly survival under harsh conditions.

Das et al. (2008) showed that PIWI in *C. elegans* is required to silence Tc3 DNA transposon but no other Tc/mariner DNA transposons; this is apparently due to the fact that Tc3 excision rates in the germline of PIWI mutants are increased at least hundredfold as compared to wild-type. Their work showed that PIWI presumably acts upstream of an endogenous siRNA pathway, suggesting that piRNA and siRNA function in *C. elegans* might be linked.

C. elegans as a model for the analysis of biological function of small ncRNAs

C. elegans is an ideal model for studying the biological role of small non-coding RNAs because miRNA expression in *C. elegans* can be inhibited by injection of 2'-oxy-methyl oligonucleotides, complementary to targeted miRNA, into developing embryos (Hutvágner et al., 2004). Previously, the technical difficulty of this procedure did not allow it to be used frequently. Several drawbacks included toxicity of modified oligonucleotides and inadequate cytoplasmic retention and tissue distribution. These problems were addressed recently (Zheng et al., 2010) by delivering the conjugated 2'-oxy-methyl antisense oligonucleotides with the polysaccharide dextran. High water solubility, increased cellular uptake, and availability increased the efficiency of the injection. Further modification of these oligos included conjugation of several molecules to a single dextran molecule. This allowed easy selection of successfully injected adult worms and embryos. Delivery of the oligos complementary to *lin-4* or *let-7* showed knockout phenotypes similar to those observed in the respective loss of function mutant strains (Zheng et al., 2010).

Indeed, progeny of animals injected with modified anti-*lin-4* oligos exhibited delayed exit in differentiation of seam cells (hypodermis) and lack of vulva formation, phenotypes typical for *lin-4(lf)*. A similar effect was observed for *let-7*; a bursting vulva phenotype with lack of alae (longitudinal ridge) formation was observed in 100% of the progeny animals. One of the limitations of the injection of dextran-conjugated antisense 2'-oxy-methyl oligonucleotides is the length of time the efficient inhibition is observed, typically limited to the first 24 hours. For this reason, the method is only effective for studies in which short-term response is sufficient. In this case, injection of modified oligos into embryos might be efficient, whereas injection of adult animals, especially for the long-term studies, associated with aging might not be efficient.

Overall, *C. elegans* is an excellent model for studying miRNA/siRNAs and their associated functions. The simplicity of various molecular and genetic manipulations in *C. elegans*, the comprehensive information over various conserved miRNAs, the availability of various miRNA deletion mutants, the detailed data for expression patterns for various miRNAs, the computational predictions of miRNA targets, and existence of genome annotation all make *C. elegans* a powerful model for studying the function of small non-coding RNAs in the variety of processes.

ncRNAs from *Drosophila melanogaster*

The fruit fly *Drosophila melanogaster* is perhaps the oldest model organism used for genetic and developmental studies. Similarly to *C. elegans*, it contains all three major types of small non-coding RNAs involved in epigenetic regulation including miRNAs, siRNAs, and piRNAs. miRNAs and siRNAs in Drosophila are produced by two specific Dicers: DCR1 for miRNA and DCR2 for siRNA production.

Exo-siRNAs

It was initially believed that siRNAs were produced from exogenous sources such as viral genomes and their amplification intermediates, exogenous double-stranded RNAs from bacteria, RNAs produced from transgenes as well as from injected artificially produced dsRNAs. Such siRNAs were named **exogenous siRNAs** or **exo-siRNAs**. Specific features of siRNAs in Drosophila include a length of approximately 21 nt, methylation at their 3' ends, and association with AGO2. Dicers in Drosophila are specialized into DCR1, which is involved in processing of pre-miRNAs, and DCR2, which functions in the RNAi pathway that defends fly against viral infection. This specialization reduces competition between the involvements of Dicers in either pathway. No viral infection has been documented in *C. elegans*, a fact that correlates with the existence of only one Dicer in the species. In contrast to Drosophila, mammals such as humans respond to viral infection via a protein-based immune system and may not use RNAi for this purpose.

Long dsRNA precursors in flies are processed in the cytoplasm into short RNA duplexes with the help of DCR2 (see Figure 10-3). The thermodynamic stability that determines the identity of the guide and the passenger strands in this duplex is sensed by the dsRNA-binding protein R2D2, which contains two dsRNA-binding domains (R2) and is associated with DCR2 (D2). R2D2 is homologous to the *C. elegans* RNAi protein RDE-4. DCR2 and R2D2 together form the **RISC Loading Complex (RLC)**, which recruits Argonaute2 (Ago2) to the duplex. Ago2 cleaves the passenger strand, releasing it, which converts pre-RISC to RISC. The guide strand is then 2'-O-methylated at the 3' end by the S-adenosyl methionine-dependent methyltransferase, Hen1. This activated RISC is then used to cleave viral mRNAs (reviewed in Okamura and Lai, 2008).

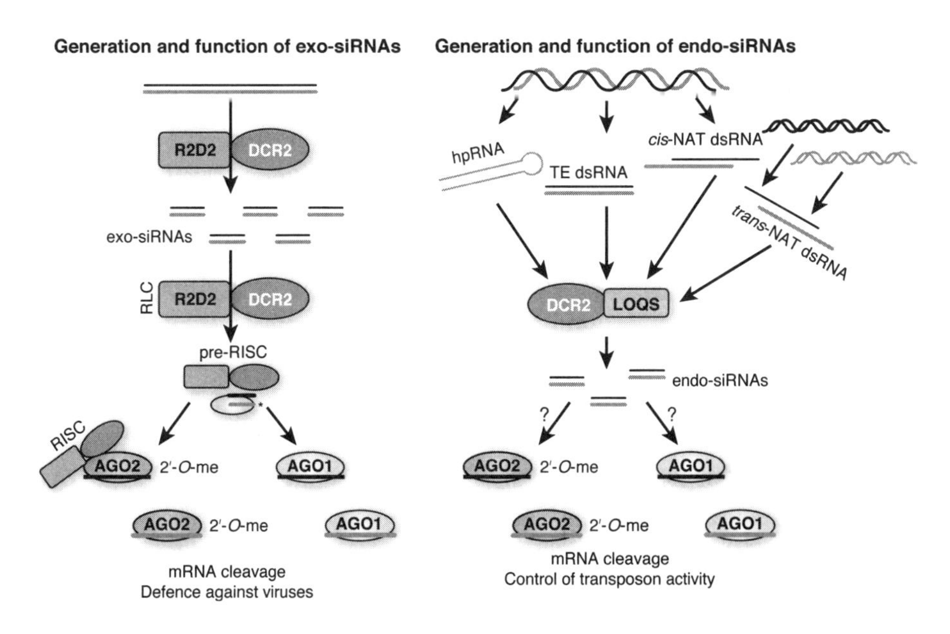

Figure 10-3 siRNA biogenesis pathways in Drosophila.

siRNAs in Drosophila are processed with the help of several sets of proteins, including RNase III enzymes Drosha, DCR1, and DCR2, **dsRNA-binding domain** (**dsRBD**) partners Pasha, Loquacious (LOQS), and R2D2 as well as Argonaute proteins AGO1 and AGO2. Exo-siRNAs are produced by DCR2/RDR2 pair (RLC) from long viral dsRNA or artificial dsRNA precursors. Binding of AGO2 to RLC forms pre-RISC which is converted to active RISC upon methylation of the 3' end of exo-siRNA. 2'-O-methyl group is added by Hen1. A relatively small fraction of exo-siRNA associates with AGO1. Endo-siRNAs are produced from several sources including hairpin RNA transcripts, dsRNAs from transposable elements (TEs), ***cis*-natural antisense transcripts (cis-NATs)**, and **trans-NATs**. Precursor dsRNAs are processed by DCR2/LOQS that pair into endo-siRNAs and are further loaded into AGO2. Similar to exo-siRNAs, a portion of endo-siRNAs is processed by AGO1.

Endo-siRNAs

Endogenous siRNAs, or **endo-siRNAs**, are produced from convergent transcripts and antisense pairs, including gene/pseudogene duplexes, repeat-associated transcripts from centromeric regions and transposons, hairpin RNAs, and **tasiRNAs** (Carthew and Sontheimer, 2009) (see Figure 10-4).

Deep sequencing of small RNAs from germ-line and somatic tissues of Drosophila and of Ago2 immunoprecipitates revealed that they are a rich source of various endogenous siRNAs. These small RNAs have features that are specific for Drosophila and mice; they are nearly always 21 nt in length; they contain equal pools of sense and antisense orientations; their 3' ends are methylated; and their 5' ends, unlike miRNAs and piRNAs, are not biased toward uracils. These features suggest that endo-siRNAs are processed from long dsRNA precursors.

Production of the 21 nt-long endo-siRNAs in Drosophila depends on DCR2, with only a small fraction of endo-siRNAs present in *dcr-2* mutants. dsRNA precursors are generated from transposon areas, repetitive sequences from heterochromatic and intergenic regions, complex long RNA transcripts, regions containing pseudogenes and processed pseudogenes, and even regular mRNA transcripts (see Figure 10-4). One of the roles of siRNAs is the silencing of transposons as demonstrated by the fact that both *dcr-2* and *ago2* mutants exhibit increased expression of transposon mRNAs.

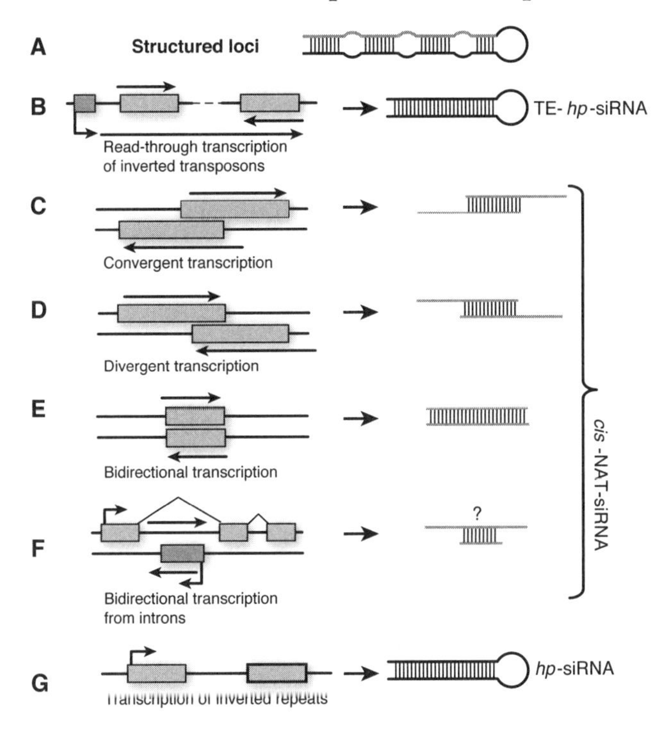

Figure 10-4 Various sources of endo-siRNA in Drosophila and mammals. A. Loci with a complex structure are the source of dsRNA precursors. B. Occasional read-through transcriptions from upstream promoters across the inverted transposons might generate dsRNAs with short-hairpins, so-called TE-hp-siRNAs (TE-siRNAs or hp-siRNAs). C. Convergent transcription from closely positioned loci with homologous sequences might result in transcripts with different length of overlap at the 5' region, leading to production of cis-NAT-siRNAs. D. Divergent transcription from closely positioned loci with homologous sequences might result in transcripts with different length of overlap at the 3' region, leading to production of cis-NAT-siRNAs. E. Bidirectional transcription of loci positioned in between two oppositely oriented promoters results in perfect long dsRNA precursor, leading to production of cis-NAT-siRNAs. F. Imperfect dsRNA precursors can be generated between the gene and pseudogene transcripts, leading to production of trans-NAT-siRNAs. G. Loci with homologous sequences located in inverted orientation serve as a template for generation of long "fold-back" dsRNAs, leading to production of hp-siRNAs.

Because endo-siRNA's production does not include the RdRP-dependent amplification step for their biogenesis, it is correct to assume that precursors of endo-siRNAs should contain long stretches of dsRNA. Indeed, siRNAs originating from mRNAs are more than 90% generated from regions with overlapping, convergent transcripts.

A small fraction of endo-siRNAs in Drosophila are generated from so called "structured loci." RNA produced from these complex loci fold into long intramolecularly paired hairpins (refer to Figure 10-4). Transposons, due to their abundance and frequent inverted orientation, are an excellent source of endo-siRNAs. Recently discovered piRNAs might be ideal candidates for the silencing of transposons in germ cells. It is possible that role of endo-siRNAs derived from transposons is to target them in somatic cells; however, it can be excluded that certain types of transposons in germline are preferentially targeted by endo-siRNAs, piRNAs, or combinations of both. In Drosophila, precipitation and deep sequencing of AGO2-associated small RNAs shows that **TE-siRNAs** of 21 nt represent a high portion of all siRNAs. Similarly, deep sequencing of various cell cultures show the presence of large number of TE-siRNAs, which correlates with significant amplification of specific LTR retrotransposons in these cells.

A large set of endo-siRNAs are produced through pairing between naturally occurring antisense transcripts, so-called *cis*-NAT. Location of these endogenes, transgenes, or transposons allows generation of long *cis*-NAT precursors through convergent, divergent, or bidirectional transcription or transcription from intronic regions (refer to Figure 10-4). Almost all Drosophila *cis*-NAT-siRNAs are generated through divergent transcription and thus are produced from the overlaps between 3' UTRs. However several highly expressed siRNAs are produced either from the overlap between 5' UTRs or from annotated introns. *cis*-NAT-siRNAs stemming from Drosophila klarsicht (klar) gene and the thick veins (tkv) gene involve overlaps with 5' exons, internal transcript exons, and/or annotated intronic regions (reviewed in Okamura and Lai, 2008). Deep sequencing of the siRNA pool showed strong bias to specific types of *cis*-NAT-siRNA with strong preference for binding to RNA that codes for nucleic acid-interacting proteins such as those functioning as transcription cofactors, DNases and RNases. At the same time, it was noted that approximately 75% of *cis*-NATs from Drosophila S2 cells did not generate *cis*-NAT-siRNAs. If only a portion of all co-expressed *cis*-NAT pairs are selected for siRNA production, is there active selection for entry of these *cis*-NATs into the RNAi pathway? Does this selection depend on possible functionality of selected *cis*-NAT-siRNAs? This is quite possible, especially given that one of the most highly expressed *cis*-NAT-siRNA loci contains the CG7739/Ago2 gene pair, suggesting that the expression of AGO2 itself can be regulated by its own siRNAs (Okamura et al., 2008a).

Transcription from inverted repeats results in generation of **hairpin RNAs (hpRNAs)**. Despite some similarity between miRNA and hpRNAs, the two are produced via different pathways; whereas miRNAs are processed by DCR1 (see later in this chapter), hpRNAs are processed by DCR2. Some of the hpRNA-producing loci in Drosophila have stems of up to 400 base pairs long. Some of the loci contain single tandem repeats, whereas others, such hp-CG4068 locus, encodes up to 20 tandem hairpins (Okamura et al., 2008b). One of the siRNAs produced from hp-CG4068 locus targets mutagen sensitive-308 (mus308), a DNA damage response polymerase. Another hpRNA, hp-CG18854, was shown to target and repress the expression of chromodomain protein CG8289.

Biogenesis of endo-siRNAs in Drosophila requires DCR2 and the dsRNA-binding protein Loquacious (LOQS). LOQS is essential for the accumulation of many endo-siRNAs, including TE-siRNAs, *cis*-NAT-siRNAs and hpRNA-siRNAs. LOQS is known to associate with DCR1 for miRNA production but earlier finding demonstrate that the protein plays a crucial role in the production of siRNAs from RNAi transgenes producing long, intramolecularly paired inverted repeat transcripts (Fürstemann et al., 2005). Proteomic analyses of DCR2 complexes suggest that the protein equally frequently interacts with LOQS and R2D2 proteins (Czech et al., 2008).

piRNA in Drosophila

PIWI protein was originally identified in Drosophila while the results of P-element enhancer trap screen were being analyzed. Insertion of P-element in the germline cells of Drosophila triggers severe defects in spermatogenesis, resulting in sterility in male flies (Lin et al.,

1997). Subsequently, it was discovered that Drosophila has three PIWI proteins: PIWI, AUBERGINE (Aub), and ARGONAUTE3 (AGO3). Male and female PIWI mutants are deficient in oogenesis and spermatogenesis; they are unable to maintain germline stem cell asymmetric division resulting in the loss of germline stem cells (Cox et al., 2000). Thus, it was suggested that PIWI protein is involved in the asymmetric division of germline stem cells (Cox et al., 2000). PIWI's function as a regulator of heterochromatic gene silencing depends on its interaction with heterochromatin protein 1a and regulation of methylation of specific regions of DNA, primarily including transposon regions (Brower-Toland et al., 2007). In the process of oogenesis, cytoplasmic fraction of PIWI interacts with DCR1 and dFMRP, and this interaction is important in the formation of the pole cells of the embryo, which are precursors of germline cells (Megosh et al., 2006).

Aub protein was originally identified because it was required for specification of the embryonic axes. In the male germ line, Aub is required for the silencing of the repetitive *Stellate* locus, which would otherwise cause male sterility. It was found that Aub requirement for suppression of *Stellate* is through processing of repetitive Suppressor of *Stellate* locus, the source of antisense piRNAs (Aravin et al., 2001). Deficiency in Aub results in the loss of anterior-posterior and dorsal-ventral patterning in embryos. This developmental defect appears to be the consequence of the double-stranded DNA breaks that occur in the oocyte in Aub absence.

Despite their diversity, piRNAs, with more than 1.5 mln of unique sequences in Drosophila, map to a few hundred genomic clusters (reviewed in Ghildiyal and Zamore, 2009). In Drosophila, the best studied cluster is the *flamenco* locus that codes for piRNAs targeting a repressor of the *gypsy*, *ZAM*, and *Idefix* transposons. In contrast to siRNAs, *flamenco* piRNAs are mainly antisense, suggesting that piRNAs might arise from long, single-stranded precursor RNAs (reviewed in Ghildiyal and Zamore, 2009). Curiously, when a transposon inserts near the 5' end of the *flamenco* locus, the production of piRNAs is blocked at the distance of up to 168 kbp downstream. Thus, piRNAs from *flamenco* locus are generated from an enormously long, single-stranded RNA transcript (Brennecke et al., 2007).

Are piRNAs also generated or at least function in the somatic cells of Drosophila? This is still debatable, but silencing of white, a gene required to produce red eye pigment, for example, depends on PIWI and Aub. The effect of piRNAs on somatic cells may be indirect through long-lived chromatin marks left by piRNAs in germ lines. Although endo-siRNAs and piRNAs are involved in transposon repression in the germline, endo-siRNA-mediated repression of transposons is vastly predominant in somatic cells. Indeed, loss in *dcr-2* and *ago2* mutants results in the increase in the expression of transposons (Chung et al., 2008). At the same time, somatic piRNA like small RNAs were found in *ago2* mutant flies. It has been suggested that, in the absence of endo-siRNAs, piRNAs might take the job of controlling transposon surveillance in somatic cells.

miRNAs in Drosophila

In Drosophila, miRNAs are approximately 22 nt long, have free hydroxyl groups at their 3' ends, and associate primarily with the Argonaute protein AGO1. miRNAs biogenesis in Drosophila occurs in a similar manner to those in mice and human as described in Chapter 7, "Non-Coding RNAs Involved in Epigenetic Processes: A General Overview." In short, miRNAs are processed by Drosha/Pasha pair in the nucleus from **primary transcript miRNA (pri-miRNA)** precursors generated from endogenous inverted repeats of less than 100 nt long. Generated molecules, named **precursor-miRNAs (pre-miRNAs)**, are transported into cytoplasm where they are further processed by DCR1 and its dsRBD partner LOQS into miRNAs (see Figure 10-5) (reviewed in Okamura and Lai, 2008). The 3' ends of these miRNAs are methylated by DmHen1 methyltransferase and then preferentially loaded into AGO1. Some miRNAs are also shown to associate with AGO2. AGO1/miRNA regulate mRNA expression via cleavage, translational inhibition, and deadenylation.

Despite LOQS being suggested as a core component of the miRNA pathway, loqs-null mutants are only mildly defective in production of mature miRNAs (Liu et al., 2007). DCR1 is capable of processing pre-miRNAs without LOQS, although the efficiency of this process is much lower.

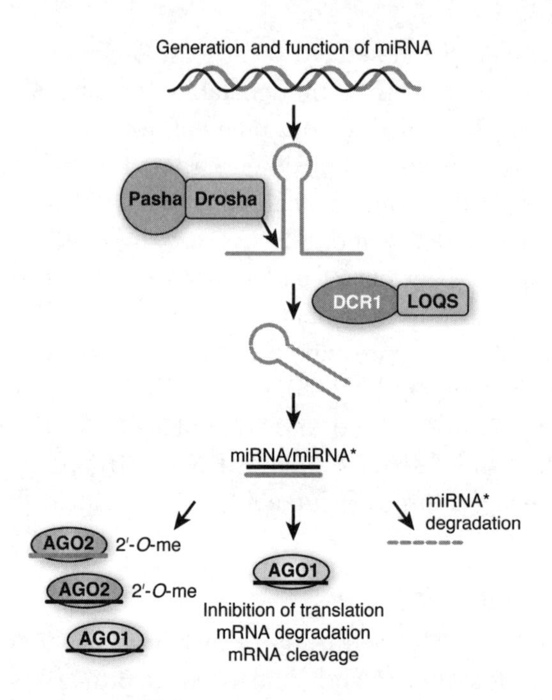

Figure 10-5 miRNA biogenesis in Drosophila. Endogenous transcripts with short inverted repeats are processed into pri-miRNA hairpins. Pri-miRNAs are processed by Drosha/Pasha into pre-miRNAs, which are sent to the cytoplasm. Pre-miRNAs in the cytoplasm are processed by DCR1/LOQS into miRNA that are methylated by DmHen1 and then loaded into AGO1 (mainly) or AGO2 (less frequently).

Activation of translation

Although small ncRNAs are largely involved in downregulation of expression and inhibition of translation, there is evidence that they are involved in activation of translation. In Drosophila, both Ago1 and Ago2 regulate translation repression. Ago1 promotes deadenylation of mRNAs, whereas recent work from Iwasaki and Tomari (2009) showed Ago2, but not Ago1, can activate translation of the target mRNAs when they lack the poly(A) tail. Therefore, activation of translation might be dependent on the length of the poly(A) tail. Being part of RISC, Ago2 represses translation of poly(A)+ target mRNAs in a cap-dependent manner, but does not induce deadenylation of the target mRNAs. The experiments showed that Ago2 stimulates translation of the poly(A)- target mRNAs. What is the mechanism of such activation? Iwasaki and Tomari (2009) suggest that Ago2-RISC binding of eIF4E can form a “closed-loop” of target mRNA even without a poly(A) tail, which might be beneficial for translation. At the same time, because translation activation was observed in both G-capped and A-capped target mRNA, to which eIF4E has a very low affinity, it is not clear how Ago2 can activate translation.

Mammals/human

Mammals contain most of the abovementioned classes of small RNAs, and their mechanisms of biogenesis are largely similar too. Some of the details of the biogenesis of miRNAs, siRNAs and piRNAs as well as their mode of action are introduced in more details under the next several subheadings.

miRNAs

There are currently more than 540 mature human miRNAs listed in the official miRNA database (miRBase) representing more than 1.0% of all the genes in the human genome and potentially targeting up to one-third of the human coding genes (Griffiths-Jones et al., 2008). miRNAs typically derive from three types of loci with annotated transcripts: the introns of protein coding genes, the exons of non-coding genes, and the introns of non-coding genes. Alternatively, these genes might derive from intergenic regions. miRNAs belonging to the class that reside within annotated transcripts are under the direct transcriptional control of their host gene. Intergenic miRNA genes, however, are under their own transcriptional control, and in most cases are thought to be transcribed using RNA polymerase II. One exception was subsequently found as a cluster situated on chromosome 19 with numerous surrounding Alu elements is transcribed by RNA polymerase III. Some of these intergenic miRNAs have shown to be regulated by some very important transcription factors such as cNF-κB, c-Myc, and p53.

Interestingly, 36% of the miRNAs within the human genome are found in clusters less than or equal to 10 kb apart (Griffiths-Jones et al., 2008), and many of these are only 100 to 1000 kb from each other. This resulted in the discovery that many miRNAs are transcribed together as a single transcriptional unit, or polycistron (Griffiths-Jones et al., 2008). The function of multi-miRNA polycistrons is thought to be efficient targeting of a single mRNA transcript, or targeting of multiple transcripts in a signal molecular pathway.

Both miRNAs and siRNAs in mammals are part of RISC-directed regulation of gene expression. miRNA biogenesis in mammals is similar to the one functioning in *C. elegans* and Drosophila, although miRNAs are not methylated in mammals in contrast to Drosophila.

The mechanism by which miRNAs regulate their target mRNAs is dependent on two factors: the type of Argonaute protein they associate with and the extent of complementarity between the miRNA and their targets. Endonucleolytic cleavage that relies on perfect or nearly perfect complementarity between miRNA and mRNA target is rare in animals, although it is common for plants. The more common mechanism represents pairing of partially complementary miRNAs with the region of complementarity being a sequence at the 5' end of the miRNA—the so-called "seed" sequence. Binding of most miRNAs in animals occurs with mismatches and bulges. This interaction results in translational inhibition of mRNA, followed by direct degradation of their mRNA targets or translocation of these mRNAs into P bodies. The seed region defines the specificity for target selection. Because the size of the seed sequence is rather small (less than 10 nt, typically 7 to 8 nt), a single miRNA can regulate the activity of many different genes (Baek et al., 2008). The miRNAs in this case work as adaptors for miRISC to specifically recognize and regulate particular mRNAs. In most of the cases, miRNA-binding sites in mammalian mRNAs are in the 3' UTRs. Often, 3' UTRs contain binding sites multiple miRNAs.

Translational inhibition might occur at different stages of translation, either initiation or elongation. In the process of translation, initiation begins with the recognition of the mRNA 5' cap nucleotide by eIF4E, a subunit of the eIF4F complex that also contains eIF4A and eIF4G. Interaction of the eIF4G protein with another initiation factor, eIF3, recruits the 40S small ribosomal subunit to the 5' end (reviewed in Carthew and Sontheimer, 2009). The small ribosomal subunit is joined by the large ribosomal subunit 60S to begin elongation from AUG codon. Interaction between 5' and 3' ends occurs through eIF4G interaction with the polyA-binding protein PABP1. This interaction circularizes the mRNA molecule, enhancing translation efficiency. Initiation factors might not always be necessary, as translation of some viral mRNAs starts at **internal ribosome entry site** (**IRES**). How do miRNAs regulate the process of translation?

According to Petersen et al. (2006), miRISC represses elongation by promoting premature ribosome dissociation from mRNAs. Three not necessarily mutually exclusive models for how miRISC represses initiation were recently reviewed by Carthew and Sontheimer (2009) (see Figure 10-6).

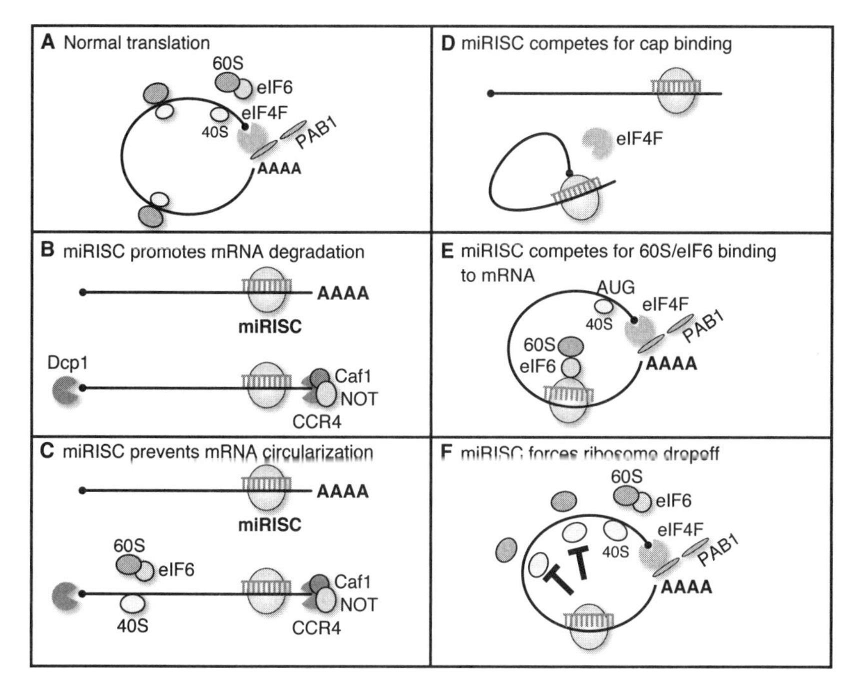

Figure 10-6 Potential mechanisms of miRNA-mediated repression of translation (reproduced with permission from Carthew RW, Sontheimer EJ. (2009) Origins and Mechanisms of miRNAs and siRNAs. Cell 136:642-55). Simplified model of translation. Unrepressed mRNAs recruit transcription factors and ribosomal subunits resulting in circularization of mRNA and enhancement of translation. miRNAs can initiate mRNA degradation by inducing deadenylation and decapping of mRNAs. miRNAs induce deadenylation of the mRNA, preventing circularization of the mRNA and deposition of ribosomal subunits. miRNAs repress initiation of translation by blocking cap recognition. miRNAs repress initiation of translation by preventing binding of 60S subunit. miRNAs prevent proper elongation steps by inducing premature drop off of ribosomes.

One model suggests that miRISC and eIF4E (part of eIF4F complex) compete for binding to the mRNA 5' cap structure (refer to Figure 10-6). This model is supported by the experiments showing that addition of purified eIF4F complex containing eIF4E into cell-free translational system inhibits miRISC binding to the cap (Mathonnet et al., 2007). Inversely, miRISC was shown to inhibit loading of the small ribosomal subunit onto mRNA (Thermann and Hentze, 2007). The exact mechanism of competition between miRISC and eIF4F complex is not clear, but it is suggested that the Mid domain of human Ago2 resembles eIF4E. Indeed, mutations replacing two phenylalanines in the Mid sequence impair the ability of Ago2 to repress translation and bind guanine cap *in vitro*. This is debatable because mutations of phenylalanines in Drosophila Ago1 protein does not inhibit binding guanine cap. Also, repression of translation was shown with only GW182 bound to mRNAs, suggesting that GW182 could be the eIF4E competitor.

A **second model** suggests that translational inhibition depends on direct deadenylation of mRNA by miRISC (refer to Figure 10-6). As a consequence, deadenylated miRNAs are not circularized, thus inhibiting translation. Indeed many repressed mRNAs were shown to be deadenylated *in vivo* and *in vitro*. In this model, too, the GW182 is suggested to promote deadenylation. The evidence against this model include the following facts: miRNAs are able to inhibit translation of nonpolyadenylated mRNAs; Ago2 seems also to be required for the process; deficiency in deadenylase enzyme does not inhibit translation repression by GW182.

A **third model** suggests that miRISC prevents association between small and large ribosomal subunits (refer to Figure 10-6). This hypothesis is supported by the fact that human Ago2 *in vitro* can physically associate with eIF6 and large ribosomal subunit (Chendrimada et al., 2007). eIF6 regulates the biogenesis and maturation of large ribosomal subunits and prevents their premature association with small ribosomal subunits. When ribosomal subunits are mature and ready to assemble at AUG site, the eIF6 has to be removed from the translation machinery. miRISCs apparently recruits eIF6 to translational machinery and prevents the subunit assembly where miRNA provides the mRNA target specificity, whereas Ago2 acts as the binding partner to eIF6. Also, depletion of eIF6 in human cells rescues mRNAs from miRNA inhibition (reviewed by Carthew and Sontheimer, 2009). This model can be challenged by the fact that depletion of eIF6 in Drosophila cells has little or no effect on silencing.

The mammalian Dicer/Ago/miRNA complex is associated with GW182, Gemin3, Gemin4, Mov10, and Imp8. Functional analysis in humans indicates that GW182 protein is both necessary and sufficient for bound Ago to silence gene expression. Thus, miRNA/Ago/GW182 might represent a minimal miRISC complex.

The ability to target hundreds of mRNAs at a time makes miRNA-RISC a powerful mechanism for regulating developmental and metabolic processes in mammals. Mutations in Dicer or miRNA-associated Argonaute proteins are either lethal or result in severe developmental defects in animals. Dicer knock-out in mice is lethal at the stage of early embryo, whereas loss of Dicer in mouse embryonic fibroblasts leads to DNA damage, apoptosis, and premature senescence (reviewed in Ghildiyal and Zamore, 2009).

Details of the role of miRNAs in various physiological and pathological processes such as pluripotency, differentiation, tumorogenesis, apoptosis, genome stability, cell identity, cell memory, organism behavior, and many others is covered in Chapters 17 through 20.

siRNAs

The mechanisms of exo- and endo-siRNA generation in mammals are similar to those in Drosophila. Lack of RdRP proteins in flies and mammals suggests that endo-siRNAs are produced from dsRNA precursors. The first mammalian endo-siRNAs was found to have a sequence homologous to LINE-1 (L1) retrotransposon in cultured human cells (Yang and Kazazian, 2006). Analysis of genomic structure of L1 showed that 5' UTR contains both sense and antisense promoters and thus may drive bidirectional transcription of L1, producing overlapping, complementary transcripts that can form dsRNA. Dicer then processes these dsRNAs into siRNAs.

Mammals produce a substantial number of *trans*-NAT-siRNAs and hpRNA siRNAs. High frequency of occurrence of pseudogenes and processed pseudogenes allows the production of *trans*-NAT-siRNA precursors with imperfect intramolecular complementarity (see Figure 10-7A). Unequal crossing-overs might also result in duplication and inversion of pseudogenes, forming long hairpin structures upon transcription (see Figure 10-7B). Indeed, a number of hpRNA-producing loci were identified in mice.

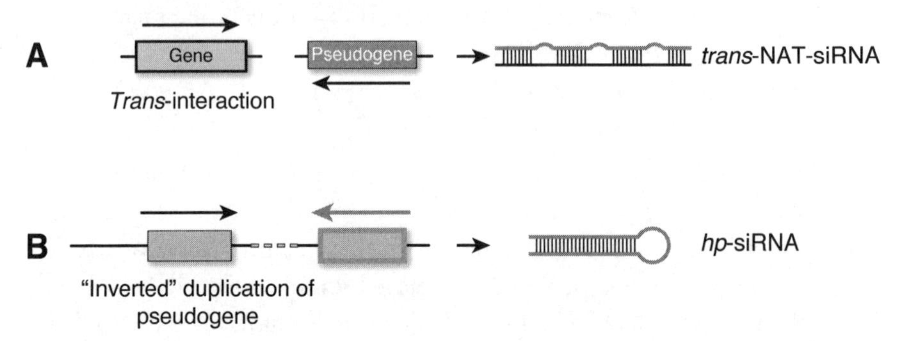

Figure 10-7 Additional endo-siRNA production mechanisms in mammals. A. Imperfect dsRNA precursors can be generated between the gene and pseudogene transcripts, leading to production of trans-NAT-siRNAs. B. Common duplication of pseudogenes followed by inversion might also serve as a template for generation of long "fold-back" dsRNAs, leading to the production of hp-siRNAs.

In humans, the mechanism of RISC assembly depends on the presence of three proteins: Dicer (a single protein in mammals), TRBP (the human immunodeficiency virus transactivating response RNA-binding protein), and Ago2. These proteins are capable of associating with each other in the absence of the dsRNA trigger. Their main function is to bind dsRNA, slice it into siRNA molecules, load these siRNAs into Ago2, and degrade the passenger strands. A number of other proteins interact with RISC but they are not necessary for RISC's functionality. Although Dicer is involved in processing endo-siRNAs, it is apparently not required for the assembly of siRNAs into functional RISCs in mice.

piRNAs—ncRNAs functioning in the germline

piRNAs are a specific class of non-coding RNAs that bind to the PIWI subfamily of Argonaute proteins. The PIWI clade is a group of animal-specific Argonaute proteins that include PIWI, Aubergine (Aub), and Ago3 in Drosophila; MILI, MIWI, and MIWI2 in mice; and HILI, HIWI1, HIWI2, and HIWI3 in humans. The main functions of piRNAs include the regulation of genome stability via repression of transposon elements and silencing of repetitive DNA sequences. The former function is regulated by shorter piRNAs of approximately 24 to 27 nt, whereas the latter is triggered by longer piRNAs of approximately 27 to 31 nt. Specific features of piRNA biogenesis that allowed them to be placed into a separate group were that they bind PIWI group of Argonautes; they do not require Dcr-1 or Dcr-2 for their production, unlike miRNAs and siRNAs; and they undergo 2'-O-methylation of the 3' like siRNAs in flies and miRNAs in plants (reviewed in Ghildiyal and Zamore, 2009).

According to one of the classifications, mammalian piRNAs are broadly divided into pre-pachytene and pachytene piRNAs, depending on the time during spermatocytes meiosis at which they are expressed. Pre-pachytene piRNAs have sequence homology with repetitive elements and silence transposons, such as IAP, LINE, SINE, and LTR. The silencing involves sequence-specific methylation of transposons early during gamete development. In male mice, methylation patterns in gametes are established during the cell cycle arrest occurring approximately between day 14.5 postcoitum and day 2 to 3 after birth. Expression of MILI and MIWI2 follows the same pattern of expression, suggesting that pre-pachytene piRNAs interacting with MILI and MIWI2 guide transposon methylation. Indeed, mice deficient in either protein are unable to properly methylate transposons. In contrast to pre-pachytene piRNAs, the pachytene piRNAs mainly arise from unannotated regions of the genome, not transposons, and their function remains unknown (reviewed in Samji, 2009).

MILI/MIWI2 and MIWI expression differs slightly however, with MILI/MIWI2 expressed at the first stage of spermatogenesis, peaking at day 9 (leptotene stage), whereas MIWI expression is highest at the second stage, lasting from pachytene until spermatide stage, reaching its peak at approximately day 17 of spermatogenesis. The expression pattern of these two protein groups follows the pattern of expression of two groups of piRNAs, those with a size of 24 to 28 nt are produced predominantly during the pre-pachytene stage, whereas those of 29 to 31 nt are produced during the pachytene and postpachytene stage. This suggests that MILI/MIWI2 proteins associate with the shorter size group, whereas MIWI associates with the larger size.

MIWI2 cellular localization in mice seems to be dependent on MILI function. MILI knockout results in re-location of MIWI2 to the cytoplasm, whereas in *miwi2* mutant, MILI is still localized in the cytoplasm (Aravin et al., 2008). Some peculiarities of PIWI function in mammals include MILI's role in translational upregulation of some mRNAs. Also, mutations of PIWI proteins in mice reduce the number of male gonads, whereas similar mutations in fly affect both gametes (reviewed in Samji, 2009).

Another difference in pre-pachytene and pachytene piRNAs includes the frequency of occurrence of uridine at their 5' end. Whereas the uridine at 5' end occurs in 81% to 87% in pre-pachytene piRNAs targeting transposons and even as rarely as 29% in those that target rRNA loci, it is found in 94% of cases at 5' end of pachytene piRNAs. Table 10-1 shows the type of loci with sequence homology to piRNAs in mammals.

Table 10-1 Comparison of potential targets of pre-pachytene and pachytene piRNAs

	Pre-pachytene		**Pachytene**	
	Occurrence, %	**5'U**	**Occurrence, %**	**5'U**
Unannotated	28%	87%	74%	94%
Transposons	35%	81%	17%	94%
LTR	33.8%			
LINE	15.8%			
SINE	49%			
others	1.4%			
gene	29%	84%	4%	94%
rRNAs	5%	29%		
other	3%		5%	94%

Data are composed from Aravin et al. (2006, 2007).

Other types of ncRNAs involved in epigenetic regulation in human

Xist and Tsix RNAs are among the most-studied long ncRNAs as they mediate X chromosome inactivation for dosage compensation in mammals. Xist RNAs are transcribed from the X chromosome that will be inactivated, and they stay closely attached. This interaction results in recruitment of chromatin-remodeling factors and histone modifying enzymes leading to H3K27 and K9 hypermethylation. In addition, Xist and Tsix RNA form dsRNA complexes that become targets to the formation of small RNAs between 25 and 42 nucleotides length that are present during X chromosome inactivation. This process seems to be important particularly on the active X chromosome, where it limits accumulation of Xist RNA (Ogawa et al., 2008).

HOTAIR RNA (Hox antisense intergenic RNA) is transcribed from human Hox loci. Orchestrated activation of Hox gene expression is fundamental to normal development. Rinn et al. (2007) have discovered numerous Hox ncRNAs and characterized one of them: HOTAIR RNA. This ncRNA is transcribed from the HOXC locus and mediates repression of the 40 kb distant HOXD locus by recruiting the Polycomb repressive complex 2 (PRC2), resulting in trimethylation of H3K27 and transcriptional silencing. A recent study showed that the 5' domain of HOTAIR interacts with PRC2, whereas the 3' domain interacts with another histone modifying enzyme LSD1 (Tsai et al., 2010). Thus, the authors of this work suggest that HOTAIR might mediate its function by building a scaffold for histone-modifying enzymes.

Interestingly, Xist RNA also seems to recruit PRC2 to its binding sites. Therefore, this might potentially be one common mechanism of long ncRNA function.

HSR1 RNA (Heat-shock RNA1) is a constitutively expressed non-coding RNA in human and rodent cells. HSR1 acts in cooperation with eEF1A (a translation elongation factor) in a ribonucleoprotein complex to activate HSF1 (Heat-shock transcription factor 1) during the heat-shock response (Shamovsky et al., 2006). This activation is mediated through formation of a complex between the ribonucleoprotein complex and HSF1. HSR1 can be negatively regulated by siRNAs, which renders cells heat-insensitive (Shamovsky et al., 2006). This RNA seems to play a structural role that affects transcription factor activity and thus transcription profile.

miRNAs in animal development

As in plants, small regulatory RNAs play essential roles in normal animal development. As would be expected, mutations to Dicer, Ago2, and Drosha are embryo lethal in mice. It has been hypothesized that one function of miRNA-mediated gene control is bestowing robustness to developmental programs, allowing for the control of such factors as leaky transcription and maintaining optimal mRNA expression levels (Stefani and Slack, 2008).

Unfortunately, most work on the developmental role of miRNAs in animals has been done in invertebrates and lower vertebrates, but work in mammals is beginning to find its way into the literature. For example, the brain-specific miR-124 in mice has been found to play a critical role in neuronal differentiation, downregulating hundreds of genes to help determine neuronal-specific gene pattern in mice. In addition, several miRNAs are involved in correct muscle and heart development, most notably miR-1, whereas others have been found to be involved in such responses as cardiac hypertrophy (Stefani and Slack, 2008). Lastly, various stages of T-lymphocyte development are characterized by a distinct miRNA expression pattern, whereas other sets mediate myeloid lineage development and macrophage function. These are but a few well-studied mechanisms in miRNA and mammalian development, but to cite further examples is far beyond the scope of this chapter.

Conclusion

Animals have three groups of small RNAs—miRNAs, siRNAs, and piRNAs (21U RNAs in *C. elegans*)—with clearly defined functions. Whereas miRNAs are primarily involved in translational inhibition, siRNAs are targeting mRNAs for degradation, and piRNAs are involved in multiple processes, including targeting transposons RNAs and promoting methylation of transposon DNA. Despite the minute differences in the mechanism of action of these groups of small RNAs, their biogenesis and their function is largely comparable among different animals.

Exercises and discussion topics

1. What are the three main categories of non-coding RNAs in animals, and which two are the most prevalent?
2. What are the three specific characteristics of ncRNA biogenesis in *C. elegans*?

3. How was sequence redundancy of miRNAs in *C. elegans* determined? Why would it be possible for sequence redundancy not to be transformed into functional redundancy?
4. Describe transitive RNAi in *C. elegans*.
5. Which Dicers are present in Drosophila and what are their specific roles?
6. What is the difference between exo-siRNAs and endo-siRNAs?
7. Which features suggest that endo-siRNAs are processed from long dsRNA precursors?
8. How are hairpin RNAs produced in Drosophila?
9. Besides for Aub, what are the other two PIWI proteins in Drosophila? What is the role of the Aub protein?
10. Do piRNAs appear to be generated or at least functional in the somatic cells of Drosophila? Why or why not?
11. What evidence is there that small ncRNAs are involved in activation of translation?
12. miRNAs typically derive from what three types of loci with annotated transcripts?
13. What are miRNA polycistrons and what is their function?
14. What are the three models that exist for how miRISC represses initiation? Provide support for each one. Are they mutually exclusive?
15. In humans, what are the three proteins on which the mechanism of RISC assembly relies? What do they do?
16. What are the main functions of piRNAs in mammals?
17. What are three differences between pre-pachytene and pachytene piRNAs?
18. What are the roles of Xist and Tsix RNAs?
19. What is the role of HOTAIR RNA?
20. What is the role of HSR1 RNA?

References

Aravin et al. (2006) A novel class of small RNAs bind to MILI protein in mouse testes. *Nature* 442:203-207.

Aravin et al. (2001) Double-stranded RNA-mediated silencing of genomic tandem repeats transposable elements in the D. melanogaster germline. *Curr Biol* 11:1017-1027.

Aravin et al. (2008) ApiRNA pathway primed by individual transposons is linked to de novo DNAmethylation in mice. *Mol Cell* 31:785-799.

Aravin et al. (2007) Developmentally regulated piRNA clusters implicate MILI in transposon control. *Science* 316:744-747.

Baek et al. (2008) The impact of microRNAs on protein output. *Nature* 455:64-71.

Brennecke et al. (2007) Discrete small RNA-generating loci as master regulators of transposons activity in Drosophila. *Cell* 128:1089-1103.

Brower-Toland et al. (2007) Drosophila PIWI associates with chromatin and interacts directly with HP1a. *Genes Dev* 21:2300-2311.

Carthew RW, Sontheimer EJ. (2009) Origins and Mechanisms of miRNAs and siRNAs. *Cell* 136:642-655.

Chendrimada et al. (2007) MicroRNA silencing through RISC recruitment of eIF6. *Nature* 447:823-828.

Chung et al. (2008) Endogenous RNA Interference Provides a Somatic Defense against Drosophila Transposons. *Curr Biol* 18:795-802.

Cox et al. (2000) Piwi encodes a nucleoplasmic factor whose activity modulates the number and division rate of germline stem cells. *Development* 127:503-514.

Czech et al. (2008) An endogenous siRNA pathway in Drosophila. *Nature* 453:798–802.

Das et al. (2008) Piwi and piRNAs act upstream of an endogenous siRNA pathway to suppress Tc3 transposon mobility in the Caenorhabditis elegans germline. *Mol Cell* 31:79-90.

Fire et al. (1998) Potent and specific genetic interference by double-stranded RNA in Caenorhabditis elegans. *Nature* 391:806-811.

Fürstemann et al. (2005) Normal microRNA maturation and germ-line stem cell maintenance requires Loquacious, a double-stranded RNA-binding domain protein. *PLoS Biol* 3:e236.

Ghildiyal M, Zamore PD. (2009) Small silencing RNAs: an expanding universe. *Nat Rev Genet* 10:94-108.

Griffiths-Jones et al. (2008) miRBase: tools for microRNA genomics. *Nucleic Acids Res* 36 (Database issue):D154-158.

Hutvágner et al. (2004) Sequence-specific inhibition of small RNA function. *PLoS Biol* 2:e98.

Ibáñez-Ventoso C, Driscoll M. (2009) MicroRNAs in C. elegans Aging: Molecular Insurance for Robustness? *Current Genomics* 10:144-153.

Iwasaki S, Tomari Y. (2009) Argonaute-mediated translational repression (and activation). *Fly (Austin)* 3:204-206.

Kato et al. (2009) Dynamic expression of small non-coding RNAs, including novel microRNAs and piRNAs/21U-RNAs, during Caenorhabditis elegans development. *Genome Biol* 10:R54.

Lee et al. (1993) The C. elegans heterochronic gene lin-4 encodes small RNAs with antisense complementarity to lin-14. *Cell* 75:843-854.

Lin H, Spradling AC. (1997) A novel group of pumiliomutations affects the asymmetric division of germline stem cells in the Drosophila ovary. *Development* 124:2463-2476.

Liu et al. (2007) Dicer-1, but not Loquacious, is critical for assembly of miRNA-induced silencing complexes. *RNA* 13:2324-2329.

Mathonnet et al. (2007) MicroRNA inhibition of translation initiation in vitro by targeting the cap-binding complex eIF4F. *Science* 317:1764-1767.

Megosh et al. (2006) The role of PIWI and the miRNA machinery in Drosophila germline determination. *Curr Biol* 16:1884-1894.

Mello CC, Conte D Jr. (2004) Revealing the world of RNA interference. *Nature* 431:338-42.

Miska et al. (2007) Most *Caenorhabditis elegans* microRNAs are individually not essential for development or viability. *PLoS Genet* 3:e215.

Ogawa et al. (2008) Intersection of the RNA interference and X-inactivation pathways. *Science* 320:1336-1341.

Okamura et al. (2008a) Two distinct mechanisms generate endogenous siRNAs from bidirectional transcription in Drosophila. *Nature Struct Mol Biol* 15:581-590.

Okamura et al. (2008b) The Drosophila hairpin RNA pathway generates endogenous short interfering RNAs. *Nature* 453:803-806.

Okamura K, Lai EC. (2008) Endogenous small interfering RNAs in animals. *Nat Rev Mol Cell Biol* 9:673-678.

Pak J, Fire A. (2007) Distinct populations of primary and secondary effectors during RNAi in C. elegans. *Science* 315:241-244.

Rinn et al. (2007) Functional demarcation of active and silent chromatin domains in human HOX loci by noncoding RNAs. *Cell* 129: 1311–23.

Ruby et al. (2006) Large-scale sequencing reveals 21U-RNAs and additional microRNAs and endogenous siRNAs in C. elegans. *Cell* 127:1193-207.

Samji T. (2009) PIWI, piRNAs, and Germline Stem Cells: What's the link? *Yale J Biol Med* 82:121-124.

Shamovsky et al. (2006) RNA-mediated response to heat shock in mammalian cells. *Nature* 440:556-560.

Stefani G, Slack FJ. (2008) Small non-coding RNAs in animal development. *Nat Rev Mol Cell Biol* 9:219-230.

Thermann R, Hentze MW. (2007) Drosophila miR2 induces pseudo-polysomes and inhibits translation initiation. *Nature* 447:875-878.

Tsai et al. (2010) Long noncoding RNA as modular scaffold of histone modification complexes. *Science* 329:689-693.

Wang G, Reinke V. (2008) A *C. elegans* Piwi, PRG-1, Regulates 21U-RNAs during Spermatogenesis. *Curr Biol* 18:861-867.

Yang N, Kazazian HHJ. (2006) L1 retrotransposition is suppressed by endogenously encoded small interfering RNAs in human cultured cells. *Nat Struct Mol Biol* 13:763-771.

Zheng et al. (2010) Inhibiting miRNA in *Caenorhabditis elegans* using a potent and selective antisense reagent. *Silence* 1:9.

11

Non-coding RNAs across the kingdoms—plants

The phenomenon of **posttranscriptional gene silencing (PTGS)** was first discovered in 1990, when the attempt to transgenically upregulate pigment production in petunias resulted in the loss of activity from both the endogene and transgene (Napoli et al., 1990). Further analysis showed that the loss of enzyme activity was due to decreased levels of mRNA without decreased levels of transcription. It was suggested that some kind of mRNA degradation process was occurring between transcription and translation, and this phenomenon was called **co-suppression**. Subsequently, the phenomenon of co-suppression was found to be regulated by small regulatory RNAs.

Since the time of PTGS discovery, many more of various **non-coding RNAs (ncRNAs)** types have been discovered. Plants have examples of almost all of the groups of regulatory ncRNAs that were discussed in the previous section, with the exception of piRNAs as they lack a PIWI homolog.

Many ncRNA populations are usually present in plant cells. They derive from different types of precursors, and have different biogenesis pathways and functions (Vazquez, 2006). Small RNAs can be classified into two large groups. The first group includes micro RNAs or **miRNAs**, the 21 to 24-nt long gene-encoded smRNAs molecules that are processed from single-stranded imperfectly folded stem-loop-like structures of **precursor miRNA (pre-miRNA)**. miRNAs play regulatory functions in many cellular processes, in particular plant development and stress response. In contrast to miRNAs, the second large group of smRNAs, called **short (small) interfering RNAs or siRNAs**, derive from long perfect dsRNA duplex precursors that can be endogenous or exogenous by origin (Vazquez, 2006). Short interfering RNAs are usually 21 to 24 nt long and are involved in the establishment of sequence-specific transcriptional and post-transcriptional gene silencing, and they can mediate cell defense against viruses.

Small RNA biogenesis in plants depends on the set of proteins common for many eukaryotes, including ARGONAUTES (AGO), DICERs (or DICER-LIKE proteins, DCLs), and RNA Dependent RNA Polymerases (RDRs). However, speciation events followed by adaptation to specific environments resulted in the specialization of many small RNA pathways. For example, in *Arabidopsis thaliana* such specialization requires 4 DCL, 10 AGO, and at least three functional RDR genes.

11-1. ncRNA biogenesis: Types of ncRNAs produced and proteins involved (as per Pikaard et al., 2008)

Polymerases involved:

Pol II: DNA-DEPENDENT RNA POLYMERASE II; main RNA polymerase, transcribing most mRNA and miRNA genes.

Pol III: DNA-DEPENDENT RNA POLYMERASE III; primarily transcribes tRNA and 5S rRNA genes.

Pol IV: nuclear RNA polymerase IV; two largest subunits are the NRPD1a and NRPD2a.

Pol V: nuclear RNA polymerase V; two largest subunits are the NRPD1b and NRPD2a.

ncRNAs produced:

l-siRNA: long siRNA of approximately 40 nt in size.

miRNA: microRNA; small RNAs transcribed from dedicated genes; processed by DCL1 and primarily mediate mRNA cleavage and degradation.

nat-siRNA: natural antisense transcripts siRNAs; typically produced from transcription of overlapping adjacent genes located on opposite DNA strands.

ra-siRNAs: repeat-associated siRNAs (ra-siRNAs); produced from repetitive genomic areas.

siRNA: small interfering RNA.

ta-siRNA: transacting siRNAs; plant endo-siRNAs generated by interaction between miRNA and siRNA pathways.

Proteins involved:

AGO: ARGONAUTE, family of 10 proteins in Arabidopsis; AGO proteins bind to both miRNAs and siRNAs; AGOs slice mRNAs into small RNAs.

CLSY1: CLASSY1, a chromatin-remodeling protein required for RdDM.

DCL1: Arabidopsis DICER-LIKE 1; responsible for miRNA biogenesis and is also required for nat-siRNA and some viRNA processing.

DCL2: Arabidopsis DICER-LIKE 2; generates 22-nt nat-siRNAs as well as viRNAs.

DCL3: Arabidopsis DICER-LIKE 3; generates 24-nt siRNAs from heterochromatic regions and repetitive elements (hc-siRNAs, ra-siRNAs).

DCL4: Arabidopsis DICER-LIKE 4; produces 21-nt ta-siRNAs and viRNAs.

DMS3: DEFECTIVE IN MERISTEM SILENCING 3; involved in DNA methylation and gene silencing, part of DDR complex.

DRD1: DEFECTIVE IN RNA-DIRECTED DNA METHYLATION 1; a chromatin-remodeling protein required for RdDM, part of DDR complex.

DRM2: DOMAINS REARRANGED METHYLTRANSFERASE 2; main Arabidopsis *de novo* DNA methyltransferase.

HEN1: HUA ENHANCER 1; methyltransferase that methylates the 2' hydroxyl groups of siRNAs and 3'-terminal nucleotide of miRNAs.

HST1: HASTY1; a homolog of exportin 5, transferring miRNAs from the nucleus to cytoplasm.

HYL1: HYPONASTIC LEAVES 1; a dsRNA-binding protein interacting with DCL1.

RdDM: RNA-directed DNA methylation; gene silencing process that includes *de novo* DNA methylation, followed by histone and chromatin modifications.

RDM1: RNA-DIRECTED DNA METHYLATION 1; involved in RdDM, part of DDR complex.

RDR2: RNA-DEPENDENT RNA POLYMERASE 2; synthesizes the second RNA copy from long ssRNA precursors; involved in the biogenesis of 24-nt siRNAs and RdDM.

RDR6: RNA-DEPENDENT RNA POLYMERASE 2; synthesizes the second RNA copy from long ssRNA precursors; generates ta-siRNAs, nat-siRNAs, l-siRNAs, viRNAs; important for long-distance silencing.

SDE3: SILENCING DEFECTIVE 3; a putative RNA helicase.

SGS3: SUPPRESSOR OF GENE SILENCING 3; a putative coiled-coil protein.

Processes involving ncRNAs:

RISC: RNA-induced silencing complex; contains ARGONAUTE protein; those involving miRNA are called miRISCs, and those including siRNAs—siRISCs.

RITS: RNA-induced transcriptional silencing complex; contains AGO4, siRNA, DOMAINS REARRANGED METHYLTRANSFERASE 2 (DRM2), DDR complex composed of DEFECTIVE IN RNA-DIRECTED DNA METHYLATION 1 (DRD1), DEFECTIVE IN MERISTEM SILENCING 3 (DMS3), RNA-DIRECTED DNA METHYLATION 1 (RDM1).

Plant miRNAs

In plants, miRNAs are single-stranded RNAs, which are typically of 21 nt in length and processed by the RNase-III-type enzyme Dicer from endogenous transcripts containing local hairpin structures.

The majority of miRNA genes in plants exist as independent transcriptional units and are transcribed by RNA polymerase II into long primary transcripts (**pri-miRNAs**).

In plants, DICER-LIKE protein 1 (DCL1) plays the roles of both Drosha and Dicer proteins in animals, converting pri-miRNAs to miRNA/miRNA* duplexes. The dsRNA arm of the hairpin loop formed by the pri-miRNA is recognized by DCL1 in the nucleus where it—with the dsRNA binding protein HYPONASTIC LEAVES1 (HYL1)—cuts the hairpin twice to excise the mature miRNA/miRNA* duplex from the stem of the hairpin with a two-nucleotide 3' overhang on each strand (Vazquez, 2006). DCL1 and HYL1 apparently localize to a distinct region of the nucleus termed the nuclear dicing body where the excision of the mature miRNA occurs. It has also been postulated that the zinc-finger containing protein SERRATE (SE) is involved in the primary processing of miRNAs; however, they are not specifically localized to nuclear dicing bodies and their function has not been characterized.

There is believed to be a mechanism by which DCL1/HYL1 preferentially cuts from the stem due to the fidelity in sequence of the mature miRNAs; however, the mechanism behind this specificity remains enigmatic. It has been speculated that it is the secondary structure of the stem and not the primary sequence that grants the specificity (Parizotto et al., 2004). Mature plant miRNAs can vary in size from approximately 20 to 24 nucleotides in length, but are most often 21 nuclotides long. The variance in size is thought to depend on the structure of the stem (Vazquez, 2006).

After it's excised from the stem-loop structure, mature miRNA/miRNA* duplex is methylated at the 2'-OH of the 3' ends of the duplex by the dsRNA methyltransferase HUA ENHANCER1 (HEN1), a process thought to impede polyuridylation and degradation. Export of either the mature duplex or a single-stranded RNA to the cytoplasm is thought to occur through the exportin5 homologue HASTY (HASTY mutants develop precociously, hence their name).

Unlike miRNAs in mammals, plant miRNAs are 2'-O-methylated at their 3' ends by HEN1. HEN1 protects plant miRNAs from 3' uridylation, and thus from possible degradation. Methylation likely occurs before miRNAs are loaded into **RNA-induced silencing complex (RISC)** because both miRNA* and miRNA strands are modified in plants.

When exported to the cytoplasm where the primary function of miRNAs occurs, the miRNAs are loaded into a RISC, which facilitates the recognition and degradation of target mRNAs. The main protein involved in **miRNA-RISC (mi-RISC)** is the ARGONAUT1 (AGO1), one of 10 members of the Argonaute protein family in Arabidopsis (Vaucheret, 2006). Argonaute proteins contain an RNA-binding PAZ domain and an RNase H-like PIWI domain. It is believed that AGO1 recognizes the miRNA/miRNA* duplex via the two-nucleotide 3' overhang, which places the 5' end of the miRNA in to the binding pocket of PAZ-domain of AGO1. The selection of the strand entering RISC is similar to animals as it follows the asymmetry rule; whichever strand is less stably paired at its 5' end incorporates into the final RISC. The passenger miRNA* strand is then released and degraded.

Being part of the RISC, miRNAs target transcripts working as a sequence-specific guide through the perfect or near-perfect pairing between the miRNA and the mRNA transcript. A high degree of complementarity triggers the RNase activity of the PIWI-domain of AGO1, which generates a single cut of the target mRNA's phosphodiester backbone. Fragments of cleaved mRNAs are then released and the RISC is available to perform another cleavage of remaining complementary mRNAs. This is in contrast to animals where miRNAs predominantly are involved in translation inhibition.

miRNA-mediated repression of translation in plants might also be possible. Early evidence suggested members of the *APETALA* family of transcription factors might be translationally inhibited; however, this finding was questioned by subsequent studies (Schwab et al., 2005). Gandikota et al. (2007) provided evidence that the 3' UTR of the Arabidopsis SBP box gene SPL3 contains a functional **miRNA-responsive element (MRE)** (also referred to as miRNA recognition element) that is used by miR156 and miR157 for translation inhibition. Later on, Brodersen et al. (2008) isolated *ago1*, *ago10*, and *katanin* Arabidopsis mutants that were defective in miRNA action, allowing them to genetically separate translational inhibition from endonucleolytic cleavage. Katanin is the microtubule-severing enzyme, and thus it is possible that cytoskeleton dynamics is important in miRNA function in translational inhibition. Recently, Beauclair et al. (2010) showed that one AGO10 in Arabidopsis is also involved in miR398-directed translational inhibition of CCS1 mRNAs, as CCS1 protein, but not CCS1 mRNAs, accumulates in *ago10* mutants.

Evolutionary conservation of miRNAs

According to Release 16 of the miRNA Registry Database (www.mirbase.org), there are 916 miRNAs from plants including Arabidopsis, rice (*Oryza sativa*), corn (*Zea mays*), poplar (*Populus trichocarpa*), as well as many others. miRNA genes are conserved among many plants. The fact that miRNAs are conserved in rice and Arabidopsis suggests that miRNA regulatory mechanism has developed before the divergence of monocot and dicot plants (reviewed in Yang et al., 2007). miR166 targets class III homeodomain-leucine zipper (HD-ZIP III) transcription factors in various plants, providing further evidence that miRNA-based regulation is one of the oldest mechanisms of posttranscriptional regulation in plants. The expression of flowering plant miRNAs—such as miR160, miR390, and miR160—were detected in gymnosperms, fern, and moss, suggesting that many plant miRNAs have remained mostly unchanged since before the emergence of flowering plants (reviewed in Yang et al., 2007). Analysis of hairpin structures of miRNAs belonging to the same gene family showed a high degree of conservation among the analyzed plants species (Yang et al., 2007). In fact, more than eight families of miRNA genes have apparently appeared before the emergence of seed plants. At the same time, most of the other miRNA families exhibit a substantial degree of divergence between species, possibly allowing to them regulate common processes that became specialized in a given plant species. This diversity is also reflected in that some of the miRNAs specific to Arabidopsis have appeared recently and are rather unique (nonconserved) for dictos. According to one of the models, miRNA genes appeared as a result of the transcription of genes that underwent inverted duplication. "Read through" transcription of such genes results in the formation of hairpin structures that are recognized by DCL1 as a pri-miRNA.

Plant siRNAs

The group of siRNAs in plants is very diverse and contains several subclasses including trans-acting siRNAs (ta-siRNAs), natural-antisense transcript-derived siRNA (nat-siRNAs), and repeat-associated siRNAs (ra-siRNAs) (Vazquez, 2006).

Transacting-siRNAs

Trans-acting siRNAs (ta-siRNAs) represent a class of plant endo-siRNAs that are generated by interaction between miRNA and siRNA pathways (see Figure 11-1). Genes coding for ta-siRNAs are transcribed by PolII from so called *TAS* loci. Generated mRNAs, which represent pri-ta-siRNAs, are targeted for cleavage by specific miRNAs (see Figure 11-1).

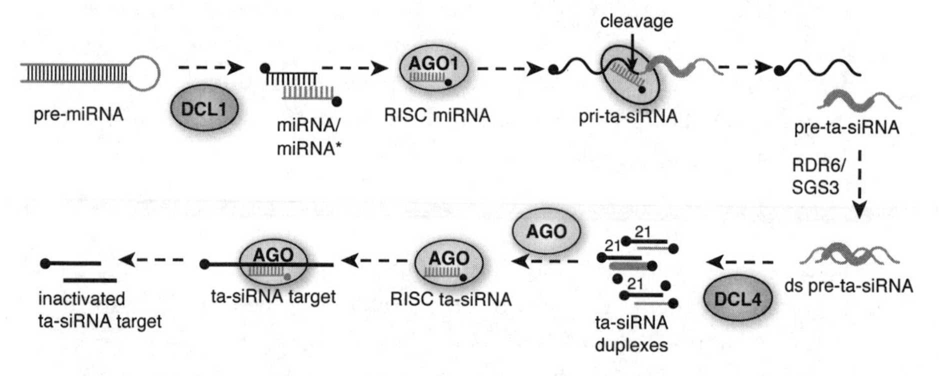

Figure 11-1 Generation of ta-siRNAs. Specific miRNAs, generated by the activity of DCL1 proteins, are forming miRNA-containing RISCs that target pri-ta-siRNAs for cleavage. This results in formation of single-stranded form of pre-ta-siRNAs that are processed to double-stranded form through the activity of RDR6/SGS3. DCL4 dice the ds pre-ta-siRNAs into duplexes containing ta-siRNAs. Together with AGO protein (not clear which one) the RISC ta-siRNA complexes are formed that find and cleave ta-siRNA targets.

The RdRP enzyme, RNA-DEPENDENT RNA POLYMERASE6 (RDR6), targets cleaved ta-siRNAs, named pre-ta-siRNAs. RDR6 together with SUPPRESSOR OF GENE SILENCING3 (SGS3) generate a second strand of RNA, using pre-ta-siRNA as a template.

The role of SGS3 might include protection of ssRNA from degradation. These long double-stranded pre-ta-siRNAs are targeted by DCL4 (together with DRB4), which dices dsRNA precursors into many 21 nt-long copies—the process called phasing. The fact that all diced dsRNAs are 21 nt in size suggests that DCL4 starts dicing precisely at the miRNA cleavage site, making ta-siRNAs every 21 nt. The phasing process is essential for ta-siRNA's target specificity. The exact site of miRNA cleavage determines the entry point for DCL4, and thus the phase of a ta-siRNA. Curiously, production of ta-siRNAs is dependent on binding of two miRNAs to pri-ta-siRNA. Whereas one miRNA promotes cleavage, the second complementary miRNA might be necessary to initiate RDR6-dependent conversion to dsRNA. Indeed, mRNA produced from *TAS3* locus contains two binding sites for miR-390. One miR-390 cleaves TAS3 whereas the second miRNA does not, but it is necessary for production of ds-pre-ta-siRNA.

Most of the ta-siRNAs in Arabidopsis derive from eight loci belonging to four families in Arabidopsis: *TAS1, TAS2, TAS3, and TAS4*. Pri-ta-siRNAs are targeted by specific miRNAs. For example, miR-173 cleaves the *TAS1* and *TAS2* transcripts and miR-390 is responsible for the targeting of *TAS3*. miR-828 promotes cleavage of the transcripts produced from the *TAS4* loci. The *TAS1* family (such as *TAS1a*, *TAS1b*, and *TAS1c*) contains three genes encoding ta-siRNAs with a high degree of similarity, such siR255 and siR480(+). These siRNAs are able to cleave four different mRNAs coding for unknown proteins. siR1511 produced from *TAS2* promotes cleavage of several mRNAs encoding pentatricopeptide repeat proteins. ta-siRNAs produced from the *TAS3* cleave several mRNAs encoding Auxin response factors (ARFs)—ARF3 (ETTIN) and ARF4—that control the juvenile-to-adult transition in leaf development. ta-siRNAs derived from *TAS4* locus target transcripts of MYB genes that are thought to control cell cycle and cell fate (Rajagopalan et al., 2006).

The ds-ta-siRNAs are methylated by HEN1, and the strand that served as a RDR6 template is then loaded into the RISC complex with one of the AGO proteins. ta-siRNA-containing RISCs direct the cleavage of specific mRNA targets. The specificity of the AGO depends on what *TAS* family was cleaved. For example, the biogenesis and function of tas3-siRNAs depend on AGO7, whereas several other ta-siRNAs involve AGO1 (Jones-Rhoades et al., 2006). The ta-siRNAs might be able to target their own transcripts for degradation. Because miRNAs are required for ta-siRNA biogenesis, it can be speculated that single miRNA can regulate the function of many ta-siRNAs, thus cleaving multiple non-homologous targets.

Natural antisense transcripts short interfering RNAs (nat-siRNA)

Nat-siRNAs are a class of endogenous ncRNAs of 21 to 24 nt in length (with 24 nt being predominant) that are generated from two overlapping and partially converging coding transcripts (Borsani et al., 2005). When nat-siRNAs are generated from transcripts transcribed from the opposite strands at the same locus, they are called ***cis*-nat-siRNAs**, and when they are produced from transcripts generated from different genomic locations, they are called ***trans*-nat-siRNAs** (Sunkar et al., 2007). Similar to ta-siRNAs, nat-siRNAs regulate targeted gene activity at the posttranscriptional level by guiding mRNA cleavage. Search for potential nat-siRNAs in Arabidopsis has identified 1,126 putative *cis*-nat-siRNA pairs3 and 1,320 putative *trans*-nat-siRNA pairs. Moldovan et al. (2010) compared the expression of the two aforementioned groups of gene pairs and identified 27 putative *cis*-nat-siRNA and 7 putative *trans*-nat-siRNA gene pairs to have an inverse expression correlation.

Nat-siRNAs were first discovered in plants to be produced in response to stress. Nat-siRNAs are generated from a pair of convergently transcribed RNAs in which one transcript has constitutive expression, whereas the complementary RNA is produced only in response to a given stress. When transcript overlap is generated, DCL2 and DRB partner target the dsRNA duplex (see Figure 11-2). The exact mechanism is unclear, but the process might include the function of RDR6/SGS3, allowing production of longer RNA duplexes. The role of Pol IV in the process is not clear but the great majority of nat-siRNAs are produced in a Pol IV-dependent manner. It is possible that Pol IV transcribe some of the nat-siRNA loci. It is also possible that Pol IV acts as an RNA polymerase, transcribing dsRNAs. The role of PolIV might also be RNA-directed DNA methylation of *cis*-nat-siRNAs. Short 24 nt siRNAs produced by DCL2/DRB from a region of overlap of the two transcripts are methylated by HEN1 and then picked up by AGO (it is not clear which one at this time) into RISCs that direct cleavage of one of the mRNAs of the pair. Cleaved mRNAs are used as templates by the RDR6/SGS3 pair to produce a long RNA duplex diced by DCL1/DRB into 21 nt-long **secondary nat-siRNAs**. After being methylated by HEN1, secondary nat-siRNAs are then used by AGO to promote the cleavage of target mRNAs (see Figure 11-2).

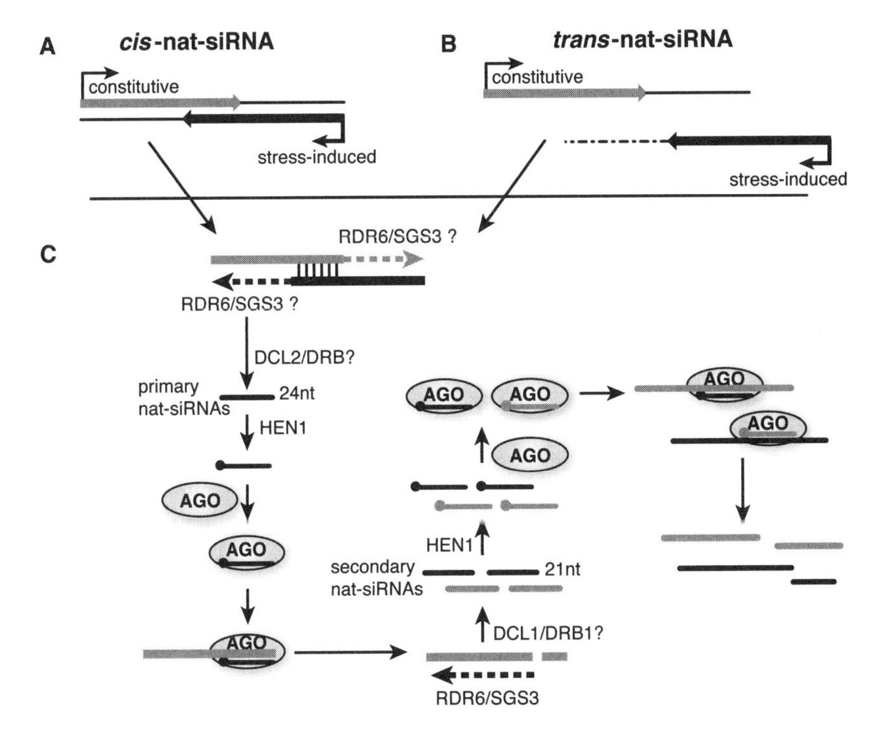

Figure 11-2 Generation of primary and secondary nat-siRNAs. Nat-siRNAs are produced from genes that produce complementary transcripts. cis-nat-siRNAs are generated from transcripts transcribed from the opposite strands of the same locus (**A**), whereas trans-nat-siRNAs are generated from transcripts produced from different genomic positions (**B**). In both cases, the pairs are formed from transcription of one gene that is constitutively expressed and one gene that is induced by stress. Next (**C**), the resulting transcript duplexes are presumably elongated with RDR6/SGS3 and processed by DCL2/DRB to primary nat-siRNAs 24 nt in length. After being methylated by HEN1, these small RNAs are picked up by AGO (not clear which one) proteins and are incorporated into RISCs to target one of the transcripts of the *cis*- or *trans*- pair. Cleavage products are processed by RDR6/SGS3 into ds form and diced by DCL1/DRB into 21 nt-long secondary ta-siRNAs involved in cleavage of secondary mRNA targets.

In addition to nat-siRNAs, **long siRNAs (lsiRNAs)** in Arabidopsis also originate from NAT pairs and are stress-induced. Unlike nat-si-RNAs, lsiRNAs are 30 to 40 nt long and for their biogenesis require DCL1, DCL4, AGO7, RDR6, and POL IV for their production. lsiRNAs are very frequent in Arabidopsis and rice, although their mechanism of action is not very well understood.

cis-acting nat-siRNAs (ca-siRNAs), heterochromatic siRNAs (hc-siRNAs), and repeat-associated siRNAs (ra-siRNAs)

In plants, a specific type of ***cis*-acting nat-siRNAs (ca-siRNAs)** originates from transposons, repetitive elements, and tandem repeats such as 5S rRNA genes. Because ca-siRNAs can also be produced from heterochromatic areas from RNA transcripts generated by PolIII, they are also referred to as **heterochromatic siRNAs** or **hc-siRNAs** (see Figure 11-3). The final group of siRNAs produced from heterochromatin and repetitive elements is **repeat-associated siRNAs (ra-siRNAs)**. Ra-siRNAs play a crucial role in maintaining gene expression via DNA methylation and histone modification and control activity of retrotransposons and repetitive DNA sequences.

Pol II and Pol III generate precursor transcripts. It is also hypothesized that Pol IV might play a unique role by transcribing methylated heterochromatic loci (Pikaard et al., 2008). Accumulation of siRNAs depends on cleavage by DCL3 assisted by DRB. Those RNA precursors that are not generated from overlapping transcripts are targeted by RDR2 to produce long RNA duplexes. One possibility is that Pol V transcribes long dsRNAs and then processes RDR2 to form dsRNAs again. These dsRNAs are diced by DCL3 into long duplexes of mainly 24 nt in length, which are then methylated by HEN1. These 24 nt-long duplexes are either picked by AGO6 (primarily) for promoting cleavage of mRNA targets as a part of RISC, or picked by AGO4 for incorporation into **RNA-induced transcriptional silencing complex** (**RITS**) to promote sequence-specific DNA methylation (refer to Figure 11-3). Besides AGO4 and siRNA, RITS involves DOMAINS REARRANGED METHYLTRANSFERASE 2 (DRM2), DDR complex composed of DEFECTIVE IN RNA-DIRECTED DNA METHYLATION 1 (DRD1), DEFECTIVE IN MERISTEM SILENCING 3 (DMS3), RNA-DIRECTED DNA METHYLATION 1 (RDM1) (Law et al., 2010), and PolV. PolV might help either in recognition of the targeted loci or in the pairing of small RNAs with DNA by transiently transcribing a given target region. Thus, ca-siRNAs and ra-siRNAs might promote heterochromatin formation by directing DNA methylation and histone modification of the loci from which they originate.

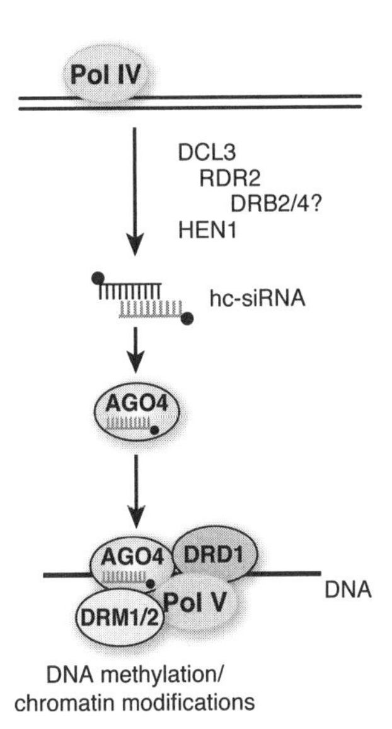

Figure 11.3 Heterochromatic siRNAs. Hc-siRNAs are produced from the areas of transposons and inverted repeats with the help of Pol IV. Sense and antisense transcripts form duplexes that are sliced by DCL3. RDR2 might be involved in elongation of partial duplexes either before or after DCL3 slicing. HEN1 methylates the short duplexes making active hc-siRNA forms that are picked up by AGO4 into RITS-type complexes that promote changes in methylation and chromatin structure at the target loci. Proteins such as DRM1/2 and DRD1 as well as the activity of Pol V are required.

It is debatable whether these three groups of siRNAs belong to the same class or not. ca-siRNAs might be involved in mRNA cleavage and DNA methylation, whereas hc-siRNAs and ra-siRNAs are primarily involved in RNA-directed DNA methylation.

Pollen siRNAs

Pollen siRNA are a recently discovered set of siRNAs that are induced in the pollen vegetative nucleus. Interestingly, transposons and transposase were found to be reactivated in the vegetative nucleus, resulting in a loss of heterochromatin structure in those sequences, but coinciding with accumulation of specific siRNAs in the vegetative nucleus as well as the gamete (Slotkin et al., 2009). This finding is interesting because it suggests that even though plants do not have piRNAs, they employ a small regulatory RNA-mediated mechanism to control transposon activity in the germline.

Long ncRNAs

One of the examples of lsiRNAs produced in plants is IPS1 RNA, a long non-coding RNA that is induced during phosphate starvation in Arabidopsis. IPS1 RNA is a target for a miRNA that is also induced under phosphate starvation—miR399. In plants, most miRNAs mediate target cleavage, showing almost perfect complementarity around the cleavage site rather than inhibition of translation. In the IPS1-miR399 duplex, there is a mismatch in the cleavage site that impedes degradation. This results in dominant negative regulation of miR399 that is bound but non-functional in complex with IPS1, but not available to bind to other target RNAs such as PHO mRNA, which consequently accumulates. PHO, or phosphate overaccumulator, was discovered in a mutant that, as its name suggests, overaccumulates phosphate. As miR399 binds to PHO mRNA, it downregulates the transcript and thereby allows phosphate accumulation.

This mechanism has important implications as it shows that ncRNAs can potentially negatively regulate other ncRNAs instead of directly regulating transcript levels of coding RNAs. Thus, ncRNAs could be important players in controlling small RNA-mediated mechanisms.

Arabidopsis has multiple redundant "back-up" mechanisms for ncRNA production

In plants—Arabidopsis in particular—various classes of ncRNAs display substantial similarities in their biogenesis. ncRNAs production in plants depends on the cleavage of dsRNA precursor molecules by a DICER protein. Four known Arabidopsis DCL proteins—DCL1, DCL2, DCL3, and DCL4—display substantial functional redundancy in biogenesis of various types of ncRNAs (Vazquez, 2006; Sunkar et al., 2007). Although DCL1's primary role is in production of miRNAs, it is also required for the processing of nat-siRNAs and—at least indirectly—required for generation of ta-siRNAs. DCL2 generates nat-siRNAs (Borsani et al., 2005) and is needed for the production of siRNAs involved in viral resistance, whereas DCL3 processes siRNAs involved in RdDM and transcriptional gene silencing. Finally, DCL4 processes ta-siRNAs, but it is also necessary for production of some viral siRNAs. Mutational studies support functional redundancy of Arabidopsis DCLs. When DCL3 is mutated, methylation marks are partially maintained in the plant genome, suggesting that DCL2 or DCL4 proteins might partially compensate for the loss of DCL3 in *dcl3* mutants (Gasciolli et al., 2005). Methylation at non-symmetrical cytosines is not significantly altered in *dcl3* mutants but is severely inhibited in *dcl2 dcl3 dcl4* triple mutants, suggesting that DCL2 and DCL4 might in fact be involved in non-CG methylation. At the same time, production of miRNAs by DCL1 is significantly increased in *dcl2* mutants, *dcl4* mutants, and *dcl2 dcl4* double mutants (Gasciolli et al., 2005).

There also exists a functional redundancy among the Argonaute. AGO proteins are an essential part of RISC and RITS. In RISC, AGOs cleave complementary mRNA in the center of the region of the mRNA/smRNA pairing, whereas in RITS, AGOs mediate RdDM. AGO1 are part of mi RISC and si-RISC-mediated mRNA degradation, although there is some functional redundancy with AGO10 protein. AGO7 is a part of ta-siRNAs-guided degradation of targeted mRNAs. However, its role in regulation of developmental timing depends on another AGO protein, possibly AGO1. AGO4 and perhaps AGO6 are involved in RdDM, chromatin modification, and epigenetic regulations at targeted sequences.

The last multi-protein family is the group of RNA-dependent DNA methyltransferases of which Arabidopsis has six known members, RDR1 through RDR6. The protein RDR6 is involved in several independent processes, including ta-siRNAs, nat-siRNAs, siRNAs involved in viral resistance, and in sense-transgene-mediated silencing. RDR1 is apparently in part able to substitute RDR6 in production of siRNAs necessary for the resistance against viral infection. In contrast, RDR2's main function is to amplify hc-siRNAs and ra-siRNAs directing silencing of repetitive DNA sequences and transposons. The function of other AGOs and RDRs is yet to be reported.

RdDM and gene silencing

RdDM pathway requires activity of many proteins, including DCL3/DRB, POl IV, RDR2, and HEN1 for the production of hc-siRNA duplexes as well as proteins such as AGO4/AGO6, Pol V, DRM2, and DDR complex involved in *de novo* DNA methylation and chromatin remodeling. Although RdDM process also promotes siRNA-regulated DNA methylation, there is a substantial difference from RNAi. In RNAi, ra-siRNAs (and other types of siRNAs) target repetitive genomic regions located in heterochromatin and promote spreading of methylation and heterochromatization in substantial (up to several thousand nucleotides) distance from the original target. In contrast, in RdDM hc-siRNA-targeted methylation does not spread and remains confined to single loci (Matzke and Birchler, 2005).

The initiating signal of RdDM is not clear; however, it might be the dsRNAs produced by RDR2 from ssRNA transcripts of various repetitive DNA sequences and transposons, or by the mRNA transcripts that are produced by the action of Pol II on genes carrying inverted repeats. The latter may not be completely impossible as the first example of endogenous promoter methylated by RdDM pathway was the promoter driving the expression of the FWA gene that contained direct repeats. RITS complexes carrying specific hc-siRNAs utilize the activity of *de novo* DNA methyltransferases such as DRM2 to promote site-specific changes in DNA methylation. *De novo* methylation in plants is established in both, CpG and non-CpG sites (see details in Chapter 4, "DNA Methylation as Epigenetic Mechanism") by METHYLTRANSFERASE 1 (MET1) and DRM2 methyltransferases, respectively. The activity of DDR chromatin-remodeling complex consisting of three enzymes—DRD1, DMS3, and RDM1—is required to facilitate the access of hc-siRNA to target DNA (Pikaard, 2006). Although the process is still not clearly understood, the **DNA-recognition** and **RNA-recognition models** seem to be favored (reviewed in Matzke and Birchler, 2005). The RNA-recognition model requires pairing of smRNA and nascent RNA transcribed from potential target locus, whereas the DNA-recognition model allows for direct DNA-RNA pairing.

DNA methylation imprints promoted by RdDM can be easily maintained at the symmetric CpG and CpNpG sites by MET1 and CHROMOMETHYLASE 3 (CMT3), respectively, even when the original hc-siRNA signal is no longer present. In contrast, maintenance of methylation of non-symmetrical CpNpN sites requires constant RdDM activity. The effect of RdDM at the CpNpN sites may be either passively lost or actively removed. The former one occurs during replication and DNA repair, whereas the latter one requires the activity of at least two DNA glycosylase-domain containing proteins DEMETER (DME) and REPRESSOR OF SILENCING 1 (ROS1) that recognize and replace methylated cytosines with non-methylated. Although no direct evidence exists, it is suggested that ROS1 and DME functions require the presence of guide ncRNA. In fact, in animals, it was shown that the active DNA demethylation by DNA glycosylases required presence of an RNA strand complementary to methylated DNA. Table 11.1 lists factors involved in RdDM.

Table 11-1 Factors involved in RdDM

Name and function	Effects on chromatin	Effects of mutation and involvement in stress response	Modification/ Transcription
RNA POLYMERASE IV (subunits NRPD1a and NRPD2a)/ Nuclear RNA polymerase	pol IV is localized in the nucleus and required for siRNA production; it acts upstream from RDR2 that uses its products as templates for dsRNA synthesis; these dsRNAs are processed by DCL3 and AGO4; pol IV might participate in a subunit exchange reaction with NRPD1b to form a functional pol V	Loss of either *NRPD1a* or *NRPD2a* leads to loss of cytosine methylation and disrupted facultative and not constitutive heterochromatin formation; *nrpd1a* and *nrpd2a* mutants display loss of siRNAs corresponding to targeted loci; both *NRPD1a* and *NRPD2a* are required for endogenous retroelements and transgenes silencing	Local, heterochromatin maintenance/ Repression

Table 11-1 Factors involved in RdDM

Name and function	Effects on chromatin	Effects of mutation and involvement in stress response	Modification/ Transcription
RNA POLYMERASE V (subunits NRPD1b and NRPD2a)/ Nuclear RNA polymerase	Presence of CTD at NRPD1b subunit might provide a platform for formation of a hypothetical NRPD1b-containing RISC complex that includes AGO4 protein associated with siRNAs; NRPD1b-containing RISC delivers siRNA to the target site, where pol V is formed by subunit exchange with pol IV; pol V initiates RNA transcription at an siRNA-targeted site that exposes DNA for epigenetic modifications; NRPD1b possibly recruits *de novo* DNA methyltransferases MET1 and DRM2 that are aided by DRD1; together with DRD1 pol V reversibly controls promoters and LTRs in euchromatin	Loss of *NRPD1b* leads to the loss of cytosine methylation mainly in non-CpG sequence context and doesn't affect heterochromatic siRNAs production; methylation loss is primarily observed on previously silenced euchromatic promoters and transposons	Local, euchromatic promoters/ Repression, possibly activation
DEFECTIVE FOR RNA-DIRECTED DNA METHYLATION1 (DRD1)/SWI /SNF-like protein	DRD1 directs non-CpG DNA methylation in response to an RNA signal; DRD1 is required to facilitate the access of RNA signals to DNA; mediates full erasure of methylation when the signal is removed; together with pol V acts downstream of siRNA biogenesis pathway; DRD possibly interacts with DNA methyltransferases and DNA glycosylases; preferentially targets promoter and LTRs in euchromatin	Mutants do not show significant defects in CpG methylation, but exhibit loss of non-CpG methylation at previously silenced euchromatic promoters and transposons; *drd1* mutation leads to down regulation of *ROS1* and *DME*	Local, euchromatic promoters/ Repression, activation

PolIV and PolV, essential components of RdDM

Polymerase IV and Polymerase V (originally named Pol IVa and Pol IVb) represent multisubunit complexes that are involved in siRNA biogenesis and RdDM process in plants (reviewed in Pikaard et al., 2008). Whereas the largest and second-largest subunits of Pol IV are NRPD1a and NRPD2a, Pol V consists of NRPD1b and NRPD2a subunits, respectively (see box 11-2 and Pikaard et al., 2008).

11-2. PolIV and PolV nomenclature

Nomenclature for genes coding for Pol IV subunits is based on the nomenclature used for *Saccharomyces cerevisiae*. In yeasts, RNA polymerases I, II, and III are designated as RPA, RPB, and RPC, respectively. To prevent confusion, in Arabidopsis, the names of polymerase subunit genes acquired an N for "nuclear" (for example, NRPA, NRPB, and so on). Arabidopsis genes encoding the largest subunits of Pol I, II, and III were designated with the number 1 (for example, NRPA1, NRPB1, and NRPC1). Similarly, the genes that encode the second-largest subunits of Arabidopsis Pol I, II, and III were designated NRPA2, NRPB2, and NRPC2. Likewise, the two related Pol IV and Pol V largest subunits were designated NRPD1a and NRPD1b, whereas the two Pol IV and Pol V second-largest subunit genes—NRPD2a and NRPD2b. It should be noted that NRPD2b is not functionally active in Arabidopsis.

Box 11-3 describes the steps of discovery of these two polymerases.

11-3. Pol IV and Pol V discovery

Genetic screen for reactivation of repressed transgene loci, performed in Craig Pikaard's lab identified several silencing-defective (sde) mutants, one of which, *sde4* turned out to be an NRPD1a allele (Herr et al., 2005). Analysis of Arabidopsis genomic sequence revealed the presence of second-largest potential subunit of Pol IV, NRPD2. It appeared that the *nrpd2a* mutant is also deficient in gene silencing. The analysis showed derepression of the silenced transgene coincided with the disappearance of 24-nt siRNAs and the loss of cytosine methylation at the transgene locus

(Herr et al., 2005). Pikaard's laboratory also showed that NRPD2 activity was not redundant with the activity of homologous subunits of Pol I, II, or III, but instead affects the heterochromatization of chromocenters (Onodera et al., 2005). The mutants, *nrpd2a* and *nrpd1a*, appeared to be impaired in methylation of these heterochromatic regions. The authors thus proposed the existence of an additional polymerase, Pol IVa (later renamed as Pol IV), the components of which turned up in another screen. Genetic screens for plants exhibiting derepression of a silenced reporter gene identified several DEFECTIVE IN RNA- DIRECTED DNA METHYLATION genes, including DRD1, a SWI2–SNF2 chromatin-remodeling protein family, as well DRD2 and DRD3, which appeared to be NRPD2a and NRPD1b genes, respectively.

Further studies confirmed the existence of yet another polymerase complex, Pol V (originally named Pol IVb) and showed that *nrpd1a* and *nrpd1b* mutants differ in their siRNA composition. Despite comparable losses of methylation, siRNAs depleted in Pol IV mutant *nrpd1a* were not abolished in Pol V mutant *nrpd1b* (reviewed in Pikaard et al., 2008). This was an initial hint that Pol IV acts upstream of siRNA biogenesis, whereas Pol V acts downstream of siRNA biogenesis in the RdDM pathway. The analysis of small RNA pools in different mutants combined with comprehensive genome-wide analyses showed that production of 94% of nearly 5,000 unique siRNAs depends on Pol IV function, with about one-third of them depending on both Pol IV and Pol V and not a single one depending only on Pol V (Mosher et al., 2008). These results further confirmed the differential roles the two polymerase complexes play in RdDM.

Several studies attempting to co-localize pol IV subunits with other proteins involved in siRNA production and RdDM resulted in the development of hypothetical model for RdDM (reviewed in Pikaard, 2006). The first study showed co-localization of Pol IV,

DRD1, and part of the NRPD1b protein pool at siRNA source/target loci (Pontes et al., 2006). At the same time, RDR2, DCL3, AGO4, and the rest of the NRPD1b pool were co-localized in the nucleolus, where they interacted with corresponding siRNAs (Pontes et al., 2006). Because Pol IV subunits co-localize with source/target loci and because in *nrpd1a* all other known components of the RdDM pathway to mislocalize, Pol IV is believed to act at an initial step of siRNA biogenesis, upstream of RDR2. Another protein that is potentially involved in RdDM is CLASSY 1 (CLSY1), an SWI–SNF family protein. It co-localizes with RDR2 at the inner perimeter of the nucleolus, and a mutation of CLSY1 disrupts RDR2 localization. Localization of Pol IV is affected to a lesser degree; therefore, CLSY1 might function at the interface between Pol IVa and RDR2, presumably facilitating the generation of dsRNAs that are diced by DCL3 and loaded into AGO4 effector complexes within the nucleolar siRNA processing center.

It is further hypothesized that Pol IV transcripts produced at siRNA source/target loci move from the nucleoplasm to nucleolar siRNA processing centers where they are converted into dsRNAs by RDR2, cleaved by DCL3 and then loaded into RISCs containing AGO4 and NRPD1b. These processing centers have several molecular markers of Cajal bodies that represent compartments in which the assembly of ribonucleoprotein complexes involved in splicing, pre-rRNA processing, and other RNA modifications occur.

The assembled RITS complex leaves the nucleolus and arrives back to the siRNA source/target site. It is possible that Pol IV donates its NRPD2a subunit to Pol V, allowing it to use siRNAs delivered by AGO4 as primers to initiate RNA transcription from an siRNA source/target site with help of DRD1 protein (reviewed in Pikaard, 2006). Nucleosome displacement caused by RNA transcription or possibly the invasion of the Pol V transcript into the DNA duplex might temporarily expose DNA to the activity of DRM2 and KYP proteins (reviewed in Pikaard, 2006). Exposed DNA is then targeted by DRM2 and KYP to promote *de novo* DNA methylation and histone H3K9 methylation, respectively. Table 11.2 lists factors involved in RISC/RNAi.

Table 11-2 Factors involved in RISC/RNAi

Name and function	Effects on chromatin	Effects of mutation and involvement in stress response	Modification/ Transcription
RNA-DEPENDENT RNA POLYMERASE2 (RDR2)/RNA-dependent RNA polymerase protein family	Synthesis of dsRNA from ssRNA templates; RDR2 provides substrates for DCL3 and AGO4 activity; might work at ssRNA templates produced by pol IV	Loss of *RDR2* results in decreased cytosine methylation at endogenous repeats and a complete loss of corresponding siRNAs	Local/Repression
DICER-LIKE3 (DCL3)/RNase III protein family	DCL3 processes dsRNA and produces RNA signals that guide siRNA-directed site-specific *de novo* DNA methylation	Loss of *DCL3* results in decreased cytosine methylation at endogenous repeats and a complete loss of corresponding siRNAs; loss of *DCL3* is partially compensated for by *DCL2* and *DCL4* genes	Local/Repression
ARGONAUT4 (AGO4)/PPD protein (contains PAZ and PIWI domains)	AGO4 functions downstream of DCL3; initiates and maintains histone H3K9 methylation, initiates *de novo* DNA methylation; might be a functional part of NRPD1b-containing RISC	Loss of *AGO4* results in decreased cytosine methylation and histone H3K9 methylation at endogenous repeats and leads to a complete loss of corresponding siRNAs	Local/Repression

Function of small RNAs in plants

A network of complex and partially overlapping pathways represents ncRNA biogenesis in plants. Various endogenous and exogenous stimuli can initiate production of dsRNAs, including various types of viruses and viroids, transgenes, repetitive DNA sequences, and particular endogenous loci.

In development

Many of ncRNAs reported to date are involved in the regulation of plant development. In fact, members of the small RNA biogenesis pathways were first identified through mutants with severe developmental defects (Jones-Rhoades et al., 2006). Among all ncRNA potentially involved in plant development, miRNAs might be the most important as *dcl1-/-* plants are embryo lethal.

Temporal and spatial expression of miRNAs enables them to target specific genes involved in cell fate and patterning. The following are just few examples of miRNAs involved in developmental processes in plants. miR-166/165 appears to target the class III HD-ZIP members *PHABULOSA*, *PHAVOLUTA*, and *REVOLUTA*, allowing for the establishment of abaxial/adaxial leaf polarity. miR-172 regulates the expression of *APETELA2* whose role is in regulation of stem cell fate in the shoot apical meristem as well establishment of floral organ identity. miR-160, miR-167, and miR-390 are involved in targeting different members of the *AUXIN RESPONSE FACTOR (ARF)* transcription factors, which play important roles in male and female reproduction, root cap formation, seed germination, and leaf polarity. miR-159, induced by abscisic acid, controls the expression of Myb transcription factors during seed germination.

Information about the role of siRNAs in development is scarce. However, recently Ron et al., (2010) showed that *cis*-nat-siRNAs have an important role in reproductive function in Arabidopsis, facilitating gametophyte formation and double fertilization. The authors identified a new sperm-specific pair of genes that give rise to the nat-siRNA pair. These genes are KPL and an inversely transcribed gene, ARIADNE14 (ARI14), which encodes a putative ubiquitin E3 ligase.

Male gametophytic *kokopelli* (*kpl*) mutants display frequent single-fertilization events. It was shown that in the absence of KPL, ARI14 RNA levels in sperm are increased, leading to defects in fertilization (Ron et al., 2010).

In response to stress

It has long been known that whole concerts of genes change in response to abiotic stresses, such as temperature, water, and light extremes, as well as soil salinity, heavy metals, radiation, and nutrient deprivation. Traditionally, changes in gene expression in response to these factors were thought to take place at a transcriptional level; however, an ever-increasing role for post-transcriptional events is beginning to emerge.

Epigenetic regulation represents a significant advantage over other types of stress response, such as permanent genetic changes because it includes fast stimulus-directed generation of new transcriptional states that are heritable and reversible. One key mechanism involved in targeting chromatin structure and modifying a gene expression pattern in response to environmental stimuli is based on the activity of ncRNAs. They were shown to guide sequence-specific gene regulation in response to various abiotic and biotic stresses and include modifications of DNA, histone proteins, and chromatin.

miRNAs play a very crucial role in response to stress. It is not surprising thus that Arabidopsis mutants *hen1-1* and *dcl1-9* (a different, viable mutant of DCL1), which are partially impaired in the production of miRNAs, were shown to be more stress sensitive (Sunkar and Zhu, 2004). miRNAs produced in plants are tissue- and organ-specific and are regulated by a number of abiotic stresses, including mechanical and oxidative stress, dehydration, salinity, cold, abscisic acid, and nutrient deprivation (reviewed in Sunkar et al., 2007), which might allow for the directing of methylation to specific loci as a result of producing stress-specific smRNAs.

Drought, cold, salinity, high light, and heavy metals all cause an increase in reactive oxygen species (ROS) in plants. ROS is a photosynthetic by-product, making ROS scavenging within plant cells a very important mechanism. CU-ZN SUPEROXIDE DISMUTASE1 and 2 (CSD1 and 2) are upregulated in response to oxidative stress and are important players in maintaining intercellular ROS homeostasis. Interestingly, despite their upregulation in response to abiotic stress, the transcriptional level of these genes remains static. Exposure to oxidative stress downregulates transcription of miR-398, which normally promotes cleavage of cytosolic (*CSD1*) and plastidic (*CSD2*) Cu-Zn superoxide dismutase gene transcripts (Sunkar et al., 2006), resulting in the rapid accumulation of *CSD1* and *CSD2* transcripts and allowing plants to resist stress.

Many miRNAs have been found to be regulated in response to abiotic stresses, but only a handful of small RNAs have been ascribed pathways in which they modulate resistance or acclimation. For example, miR-399(a-f) induction upon phosphate starvation targets a ubiquitin-conjugating enzyme (PHO2) releasing the regulation of a number of phosphate homeostasis-mediating genes; miR-395 is attenuated during sulfur starvation to target ATP sulfurylases (APS1, APS3, and APS4) and a low-affinity sulfate transporter (AST68) potentially modulating the rates of sulfate translocation and assimilation (Sunkar et al., 2007). Less well understood are the roles of miR-417 in affecting seed germination under salt stress and the severe upregulation of miR-393 and downregulation of miR-389a under cold, drought, and salinity stress (Sunkar and Zhu, 2004). Further, microarrays have revealed a number of miRNAs up- and downregulated by stress in response to UV treatment and cold, drought, and salt stress. With the advances in technology that might alleviate some of the technical and monetary problems associated with small RNA microarrays, we can expect this field to expand rapidly.

There are a large number of non-conservative miRNAs that are available in some species but absent in others. This might support a hypothesis that the development of a specialized miRNAs network was driven by physiological and stress conditions specific for each species (reviewed in Lu et al., 2005). Identification of 22 miRNAs from developing secondary xylem of *P. trichocarpa* stems (Lu et al., 2005) further confirmed that species-specific miRNAs contribute to regulation of gene expression associated with specific growth/stress conditions. The expression of many *P. trichocarpa* miRNAs (ptr-miRNAs) was induced in the developing xylem of stems in the presence of gravitropism-mediated mechanical stress (Lu et al., 2005). This stress increases ptr-miR408 expression thus regulating the plastocyanin-like protein involved in lignin polymerization. Mechanical stress also downregulates the expression of ptr-miR164 and ptr-miR171. These miRNAs target genes that are involved in cell division and elongation in response to gravitropism. Some of the non-conservative ptr-miRNAs induced by abiotic stress in *P. trichocarpa* are ptr-miR473, ptr-miR482, ptr-miR472, and ptr-miR159. ptr-miR472 and ptr-miR159 regulate the activity of putative disease resistance gene and cell wall polysaccharide synthesis (Lu et al., 2005 and references within), whereas ptr-miR473 and ptr-miR482 target synthesis of cell wall polyphenolics via the UV-B resistant gene (*UVR8*) and three putative disease resistance genes, respectively.

Some miRNAs induced by stress are able to target mRNAs of genes involved in epigenetic regulations themselves. These miRNAs include miR407 and miR402, which are regulated by dehydration, salinity, cold, and abscisic acid. miR407 targets a SET domain protein functioning in histone *Lys* methylation, whereas miRNA402 targets ROS-like DNA glycosylase (Sunkar and Zhu, 2004). Many stress-regulated miRNAs have multiple target sites within the same gene, suggesting that degradation might be achieved through binding to any independent binding site in the mRNA, or perhaps through some kind of translation inhibition.

Beside stress type specificity and time of the induction, stress-regulated miRNAs also exhibit tissue-specific expression patterns by regulating organ-specific functional and metabolic differences in response to stress. For example, miR393 downregulates *TIR1,* a positive regulator of auxin signaling, and has its strongest expression in the inflorescence under physiological conditions. When miR393 is induced by stresses such as salinity, dehydration, cold, or exposure to abscisic acid, it results in inhibition of plant growth, most pronounced at the apical meristem and developing inflorescence (Sunkar and Zhu, 2004).

There is less information on the involvement of siRNAs in stress response. However, most of the nat-siRNAs found to date were shown to be induced by salt stress. Borsani et al. (2005) showed that activation of the expression of one of the genes involved in an antisense pair by stress leads to production of nat-siRNA, which guides the cleavage of other gene transcripts. The SRO5-P5CDH nat-siRNA, which are siRNAs produced from the overlapping *SRO5* and

P5CDH transcripts, contribute to a regulatory loop that controls ROS production and proline homeostasis, leading to salt tolerance.

Another specific nat-siRNA was activated by *Pseudomonas syringae* infection, allowing infected plants to resist the pathogen. This mechanism might play an important role because 26% of human genes, 22% of mouse genes, 16% of Drosophila genes, and 9% of Arabidopsis genes are grouped in antisense overlapping pairs and might form *cis*-nat-siRNAs (Ron et al., 2010).

Finally, siRNAs produced from the genome of viruses infecting plants, called **viRNAs**, play an important role in protection of systemic uninfected plant tissue against incoming infection. Detailed description of viRNA production, mechanisms of virus-induced gene silencing (VIGS), and the function of viral suppressors is presented in Chapter 15.

Conclusion

Arguably, plants possess the most versatile mechanisms of biogenesis of ncRNAs. Small interfering RNAs alone consist of a number of families, including *natural-antisense transcript-derived siRNA (nat-siRNAs), cis-acting nat-siRNAs (ca-siRNAs), trans-acting nat-siRNAs (trans-nat-siRNAs), trans-acting siRNAs (ta-siRNAs), heterochromatic siRNAs (hc-siRNAs), repeat-associated siRNAs (ra-siRNAs), long siRNAs (lsiRNAs)*, and *viral induced siRNAs (viRNAs)*. This great abundance of siRNAs requires several partially independent Dicer-like proteins (DCL2, DCL3, and DCL4) for their biogenesis, whereas miRNA biogenesis relies on separate DCL, DCL1. Small RNAs in plants interact with their targets through two complexes, RITS and RISC involved in regulation of gene expression at transcriptional and posttranscriptional levels. Finally, plants exhibit sophisticated mechanisms of self-defense against invading viruses, virus-induced gene silencing, or VIGS.

Exercises and discussion topics

1. What is posttranscriptional gene silencing (PTGS), and how was it first discovered?
2. Which group of regulatory ncRNAs is missing in plants?

3. Which two large groups can small RNAs be classified into, and what is the difference between them?
4. Name five of the many proteins involved in ncRNA biogenesis as well as their function in plants.
5. What role does DICER-LIKE protein 1 play in plants?
6. What is RISC, and how does it work?
7. Provide evidence for whether or not miRNA-mediated repression of translation occurs in plants.
8. What evidence is there that miRNA-based regulation is one of the oldest mechanisms of posttranscriptional regulation in plants?
9. What are the three subclasses of siRNAs in plants? How is each subclass generated?
10. What is the role of pollen siRNAs?
11. What is an example of an l-siRNA produced in plants, and how does it work?
12. How many DICER proteins are there in Arabidopsis, and what role does each of them perform?
13. What is the role AGO proteins?
14. How many different RNA-dependent DNA methyltransferases are present in Arabidopsis? Which of these have known functions and what are they?
15. How are RdDM and RNAi similar? What is the substantial difference?
16. How do the DNA-recognition and RNA-recognition models work?
17. How does maintenance methylation of non-symmetrical CpNpN sites occur? How is it different from the symmetric CpG and CpNpG sites?
18. How are the roles of Pol IV and Pol V in RdDM different?
19. Among all ncRNA potentially involved in plant development, which is thought to be the most important and why?
20. What role can miR393 play in response to stress?
21. What kind of involvement in stress response do siRNAs have?

References

Beauclair et al. (2010) microRNA-directed cleavage and translational repression of the copper chaperone for superoxide dismutase RNA in Arabidopsis. *Plant J* 62:454-462.

Borsani et al. (2005) Endogenous siRNAs derived from a pair of natural cis-antisense transcripts regulate salt tolerance in Arabidopsis. *Cell* 123:1279-1291.

Brodersen et al. (2008) Widespread translational inhibition by plant miRNAs and siRNAs. *Science* 320:1185-1190.

Gandikota et al. (2007) The miRNA156/157 recognition element in the 3' UTR of the Arabidopsis SBP box gene SPL3 prevents early flowering by translational inhibition in seedlings. *Plant J* 49:683-693.

Gasciolli et al. (2005) Partially redundant functions of Arabidopsis DICER-like enzymes and a role for DCL4 in producing trans-acting siRNAs. *Curr Biol* Aug 15:1494-1500.

Herr et al. (2005) RNA polymerase IV directs silencing of endogenous DNA. *Science* 308:118-120.

Jones-Rhoades et al. (2006) MicroRNAS and their regulatory roles in plants. *Annu Rev Plant Biol* 57:19-53.

Law et al. (2010) A protein complex required for polymerase V transcripts and RNA-directed DNA methylation in Arabidopsis. *Curr Biol* 20:951-956.

Lu et al. (2005) Novel and mechanical stress-responsive MicroRNAs in Populus trichocarpa that are absent from Arabidopsis. *Plant Cell* 17:2186-2203.

Matzke MA, Birchler JA. (2005) RNAi-mediated pathways in the nucleus. *Nat Rev Genet* 6:24-35.

Moldovan et al. (2010) The hunt for hypoxia responsive natural antisense short interfering RNAs. *Plant Signal Behav* 5:247-251.

Mosher et al. (2008) PolIVb influences RNA-directed DNA methylation independently of its role in siRNA biogenesis. *Proc Natl Acad Sci USA* 105:3145-3150.

Napoli et al. (1990) Introduction of a Chimeric Chalcone Synthase Gene into Petunia Results in Reversible Co-Suppression of Homologous Genes in trans. *Plant Cell* 2:279-289.

Onodera et al. (2005) Plant nuclear RNA polymerase IV mediates siRNA and DNA methylation-dependent heterochromatin formation. *Cell* 120:613-622

Parizotto et al. (2004) In vivo investigation of the transcription, processing, endonucleolytic activity, and functional relevance of the spatial distribution of a plant miRNA. *Genes Dev* 18:2237-2242.

Pikaard et al. (2008) Roles of RNA polymerase IV in gene silencing. *Trends Plant Sci* 13:390-397.

Pikaard CS. (2006) Cell biology of the Arabidopsis nuclear siRNA pathway for RNA-directed chromatin modification. *Cold Spring Harb Symp Quant Biol* 71:473-480.

Pontes et al. (2006) The Arabidopsis chromatin-modifying nuclear siRNA pathway involves a nucleolar RNA processing center. *Cell* 126:79-92.

Rajagopalan et al. (2006) A diverse and evolutionarily fluid set of microRNAs in Arabidopsis thaliana. *Genes Dev* 20:3407-3425.

Ron et al. (2010) Proper regulation of a sperm-specific cis-nat-siRNA is essential for double fertilization in Arabidopsis. *Genes Dev* 24:1010-1021.

Schwab et al. (2005) Specific effects of microRNAs on the plant transcriptome. *Dev Cell* 8:517-527.

Slotkin et al. (2009) Epigenetic reprogramming and small RNA silencing of transposable elements in pollen. *Cell* 136:461-472.

Sunkar et al. (2007) Small RNAs as big players in plant abiotic stress responses and nutrient deprivation. *Trends Plant Sci* 12:301-309.

Sunkar et al. (2006) Posttranscriptional induction of two Cu/Zn superoxide dismutase genes in Arabidopsis is mediated by downregulation of miR398 and important for oxidative stress tolerance. *Plant Cell* 18:2051-2065.

Sunkar R, Zhu JK. (2004) Novel and stress-regulated microRNAs and other small RNAs from Arabidopsis. *Plant Cell* 16:2001-2019.

Vaucheret H. (2006) Post-transcriptional small RNA pathways in plants: mechanisms and regulations. *Genes Dev* 20:759-771.

Vazquez F. (2006) Arabidopsis endogenous small RNAs: highways and byways. *Trends Plant Sci* 11:460-468.

Yang et al. (2007) Functional diversity of miRNA in plants. *Plant Science* 172:423-432.

12

Non-coding RNAs—comparison of biogenesis in plants and animals

Regulation of gene expression via small non-coding RNAs (ncRNAs) is one of the most ancient mechanisms, and it appears to exist throughout all domains and kingdoms of life. In most of the species analyzed to date, small silencing RNAs are characterized by their short length (approximately 20 to 30 nt), their association with members of the Argonaute (Ago) family of proteins, and their capacity to reduce gene expression by several independent mechanisms. At the same time, there exist longer small ncRNAs. The expression of ncRNAs might also lead to an increase in gene expression. Some ncRNA classes might have common regulatory pathways. Moreover, similar types of ncRNAs may be derived from single- or double-stranded RNA precursors, and different types of ncRNAs may be derived from either single-stranded or double-stranded RNAs. Typically, the processing machinery of ncRNAs of the same class is very similar; however, there are many organism-specific differences. Also, new classes of small regulatory RNAs are constantly discovered; therefore, our understanding of modes of action and biogenesis pathways is far from being complete.

This chapter compares the mechanism of small RNA biogenesis in various organisms. First, we summarize the most commonly occurring small non-coding RNAs, their origin, and their functions. It appears that plants are the most versatile organisms in their ability to generate different types of small RNAs.

Table 12-1 summarizes various types of small ncRNAs involved in silencing (reviewed in Ghildiyal and Zamore, 2009).

Table 12-1 Small ncRNAs involved in silencing

Type of ncRNA (size, nt)	Organisms	Origin	Proteins involved	Function
miRNA (20 to 25)	Protists, algae, plants, animals	Inter- and intra-genic; pri-miRNA	DCL1 in plants; Drosha, Dicer in animals	mRNA cleavage and transcription inhibition
exo-siRNA (21; 24 in plants)	Protists, fungi, plant, animals	Transgenes and viral RNAs	Dicer (various DCLs in plants)	mRNA cleavage; defense against viruses
endo-siRNA (21)	Protists, algae, fungi, plant, animals	Gene/pseudo-gene pairs, bidirectional transcription, structured loci	Dicer and RdRP-dependent in animals	mRNA cleavage; DNA methylation
ca-siRNA (24)	Plants	transposons	DCL3	mRNA cleavage; DNA methylation
hc-siRNA (21-24)	Plants	heterochromatic regions	DCL3	DNA methylation; heterochromatin formation
hp-siRNA (21-24)	Animal	inverted transposons	Dicer	mRNA cleavage; heterochromatin formation
ta-siRNA (21)	Plant	miRNA-cleaved TAS RNAs	DCL4	mRNA cleavage
nat-siRNA (24)	Plant (primary)	Bidirectional transcripts	DCL2	Cleavage of stress-induced mRNAs
nat-siRNA (21)	Plant (secondary)	Secondary nat-siRNA	DCL1	Cleavage of stress-induced mRNAs
ra-siRNA (21; 24 in plants)	Protists, fungi, plant, animals	Repeat-associated	Dicer (various DCLs in plants)	DNA methylation; heterochromatin formation
piRNA (24-30)	Drosophila, zebrafish, mammals	Long primary transcripts	Piwi/Aub/Ago	Transposon silencing; RNA cleavage
21U-RNA (piRNA) (21)	*C. elegans*	Individual piRNA transcripts	PRG1	Transposon silencing

miRNA (microRNA); ca-siRNA (*cis*-acting nat-siRNA); hc-siRNA (heterochromatic siRNA); hpsiRNA (hairpin siRNA (also known as TE-hp-siRNA, transposons element hairpin siRNA)); ta-siRNA (transacting siRNA); nat-siRNA (natural antisense transcript-derived siRNA); ra-siRNA (repeat-associated siRNA); exo-siRNA (exogenous siRNA); endo-siRNA (endogenous siRNA); piRNA (PIWI-interacting siRNA); 21U-RNA (piRNA of 21 nt in length)

Comparison of miRNAs biogenesis and function in plant and animals

Because no common miRNAs have been found in plants and animals, it is assumed that miRNA biogenesis pathways have evolved independently in these two kingdoms. However, the report by Arteaga-Vazquez et al. (2006) demonstrated that Arabidopsis miR854 has a single nucleotide mismatch with miRNA homologs found in four animals including humans. This miRNA not only possesses homologs in different animals, it apparently has a target conserved among plants and animals. In Arabidopsis, the 3' UTR of the *UBP1*-like gene that encodes an hnRNP-like protein contains several binding sites for miR854. Thus, this miRNA appears to be one of just a few miRNAs capable of translation inhibition in plants. Similarly, in animals, the 3' UTR of the *UBP1* homologs also contains binding sites for miR854, although translation inhibition by this miRNA in animals remains to be shown.

Table 12-2 summarizes differences between plants and animals as to the origin, biogenesis, and function of miRNAs.

Table 12-2 Comparison of the mechanisms of miRNA biogenesis in plants and animals

	Plants	Animals
Origin	Mostly intergenic regions, mostly individual	Exons of non-coding genes, introns of coding and non-coding genes, intergenic regions, often clustered, polycistronic
Biogenesis	Larger and more complex dsRNA loops	More simple structures of precursors
	miRNA precursors are processed twice in the nucleus	miRNAs are processed first in the nucleus and then in the cytoplasm
	miRNAs/miRNAs* are formed in the nucleus	miRNAs/miRNAs* are formed in the cytoplasm
	Single miRNA is processed from one precursor	Multiple miRNAs are processed from one precursor
	2'-O-methylated at their 3' ends	2'-O-methylation only in Drosophila
Editing	The phenomenon is not observed	miRNAs can be edited

Table 12-2 Comparison of the mechanisms of miRNA biogenesis in plants and animals

	Plants	**Animals**
Function	Targeted mRNAs are cleaved	Translation is inhibited in target mRNAs
	A single miRNA targets mRNA	Multiple miRNAs target mRNA
	A single miRNA targets a single mRNA	A single miRNA can target many mRNAs
	A high degree of homology to mRNA	A lower degree of homology to mRNA
	Mainly target transcription factors	Target various mRNAs
	Mature miRNAs are formed in the nucleus	Mature miRNAs can be transferred to the nucleus from the cytoplasm
Fate of repressed mRNAs	The phenomenon is less common in plants	Repressed mRNAs are stored in P bodies
Conservation	Highly conserved within the kingdom	Highly conserved within the kingdom

In the plant genome, miRNAs are primarily transcribed from intergenic regions (found between protein-coding genes) as individual genes. In contrast, in animals, miRNAs are often found in the introns of coding and non-coding genes, in the exons of non-coding genes, and in the intergenic regions. In animals, such miRNAs are often polycistronic. Thus, in animals, the loci carrying miRNAs are co-transcribed into multiple pri-miRNAs, each capable of producing single mature miRNAs. In plants, with rare exceptions, the miRNA-containing loci are transcribed into single pri-miRNAs, thus giving rise to single miRNAs. Also, it appears that in plants, the stem-loop structures of pre-miRNAs are larger and more variable (see Figure 12-1). In unicellular organisms, the miRNA-containing loci are also transcribed into single stem-loop precursors. In contrast to plants and animals, these structures are much larger and can produce multiple mature miRNAs (see Figure 12-1).

In animals, precursors of miRNAs contain several loops with potential miRNAs. In plants, the stem-loop structure carries only a single miRNA. In algae, the loops are much larger and can contain more than one potential miRNA.

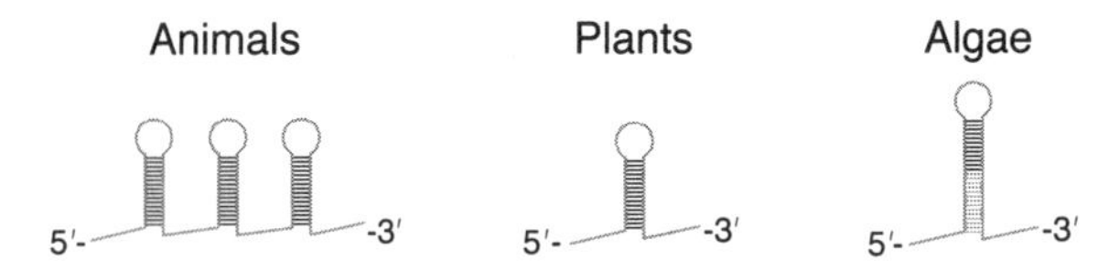

Figure 12-1 Comparison of stem-loop structures among different organisms (adapted from original figure from Naqvi AR, et al. The Fascinating World of RNA Interference. Int. J. Biol. Sci. 2009; 5:97-117. Available from http://www.biolsci.org/v05p0097.htm).

The difference between the biogenesis of plant and animal miRNAs lies in the fact that in plants, miRNA maturation occurs entirely in the nucleus, and the cleavage of RNA precursors is performed with the help of one protein, DCL1 (see Figure 12-2); in contrast, in animals, the precursor molecules are first cleaved by the Drosha/DGCR8 pairs, and the actual maturation of miRNAs occurs in the cytoplasm (see Figure 12-2). The steps of miRNA maturation in plants also include 2'-O-methylation at the 3' ends.

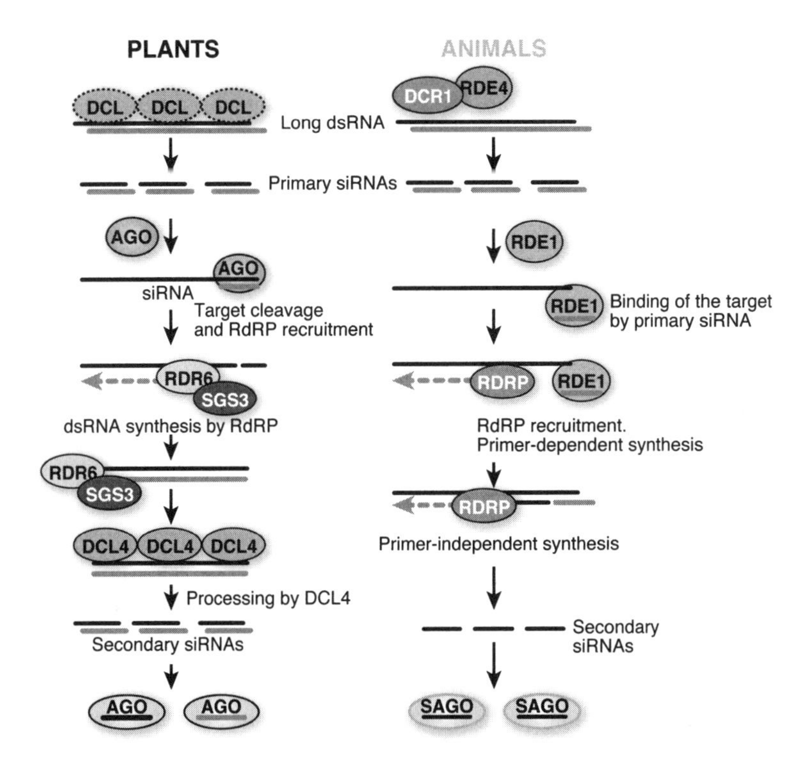

Figure 12-2 Comparison of miRNA biogenesis in plants and animals.

Further differences between miRNAs in plants and animals are obvious from the analysis of their influence on gene expression. In plants, miRNAs regulate single targets, whereas in animals, they can target multiple mRNAs. Also, plant miRNAs primarily bind to the coding regions of mRNAs, whereas in animals, miRNAs bind to the 3' UTR regions, multiple molecules at a time. Whereas miRNAs in animals are involved in translation inhibition, miRNAs in plants regulate transcription via transcript cleavage. However, this is not a firm rule. For example, in plants, Apetala, a transcription factor, is translationally repressed by miR-172, whereas in animals, miR-196 is able to cleave a HOXB transcript. Also, it was recently shown that translational repression by miRNAs in plants is quite frequent and might involve P-body function (Broderson et al., 2008).

In plants, miRNA-directed cleavage requires nearly perfect homology to the target mRNA, whereas in animals, translational repression might occur if there are a few mismatches to the target mRNA. It should be noted, however, that homology between seed sequence and target mRNA should be nearly perfect, with a maximum of a single mismatch allowed. The requirement of a high degree of homology between plant miRNAs and the targeted coding sequences suggests that in animals the function of plant miRNAs resembles the function of siRNAs rather than that of miRNAs.

The mechanisms of miRNA-mediated translational inhibition might be similar in plants and animals

How are the mechanisms regulating transcript stability controlled in plants? In the past, it was suggested that these mechanisms are mutually exclusive and are primarily regulated by the presence of mismatches between miRNAs and their targets. Perfect and near-perfect complementarity between miRNAs and mRNA targets presumably excludes translational repression because it activates the cleavage of mRNA by AGO1. Conversely, an imperfect pairing (especially at the seed region) typically results in translational repression because the target mRNAs avoid cleavage. The aforementioned observations have contributed to the notion that the mechanism of miRNA function in plant and animal miRNAs is totally different. Recent findings by Brodersen et al. (2008) suggest that this might not be the case. The authors argue that the mechanisms of translational inhibition in animals and plants are common. Indeed, translational repression by near-perfectly matched miRNAs has been reported several times in Arabidopsis (Aukerman et al., 2003). Also, the complete reduction of mRNA levels in RNAi promoted by artificial miRNAs is not necessary for producing phenotypes that are indistinguishable from genetic knockouts of their targets. Finally, experiments with the inducible RNAi construct showed that targeted mRNA levels return to normal, whereas the protein's activity remains low.

The authors propose that translational repression is the default mechanism by which small RNAs silence messages, both in plants and animals (Brodersen et al., 2008). In addition to that, if miRNAs have near perfect homology with mRNAs, they might regulate transcription through cleavage of their targets. However, translational repression might be more common for so-called young miRNAs. For example, non-conserved miR834 targets the COP1-interacting protein 4 (CIP4) with near-perfect matches; however, this is not sufficient for cleavage of the target. Many young miRNAs are presumed to be nonfunctional because in *dcl1* and *hen1* mutants—which are responsible for processing and methylation of miRNAs, respectively—the putative target miRNA levels are unchanged (reviewed in Brodersen et al., 2008).

Future research will need to substantiate these findings. Importantly, since the discovery of novel miRNAs and prediction of their targets have been reported, it has also become critical to investigate the degree of base-pairing and mismatches between miRNAs and their potential targets. This will also allow the prediction of mRNAs that are potential targets of translational inhibition in plants.

It remains to be shown whether AGO1 proteins, which are essential components of RISC, may be simultaneously involved in cleavage and translational inhibition of the same miR-mRNA duplex. It is possible that these two mechanisms coexist within the same cells, and some third-party factors might promote cleavage or inhibition. However, it is also possible that cleavage and translational inhibition are spatially and/or temporally separated. The same miRNAs might slice mRNA targets during one developmental stage or in a particular tissue type and inhibit translation during another developmental stage or in a different tissue type.

The mechanism of miRNA-mediated translational repression might be similar in plants and animals. In both organisms, translational repression requires AGO1, although in plants, it can also occur through the function of AGO10. Indeed, *ago1* and *ago10* mutants show overlapping developmental defects, suggesting their involvement in the similar pathways. Repression might also require the activity of the microtubule severing enzyme KATANIN (KTN1), implicating cytoskeleton dynamics in miRNA action in plants and animals (Brodersen et al., 2008). Indeed, the authors demonstrated that *mad5* plants mutated in KTN1 have both the normal activity of miRNA-induced mRNA cleavage and a substantial reduction in translational inhibition. Because KTN1 functions as a microtubule-severing enzyme, it can be inferred that the reorganization of the microtubule network is a requirement for miRNA-directed translational repression in plants. Moreover, it has been suggested in other organisms that microtubules play a role in miRNA function. For example, depletion of tubulin protein in *C. elegans* causes the impaired function of several different miRNAs (Parry et al., 2007). Also, in Drosophila, the assembly of functional RISC requires a microtubule-associated protein, Armitage, and miRNA-mediated translational activation requires another microtubule-associated protein, FMR (reviewed in Brodersen et al., 2008).

Another similarity of miRNA function between animals and plants includes the possible fate of translationally repressed mRNAs. In animals, three proteins—DCP1, DCP2, and Ge-1—which compose the decapping complex, are necessary for miRNA-guided translational repression. Although mutations in the homologs of these genes in Arabidopsis ecotype *Columbia* (*Col-0*) plants are lethal, mutations in the Ge-1 homolog *VARICOSE* (*VCS*) in Arabidopsis ecotype *Landsberg erecta* (*Ler-0*) are not lethal. Brodersen et al. (2008) were able to analyze the role of VCS in translational inhibition and showed that *vcs* plants are impaired in this process. Previously it was shown that the *vcs* mutant accumulates capped mRNAs and arrests the decay of many mRNAs (Goeres et al., 2007). Thus, it can be suggested that the decapping component Ge-1/VCS plays a similar role in translational inhibition in animal and plant kingdoms (Brodersen et al., 2008).

The connection between the activity of the decapping complex and cytoskeleton was demonstrated in yeasts. Many factors involved in mRNA decay co-localize with the decapping complex in P-bodies and interact with tubulin and microtubule polymers in yeast (Gavin et al., 2008). It remains to be shown whether a similar parallel can be drawn for plants and animals.

Loss of miRNA biogenesis results in severe developmental phenotypes in all organisms

In animals, mutations in Dicer and Argonaute proteins associated with miRNA biogenesis are almost always lethal and cause severe defects in development. In the *dcr-1* mutant of Drosophila, the germline stem cell clones divide slowly, whereas the *dcr-1* mutant of *C. elegans* displays defects in germline development and embryogenesis. Zebrafish mutants that are deficient in maternal and zygotic Dicer activity are also defective in embryogenesis. Mice lacking Dicer are embryonic lethal and deficient in stem cells (reviewed in Ghildiyal and Zamore, 2009). In Arabidopsis, the *dcl1* mutant also exhibits developmental defects and is infertile. It should be noted, however, that the *dcl1* mutant is not completely devoid of miRNA, suggesting that other DCL proteins can, in part, substitute for DCL1 function.

Comparison of siRNA biogenesis and function in plant and animals

Plant siRNAs appear to be more diverse than animal siRNAs (see Table 12-3). This might be the consequence of the sessile lifestyle of plants, which cannot utilize the mechanism of escape in response to abiotic and biotic stress. For details on biogenesis of various siRNAs, refer to Chapters 7, 10, and 11. Here, we only mention that siRNAs in plants are methylated in a similar manner as miRNAs. In animals, methylation of siRNAs occurs only in Drosophila. In general, siRNAs in animals are processed using a single Dicer protein, whereas plants use three different DCL proteins—DCL2, DCL3, and DCL4—which are very redundant in their function. In plants, evolution of multiple DCL proteins might be a reflection of the need to protect their sedentary life. Table 12-3 compares the mechanisms of siRNA biogenesis in plants and animals.

Table 12-3 Comparison of the mechanisms of siRNA biogenesis in plants and animals

	Plants	Animals
Type	ta-siRNAs, nat-siRNAs, ca-siRNAs, ra-siRNAs, l-siRNAs, hc-siRNAs, viRNAs	*cis*-NAT-siRNAs, *trans*-NAT-siRNAs, hp-siRNAs, TE-siRNAs, piRNAs
Biogenesis	siRNAs are 2'-O-methylated at their 3' ends	2'-O-methylation at their 3' end only in Drosophila
	DCL2, DCL3, and DCL4 process different siRNAs	Single Dicer processes all siRNAs
	Substantial redundancy in function of DCL and AGO proteins	Limited functional redundancy
Function	Targeted mRNAs are cleaved	Targeted mRNAs are cleaved
	Systemic spread of silencing	Systemic spread of silencing
	Formation of RISC and RITS	Formation of RISC and RITS
	DNA methylation through RdDM is more common	DNA methylation through RdDM is less common
	No such phenomenon reported	siRNAs direct DNA elimination in ciliates

In plants and animals, siRNAs are involved in cleavage of target mRNAs and DNA methylation, and thus they are part of two main complexes: RISC and RITS. siRNA-mediated methylation is more common in plants. In ciliates, siRNAs have been shown to promote DNA elimination (see Chapter 9, "Non-Coding RNAs Across the Kingdoms—Protista and Fungi"). Also, in humans and *C. elegans*, a specific type of siRNAs is produced—Piwi-associated siRNAs or piRNAs (refer to Table 12-2). Although plants and animals (excluding human and worms) do not produce piRNAs, their function may be maintained by ra-siRNAs and hc-siRNAs in plants and by hp-siRNAs and TE-siRNAs in animals. In plants and animals, siRNAs are involved in short- and long-distance signaling.

Table 12-4 compares the mechanisms of silencing spread in worms and plants.

Table 12-4 Comparison of the mechanisms of amplification of silencing in worms and plants

Target	Plants	Worm
long dsRNA	Processed by DCL2, DCL3, or DCL4.	Processed by DCR1/RDE4.
mRNA	AGO complex recruits primary siRNAs to mRNA.	RDE1 complex recruits primary siRNAs to mRNA.
mRNA/siRNA	RDR6, assisted by SGS3 amplifies second strand.	RDRP, assisted by RDE1 performs primer-independent synthesis of secondary siRNAs. The mechanism is not entirely clear.
dsRNA	DCL4 processes dsRNA into secondary siRNAs.	
secondary siRNAs	AGO complexes incorporate secondary siRNAs to target mRNA.	Secondary AGO (SAGO) such as CSR-1 associate with secondary siRNAs to target mRNA.

The silencing response in plants and worms requires the amplification of primary siRNAs through the activity of RNA-dependent RNA polymerases (RdRPs). Primary siRNAs are produced from endogenous or exogenous double-stranded RNA targets, such as complementary mRNAs or replicating viral genomes. In both plants and worms, Dicers process dsRNA precursors into primary siRNAs. In plants, primary siRNAs direct cleavage of target mRNAs by AGO1; these cleaved transcripts are then used by the RdRP enzyme RDR6 to synthesize long dsRNAs. The newly amplified long dsRNAs are then diced into secondary siRNAs with the help of DCL enzymes (reviewed in Ghildiyal and Zamore, 2009). Because secondary siRNAs in plants can be produced from either the 5' or 3' ends of the cleaved transcript, it can be suggested the signal for amplification of primary siRNAs requires the initial mRNA cleavage rather than priming of RDR6 by primary siRNAs. However, it is suggested that transitivity in Arabidopsis can sometimes be primed (Moissiard et al., 2007). In plants, the RDR6-dependent amplification of secondary siRNAs plays an important role in the protection of plants against viruses.

In *C. elegans*, secondary siRNAs are produced from primary siRNAs through a different mechanism. The Argonaute protein RDE-1 binds primary siRNAs and guides them to the target mRNAs. Binding of the target recruits the RdRP protein that amplifies secondary siRNAs (reviewed in Ghildiyal and Zamore, 2009). Two evidences suggest that secondary siRNAs in *C. elegans* are produced by reverse transcription rather than dicing. First, worm secondary siRNAs have di- or triphosphates at the 5' end, indicating that they are produced by transcription. Second, *in vitro*, secondary siRNAs are produced in the absence of Dicer protein. Because secondary siRNAs are reverse-transcribed in *C. elegans*, they are in antisense to the target mRNA. Curiously, in Neurospora, QDE1 can transcribe approximately 22 nt small RNA oligos from a long RNA template *in vitro*, suggesting that secondary siRNAs are also produced in fungi.

Secondary siRNAs can be amplified in the same cells as primary siRNAs or in distal cells. Regardless, secondary siRNAs are bound to secondary Argonautes (such as CSR-1) that promote cleavage of target mRNAs. Amplification of siRNAs can also occur in flies that have RdRP activity (Lipardi and Paterson, 2009). However, extensive biochemical and genetic studies suggest that the RNAi pathway in Drosophila is independent of RdRP (reviewed in Ghildiyal and Zamore, 2009). Also in mammals, the RNAi mechanism can be used without RdRP. It is possible that the development of a sophisticated immune system in mammals substitutes the need for RdRP-dependent amplification. The absence of amplification of secondary siRNAs allows for an efficient allele-specific RNAi in cultured mammalian cells and will, in the future, allow for efficient and specific methods for the correction of various diseases in humans.

Summary of various non-coding RNAs produced in different organisms

Table 12-5 summarizes various types of non-coding RNAs produced in different organisms and common proteins involved in the mechanisms of their biogenesis.

Table 12-5 Summary of various non-coding RNAs produced in different organisms

	miRNA	siRNA	piRNA	Unique RNA types	Dicer	Argonaute	RdRP
Bacteria	-	-	-	Protein-binding, 6S RNA, *cis-* and *trans-* sRNAs, CRISPRs	-	-	-
Archaea	-	-	-	cis-sRNAs and CRISPRs	1 (Dcl1)	-	-
Protozoa (*Tetrahymena thermophila; Trypanosoma brucei*)	+	+	-	scnRNA	3 (Dcl1, Dcr1, Dcr2)	Ago1	-
N. crassa	+	+	-	qiRNAs, milRNAs, disiRNAs	DCL1, DCL2	2 (QDE-2, SMS2)	3 (QDE-1, SAD-1, RRP-3)
S. cerevisiae	-	+	-	CUTs, MUTs, ncRNAs	-	-	-
S. pombe	+	+	-		Dcr1	Ago1	Rdp1
C. elegans	+	+	+	21U-RNAs	Dcr1	27 (Primary and secondary, SAGO)	4 (EGO1, RRF-1, RRF-2, RRF-3)
Drosophila	+	+	+		2 (DCR1, DCR2)	5 (Ago1, Ago2...)	-

Table 12-5 Summary of various non-coding RNAs produced in different organisms

	miRNA	siRNA	piRNA	Unique RNA types	Dicer	Argonaute	RdRP
Humans	+	+	+	Xist, Tsix, HOTAIR, HSR1	1	2 (Ago1, Ago2)	-
Plants (Arabidopsis)	+	+	-	ra-siRNAs, ta-siRNAs	4 (DCL1–DCL4) 10 in rice	10 (Ago1–Ago10)	7 (RDR2, RDR6...)

In Arabidopsis, AGO4 and AGO6 mediate DNA methylation, and AGO7 is involved in biosynthesis of some transacting siRNAs, whereas AGO1 can function as a miRNA-guided slicer. AGO2, AGO5, and AGO7 do not interact with most miRNAs. The knowledge about function of various AGO proteins found in *C. elegans* is scarce.

Conclusion

There are substantial similarities and great differences in type of ncRNAs produced and their mechanisms of biogenesis among eukaryotes. Despite substantial similarity in structure of miRNAs between plants and animals, it is hypothesized that miRNAs evolved independently in these kingdoms. miRNAs in plants and animals are processed by different proteins and even similar steps are carried on in different cell compartments. Moreover, despite substantial conservation of miRNA sequences among plant species, there is little to no conservation (as far as a target sequence is involved) between plants and animals. Plants possess more types of ncRNAs, including ca-siRNAs, ta-siRNAs, and nat-siRNAs, although ca-siRNAs might be similar to hc-siRNAs in plants and hp-siRNAs in animals. Animals also have their own unique class of siRNAs—piRNAs—utilized for targeting transposon sequences for degradation and methylation-dependent silencing in the germline. At the same time, several groups of siRNAs, endo-siRNAs, exo-siRNAs, and ra-siRNAs exist in nearly all eukaryotes, plants, animals, fungi, and protists.

Exercises and discussion topics

1. What are the three main characteristics of small silencing RNAs that are similar in most of the species?
2. Are there common miRNAs among plants and animals?
3. In which organisms are piRNAs produced?
4. What are the main differences in the origin, mechanisms of biogenesis, and function of miRNAs between plants and animals?
5. Are there any similarities in mechanisms of miRNA-mediated translational inhibition between animals and plants?
6. What is the possible fate of translationally repressed mRNAs in plants and animals?
7. What are the similarities and differences of siRNA biogenesis in plants and animals?
8. Compare the mechanisms of silencing spread in worms and plants.
9. List organisms that do or do not have active RdRP mechanisms.
10. What are the unique types of RNAs produced in plants and animals?

References

Brodersen et al. (2008) Widespread translational inhibition by plant miRNAs and siRNAs. *Science* 320:1185-1190.

Ghildiyal M, Zamore PD. (2009) Small silencing RNAs: an expanding universe. *Nat Rev Genet* 10:94-108.

Goeres et al. (2007) Components of the Arabidopsis mRNA Decapping Complex Are Required for Early Seedling Development. *Plant Cell* 19:1549-1564.

Lipardi C, Paterson BM. (2009) Identification of an RNA-dependent RNA polymerase in Drosophila involved in RNAi and transposon suppression. *Proc Natl Acad Sci USA* 106:15645-15650.

Moissiard et al. (2007) Transitivity in Arabidopsis can be primed, requires the redundant action of the antiviral Dicer-like 4 and Dicer-like 2, and is compromised by viral-encoded suppressor proteins. *RNA* 13:1268-1278.

13

Paramutation, transactivation, transvection, and cosuppression—silencing of homologous sequences

Epigenetic regulation of gene expression involves mitotically and meiotically stable but potentially reversible modifications not including changes in DNA sequence. **Paramutation** is one such epigenetic phenomena that involves changes in expression of one allele upon interaction with another allele. The interaction involves transmission of an epigenetically regulated expression state from one homologous sequence to another. Paramutation was first described in maize but has also been reported in other organisms as well as in fungi, tomato, pea, and mouse. The exact mechanisms triggering paramutation are not known. In this chapter, two major models are presented: the **physical interaction model** and the **small noncoding RNA-mediated model**.

We are used to the fact that gene expression is faithfully inherited from cell to cell. Cells belonging to the same tissue acquire the same phenotypic characteristics that depend on the expression pattern of a given gene at a given time. The flow of genetic information follows the Mendelian laws of segregation. These patterns are rarely altered by various types of mutations that cause an increase or a decrease in the expression of the mutated allele.

Previous chapters have discussed the mechanisms of transcriptional and post-transcriptional gene silencing. Transcriptional gene silencing leads to inactivation of previously active alleles. Paramutation is a somewhat related mechanism. It also results in gene inactivation that is triggered by *in trans* interactions between a paramutagenic silent allele and a paramutable active allele. Such interactions lead to inactivation of the paramutable allele that acquires the characteristics of the paramutagenic allele. Changes in the paramutable allele are mitotically and meiotically stable; segregation of the paramutagenic and paramutable alleles does not activate the expression of the paramutable allele. Curiously, this phenomenon is "contagious" as crosses between plants carrying the newly converted paramutable allele, which became paramutagenic, and plants carrying the normally expressing allele results in silencing and thus, conversion of a new paramutable allele. This chapter introduces the phenomenon of paramutation and discusses possible mechanisms behind it. Additionally, we briefly introduce such phenomena as transactivation, transvection, and cosuppression to show the versatility of the regulation of the expression of homologous sequences and allelic interactions.

Paramutation as an allelic interaction

Paramutation is most frequently observed between homologous DNA sequences present in an allelic position. It should be noted, however, that paramutation can occur between homologous sequences at non-allelic positions. This chapter, however, refers to all sequences that involve paramutation as "alleles" simply to indicate their participation in the paramutation process.

Paramutation involves two different types of alleles: one that triggers paramutation, and thus is called **paramutagenic**, and another allele that acquires paramutagenic state and is called **paramutable**. The state of paramutation is typically low expression of the allele. Thus, the process of paramutation is characterized by the interaction between low expressing paramutagenic allele with high expressing paramutable allele that leads to acquisition of low expression state by the paramutable allele. At the same time, the alleles that do not participate in paramutation are referred to as neutral or **non-paramutagenic** (Chandler, 2004). Changed paramutable alleles become paramutated and are designated with an apostrophe, as in the case of the *R-r'* alleles responsible for anthocyanin pigmentation.

Because paramutable alleles also have the ability to become paramutagenic and the paramutated state is heritable, it is important to differentiate between the initiation and maintenance steps of paramutation as well as between primary and secondary paramutations. The initial transfer of distinct epigenetic states is referred to as the **establishment of paramutation**, whereas the transmission of this state through mitotic and meiotic cell divisions is called **maintenance of paramutation**. The frequency with which paramutation is manifested in the progeny (F1 or any generation in which the transgene and the endogene are simultaneously present in the genome) is called **penetrance**; it is expressed as a percentage and varies from 0% to 100%. And finally, the stability of the paramutated state also varies; there is a possibility that the paramutable allele loses its newly acquired epigenetic state and returns to the original chromatin structure and gene expression. The stability of the paramutated state can be influenced by the presence of other alleles, such as the neutral allele *in trans*.

Paramutation as a process could be described as *in trans* interactions between homologous sequences that lead to the establishment of distinct epigenetic states that are heritable in nature. As mentioned before, homologous sequences might represent two alleles of the same endogene, two alleles of the transgene or an allele of the transgene, and an allele of the endogene.

Historical overview

Perhaps, the first documented paramutation-like phenomenon was reported in 1915 (Bateson and Pellew, 1915). The scientists observed an unusual phenotype of garden pea *Pisum sativum* that had narrow leaves and petals. Normal and rogue plants formed hybrids that had an intermediate phenotype at the base of the plant. During development, the hybrids turned more rogue-like in appearance, and their offspring displayed only the rogue phenotype.

It was not until 1950s when more details about possible influence of alleles on each other was described. In 1956, Alexander Brink described the phenomenon of paramutation at the *r1* locus (Brink, 1956). Brink observed that one allele encoding the synthesis of purple anthocyanin pigments in kernels of maize had an influence on another allele regulating the synthesis of light-colored pigments, which resulted in 100% conversion of color. At the same time, the stability of the change was much lower than if it were a true genetic mutation. A few years later, Edward Coe, Jr., reported that the *b1* locus in maize underwent similar changes in heritability (Coe, 1959). Finally, Rudolf Hagemann discovered similar interactions at the *sulfurea* locus of *Lycopersicon esculentum* (tomato) (Hagemann, 1958). Although since then other examples of paramutation have been reported in maize and other species, paramutation in the two maize loci, *r1* and *b1,* remains by far the most described and understood.

Other examples of similar phenomena include the cruciata character, which describes the smaller cross-shaped flowers observed in hybrids of several Oenothera species (evening primrose), reversion of phenotypic changes in transgenic *Arabidopsis thaliana* (Arabidopsis) plants expressing GFP-COP1 transgenes, and so on. (See a detailed list in Chandler and Stam, 2004.) In the past, similar phenomena were referred to as mass somatic mutation, somatic conversion, conversion, and conversion-type phenomenon. Finally, in 1968 at the International Congress of Genetics in Tokyo, the term paramutation was proposed by Alexander Brink, Rudolf Hagemann, and Edward Coe to be used for describing similarities among the aforementioned phenomena (Chandler and Stam, 2004).

Until the appearance of molecular tools, the phenomenon of interactions between alleles was studied by careful genetic analysis and observations of a particular phenotype. The ability to perform molecular analysis and look at the chromatin structure of interacting alleles, methylation, and histone modification patterns and the gene expression has dramatically advanced this field. According to PubMed database, the search for "paramutation" reveals 119 articles (April, 2012), with just 20 papers being published in the first 35 years after the initial report, and practically all of them being authored or co-authored by Dr. Brink. In the last 15 years or so, the scientific community witnessed a real splash of interest in the phenomenon of paramutation; nearly 100 articles describing it were published, with 37 of them appearing in print in the last three years. More importantly, currently there are many more labs studying the phenomenon around the world, and more new facts will be discovered every day.

Plant transgenesis was another revolutionary gadget that produced more hints on paramutagenic activity. Because transgenes are often chosen for giving rise to a distinct phenotype, and the sequence of the transgene and often the genomic position can be identified, changes in chromatin structure and gene expression are relatively easy to observe. Interactions *in trans* were described between two homologous transgenes as well as between transgenes and endogenous genes. Indeed, such interactions typically resulted in altered transcription and changes in chromatin organization observed as changes in DNA methylation or chromatin compactness leading to differential sensitivity to nucleases noticeable at both interacting homologous sequences (Chandler and Stam, 2004).

Although paramutation has been initially described in plants, it is apparently not limited to plants as similar phenomena have been described in human and mouse.

13-1. Plant transgenesis

Transgenesis—or transformation or genetic modification—is the process describing the generation of transgenic (genetically modified) plant using a variety of techniques with the main ones being Agrobacterium-mediated and biolistic. Transgenes are the genes that are integrated into plants. Transgenes may be the genes from the same plant species, from different plant species, or even from a different kingdom. Transgenic plants are primarily used in research as a powerful forward and reverse genetic tool but are also used for improving food and feed quality.

Plant paramutation

As we have already mentioned, the earliest example of paramutation as a phenomenon dates back to 1915, although the work reporting phenotypes of garden pea was rather descriptive and was never followed up. Thus, the systematic analysis of paramutation as well as the introduction of this term itself were not done until the late 1950s, when Alexander Brink described a somewhat puzzling and controversial phenomenon of the inheritance of the *Red 1* (*r1*) locus in maize

(Brink, 1956). It was observed that the spotted seed allele (*R-st*) was able to transform the *R-r* phenotype allele (purple color seeds) into a colorless seed phenotype in subsequent generations. As a result of this cross, all seeds of the F_2 progeny had a reduced content of anthocyanin, contrary to Mendel's law of segregation. The *R1* locus encodes a transcription factor involved in the induction of anthocyanin pigment genes. In this respect, the *R1* transcription factor is similar in function to the *b1* transcription factor described by Ed Coe, Jr. It is very much possible that these genes appeared as a result of duplication events due to ancient **allotetraploidization** during maize evolution. At the *R1* locus, paramutation is observed between two distinct alleles. The term **allele** is normally used if there is a high level of homology between paramutagenic and paramutable sequences. It appears that in case of the *R1* locus of maize, the two sequences are very heterogenic; the paramutation process at the *R1* locus involves different genes and a different number of these genes. Partners that are involved in paramutation are referred to as **haplotypes** rather than alleles.

13-2. Allotetraploidization

Allotetraploidization is the process upon which the crosses between two closely related species results in formation of a hybrid individual having two sets of chromosomes derived from both species. Yet another example of alloploidy is Allohexaploidization of wheat. These events typically result in the increase in the nucleus size and are believed to give adaptive advantage for plants.

Paramutation at the R1 locus in maize

The paramutable *R-r* locus consists of four coding genes: *P*, *q*, *S1*, and *S2* (see Figure 13-1A). The *S1* and *S2* loci are functional coding sequences representing inverted repeats arranged in a head-to-head orientation. *S1* and *S2* genes regulate anthocyanin expression in the aleurone layer of maize seeds and are highly receptive to paramutation. The *P* locus encodes for a functional gene expressing anthocyanin in somatic tissues and is marginally influenced by paramutation. The fourth coding sequence *q* is nonfunctional and apparently is not a target of paramutation.

The paramutagenic *R-st* haplotype consists of three coding-sequence repeats: *Nc1*, *Nc2*, and *Nc3*. All three sequence repeats result in a very low level of expression; hence, seeds produce almost no anthocyanin and are colorless. The fourth coding sequence (*Sc*) is present upstream of *Nc1*, *Nc2*, and *Nc3* sequences. The gene is highly expressed, and it determines a spotted (stippled) phenotype of anthocyanin in seed expression.

Crosses between *R-r* and *R-sc* produce F1 seeds that all have the gene expression level similar to that of *R-r* (see Figure 13-1B). At this stage, it can be concluded that a high level of expression of *S1* and *S2* persists in the F_1 generation. When F_1 plants are further used for crosses with plants carrying the recessive neutral (nonparamutagenic) allele *r* coding for colorless seeds, the resulting F_2 generation loses the *R-r* phenotype. The expression analysis confirms that loss of the *R-r* phenotype is associated with reduced expression of *S1* and *S2* genes. The result of crosses depends on whether male or female F_1 gametes were used. When pollen (male gametes) of F_1 plants is brushed onto the stigma of emasculated *r* plants, a complete absence of pigmentation in F_2 seeds is observed; and conversely, when pollen from *r* plants is brushed onto the stigma of emasculated F1 plants, the expression of *R-r* in the form of occasional spots and anthocyanin expression in the aleurone layer of seeds can be seen. This indicates that paramutation is germline-specific.

The sequence of the *R-r* locus has revealed a 387 bp region located between the *S1* and *S2* genes. The sequence, named σ (sigma), is derived from rearranged remnants of the Doppia transposable element belonging to the CACTA family. The σ sequence acts as a promoter for the *S1* and *S2* genes (note that these genes represent inverted repeats). The Doppia element is also found distal to the *q* gene, but no Doppia is found close to the *P* gene. The presence of the σ sequence in the **paramutable haplotype** has been shown to be very important for paramutation. All three *Nc* genes in the *R-sc* haplotype also contain similar sequences in the upstream regions.

A similar phenomenon of paramutation is observed upon crosses of the paramutable haplotype *R-r* with another paramutagenic allele, *R-marbled* or *R-mb*. The only difference is that instead of appearance of a spotted kernel phenotype, there is appearance of a marbled phenotype—that is, a combination of kernels with different degrees of pigmentation in the aleurone layer. It is curious to note, however, that the σ sequence is not present the *R-mb* haplotype. This haplotype consists of the *S* gene and two *Lcm1* genes.

All haplotypes tested to date and involved in paramutation consist of multiple copies of *R1* genes in various combinations of *S*, *Nc*, or *Lcm* genes. The number of *R1* genes can vary from two to five. Thus, the analysis of loci shows no significant structural similarity between paramutagenic haplotypes. The same is true for the paramutable and neutral haplotypes. The only obvious difference found between the paramutagenic and paramutable haplotypes is in the level of DNA methylation. The paramutable sequences showed high levels of DNA methylation in the upstream region of the *R1* genes, whereas the paramutagenic and neutral sequences exhibited low DNA methylation levels. Hence, it can be suggested that paramutation is rather structure-independent, but it depends more on chromatin organization of the haplotype.

The differential stability of paramutation at the *R1* locus adds additional complexity to this picture. Each individual paramutagenic *R-st* haplotype has a different **paramutagenic potential**. As a result, the ability to reverse silencing of the *R-r* haplotype is directly correlated with this potential. Also, the number of generations for which the *R-r* haplotype coexists with the paramutagenic *R-st* haplotype also determines paramutation stability. These results suggest that paramutation has quantitative characteristics.

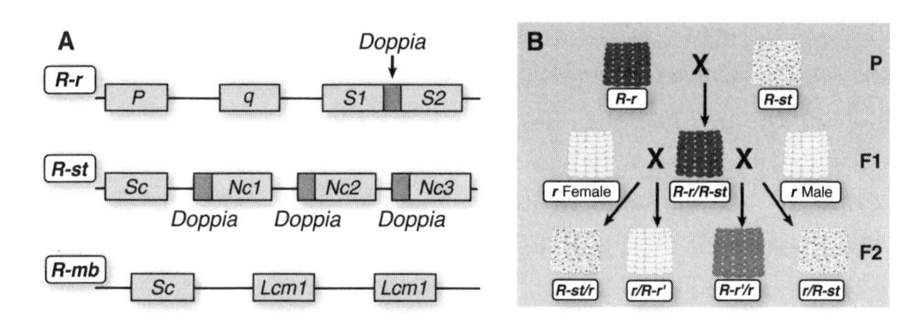

Figure 13-1 Paramutation at the r1 locus. **A**) Structural representation of various R1 haplotypes. All three haplotypes contain different sequences with the various number of r1 genes. Two of the alleles, R-r and R-st, contain the Doppia sequences. **B**) Crosses between two parents, R-r and R-st, result in the entire generation of seeds being purple (R-r/R-st). Subsequent crosses between F1 plants and r plants segregate into spotted plants R-st/r and paramutated plants that are either completely colorless or have a substantially reduced pigmentation, depending on whether male or female r plants were used for crosses, respectively. Thus, paramutation at the R1 locus is visible only in the F2 and subsequent generations.

Paramutations at the b1 locus in maize

The *b1* locus is also one of the earliest studied systems of paramutation. The system was first introduced by Ed Coe, Jr. in 1966. Similarly to *r1*, the *b1* locus also encodes a transcription factor that regulates the expression of genes involved in anthocyanin biosynthesis. High expression from the *b1* locus results in purple pigmentation easily noticeable in young seedlings and mature plant tissues. There are many reported alleles that are responsible for this phenotype, although paramutation is triggered only upon interactions between two specific alleles. The paramutable *B-I* allele expresses a transcription factor at high levels; hence, it leads to purple pigmentation. On the contrary, the paramutagenic *B'* allele is silent, the transcription factor is not produced, and hence, pigmentation is absent.

In nature, a spontaneous conversion of the *B-I* allele into the *B'* allele occurs in the frequency range of 1% to 10%. Crosses between plants carrying the *B-I* allele and plants carrying the *B'* allele result in heterozygous plants in which the *B-I* allele is converted into *B'* (Brink, 1956). The results are independent of whether a source of the silenced allele is paternal or maternal; in both cases, conversion is very stable (Patterson et al., 1993) (see Figure 13-2A). Screening of 60,000 plants of the *B-I/B'* population did not identify a single plant showing the reversal of *B'* into *B-I*. Thus, the penetrance of paramutation established is very high, presumably 100%. As it was described before, the paramutated *B-I* allele, marked as B'^{*}, becomes paramutagenic and in subsequent crosses, it is able to convert the *B-I* allele into the B'^{*} allele.

The newly formed B'^{*} allele is indistinguishable from *B'* in its ability to paramutate naive *B-I* alleles in subsequent generations. Thus, it is not possible to distinguish **primary paramutation** from **secondary paramutation** at the *b1* locus. This drastically contrasts with paramutation at the *r1* locus in which the full penetrance of *R-r'* secondary paramutation (the ability of newly converted *R-r'* to paramutate a naive *R-r* allele) is observed only when heterozygous *R-r'/R-st* plants have been propagated for multiple generations (Brown and Brink, 1960). The ability of *R-r'* plants to convert *R-r* to *R-r'* (after being crossed with it) is referred to as secondary paramutation, which is different from the primary paramutational interaction between *R-st* and *R-r*.

The first molecular experiments describing *b1* paramutation were carried out by Garth Patterson. The sequencing analysis showed that the *B-I* and *B'* alleles are identical in coding and flanking regulatory sequences. The analysis of transcription showed a ten- to twenty-fold reduction in the transcription rate of the *B-I* allele due to paramutations (Patterson et al., 1993). This, however, does not abolish transcription completely as the proteins produced from either the *B-I* or *B'* allele are functional and are able to induce anthocyanin production. It can be hypothesized that the amount of the produced transcription factor is not sufficient to result in an anthocyanin accumulation phenotype.

The results of the experiments on the *r1* haplotypes showed the importance of certain sequences, such as the σ sequence, for the establishment of efficient paramutation. Thus, the experiments were carried out to identify whether such sequences existed at the *b1* locus. The fact that paramutation at the *b1* locus exhibited the full penetrance and extreme stability allowed the genetic analysis of paramutation at the *b1* locus. Fine-structure recombination mapping indicated that a 6 kb region located 100 kb upstream of the *b1* transcription start site is essential for paramutation (Stam et al., 2002). In case of the *B-I* and *B'* alleles, this region contains seven direct repeats of an 853-bp sequence. The number of these direct repeats determines the extent of paramutation at the *b1* locus (see Figure 13-2B). If the number decreases to less than five, the paramutation-like interaction at the *b1* locus is significantly reduced. The reduction of the number of these repeats down to one renders the allele to become neutral (non-paramutagenic). The number of copies of these repeats also influences the expression level of the *b1* locus (Stam et al., 2002). The alleles such as the *B-I* ones that contain seven repeats are highly expressing alleles, and a reduction in the number of direct repeats correlates with a decrease in expression.

As was suggested earlier, the *B-I* and *B'* alleles are **epiallels** because they are identical in sequence but different in the level of expression. The analysis of epigenetic changes in these alleles showed the following picture. Whereas the methylation levels at the promoter and coding regions of these two alleles did not differ, the *B-I* repeats were significantly more methylated as compared to the *B'* repeats. Also, the sequences of the repeats of the *B-I* allele were more nuclease sensitive than those of the *B'* allele. Hence, at this point, it can be assumed that alterations in epigenetic status cause paramutation at the *b1* loci.

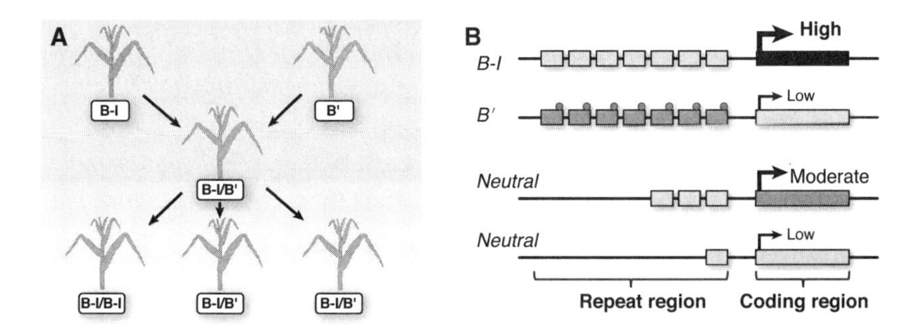

Figure 13-2 Paramutation at the b1 locus. **A**) Segregation of the B-I allele after paramutation. **B**) Structural representation of various b1 alleles and a correlation between the number of repeats and expression of the b1 gene.

Other less-described examples of paramutation in plants include *pl1 (purple plant 1)* and *p1 (pericarp color 1)* loci in maize (Zea *mays*), the maize *A1* transgenic locus in petunia (*Pitunia hybrida*), and the 271 locus in tobacco (*Nicotiana tabacum*).

The maize *p1* gene is a transcription factor mediating phlobaphene biogenesis, resulting in red ear and cob pigmentation. Upon exposure to a transgene carrying specific *P1-rr* regulatory sequences, the *P1-rr* allele becomes paramutated. The newly acquired transcriptionally silenced state, termed *P1-rr*', becomes paramutagenic in subsequent generations. The delivery of transgenes carrying various regulatory regions of *p1* identified a specific 1.2 kb enhancer fragment that was required and sufficient for the establishment of paramutation (Stam et al., 2002). Similarly to the previously described loci involved in paramutation, the *p1* locus is also flanked by a direct repeat carrying the enhancer fragment. Subsequent crosses showed that the *P1-rr*' phenotype is heritable but not fully penetrant.

Paramutation is an epigenetically regulated phenomenon

Paramutation is an epigenetically regulated phenomenon. Although epigenetics itself does not include permanent changes to DNA sequence, the mechanisms involved are established through gene function. Thus, it can be predicted that genetic components regulating the processes of establishment and maintenance of paramutation can be identified in a mutagenesis screen. Indeed, the mutagenesis experiments have identified such components (see Figure 13-3A). In the first experiment, a cross was performed between *B-I* plants carrying a *Mutator transposone* and *B*' plants (Dorweiler et al., 2000). It was hypothesized that frequent excision and re-insertion of the transposone would eventually inactivate one of the genetic components responsible for paramutation. The progeny of *B-I* and *B*'plants was screened for the presence of anthocyanin pigmentation. As a result, a plant, *mediator of paramutation 1 (mop1)*, that expressed anthocyanin was identified; it appeared to be a **recessive mutation** that increased the transcription at the *B*' locus (see Figure 13-3B). The *mop1* plant was then further used for the analysis of paramutation mechanisms.

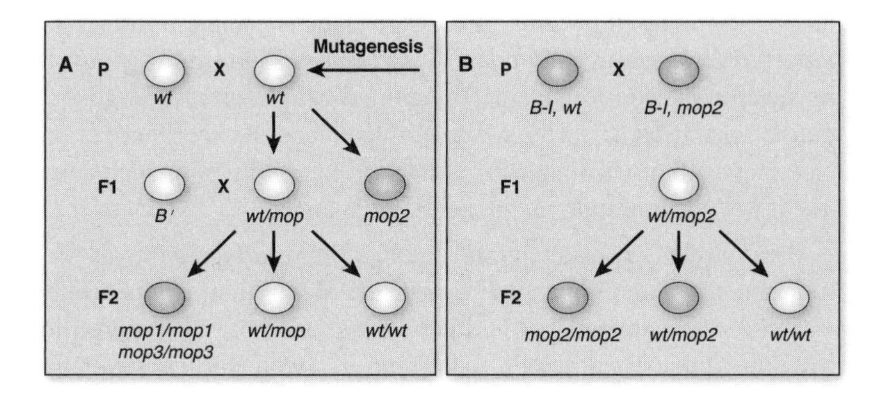

Figure 13-3 Experiments for the identification of genes involved in paramutation. **A**) The mutagenesis experiment carried out for the identification of genes involved in the establishment and maintenance of paramutation. **B**) The ability of mop2 mutants to reverse paramutation is limited to plants containing mutation. In subsequent generations, wild type plants not only reveal gene silencing but also appear to be paramutagenic.

The mutation in *mop1* plants appeared to be in a gene coding for an RNA-dependent RNA polymerase (RdRP) (Alleman et al., 2006). As RdRP is required for the production of siRNAs, the establishment of paramutation might be dependent on specific siRNAs. Both the establishment of paramutation and the maintenance of transcriptional silencing of *B'* absolutely depend on *mop1* (Dorweiler et al., 2000), thus suggesting a role for siRNAs in both processes. Interestingly, when a *mop1* plant (from the *B-I/B'* background) is crossed to a *B-I* plant, the F_2 population with a reintroduced wild-type copy of MOP1 protein is paramutated again exhibiting *B'* expression levels (Dorweiler et al., 2000). This experiment clearly shows that the MOP1 protein has the ability to resilence the activated epiallele transmitted from the *mop1* mutant plant. Another possibility suggested by Vicki Chandler is that this activated allele carries some kind of heritable mark that labels it as a *B'* allele. That mark would have to be transmitted through meiosis in order to resilence the B' allele from the activated allele transmitted from the *mop1* mutant to the silent B' allele in the presence of the MOP1 protein. Vicki Chandler's group reported a rare occurrence of changes from *B'* to *B-I* observed among thousands of *mop1* plants in which a new state is heritable and immune to resilencing upon re-introduction of the MOP1 protein. This finding is really fascinating as nothing like this ever happened to the *B'* allele in the wild-type background. A similar finding was reported for a transcriptionally silent transgene; the transgene was reactivated in *mop1* plants and remained active for several generations after the MOP1 protein was re-introduced. Both of these examples suggest that chromatin structure can be modified in such a way that it remains immune to epigenetic regulation.

Subsequent experiments showed that the wild-type *mop1* allele is required for maintaining low expression states that are associated with the paramutagenic *b1* and *pl1* alleles, but it is not required for sustaining the low expression state associated with *R-r'*. However, it is absolutely required to establish paramutation at three loci: *b1, pl1,* and *r1*. The effect of the *mop1* mutation on the establishment of paramutation at the *p1* locus is currently unknown.

The MOP1 gene seems also to be required for maintaining the extensive DNA methylation that silences the MUTATOR transposons. At the same time, the *mop1* mutation does not cause global hypomethylation of repeated sequences at centromeric or rDNA loci. According to Vicki Chandler, the *mop1* mutation also reactivates some transcriptionally silent transgenes tested. The observation that *mop1* does not affect all aspects of paramutation at each locus tested and it does not reactivate all transcriptionally silent transgenes tested indicates that multiple mechanisms might be involved in these different examples of silencing.

Vicki Chandler and Mary Alleman suggest several potential explanations for the inability of *mop1* to maintain the low expression state associated with the *R-r'* phenotype. Because MOP1 is an RdRP enzyme, it is possible that an RNA mechanism might not be involved in maintaining the silencing of *R-r*. Another possibility could be that the *R-r'* allele does not require the function of RdRP for maintaining an RNA signal. In this case, the inverted repeats at the *R-r* locus might produce a sufficient amount of hairpin RNA.

It remains to be established whether other mutations impaired in the RNAi pathway are impaired in maintaining *R-r'* silencing. The fact that paramutation at *r1* is much more efficiently established through the paternal allele that is subject to **genomic imprinting** further complicates the matter. DNA methylation observed within the inverted repeats in *R-r* correlates with its silencing, thus just this methylation might be responsible for the maintenance of silencing in *mop1* plants. Another plausible hypothesis is that there exist two silencing mechanisms that are **epistatic** to each other, the silencing mediated by methylation and imprinting and the silencing mediated by MOP1. As a result, the absence of MOP1 does not influence the maintenance of silencing.

13-3. Epistatic interaction

Epistatic interaction or epistasis is the interaction between two or more genes to control a single phenotype. An epistatic interaction between two or more genes occurs when the phenotypic effect of one gene depends on another gene, often exposing a functional association. Sometimes epistatic interactions are based of physical interaction between gene sequences located on interacting chromatin loops. The absence of one of the genes gives rise to a distinct phenotype.

A second screen for mutants impaired in the establishment of paramutation was done using the chemical mutagen ethyl methanesulfonate (EMS). The *B'* pollen was exposed to EMS and then brushed onto the stigma of *B-I* plants. The F_1 and F_2 populations from crosses were screened for the presence of plants expressing anthocyanin. The screen identified several mutants—*mop2*, *mop3*, *rmr1*, *rmr2*, *rmr6*, and *rmr7*. The *rmr6* gene encoded a large subunit that was probably a homolog of a large subunit of Pol IV, NRPD1 in Arabidopsis. The *mop2/rmr7* mutant showed the highest homology to the gene that encoded NRPD2/E2, the second largest subunit involved in both Pol IV and Pol V complexes in Arabidopsis. The *rmr1* gene encodes a putative chromatin-remodeling protein with a Snf2 domain that shares similarity to the helicase domain of DRD1; *rmr1* is also required for the accumulation of siRNAs in maize. In Arabidopsis, RDR2, NRPD1, NRPD2/E2, and DRD1 are the main components of an RNA-dependent DNA methylation (RdDM) pathway that mediates sequence-specific, chromatin-based transcriptional gene silencing. Thus, RdDM might be the main mechanism involved in paramutation in maize. It has to be noted, however, that there are also the unique and distinct properties of paramutation in maize that might be dissimilar to the RdDM pathway in Arabidopsis. For example, heritability of RdDM gene silencing in Arabidopsis is not so common, whereas heritability of paramutation is an established fact. Moreover, RdDM is not heritable if the targeted allele is separated from a locus producing siRNA. Perhaps, the most important difference is that genes that are targeted by the RdDM-dependent silencing mechanism do not acquire the ability to silence similar genes at other loci (or other plants), and, therefore, they do not become paramutagenic.

Arabidopsis represents a unique system for studies of paramutations. It has a small-sequenced genome and a great number of well-characterized mutants. Unfortunately, the phenomenon of true paramutation is yet to be discovered in Arabidopsis. Previous studies, however, report several phenomena that resemble paramutation. Arabidopsis mutants *cpr1* (constitutive expression of the PR1 gene) have dwarf phenotype; however, in the F2 generation of these plants, there is a 20% reversion of dwarfism with some indication of heritability. In another example, tetraploidization of Arabidopsis plants that carry a gene of resistance to hygromycin results in the loss of resistance to hygromycin observed in 100% of plants from the F2 generation; this effect is 100% heritable. In both of these cases, no evidence of the ability to convert other plants to the same phenotype (the effect of **paramutagenecity**) has been observed. In yet another example of so-called ***trans*-silencing**, plants transgenic for GFP-COP1 (constitutive photomorphogenesis protein 1) act as *trans*-silencers; they are able to silence other similar loci when crosses are performed. The most well-described *trans*-silencer locus, C73 carries three copies of a 35S-driven GFP-COS1 fusion that are transcriptionally silenced but show low levels of GFP expression in roots. Exposure to this locus results in the loss of GFP expression and dwarfism due to acquisition of a *cop1* phenotype. Thus, the C73 locus represents a paramutagenic locus that results in 100% conversion and is 100% heritable. Paramutated loci carrying GFP transgenes also become paramutagenic when used for crosses with naive plants.

Yet another example of paramutation-like phenomenon is described in the Wassilewskija (WS) ecotype of Arabidopsis. Plants of this cultivar carry four copies of the phosphoribosylanthranilate isomerase (*PAI*) gene located at three unlinked sites in the genome. All four genes are heavily cytosine-methylated over their regions of shared DNA sequence similarity, and their combined expression provides a sufficient amount of protein for a normal plant phenotype. If two tandemly organized PAI genes—*MePAI1* and *MePAI4*—are deleted, the other two genes provide insufficient amounts of protein; and the Δ*pai1–pai4* deletion mutant becomes blue fluorescent under UV light, caused by accumulation of early intermediates in the tryptophan pathway, anthranilate, and anthranilate-derived compounds. These plants undergo reversion at the frequency of 1% to 5%, and this reversion is associated with substantial hypomethylation at the PAI loci. The research shows that the locus containing the *MePAI1* and *MePAI4* genes functions as a paramutagenic allele and is able to silence the *MePAI2* and *MePAI3* alleles and *trans*-methylate them. This change becomes heritable, and thus, the phenomenon satisfies the requirement to be called paramutation. It is important, however, to mention that the transmethylated *PAI2* and *PAI3* genes do not transfer their methylation status to naive singlet genes; therefore, they themselves do not become paramutagenic.

13-4. *Trans*-silencing

In contrast to paramutation, *trans*-silencing represents meiotically heritable changes of phenotype occurring due to non-allelic interactions. Such interactions are most well described for transgenic loci; in this case one master locus tends to suppress the expression of its target locus, hence "*trans*-silencing." Paramutation and *trans*-silencing are related processes. For example, an inverted repeat allele carrying *PAI1/PAI4* genes in the WS cultivar can silence homologous PAI alleles as well as unlinked PAI genes from the Columbia ecotype, an example of allelic *trans*-silencing or paramutation. Similarly, synthetic transgenes consisting of PAI inverted repeats are also able to *trans*-methylate homologous, but non-allelic, target loci.

What are other proteins potentially involved in the establishment and maintenance of paramutation? Mutations in three different genes (*ddm1*, *met1*, and *cmt3*) that affect DNA methylation patterns through distinct mechanisms were tested for their ability to reduce methylation and increase expression of the silenced *PAI* loci in *Arabidopsis thaliana*. All three mutations caused partial demethylation and increased expression of the singlet *PAI* locus, although not in the same manner. Whereas the *cmt3* mutation altered methylation of both the inverted repeat and the singlet locus, the *ddm1* mutation affected the singlet gene more strongly than the inverted repeat, yet *met1* affected the inverted repeats more strongly than the singlet gene. Mutations in *SUVH4*, a SET domain protein with histone H3Lys9 methyltransferase activity, showed reduced cytosine methylation on the singlet gene *PAI2*, but they did not affect methylation on the PAI inverted repeat. The comparison analysis performed between the *suvh4* and *cmt3* mutants showed that an unmethylated singlet *PAI2* gene was not methylated *de novo* in *cmt3*, whereas it was methylated *de novo* in *suvh4*. It can thus be suggested that SUVH4 is involved in maintenance but not establishment of the inverted-repeat-induced methylation, whereas CMT3 is involved in both.

In another set of experiments, MOM1 and DDM1 were tested for their role in the maintenance of a paramutation-like phenomenon that occurs between transgene alleles in tetraploid but not in diploid plants. The *mom1* mutation had no effect; the *ddm1* mutation had only a weak effect on maintenance, in which the silent transgene started expressing only after several generations, in contrast to changes being observed more quickly in the PAI system. Unfortunately, the effects of *ddm1* and *mom1* mutations on the establishment of ploidy-related transgene silencing were not tested. Because *ddm1* and *mom1* can reactivate silenced repetitive transgenes in diploid Arabidopsis in which silencing of transgenes does not occur, it seems unlikely that they function directly in the *trans* phenomenon found in tetraploids. The fact that DDM1 is involved *in trans* activity at the PAI loci but not at the transgenic loci in tetraploids makes it possible to suggest that there are several different responsible mechanisms.

Paramutation might occur at far greater rate and at many more loci than reported to date. It is the ability to have an easily observed phenotype such as the one characterized by changes in anthocyanin production in maize kernel that allowed substantial progress in understanding the phenomenon. All four loci in maize mentioned previously (*R1*, *b1*, *pl1*, *p1*) encode transcription factors involved in pigment biosynthesis in plants. Hence, paramutation in maize was easy to observe as silencing of a transcription factor would result in a reduction of pigment accumulation in various plant tissues.

The most intriguing question about paramutation still is how the phenomenon occurs. What is/are the underlying molecular mechanism(s)? Experiments to investigate these matters and address these questions lasted over a period of time. However, the answer that explains different types of paramutation occurring at different loci has yet to be obtained. Experiments indicate that there are at least two possible mechanisms of paramutation: an RNA-dependent chromatin modification model and a direct physical interaction model.

Functional models for paramutation

There could be several different mechanisms responsible for interactions between alleles and haplotypes in the process of paramutation. The two major possibilities include the **RNA-based model** or the **DNA pairing model** (reviewed in Chandler and Stam. 2004). From the early days when the phenomenon of paramutation was discovered, scientists tried to use available genetic approaches for understanding the responsible mechanisms. Brink and Coe, two investigators who studied the phenomenon starting in the late 1950s, assumed that paramutation required a direct or an indirect contact between interacting paramutagenic and paramutable alleles. Brink hypothesized that upon paramutation, *R-st* and *R-r* alleles communicate early in development in newly formed *R-st/R-r* heterozygotes. To prove that a direct chromosomal interaction is needed for the establishment of paramutation, he attempted to disrupt pairing between two loci by the use of translocated chromosomes. The experiment did not lead to conclusive results. Later on, Coe showed that translocation of one of the interacting alleles (*B-I* or *B'*) to the other chromosome arms did not disrupt paramutation. This experiment was not perfectly clean as the translocated regions were large and could still potentially allow for somatic or meiotic pairing.

Another hypothesis tested by Brink was the existence of a "communication molecule" called by Brink a "cytoplasmic particle" acting *in trans*. The communication molecule might potentially be produced by the *R-st* allele and transmit paramutation to the *R-r* allele. Again, Brink found no evidence for the existence of such a molecule. Coe also hypothesized that the paramutagenic allele transferred a physical entity, which he thought could be either DNA or RNA. Current studies also show that there are no genetic differences between the *B-I* or *B'* alleles. However, there is evidence that RNA signals are indeed involved in these allelic interactions.

Repeat-counting mechanisms

Another feature of the *b1* locus that deserves our attention is the dependence of the penetrance and stability of paramutation on the number of **tandem repeats**. Because the locus containing three repeats is much less penetrant than that containing seven repeats, it can be suggested that cells possess a mechanism for counting the numbers of repeats. Therefore, it can be hypothesized that the number of specific chromatin marks—such as cytosine methylation, histone modifications, the binding of specific proteins, as well as the number of small RNA molecules produced (the level of which also depends on the number of repeats)—can influence the outcome of paramutagenicity.

In 1968, Brink and colleagues suggested that the *r1* haplotypes exist in a wide range of states and that the ability to be in a particular state depends on the nature of the haplotype itself, but at the same time, this ability is strongly influenced by the nature of the other allele present. Thus, Brink proposed that the *r1* locus had two components: the haplotype encoding the protein regulating anthocyanin synthesis and a heterochromatic region consisting of varying numbers of repeats, called **metameres**, which functioned to repress *r1*. He also proposed that the degree of repression was proportional to the number of metameres and this number could change through misreplication during somatic mitosis.

Brink seemed to be influenced by earlier work of Lewis (1950) describing the position-effect variegation (PEV) in Drosophila (Lewis, 1950). The mosaic pattern of PEV in Drosophila eyes was strikingly similar to the spotted phenotype of paramutant *R-r'* kernels; thus, Brink hypothesized that PEV and paramutation were both caused by aberrant gene expression.

He suggested that a special chromatin structure, which he called **parachromatin**, was responsible for communication of the state of chromatin condensation and gene silencing. Potentially, the larger number of tandem repeats at these "parachromatic" areas facilitates the communication with the inverted repeats at the paramutable locus, either directly through pairing or through some type of RNA-signaling molecule (see Figure 13-4). The inverted repeat nature of the paramutable haplotypes possibly makes them more receptive to the signal.

Multiple tandem repeats are the most important sequences required for paramutation at the *b1* locus. These tandem repeats contain an enhancer that increases the expression of *b1* genes *in cis*. The differences in chromatin structure of tandem repeats of *B-I* and *B'* loci might determine whether the enhancer can function to increase *b1* transcription. Also, the repeats are necessary to mediate *trans*-interactions that establish and maintain paramutation. Maike Stam's group used chromosome conformation capture (3C), to show that the hepta-repeat physically interacts with the transcription start site region in a tissue- and expression level–specific manner. The interaction frequency was shown to be high in *B-I* and *B'* husk tissue and relatively low in inner tissue; in *B-I* husk, it was significantly higher than in *B'* husk. Multiple repeats are required to stabilize this interaction. Three other regions of the b1 locus, ~15, ~47, and ~107 kb upstream of the transcription start site, interact with the transcription start site region in an epiallele- and expression level–specific manner. The interactions are observed only in high expressing *B-I* husk tissue, which indicates that, besides the hepta-repeat, the region upstream of the *b1* transcription start includes several other regulatory sequences involved in inducing high expression of *B-I*. Although the 3C analysis indicated additional interactions that potentially generate more loops in *B-I* compared to *B'*, there are no proofs that the *b1* repeat siRNAs are involved in enhancer activity because there is no sequence homology between the *b1* promoter region and the *b1* tandem repeat siRNAs. Thus, in *cis*, paramutation activity of the repeat region is highly unlikely.

13-5. Chromosome conformation capture (3C)

Chromosome conformation capture, or the 3C, methodology involves the use of formaldehyde to cross-link DNA and proteins conserving the three-dimensional chromosomal conformation. The cross-linked chromatin is then subjected to restriction digestion followed by intramolecular ligation. After reversing the cross-links, the DNA is purified and analyzed by quantitative PCR. The quantity of specific ligation fragments is a measure of the frequency of interactions between chromatin regions *in vivo*.

The hypothesis of the importance of the number of repeats for the establishment and maintenance of paramutation is a plausible one, but, nevertheless, studies in both maize and Arabidopsis showed that being epigenetically silenced and carrying tandem repeats was not always sufficient for paramutation to occur. In maize, the *P1-wr* locus containing multiple epigenetically silenced tandem repeats of large coding sequences does not participate in paramutation with *P1-rr* or *P1-rr'* (Stam et al., 2002). Similarly, at the **FWA locus** in Arabidopsis, transcription of two tandem repeats and generation of siRNAs are also not sufficient for *trans* silencing of this locus.

Figure 13-4 shows a possible model of repeat-counting mechanism.

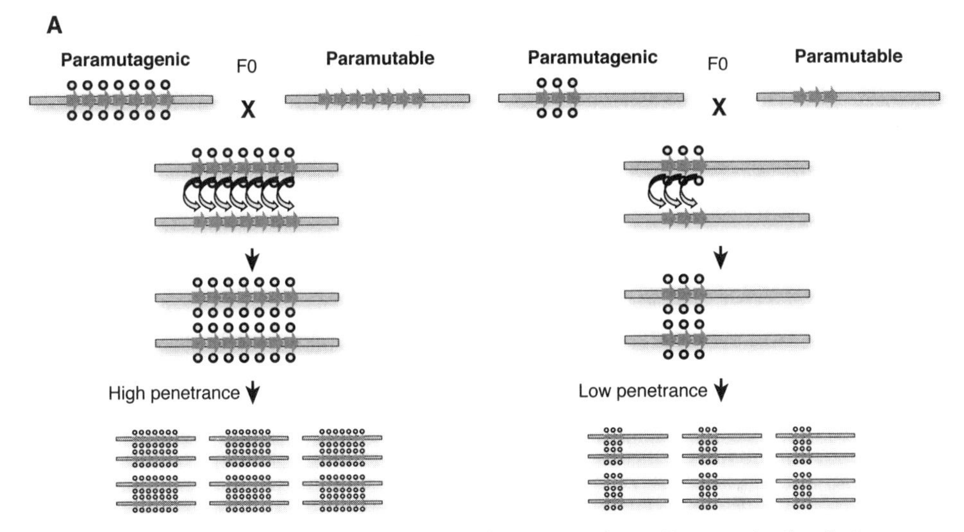

Figure 13-4A Repeat-counting mechanism of paramutations. Communication between paramutagenic and paramutable alleles may depend on the number of tandem repeats at the "parachromatic" areas. Larger number of repeats might facilitate more efficient communication with the inverted repeats at the paramutable locus, either directly through pairing or through some type of RNA-signaling molecule. **A**. Physical interaction model. Physical interaction between a large number of repeats generates more efficient silencing, possibly through larger stretches of methylation. This results in higher penetrance of silencing.

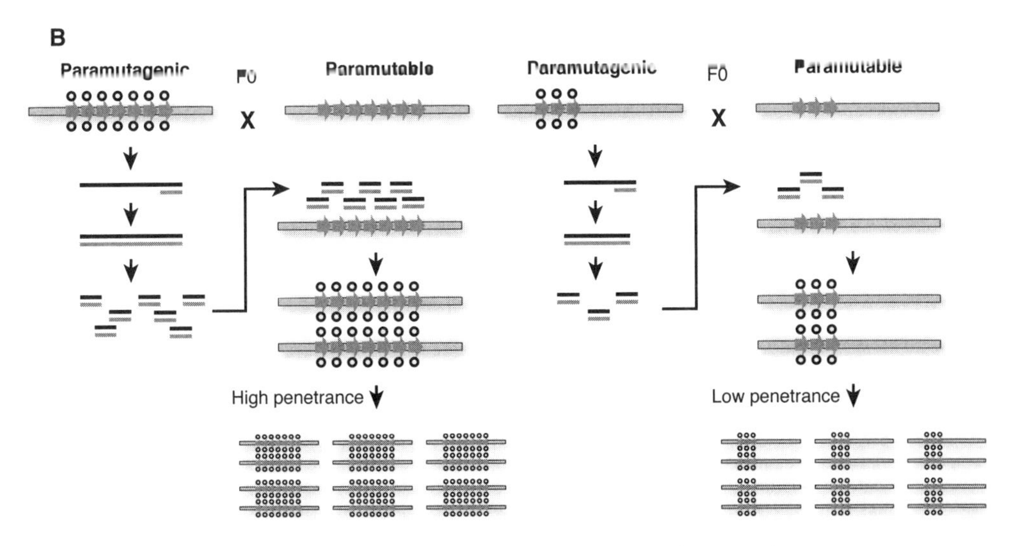

Figure 13-4B snRNA interaction model. Small RNAs generated from tandem repeats at paramutagenic loci interact with tandem repeats at paramutable loci and trigger hypermethylation. Larger number of repeats might generate more snRNAs, resulting either in higher methylation levels or a larger number of tandem repeats methylated at paramutable allele. As a result, a higher level of penetrance is achieved.

An RNA model of paramutation

RNA appears to play an important role in two distinct paramutation-like phenomena. In *P. infestans*, a sequence-specific, diffusible factor is required for *trans*-inactivation. Although not proven, the RNA molecule could be involved in this process. In Arabidopsis, dsRNA produced from the PAI1-PAI4 inverted repeat locus mediates DNA methylation. Based on the molecular characteristics of mutants impaired in various steps of paramutation (earlier in this chapter), an RNA model seems to be the most logical explanation of the phenomenon of paramutation (see Figure 13-5A).

In case of the *b1* locus, both strands of tandem repeats that are critical for paramutation at the *b1* locus are transcribed. Chandler's group showed that siRNAs (*b1TR*-siRNAs) are produced from multiple *b1* alleles: *B-I* and *B'*, the alleles carrying seven tandem repeats and undergoing paramutation and a single copy of the neutral allele (*b*) that is not involved in paramutation. Consistent with the siRNA requirements of paramutations, the *b1* tandem repeat siRNAs are dramatically reduced in *mop1* and *mop2* mutants. Because a decrease in the amount of siRNAs could occur due to either a poor expression of DNA loci or an insufficient functioning of RDR, the transcription run-on experiment was performed. Chandler's group found that wild-type and *mop1* plants had a similar number of generated transcripts. Thus, a lower number of siRNAs is due to the lack of sufficient RDR activity in *mop1* plants. The experiments with *mop2* mutants showed that a dramatic decrease in *b1TR*-siRNAs does not result in a decrease in *B'* silencing, thus indicating that siRNAs are not required to maintain silencing. These data show that the steps involved in the establishment and maintenance of a silent state can be separated, suggesting different mechanisms of occurrence of two processes. The results obtained from the experiments are similar to the observations with FWA where siRNAs are necessary but not sufficient for silencing and are required to establish, but not maintain, silencing. It is possible that *b1TR*-siRNAs are mediating the establishment of paramutation at the very early steps in development, perhaps just after fertilization or in early embryogenesis; unfortunately, these tissues are not easy to analyze biochemically. The fact that *b1TR*-siRNAs are also produced from loci that do not undergo paramutation suggests that the presence of these siRNAs might be necessary for paramutation to occur, but it is not sufficient. One possible explanation could be that it is a longer transcript that actually triggers paramutation despite the fact that siRNAs are produced from all alleles. Longer transcripts are unlikely to affect transcriptional gene silencing if not for the fact that they would produce more siRNA molecules. Thus, the longer the transcript, the more siRNAs it would produce.

The importance of the *b1* tandem repeat siRNAs in establishing paramutation was proved in the experiment with transgenic plants that produced *b1TR*-siRNAs by placing the *b1* repeat unit under the control of the strong constitutive 35S CaMV promoter. These transgenic plants behaved as paramutagenic plants. They silence a naive *B-I* allele; the change was heritable and the newly converted allele remained paramutagenic even after segregation of the transgene. Thus, it can be hypothesized that either *b1TR*-siRNAs or dsRNA templates mediate the *trans*-communication between the alleles, which allows establishing paramutation.

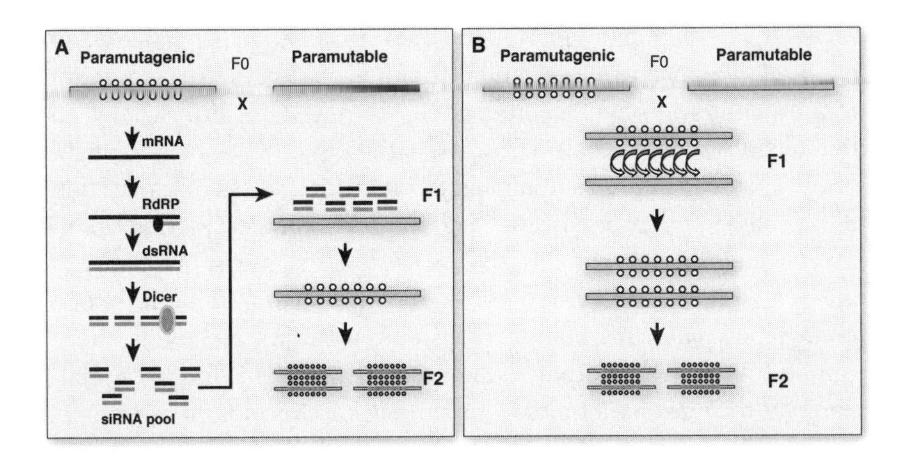

Figure 13-5 RNA and physical interaction models for paramutation. **A**. RNA model. Various steps of siRNA synthesis from a paramutagenic allele. siRNAs produced through a dsRNA intermediate lead to changes in either methylation or chromatin structure in the paramutable allele. **B**. Physical interaction model. In the F_1 generation, epigenetic information is transferred from the paramutagenic to the paramutable allele if they physically interact with each other. It leads to altered expression of the paramutable allele, which is heritable in further generations.

What is the story of paramutation at the *R1* locus? As you remember, the R1 locus contains two reverse-oriented genes: *S1* and *S2*. It can be hypothesized that upon transcription, these inverted repeats may form double-stranded RNA that can give rise to siRNAs; the latter, in turn, can promote chromatin changes in the paramutable allele. The stability of paramutation at the *R1* locus seems to be quantitative and correlated with the duration of exposure to paramutagenic sequences. It is possible that prolonged exposure causes an increase in DNA methylation. The more exposure, the more DNA is methylated. Hence, the RNA model is the best explanation of paramutation at the *R1* locus. Unfortunately, both models still remain to be elucidated.

A physical interaction model

Although much has been learned about the involvement of siRNAs in paramutation, many questions still remained unanswered. The fact that siRNAs are not required for the maintenance of paramutation suggests the involvement of other mechanisms, including changes in chromatin structure. One of the alternative models of a direct physical interaction might theoretically provide an explanation of paramutation occurrences (refer to Figure 13-5B). An important requirement for physical interactions is physical pairing between homologous sequences. In Drosophila, the physical interaction might occur in somatic cells as homologous chromosomes can be paired in these cells. In other organisms, such as plants, physical pairing is limited to meiosis. During this stage, two homologous sequences can interact with each other, and epigenetic information in the form of a methylation pattern or binding of histones with specific modifications can be transferred to other sequences. This model is based on phenomena that involve homologous pairing of chromosomes, such as **transvection** (later in this chapter) in Drosophila and X chromosome inactivation in humans. The fact that repeats play an important role in paramutation fits well with pairing models. Indeed, repeated sequences are known to pair more often than single-copy sequences. Moreover, DNA hypermethylation at repetitive sequences increases the frequency of homologous pairing. Knowing that paramutagenic alleles are hypermethylated makes it tempting to speculate that this allows for an enhanced pairing frequency, which might possibly facilitate paramutation.

The physical interaction model seems to be appropriate for the *b1* locus in which repeated sequences undergo an increase in methylation. Moreover, at the *R1* locus, *R-r* expression is silenced only in F_2 and subsequent generations, indicating that *R-r* silencing requires meiosis. Both of these evidences suggest that the physical interaction model can be a viable alternative for paramutations. At the same time, it cannot explain the participation of *mop2* and the absence of meiotic linkage during many paramutations, including the one at the *b1* locus.

The inability of any of the models to explain the outcomes of paramutations might be due to the fact that more than one model works in the process of conjugation. The previously mentioned RNA and the physical interaction model might work in parallel. This can explain variations among different types of paramutations. However, exact mechanisms of their occurrence still remain unexplained.

There are several reasons to favor the siRNA model of paramutation. Transcription from the heptamer is bidirectional. It occurs at similar rates in both *B-I* and *B'* and is most likely mediated by RNA Pol II (because it is blocked by alpha-amanitin concentrations that block Pol II). Also, mutants in *mop1* and *mop2* show no reduction in repeat transcription. Paramutation and production of b1 repeat siRNAs depend on the function of Pol IV/Pol V-like enzymes (MOP2/RMR7 and RMR6). The produced siRNAs are apparently further amplified by RDR (MOP1) through an intermediate dsRNA form, and this dsRNA is processed into siRNAs by yet an unknown enzyme. Then these siRNAs presumably induce chromatin modifications within the *b1* repeats, including changes in methylation and histones modifications. The mechanism of these changes is yet to be elucidated not only in maize but in Arabidopsis as well. Chandler and group hypothesized that the *B-I* allele is relatively immune to RNA-directed DNA and chromatin modifications due to a specific structure of already-existing chromatin. As a result, it remains in an actively transcribed state in spite of producing similar levels of siRNAs as *B'*.

In trans communication between *B-I* and *B'* might occur at several levels, including b1TR-siRNAs, DNA/DNA, or DNA/Protein/DNA interactions. All these enable transferring the *B'* chromatin structure to *B-I*. Proteins that bind to the *b1* tandem repeats could form higher order complexes and bring the *B'* and *B-I* alleles in close proximity. A similar mechanism has been proposed for the role of Zeste in *trans*-communication (transvection) in Drosophila, and it has been observed with transgene repeat arrays in Drosophila. Such *in trans* communication might result in chromatin marks being passed from the *B'* to *B-I* allele. The analysis of histone modifications showed that both alleles have H3ac, H3K4me2, and H3K27me3 modifications at the coding regions. In contrast to *B-I*, *B'* carries H3K27me2, which is a hallmark for heterochromatin. It is possible that H3K27me2 modification at the *B'* transcription unit plays a role in the heritable maintenance of the low expression state at the *B'* locus. The silent chromatin state at the *B'* hepta-repeat most probably prevents the hepta-repeat enhancer function from elevating the *b1* transcription level.

The importance and significance of paramutation in plants

Why does paramutation exist? Paramutation is one of many homology-dependent gene-silencing mechanisms that can be utilized by plant cells to counter expression of foreign invasive DNA or RNA molecules stemming from viruses and transposons. Because paramutation requires repeats that are structurally similar to repeats at centromeres, which are also regulated by RNAi processes, it is possible that the loci that carry repetitive elements and are involved in paramutation might simply be "bystanders" in the war of cells against invading nucleic acids. The fact that paramutation is an extremely rare process suggests that the number of involved genes is very small, thus confirming the idea of an aberrant unintentional mechanism.

Paramutation might function in the transmission of the environmentally adapted expression states that were established in somatic cells to progeny. Changes in temperature appear to influence paramutation in two different cases: paramutation at the *r1* locus in maize and paramutation at the *A1* transgene in petunia. Petunia plants that express the maize A1 gene encoding dihydroflavonol 4-reductase, an enzyme involved in anthocyanin biosynthesis, have red flowers. The appearance of white-flowered plants that are transcriptionally silenced and have paramutagenic activity was shown to be dependent on the season of plant growth; the later during the growth season the plants are grown, the more white flowers appear. Moreover, apparently the age of the plants used for crossing also has the influence on the outcome of paramutation.

Paramutation might function as a regulatory process that is used to control gene expression in polyploids. The fact that the discovery of the phenomenon and the main progress in the analysis of paramutation was achieved in maize, a polylploid plant, supports this hypothesis. Moreover, occurrence of other paramutation-like phenomena reported in tetraploid but not diploid Arabidopsis further strengthens this idea. Paramutation that involves *trans*-interactions among homologous sequences might also provide a mechanism for generating functional homozygosity in polyploids.

Paramutations in other organisms

In the plant pathogen *Phytophtora infestans*, a paramutation-like phenomonon was described involving *trans*-interactions between non-allelic *inf1 trans*- and endogenous genes. The *inf1* gene is highly expressed and the protein is easily detectible. INF1 is a member of the elicitin family inducing defense responses in *Phytophtora infestans*. When another copy of the *inf1* gene is inserted, the transgene becomes transcriptionally silenced and behaves as a paramutagenic locus, heritably *trans*-inactivating the endogenous, paramutable *inf1* gene. The paramutagenic ability does not depend on whether a transgene is in a sense or an antisense orientation as well as whether a transgene is with or without a promoter. Secondary paramutation for the gene *inf1* gene has not been reported.

Another example of a paramutation-like phenomenon in fungi was described in the ascomycete fungus *Ascobulus immerses*. A methylated, inactivated *b2* gene gives rise to a white-spore phenotype, whereas an unmethylated active *b2* gene results in dark brown pigmented spores. The active (paramutable) recipient *b2* gene is *trans*-activated by the meiotic transfer of DNA methylation from the inactive (paramutagenic) donor allele. Several indications, such as the occurrence of methylation transfer only during meiosis and 5' to 3' polarity of methylation transfer that is similar to the mechanism of gene conversion, suggest that *trans*-regulations observed occur through an earlier intermediate in the recombination process.

Paramutation in animals

Paramutations in animals is less well described phenomenon. Especially in mammals, because of the small size of the litter, it is more difficult to observed different phenotypes and to link them to non-genetic mechanisms of inheritance. Here you are introduced to one specific case of paramutations in animals, changes in coat color.

An example of a white-tailed phenotype

It was possible to discover and characterize paramutation in plants because of the easily observable phenotype: kernel color. Similarly, in animals, paramutation was also described based on the observation of another easily identifiable phenotype: coat color. Coat color is perhaps one of the oldest phenotypic traits used for the selection of various new traits. In mice, mutations at the *W* locus in the heterozygote leads to a visible phenotype known as the white spotting phenotype that is represented by coat discoloration, such as whitening of the belly, feet, and tails in mice. The phenotype is associated with mutation of the c-kit proto-oncogene that encodes a transmembrane tyrosine protein kinase receptor. Mutation at the *W* locus also leads to multiple developmental problems, including sterility and anemia. The effects are cell autonomous and intrinsic, and they typically result in different degrees of severity. A white tail tip in adults is caused by a defect in the migration of melanocyte progenitors responsible for pigmentation. The discoloration varies quite a bit with the size and area of white pigmentation depending on the strain of mice, although loss of pigmentation in the tail is a common characteristic of all mutants. A white tail is inherited and visible as early as one week after birth. Additionally, in 1% to 2% of cases, there appears to be an infrequent occurrence of a short white tail tip that is not systematically transmitted to the next generation, with the white portion being generally light and short.

The specific discovery that we would like to discuss is the case of transmission of a white tail of *W* mutants that contain deletions in the first exon of the c-kit gene (Rassoulzadegan et al., 2006). Because KIT is needed for hematopoiesis, $Kit^{tm1Alf/tm1Alf}$ mutants that are homozygous for the deletion die shortly after birth. However, heterozygotes survive, but because their reduced levels of KIT impair melanocyte development, they display a white belly spot of variable size, white feet, and a white tail tip usually several centimeters long.

When heterozygous animals were crossed with one of the strains of a wild-type partner (either C57Black/6, B6/D2, or 129/sv), the heterozygote progenies have the Mendelian ratio of segregation of the white tail tip phenotype. In contrast, when animals heterozygous for mutations at the *W* locus were crossed with the similar $Kit^{tm1Alf/+}$ animals, a deviation from the Mendelian law of segregation was observed. Homozygous mice ($Kit^{tm1Alf/tm1Alf}$) are dead at birth and thus cannot be used for the studies. The phenotypic analysis of the live progeny of a heterozygous cross showed that the majority of animals with the wild-type genotype mimic the phenotype of the heterozygous parents with a short but bright white tail tip. The animals with wild-type genotype but white-tail phenotype were referred to as Kit*. Curiously, it was found that the white tail tip in Kit* was shorter than in the heterozygous $Kit^{tm1Alf/+}$ mutants, but the length of a white tail increased in the progeny, with the increase being directly proportional to the number of intercrosses between Kit*and $Kit^{tm1Alf/+}$.

Because animals homozygous for the *W* mutation are dead at birth, it was expected that the frequency of occurrence of Kit* animals among all live animals would be around 33.3%, with 66.7% being $Kit^{tm1Alf/+}$animals. Surprisingly, the analysis showed that approximately 60% of animals alive at birth were animals with wild-type genotype. This dramatic overrepresentation of wild-type animals could possibly be due to problems with the development experienced by the heterozygous $Kit^{tm1Alf/+}$ animals, although no $Kit^{tm1Alf/+}$ animals were found dead at birth. The Kit* animals represented the majority of all animals that had wild-type genotype, with 54 out 65 animals being Kit* animals and the remaining 11 animals being genetically and phenotypically wild-type animals with a full black tail. Each subsequent intercrossing of the heterozygous $Kit^{tm1Alf/+}$ mutants to other heterozygous $Kit^{tm1Alf/+}$ mutants increased the frequency of occurrence of the white tail phenotype and, beyond six intercrosses, a white coat color extended to the rest of the body of mutants. A similar picture was observed regardless of whether the source of the $Kit^{tm1Alf/+}$ allele was maternal or paternal.

Kit* animals that were obtained from crosses between $Kit^{tm1Alf/+}$ and $Kit^{tm1Alf/+}$ animals had a clear quantitative variation in tail length and a strong white tail tip phenotype. Curiously, when Kit* animals were crossed with $\mathrm{Kit}^{+/+}$ animals exhibiting a normal tail color, the white tail phenotype was inherited by their progenies at an average frequency of 80% to 90%. Further transmission of the phenotype showed a decrease in frequency, and in the following generations, phenotypic changes would eventually disappear. In contrast, intercrossing of Kit* × Kit* resulted in the maintenance of the white tail tip phenotype through a number of generations with 90% efficiency.

The described changes in tail color resemble the phenomenon of paramutation in plants. All components of paramutation seem to be in place, including the presence of paramutagenic and paramutable loci, heritability, and contagiousness of paramutation.

RNAs are involved in phenotypic epigenetic inheritance in mice

The authors of the aforementioned work (Rassoulzadegan et al., 2006) hypothesized that small non-coding RNAs could be ideal candidates for acting as a signal for the transgenerational maintenance of epigenetic changes. RNA extracted from sperm was thus analyzed for its ability to induce a white tail tip phenotype after microinjection into fertilized mouse eggs. The analysis of the abundance of c-kit transcripts of RNA prepared from the sperm of $Kit^{tm1Alf1/+}$ heterozygotes, Kit* paramutants, and wild-type animals revealed higher number of short fragments of c-kit transcripts in the sperm of Kit* animals. Several cytological assays confirmed that the sperm heads of $Kit^{tm1Alf1/+}$ and Kit* animals had more RNAs than those of wild-type $\mathrm{Kit}^{+/+}$ animals. Next, the authors microinjected total RNA prepared from various tissues (brain, testes, sperm) of $Kit^{tm1Alf1/+}$ mutants, Kit* and wild-type mice into fertilized zygotes; they found that microinjection of $Kit^{tmlAlf1/+}$ and Kit* but not wild-type RNA resulted in the birth of animals with a white tail in approximately 50% of cases. No difference in paramutation frequency was found either between injection of RNA isolated from sperm and from somatic tissue or between brain somatic tissue of male and female. The paramutant animals with a white tail tip born after RNA microinjection transmitted a paramutagenic state to their progeny efficiently and independently of animal gender.

Because the authors observed an increased pool of c-kit RNA fragments in $Kit^{tmlAlf1/+}$ animals, they hypothesized that such RNA degradation could be an action of specific micro-RNAs. To test the hypothesis, they microinjected two microRNAs carrying homology to c-kit mRNA (miR-221 and miR-222) into fertilized eggs of wild-type animals. The effect of paramutagenesis was observed in 40% of all animals born after microinjection of miRNAs. There was a substantial variation in the maintenance and transmission of the white tail phenotype in the progeny of these animals. These variations are very similar to those described for the phenomenon of plant paramutation, which supports the **rheostat hypothesis** of gradual changes rather than actual "on-off switch" changes. If such an "RNA-based rheostat" exists, it can potentially be maintained and transferred to the future generations. As the aforementioned studies showed, backcrosses to animals that are heterozygous for a mutant allele are apparently required to propagate this state possibly indefinitely. Thus, although the propagation of the phenotype (and small RNAs?) over several generations is possible, it is lost unless "fresh" small RNAs are produced upon crosses. The "molecular rheostat" has been originally proposed for paramutation of the L1 gene in plants. The transmission and propagation of small interfering RNAs in plants depend in part on the function of the RNA-dependent RNA polymerase (RdRP). The fact that such enzymes are absent in mammals proves the existence of an alternative mechanism of RNA amplification and/or RNA-mediated DNA methylation and chromatin modification. Moreover, because no changes in methylation were observed at the locus, it is quite possible that the mechanism operates through histone modifications or perhaps other mechanisms of chromatin changes.

The fact that no changes in the primary DNA sequence of the Kit locus were found only proves additional similarities with plant paramutation. It should be noted, however, that plant paramutation is frequently associated with changes in methylation at the locus, whereas no such changes and repetitive elements were observed at the Kit locus.

The work published by Rassoulzadegan and colleagues raised many questions. It is curious to know whether the process of paramutation was unique to the Kit locus of the mouse. This is apparently not the case; the recent work by this group reported that a variety of phenotypes were observed in mouse families generated after RNA microinjection in fertilized eggs. This might or might not necessarily be the events of paramutation as RNAs do not have to be in perfect homology to their genomic targets, and thus it would be difficult to establish an actual allelic interaction. The idea that small RNAs that are present in the sperm/ovum or fertilized eggs influence the phenotypic appearance of the embryo is fascinating. We discuss the implications of this phenomenon in the Chapter 16 "Epigenetics of Germline and Epigenetic Memory," covering transgenerational responses and transgenerational inheritance of epigenetic changes.

A critical analysis of Rassoulzadegan's work by Heinz Arnheiter (2007) suggests that the reported paramutation phenomenon had many holes in it. Heinz Arnheiter thinks that the statement that the paramutated offspring had "the white patches characteristic of the parental heterozygotes" is an exaggeration as there were no white feet or white belly spots typically observed—only small white tail tips were seen. The author further suggests that according to an animal breeding facility (Jackson Laboratories, Bar Harbor, Maine, USA), however, white tail tips are common among normal laboratory mice, including standard C57BL/6J mice (Jackson stock no. 000664). He writes: "Indeed, in at least six of 10 such C57BL/6J mice recently shipped to our laboratory from Jackson, there were sizable white tail tips and patches. Similar observations were made with mice of a different stock (C57BL/6NCrl) purchased from another breeding facility, Charles River (Lyon, France)." Because it is safe to assume that none of the ancestors of these mice ever carried the Kit^{tm1Alf} allele, tail discoloration was unlikely to be triggered by paramutation in these animals. In his work, Rassoulzadegan claims that the strain of mice they used (129/Sv, C57BL/6, and C57BL/6 × DBA) is not prone to have white tail tips. Arnheiter suggests that several other explanations should be taken into consideration, including genetic and, most likely, some environmental and nutritional ones. Arnheiter further questions the involvement of specific RNAs and even the link between the Kit locus and the phenomenon observed. To better understand the controversy, please read the paper by Heinz Arnheiter (2007).

Because the aforementioned case of paramutation at the Kit locus was the only report in animals, doubts raised by Heinz Arnheiter would undermine the belief that paramutation exists in animals. However, as it has been recently reported, *trans* allele methylation is yet another case of paramutation or a paramutation-like effect observed in mice. In mammals, there are many imprinted genes with methylation and gene expression patterns defined by the parental origin of inherited alleles. At *Rasgrf1* (encoding RAS protein-specific guanine nucleotide-releasing factor 1), a repeated DNA element is needed to establish methylation of a differentially methylated domain (DMD), a methylation-sensitive enhancer-blocking element, which, together with a repeated DNA element, functions as a binary switch that regulates imprinting and expression of the active paternal allele. Sequences that regulate methylation of the maternal allele have been also identified for *Igf2r* (encoding insulin-like growth factor 2 receptor); an intron, a sequence called region 2, is needed for methylation of the active maternal allele.

Herman et al. (2003) analyzed whether the activities of the two identified regulatory elements overlap and can be replaced with each other. They generated transgenic mice containing *Igf2r* region 2 in place of the *Rasgrf1* repeats and performed reciprocal crosses between heterozygous transgenic animals and wild-type animals. Maternal transmission of the *Rasgrf1*$^{tm3.1Pds}$ allele (*Rasgrf1*$^{-/+}$) had no effect on methylation or expression of the locus, which remained paternal allele–specific and was expressed at levels comparable to those in wild-type mice. In contrast, mice with a paternally transmitted transgenic allele (*Rasgrf1*$^{+/-}$) lacked both methylation and expression of the paternal allele, and the locus was expressed at lower levels than in wild-type mice. Moreover, paternal transmission of the mutated allele also induced methylation and expression *in trans* of the normally unmethylated and silent wild-type maternal allele. Once activated, the wild-type maternal *Rasgrf1* allele maintained its activated state in the next generation independently of the paternal allele. Thus, the expression of the mutated paternal allele in *Rasgrf1*$^{+/-}$ mice showed that region 2 was able, in part, to replace the function of the *Rasgrf1* repeats. The authors proved specific requirements of region 2 sequence by replacing it with an unrelated sequence of similar size (allele *Rasgrf1*$^{+/tm2Pds}$). None of the heterozygous progeny of *Rasgrf1*$^{+/tm2Pds}$ males expressed *Rasgrf1* from the maternal allele, confirming the requirement of *Igf2r* sequence for this paramutation-like process.

Trans activation of the normally silent maternal allele in *Rasgrf1*$^{+/-}$ transgenic animals resembles two phenomena: transvection in *Drosophila melanogaster* and paramutation in maize. In paramutation, expression of a paramutable allele altered by a paramutagenic allele persists through meiosis independently of the presence of a paramutagenic allele. In maize, the affected paramutable allele behaves as a paramutagenic allele in the next generation, representing secondary paramutation events. Crosses performed by Herman et al. (2003) showed that transmission of the *Rasgrf1*$^{tm3.1Pds}$ allele from father to daughter modified the daughter's *Rasgrf1*$^{+d}$ wild-type allele in a manner that allowed it to affect expression of both parental alleles in the grandchildren. This recapitulates the two key properties of paramutation: regulation of expression of one allele by the other and stability of this phenotype through meiosis. The fact that offspring with a silenced paternal allele showed secondary paramutation confirms a third property of paramutation. The fact that Herman et al. (2003) observed paramutation-like effects only when the *Rasgrf1*$^{+d}$ allele was maternally transmitted proves a fourth feature of paramutation noted at the *r1* locus in maize, where paramutation was observed only on paternal transmission of the paramutated allele. In mice, transmission of the wild-type allele by *Rasgrf1*$^{+/-}$ males led to normal monoallelic expression from the paternal allele.

The phenomenon of *trans* allele methylation was also reported in the past for the fungus *Ascobolus immersus* and maize. When a methylated allele and an unmethylated allele of the *Ascobolus b2* spore color gene were brought together in individual meiotic cells, a frequent transfer of methylation to the unmethylated allele was observed. In maize, at the *Rosa26* locus modified with a *lox*P-stop-*lox*P cassette, Cre-mediated recombination led to methylation of the modified locus that was transferred to the unrecombined *lox*P-stop-*lox*P allele on the homologous chromosome in the next generation. Transferred methylation was stable for multiple generations.

Transvection

Transvection is yet another epigenetic process that describes interactions between two homologous alleles or even between non-allelic regions. These interactions lead to changes in gene expression, either activation or repression. The phenomenon was initially described in the 1950s by Edward B. Lewis (1918–2004) for the bithorax complex in Drosophila. He received the Nobel Prize in 1995 for the discovery and description of functions of the bithorax complex in Drosophila. Lewis believed that transvection only occurred when there was a pairing between certain genomic regions. He wrote, "Operationally, transvection is occurring if the phenotype of a given genotype can be altered solely by disruption of somatic (or meiotic) pairing. Such disruption can generally be accomplished by introduction of a heterozygous rearrangement that disrupts pairing in the relevant region but has no position effect of its own on the phenotype."

The work of James Morris and his research group showed that the structure of paired homologues played an important role in regulating transvection and gene expression (Morris et al., 1998). They proposed that homologue pairing promotes at least two forms of transvection at the Drosophila *yellow* locus. In one case, gene expression is directed by the *trans* action of genetic elements such as activation of the promoter by enhancer elements, whereas in the other case gene expression is regulated by enhancer bypass of a chromatin insulator triggered by the presence of a structurally dissimilar homologue. Using the protocols established by Lewis, Morris's group provided a formal demonstration of the pairing-dependent nature of *yellow* transvection and suggest that *yellow* pairing, as measured by transvection, reflects the extent of contiguous homology flanking the locus (Ou et al., 2009).

Transvection can also occur by the action of silencers *in trans* or by the spreading of position effect variegation from rearrangements having heterochromatic breakpoints to paired unrearranged chromosomes. Other cases of transvection might involve the production of joint RNAs by *trans*-splicing, although this still has to be demonstrated. In Drosophila, some cases of transvection require Zeste, a DNA-binding protein; this may facilitate homologue interactions by self-aggregation. Transvection appears to require synapsis of homologues but can also occur without it. For more details on the mechanisms of transvection in Drosophila, see the review by Duncan et al. (2002).

Since then, transvection has been observed at a number of additional loci in Drosophila—including white, decapentaplegic, eyes absent, and vestigial—as well as in other species, including fungi, nematodes, insects, mice, humans, and plants. In light of these findings, transvection might represent a potent and widespread form of gene regulation.

Cosuppression

Cosuppression is a form of post-transcriptional homology-dependent gene silencing that describes the loss of expression of a transgene and related endogenous or viral genes in transgenic plants. Most frequently, cosuppression occurs due the abundance of transgene transcripts. The process is regulated at the level of mRNA and includes processing, localization, and/or degradation of mRNA. A rapid turnover of endogenous transcripts in the presence of homologous transgenes is a **nonlinear response**. A small increase in gene expression or dosage results in an **inversely amplified response** and dramatic phenotypic alterations.

There are many reports of cosuppression in different organisms, with the two better-known examples including suppression of chalcone synthase (*Chs*) by the transgene presence in petunia and the infection of plants with RNA viruses carrying homology to endogenous plant genes. Suppression of *Chs* in transgenic petunia reported by Richard Jorgensen in 1990 is the first well-documented event of silencing (Napoli et al., 1990). A second experimental system that has been very useful for the analyses of cosuppression is the generation of resistance to infection by RNA viruses (such as tobacco mosaic virus and tobacco etch virus) in tobacco and related species. This is achieved most often through the expression of a viral coat protein (CP) transgene.

For more information on possible mechanisms of cosuppression, see Chapter 15, "Gene Silencing—Ancient Immune Response and a Versatile Mechanism of Control over the Fate of Foreign Nucleic Acids."

Conclusion

Paramutation results from an interaction between two alleles that have different chromatin structure and DNA methylation. Sequences involved in paramutation undergo silencing of genes present within these sequences. Structurally, the sequences involved can vary from highly homologous to very diverse non-homologous ones. Many of the paramutation systems consist of repeat sequences and/or multiple copies of the gene. The stability of paramutation can be very high at the *b1* locus or variable at the R1 locus. The siRNA-based model is one of the prominent candidates for being the mechanism underlying paramutation. However, this model can work in conjugation with other mechanisms, including the physical interaction model. The exact pathway of paramutation still remains inconclusive. Evidently, more experiments are needed to understand paramutation better. The presence of siRNA or other RNA species derived from paramutation sequences needs to be identified and characterized. Molecular components of siRNA and the physical interaction pathway also need to be identified. At present, the study of paramutation is limited to various visible traits. Studies focused on identification of new paramutation systems should be carried out. They might lead to more in-depth knowledge of the phenomenon and better comprehension of its relevance to the survival of organisms and evolution.

Exercises and discussion topics

1. What are the two major models of paramutation?
2. What is called a penetrance of paramutation?
3. What are the differences between the paramutagenic and paramutable haplotypes?
4. Are paramutations reversible or stable? What is the difference between mutation and paramutation?
5. Compare gene silencing, genomic imprinting, and paramutations.
6. What is the difference between primary and secondary paramutation?
7. What are the similarities and the differences between paramutation and transvection?
8. What is a connection between siRNAs and paramutation?
9. Are siRNAs required for the establishment and maintenance of paramutation?
10. Why are tandem repeats important for paramutation?
11. Why is the mechanism of paramutation important for plants?
12. Is a paramutation phenomenon observed only in plants?
13. How do paramutations differ in animals and in plants?
14. What is the significance of paramutations in evolution?

References

Alleman et al. (2006) An RNA-dependent RNA polymerase is required for paramutation in maize. *Nature* 20:295-298.

Arnheiter H. (2007) Mammalian paramutation: a tail's tale? *Pigment Cell Res* 20:36-40.

Bateson W, Pellew C. (1915) On the genetics of 'rogues' among culinary peas (*Pisum sativum*). *J Genet* 5:15-36.

Brink RA. (1956) A genetic change associated with the *R* locus in maize which is directed and potentially reversible. *Genetics* 41:872-889.

Brown DF, Brink RA. (1960) Paramutagenic Action of Paramutant R and R Alleles in Maize. *Genetics* 45:1313-1316.

Chandler VL, Stam M. (2004) Chromatin conversations: mechanisms and implications of paramutation. *Nat Rev Genet* 5:532-544.

Chandler VL. (2004) Poetry of b1 paramutation: cis- and trans-chromatin communication. *Cold spring harbor symposia on quantitative biology*. Cold Spring Harbor laboratory press. Volume LXIX.

Coe EH. (1959) A regular and continuing conversion-type phenomenon at the B locus in maize. *Proc Natl Acad Sci USA* 45:828-832.

Dorweiler et al. (2000) *mediator of paramutation1* is required for establishment and maintenance of paramutation at multiple maize loci. *Plant Cell* 12:2101-2118.

Duncan IW. (2002) Transvection effects in Drosophila. *Annu Rev Genet* 36:521-556.

Hagemann R. (1958) [Somatic conversion in Lycopersicon esculentum Mill.] *Z Vererbungsl* 89:587-613.

Herman et al. (2003) Trans allele methylation and paramutation-like effects in mice. *Nat Genet* 34:199-202.

Lewis EB. (1950) The phenomenon of position effect. *Adv Genet* 3:73-115.

Morris et al. (1998) Two modes of transvection: enhancer action in trans and bypass of a chromatin insulator in cis. *Proc Natl Acad Sci USA* 95:10740-10745.

Napoli et al. (1990) Introduction of a Chimeric Chalcone Synthase Gene into Petunia Results in Reversible Co-Suppression of Homologous Genes in trans. *Plant Cell* 2:279-289.

Ou et al. (2009) Effects of chromosomal rearrangements on transvection at the yellow gene of Drosophila melanogaster. *Genetics* 183:483-496

Patterson et al. (1993) Paramutation, an allelic interaction, is associated with a stable and heritable reduction of transcription of the maize *b* regulatory gene. *Genetics* 135:881-894.

Rassoulzadegan et al. (2006) RNA-mediated non-mendelian inheritance of an epigenetic change in the mouse. *Nature* 441:469-474.

Stam et al. (2002) The regulatory regions required for *B'* paramutation and expression are located far upstream of the maize *b1* transcribed sequences. *Genetics* 162:917-30.

14

Bacterial adaptive immunity—Clustered Regularly Interspaced Short Palindromic Repeats (CRISPR)

Clustered regularly interspaced short palindromic repeats (CRISPR) are the bacterial loci that encode regulatory RNAs conferring sequence-directed immunity against phages. CRISPR is believed to be one of the most ancient bacterial adaptive mechanisms of protection against phages. This adaptive immunity involves an active process of integration of short fragments of foreign nucleic acids into clusters of CRISPRs, followed by expression of small RNAs from the loci. These RNAs target pathogen genomes for degradation.

Bacteria and Archaea lack bona fide sexual reproduction. Some of the mechanisms that they use instead for adaptation to stress and genome diversification include **phage transduction**, **transformation**, and **conjugation**, commonly known as mechanisms of horizontal gene transfer (HGT). Expression of these chromosomal or extrachromosomal genes acquired from without enable bacteria to thrive in the process of adaptive evolution. It would thus seem logical if bacteria would have special mechanisms allowing them to acquire and incorporate the external DNA quite efficiently. Indeed, some of the mechanisms, such as **integrons**, might assist bacteria to acquire antibiotic resistance. Integrons are genetic elements that, although being unable to move themselves, carry gene cassettes that can be mobilized either to other integrons or to alternate sites in the bacterial genome.

14-1. Horizontal gene transfer

Horizontal gene transfer (HGT), also known as lateral gene transfer, is the transfer of genetic material from one organism to another without reproduction—that is, when the receiving organism does not acquire genetic material through cell division.

HGT is the main source of genetic diversity in bacteria. It involves three main processes: transformation, conjugation, and transduction.

Transformation is the uptake of naked DNA upon which short DNA fragments are "picked-up" by bacteria from the environment. Conjugation is the transfer of DNA from one bacterial cell to another with the aid of conjugal plasmid. Conjugation can occur between similar and distantly related bacteria and long DNA fragments may be transferred. Transduction is the transfer of DNA from donor to recipient cell with the aid of a bacteriophage. This mechanism is limited to closely related bacterial species as both the donor and recipient cells need to have the same cell surface receptors. DNA molecules of various lengths can be transferred, but they are typically limited by the size of the phage head.

Bacteria and Archaea have quite a lot of mechanisms of stress response, including **SOS DNA damage response**. Damaged DNA stalls replication processes, resulting in overaccumulation of single stranded DNA (ssDNA). RecA protein, a recombinase functioning in recombinational DNA repair, has a high affinity to the ssDNA molecules. A RecA association with ssDNA releases the repressive binding of LexA to SOS responsive genes. This results in an increase in the expression of three nonessential DNA polymerases that were shown to be required for generations of mutations in bacteria: Pol II (encoded by the gene polB), Pol IV (encoded by dinB), and Pol V(encoded by umuD and umuC). Together these polymerases promote nucleotide excision repair and translesion synthesis DNA repair, leading to accumulation of mutations in bacterial DNA. As a result, bacteria gain the necessary genetic diversity to adapt to new environmental conditions. You can find more detailed information about SOS mechanisms from reviews on the topic.

Despite the fact that acquisition of foreign DNA might be beneficial to bacteria, it also possesses certain dangers, such as integration of DNA from lytic phages. One of the common mechanisms of protection against bacteriophage DNA and DNA from other bacteria, which might cause a decrease in fitness, is the function of specific methylation-sensitive restrictases, the so-called **restriction-modification system (RMS)**. In the process of RMS, bacteria protect their own DNA by methylation of cytosines at specific sites recognized by restriction enzymes. As a result, only foreign unmethylated DNA is degraded by restrictases.

Phages are the most common and ancient enemies of bacteria. In some growth environments, these viruses can outnumber bacterial cells ten to one. Not surprisingly the mechanisms of phage resistance are versatile and include adsorption resistance, restriction mechanisms, and abortive infections. Adsorption resistance includes loss of host phage receptor molecules and physical barriers such as capsules hiding receptor molecules, which results in a reduced interaction between phage and bacterium. The abortive infection system functions by promoting "altruistic suicide" of the infected bacterial cell, upon which both bacterium and phage die. In contrast, the restriction mechanisms include phage-genome uptake blocks, superinfection immunity, restriction modification, and CRISPR. You can find more details on the mechanisms of protection that bacteria employ against viruses in the review by Hyman and Abedon (2010).

One of the most interesting mechanisms of protection, as far as research in epigenetics is concerned, is based on the function of regulatory RNAs encoded by CRISPR loci to confer sequence-directed immunity against phages. This process represents an active protection mechanism by which prokaryotes protect themselves against phages by integration of short fragments of phage/viral nucleic acids into clusters of CRISPRs. As such, this process represents an ancient adaptive mechanism of protection against pathogens.

Structure of CRISPR-associated genes

The first report about the existence of CRISPR arrays dates to 1987 (Ishino et al., 1987) when a group of 14 29 bp-long repeats separated by non-repetitive elements—32 to 33 bp in length—in *E. coli* were identified. Similar structures were later on found in *Thermotoga maritima*, *Methanocaldococcus jannaschii*, *Haloferax mediterranei*, *Mycobacterium tuberculosis*, and other bacteria. To date, such arrays were identified, appearing in 40% of sequenced eubacterial genomes and in 90% of sequenced archaeal genomes.

A more detailed analysis of the structure of these arrays revealed that they are highly conserved and consist of 24- to 47-bp repeats separated by variable, often unique spacer sequences derived from phage or plasmid replicons (see Figure 14-1; reviewed in Vale and Little, 2010). The 24 to 47 nt repeats can be repeated up to 250 times, and the loci are almost always accompanied by multiple (25 to 45) **CRISPR-associated (*cas*) genes**. These genes are found in proximity to the CRISPR arrays, typically upstream, and those loci that do not contain these arrays normally do not contain *cas* genes. The proteins encoded by *cas* genes show high sequence and structural similarity to endonucleases, helicases, polymerases, and DNA- and RNA-binding proteins (Jansen et al., 2002; Wiedenheft et al., 2009). That is why it was proposed that their function is associated with the function of the RNA-interference (RNAi) system (reviewed in Marraffini and Sontheimer, 2010a). However, this might not be the case because unlike RNAi, **CRISPR-Cas** complexes appear almost exclusively to bind to DNA (Mojica et al., 2009).

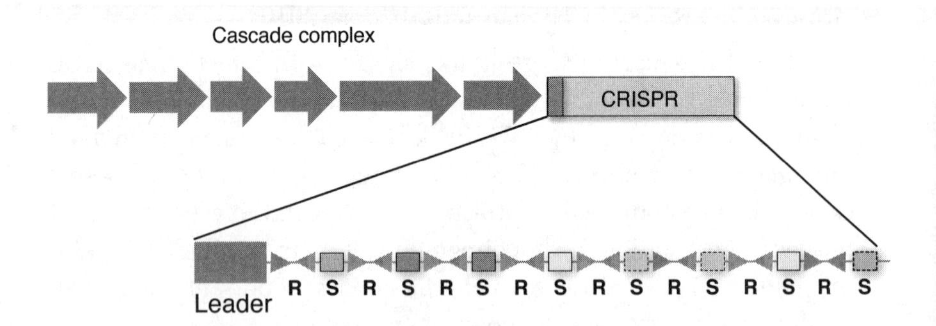

Figure 14-1 Organization and processing of the CRISPR elements. CRISPR loci consist of inverted repeats (R) and spacers (S) preceded by the leader sequence. CRISPR loci contain genes of the cascade complex, the cas genes.

CRISPR loci are classified into 12 major groups—CRISPR-1 to CRISPR-12—based on sequence similarities (Kunin et al., 2007).

Depending on gene content and gene order, CRISPR loci are classified into eight CRISPR/*cas* subtypes. The arrays of the same repeats are sometimes followed by an AT-rich sequence, referred to as a leader. In the arrays containing degenerated terminal repeats, the leader is often located upstream of the CRISPR locus. The role of leaders is not clear, but it is possible that they promote transcription through repeats or alternatively provide a binding site for the proteins (could be Cas proteins) responsible for repeat duplication and/or spacer acquisition. A fully functional CRISPR/*cas* system thus is composed of CRISPR, Cas proteins, and the leader sequence (see Figure 14-1).

Despite their sequence diversity, a majority of CRISPR repeats are partly **palindromic** and have the potential to form hairpin structures. The stem of the hairpin structure formed contains conserved G-U pairs. This implies that the hairpin structures can be formed in the transcriptional products of the locus. Such conservation of the stem structure is not a conserved property of all CRISPR repeats.

An additional group of proteins whose function is potentially associated with CRISPR is the **RAMPs (Repeat-Associated Mysterious Proteins) family**. The genes coding for these proteins are present in CRISPR-containing genomes adjacent to or distant from the repeats themselves. A subset of the RAMP genes (*cmr1-6*) is found in a well-conserved cluster called the "RAMP module" or "polymerase cassette" and is associated with several cas system subtypes (Haft et al., 2005). CRISPR-related proteins of known function or activity are summarized in Table 14-1.

Table 14-1 CRISPR-related proteins (reproduced from Karginov FV and Hannon GJ (2009) The CRISPR System: Small RNA-Guided Defense in Bacteria and Archaea. Mol Cell 37:7-19).

Organism	Name	Function/activity
E. coli	**Cascade complex**	Endonucleolytic cleavage of precrRNA
	cse1 (casA)	
	cse2 (casB)	
	cse3 (casE)	Catalytic subunit
	cse4 (casC)	
	cas5e (casD)	

Table 14-1 CRISPR-related proteins (reproduced from Karginov FV and Hannon GJ (2009) The CRISPR System: Small RNA-Guided Defense in Bacteria and Archaea. Mol Cell 37:7-19).

Organism	Name	Function/activity
P. furiosus	**RAMP module complex**	crRNA-guided endonucleolytic cleavage of RNA targets
	cmr1	
	cmr2	
	cmr3	
	cmr4	
	cmr5	Dispensable for activity
	cmr6	
P. aeruginosa	*cas1*	Acquisition of new spacers / ss/dsRNA endonuclease
S. solfataricus	*cas2*	ssRNA endonuclease
E. coli	*cas3*	crRNA-guided degradation of invading NAs
P. furiosus	*cas6*	Endonucleolytic cleavage of precrRNA
S. thermophilus	*csn1* (*cas5*)	Phage resistance using existing spacers
S. thermophilus	*cas7*	Acquisition of new spacers?
S. solfataricus	*SSO0454*	Specific binding of CRISPR repeat DNA

The proposed function of CRISPR loci

The function of CRISPRs remained unknown for a long time. Shortly after being discovered, CRISPRs were suggested to participate in adaptation to temperature stress, replicon partitioning, DNA repair, and transposition. Later on, in 2005, several research groups discovered that spacer sequences consisted of DNA derived from phages and plasmids, and thus, for the first time, it was suggested that CRISPRs were involved in bacterial immunity against invading foreign DNA molecules (Bolotin et al., 2005). The correlation between the number of these repeats and bacterial tolerance to phage infection was an indirect confirmation of the immunity hypothesis.

Recent research allowed us to gain more understanding toward the potential function of CRISPRs. It was shown that:

- Bacterial cells are able to incorporate phage genetic material into CRISPRs as spacers.
- Newly acquired spacers provide bacteria with resistance against phages that carry sequences homologous to the incorporated sequence.
- Resistance to phages is lost when spacers are removed.
- Single point mutations of specific short motifs in the phage genome are sufficient for loss of recognition (reviewed in Vale and Little, 2010).

The main function of CRISPRs thus appears to be limiting horizontal gene transfer by targeting DNA (discussed in Marraffini and Sontheimer, 2008). Based on computational analysis, Makarova and colleagues proposed that the CRISPR-Cas system (CASS) is a prokaryotic system of defense against phages and plasmids that function via a gene silencing RNAi mechanism (Makarova et al., 2006). They also proposed that the function of CASS involves integration of fragments of foreign genes into bacterial and archaeal genomes that results in heritable immunity to the respective agents.

It becomes increasingly clear that despite some similarities to RNAi, **CRISPR RNAs (crRNAs)**, and eukaryotic RNA silencing seem unlikely to be homologous processes (see Table 14-2).

Table 14-2 Comparison of CRISPR and RNAi

	CRISPR	RNAi
Protein machinery	Different	Different
Precursors	crRNAs are made of single-stranded precursors	miRNAs/siRNAs arise from double-stranded precursors
Amplification	crRNAs are not posttranscriptionally amplified	miRNAs and siRNAs can be amplified
Targets	crRNAs target DNA*	miRNAs/siRNAs target mRNA (although they can also target DNA for methylation)

Table 14-2 Comparison of CRISPR and RNAi

	CRISPR	RNAi
Function	Only for the defense against phages	Functions for the defense against invaders as well as for the regulation of the endogenes
Capacity to evade	Highly likely because the defense occurs only after the insertion of invader DNA sequences into CRISPR; meanwhile, phages can mutate	Very unlikely as the small guide RNA can be picked up from the invader RNA

* *In vitro* targeting of RNAs by a *Pyrococcus furiosus* **crRNA–Cas ribonucleoprotein (crRNP)** complex has been documented (reviewed in Marraffini and Sontheimer, 2010a), raising the possibility that RNAi may have greater functional analogies with some CRISPR systems than with others.

Although the molecular details are still being uncovered, the immune response of the CASS system presumably has three distinct stages (reviewed in Marraffini and Sontheimer, 2010a). In the initial stage (Stage 1), infection with bacteriophage activates the cascade complex, and the Cas proteins target and cleave short recognition motifs in the phage genome (Deveau et al., 2008) and incorporate these so-called proto-spacers into the host genome at the 5' leader sequence of the CRISPR locus (Barrangou et al., 2007). The proto-spacers are incorporated as short sequences of 23 to 30 nt (see Figure 14-2).

In Stage 2, the incorporated spacers are transcribed as CRISPR RNAs (crRNAs) that contain the spacer sequence flanked on either side by partial repeat sequences (see Figure 14-3) (Brouns et al. 2008). As a result, the spacer sequences are amplified at a large scale (for example, Lillestøl et al., 2009).

Finally, Stage 3 includes Cas-protein complex-mediated sequence-specific interference of phage amplification (see Figure 14-4). In this process, crRNAs serve as templates to target conserved bacteriophage motifs in future infections (Brouns et al., 2008).

Stage 1

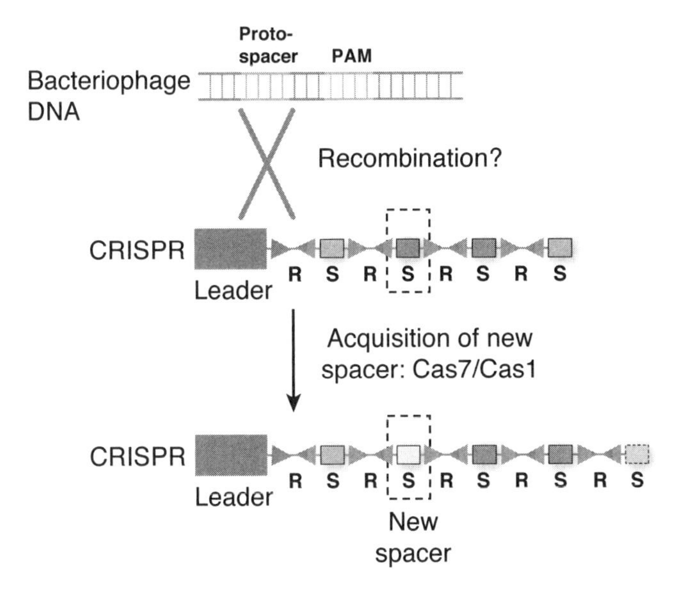

Figure 14-2 The acquisition of a new spacer into the array of the CRISPR elements. The acquisition of new spacer elements into CRISPR might occur through recombination between areas of homology in bacteriophage and CRISPR loci.

Stage 2

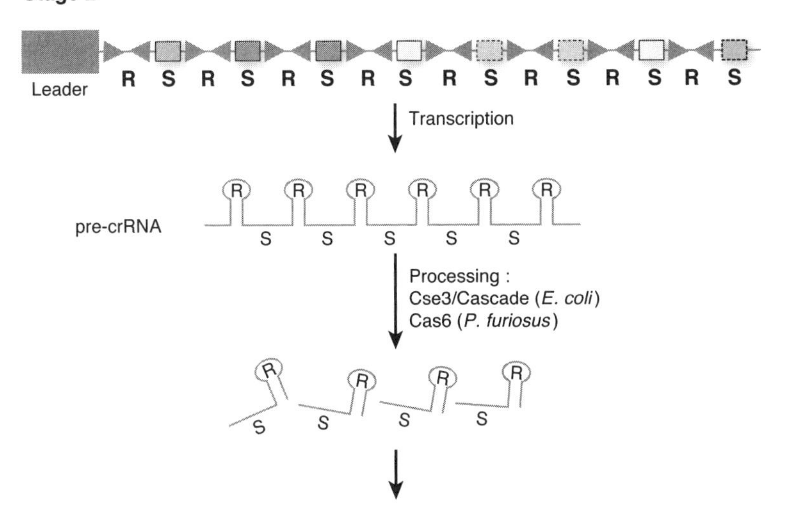

Figure 14-3 Stage 2. Processing of spacers. Transcribed pre-crRNAs are processed via Cas proteins.

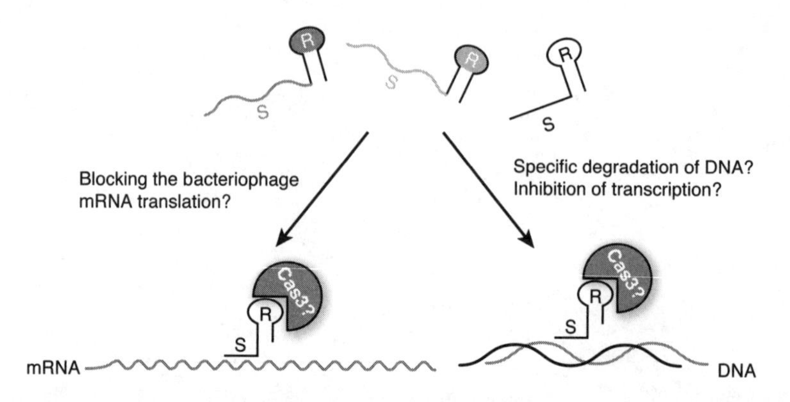

Figure 14-4 Stage 3. Mechanism of action of crRNAs. crRNAs either target mRNA or the DNA of the phage. Cas3 protein might be necessary in the process.

It was shown that a complex of five Cas proteins—CasA, CasB, CasC, CasD, and CasE, which are collectively termed Cascade—cleaves a CRISPR RNA pre-crRNA precursor in each repeat and retains cleavage products containing virus-derived sequences. Resulting mature crRNAs consist of the antiviral spacer unit flanked by short RNA sequences derived from the repeat on either side termed the 5' and 3' handle. These handles might serve as conserved binding sites for Cascade proteins. The helicase Cas3 presumably assists the mature crRNAs to serve as guide RNAs that direct the complex to viral RNAs and interfere with virus proliferation (see Figure 14-4). Although this is a hypothesis, the observed fact that anti-λ; CRISPRs of both polarities lead to a reduction of *Streptococcus thermophilus* sensitivity to phage lambda supports this hypothesis. Indeed, previous reports showed that virus-derived sequences can integrate into CRISPR loci in both orientations (Makarova et al., 2006; Bolotin et al., 2005; Deveau et al., 2008). On the contrary, a recent report by Touchon and Rocha (2010) supports the idea that new spacers are acquired in a polarized fashion, with new units being added at the 5' leader end of the CRISPR.

A decreased sensitivity to lambda phage has also been reported for *E. coli* strains carrying artificial CRISPR/*cas* systems with spacers targeting essential genes of the virus (Brouns et al., 2008). It has been also shown that CRISPR/*cas* systems can limit plasmid conjugation in

Staphylococcus epidermidis (Marraffini and Sontheimer, 2008), demonstrating a broader role for CRISPR in the prevention of HGT.

CRISPR/*cas* systems have several analogies with the way eukaryotic small RNA-based systems function, particularly with the function of **PIWI-interacting RNAs (piRNAs)**. Similarly to prokaryotic repeats, piRNAs are primarily arranged into a limited number of loci in the genome. Also, precursors of piRNAs are long single-stranded RNA molecules that are cleaved to produce functional piRNAs involved in silencing. CRISPR loci are also transcribed into a single molecule that is further processed into smaller discrete RNAs, each with the size of a repeat-spacer unit.

Transcription of the CRISPR repeats initiates in the leader sequence. The generated long pre-crRNA precursors often span the entire locus (Lillestøl et al., 2009). The pre-crRNAs are then cleaved into mature products that consist of short fragments corresponding to the spacers (intervals between the repeats). It should be noted however that irregular patterns of transcripts have also been detected from the opposite strand in *S. acidocaldarius* but not in *S. epidermidis*, *P. furiosus*, or *E. coli* (reviewed in Karginov and Hannon, 2010).

More detailed information exists for *E. coli*; Cascade produces 57 nt units from the multimeric precursor transcripts. The cascade consists of Cse1, Cse2, Cse4, Cas5e, and Cse3 proteins (also known as CasA-CasE). Within the complex, Cse3/CasE is required and sufficient to define the 5' end of the product. Processing of 3' includes the cleavage of two nucleotides by yet unknown mechanism(s). In *P. furiosus*, the cleavage of the precursor occurs through the function of the Cas6 protein that is apparently not homologous to Cse3/CasE or any other *E. coli* proteins. The product of Cse3 or Cas6 is an RNA consisting of an 8 nt repeat sequence "tag" and the spacer sequence (see Figure 14-5).

In *P. furiosus*, there is an additional exonucleolytic processing step leading to two discrete species of mature psiRNA, 38 to 45 nt and 43 to 46 nt, depending on the spacer length. The resulting RNAs maintain the 5' repeat tag. In *S. acidocaldarius*, CRISPR-derived endonucleolytically cleaved small RNAs appear to be products between 35 and 52 nt in length. The precursors and mature small RNAs are found in distinct **ribonucleoprotein complexes (RNPs)**, suggesting that there indeed exist two independent steps of processing and assembly pathways. Table 14-3 summarizes crRNA processing in different bacteria.

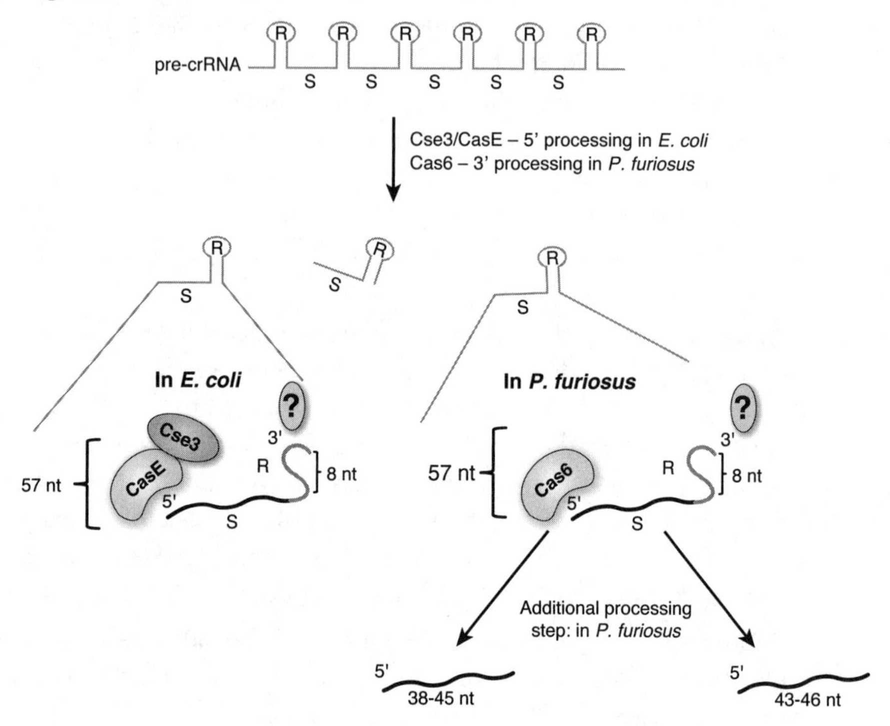

Figure 14-5 Processing crRNA loci in *P. furiosus*. Processing crRNA loci leads to generation of 57 nt crRNAs in E. coli and pre-psiRNA in *P. furiosus*. Processing of 5' end occurs through CasE/Cse3 in E. coli and Cas6 in *P. furiosus*. The mechanism of 3' processing is not known. (R represents repeat sequence; S represents spacer sequence.) Additional processing steps in *P. furiosus* generate two classes of psiRNA 38 to 45 and 43 to 46 nt.

Table 14-3 Different type of crRNAs produced in different bacteria

	E. coli	*P. furiosus*	*S. acidocaldarius*
Processing of 5' end	CasE/Cse3	Cas6	?
Processing of 3' end (trimming of 2 nt)	?	?	?
Further processing	-	+	-
Mature ncRNAs	crRNA ~ 55 nt	psiRNA: 35 to 45 nt and 43 to 46 nt	crRNA ~35 to 52 nt

The current model of the function of crRNAs suggests that they serve as sequence-specific guides during the effector stage of resistance against invading phage elements. This was demonstrated by reconstitution experiments in which the functioning CRISPR system containing the Cascade complex, Cas3, and a modified CRISPR locus with spacer sequences targeting phage lambda was incorporated in *E. coli* BL21 (DE3), which lacked the endogenous *cas* genes. It was sufficient to trigger *de novo* resistance to the phage (Brouns et al., 2008). It was further shown that Cas3 was not required for the generation of crRNAs, but it was necessary for phage resistance, suggesting that it might catalyze crRNA-guided destruction of foreign nucleic acids. Cas3 has an HD nuclease domain fused to a DExD/H helicase module and the nucleolytic activity, with substrates being dsDNA or dsRNA. In substantial analogy to eukaryotic RNAi-related pathways in the way crRNAs are produced, the initially proposed model suggested that crRNAs guide targeting of mRNAs derived from the invader (Makarova et al., 2006). Recent results demonstrate a capacity for RNA targeting in CRISPR systems containing the RAMP modules. *P. furiosus* contains a ribonucleoprotein complex consisting of six RAMP module proteins and the mature 39 nt and 45 nt psiRNAs with the shared 8 nt 5' repeat tag. This complex was able to endonucleolytically cleave complementary mRNA targets. The authors reconstituted the complex *in vitro* and demonstrated that the recombinant complex consisting of either 39 nt or 45 nt psiRNAs cleaved mRNA targets.

Although mRNA targeting is possible, other experiments suggest that there is also the direct recognition of the foreign DNA rather than targeting mRNA. The evidences include the facts that, to date, sequence analyses have only detected spacers from phage with DNA genomes (Mojica et al., 2009; Wiedenheft et al., 2009). The debate on whether DNA or RNA is being targeted goes on. Detailed analyses of bacteria and Archaea show that spacers encode crRNAs corresponding to both the coding and template strands of the phage DNA without a preference for any particular region. A bias toward the coding strand was observed upon examination of spacers arising in experimentally induced phage-resistant mutants of *S. thermophilus* (Barrangou et al., 2007).

The experiment with the use of engineered spacers in *E. coli* demonstrated that effective crRNAs could be produced from either the coding or template strand of lambda phage, thus further supporting the DNA as a target of crRNAs (Brouns et al., 2008). Further support for DNA targeting was provided by a recent study of CRISPR activity in *S. epidermidis* (Marraffini and Sontheimer, 2008). The CRISPR locus of the *S. epidermidis* isolate RP62a contains a spacer (*spc1*) against the *nickase* (*nes*) gene of staphylococcal conjugative plasmids. Because nickase activity is required for conjugation only in donor cells that deliver the plasmid, mRNA targeting by crRNAs would destroy the function of *S. epidermidis* isolate as a donor but not as a recipient. The abolishment of the activity of both a donor and a recipient and the observation that the targeted region was equally effective in either orientation within the plasmid supported the DNA targeting model (for details see Marraffini and Sontheimer, 2008). Finally, a solid proof that DNA was indeed a target was obtained by the experiment with *S. epidermidis* isolate RP62a in which Marraffini and Sontheimer (2008) designed a plasmid variant in which the proto-spacer was interrupted by the intron sequence. As a result, the intron split the DNA target but was removed from the mRNA target in the process of transcription/splicing. This artificial plasmid allowed conjugation, indicating that crRNAs were not able to target DNA and did not target the mRNA substrate (see Figure 14-6).

So far, both targeting DNA and targeting RNA molecules are possible as models of CRISPR activity. The RNA targeting model might exist only for those CRISPR systems that encode a RAMP module. Alternatively to targeting phage mRNA, the RAMP module of the CRISPR system might also be involved in targeting mRNAs that are important for endogenous cellular processes in bacteria.

Future experiments will have to clarify whether the RNA cleavage activity in *P. furiosus* relates to DNA targeting. The analysis of the CRISPR activity in *E. coli*, *S. thermophiles*, and *S. epidermidis* does not allow such study as these bacteria do not possess a RAMP module. The analysis of *P. furiosus,* on the contrary, would allow searching for both mRNA and DNA targeting activities. Additionally, *S. solfataricus* and *B. halodurans* might also be useful for finding out whether these two mechanisms exist as these bacteria contain spacers that are both sense and antisense to extrachromosomal elements, suggesting that DNA targeting is active in these organisms (Makarova et al., 2006).

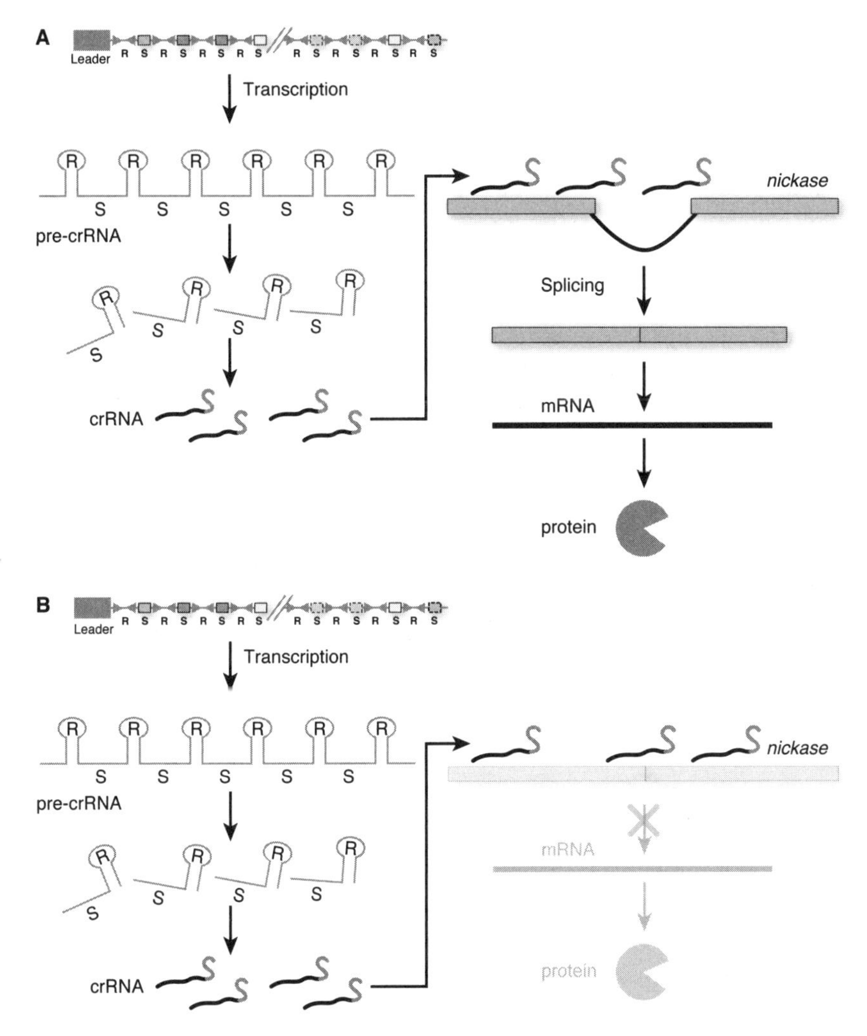

Figure 14-6 Targeting of phage nickase gene by S. epidermidis isolate RP62a spc1 crRNA. **A.** Artificial intron prevents *spc1* crRNAs from targeting the *nickase* locus. Produced mRNAs are not targeted by *spc1* crRNAs, confirming that mRNA is not the target but DNA is. **B.** When the intron is not present, *spc1* crRNAs efficiently target DNA locus and prevent production of *nickase* mRNA.

Incorporation of new sequences into CRISPR loci

If proto-spacers function as sequence-specific recognition modules then how do they evolve? How do bacteria stay tuned to changes in phage genomic DNA, and how do they "learn" the sequence composition of phage genomes?

One of the most obvious models suggests that bacteria are able to integrate phage DNA into their genome and thus use it as a template for targeting phages in subsequent infections. The incorporation of new sequences into CRISPR loci was indeed suggested by many researches. The analysis of CRISPR sequences in different *Yersinia pestis* strains allowed Pourcel et al. (2005) to hypothesize that these bacteria acquired new spacers very recently. It was then proposed that CRISPR spacers could have derived from preexisting sequences such as genetic elements of chromosomes and plasmid molecules transmitted by bacteriophages or conjugative plasmids (Mojica et al., 2005). A parallel was found that showed the inability of bacteriophages to infect strains carrying specific spacers, suggesting the existence of a correlation between CRISPR and immunity against targeted DNA (Mojica et al., 2005).

In 2007, it was shown that S. *thermophilus* bacterial strains carrying CRISPR became resistant to phage infection after acquiring new spacers derived from the virus. Barrangou and colleagues infected the S. *thermophilus* strain DGCC7710 with the virulent *pac*-type phages 2972 and 858 and observed the appearance of the development of **bacteriophage-insensitive mutants (BIMs)** (Barrangou et al., 2007). They found that BIMs consisted of bacteria that had new spacers derived from the phage genome integrated into CRISPR loci, thus generating a phage-resistant phenotype. The specificity of the resistance was thus determined by the identity between newly acquired spacers and the actual phage sequences (Barrangou et al., 2007). Curiously, a small population of phages was able to infect BIM cells, thus suggesting that both CRISPR loci and phage genomic regions are subject to rapid evolutionary changes.

Acquisition of phage sequences can be either a random or a directed process. The integrated sequences may be randomly picked from the cytoplasm, or there possibly exists a direct mechanism of selection of sequences that would be the most useful to integrate. If the latter is true, then it could be hypothesized that the CRISPR-leader system is capable to recognize the potential spacer sequence or the proto-spacer (Deveau et al., 2008). Do all phages have the potential to serve as a source of proto-spacers? Because a direct evidence of acquisition of proto-spacers exists only for dsDNA phages that have no RNA stage (phages 858 or 2972), it is possible that only genomic DNA of phages can be integrated. It cannot be excluded, however, that the process of lysogeny, phage integration into bacterial chromosomes, is an initial step of CRISPR formation. In this case, RNA phages can also be a source of CRIPSR. It is also possible that DNA synthesized from RNA via function of RNA-dependent DNA polymerase might also serve as proto-spacer.

Thus, a more detailed analysis of the mechanism of proto-spacer's acquisition might include the analysis of different types of bacteriophages (DNA or RNA phages). There is also not much known about the mechanism of the integration of proto-spacers, but some models have been proposed. The computational analysis by Mojica et al. (2009) suggested the importance of **proto-spacers adjusting motifs (PAMs)** in the process of integration. While analyzing several groups of phage sequences containing homology to CRISPR, the authors found the conservation of di- and trinucleotides starting immediately or one position after the proto-spacer. The authors further suggested a relationship between CRISPR and spacer/PAM orientation (Mojica et al., 2009). Specifically, the computational analysis showed that the spacer ends equivalent to the proto-spacer edges adjacent to PAM (referred to as **PAME, proto-spacer adjacent motif end**) were oriented toward the leader irrespective of the location of the *cas* genes. PAME thus may direct the orientation of spacers with respect to both repeats and the leader. Together with the preferential incorporation of new spacers at the leader's end of the CRISPR array, this suggests that CRISPR and the leader could participate in PAM recognition. Correctly oriented proto-spacers are then possibly base-paired with CRISPR and inserted into the bacterial genome through the process of homologous recombination (Mojica et al., 2009) (see Figure 14-2).

Other potential functions of CRISPRs

In silico phylogenetic analysis suggests that spacers are chronological records reflecting previous encounters with mobile genetic elements. The analysis showed the loss of one or more repeat-spacer units. This proposes that the growth in the number of spacers in CRISPR is controlled. It would make sense that the older spacers should be more frequently deleted because they have been inserted for a longer time. The analysis by Touchon and Rocha (2010) showed, however, that some of the most ancient spacers were highly persistent and shared by nearly all CRISPRs of the given species. This might indicate a critical yet unknown function in the activity of CRISPR elements.

CRISPRs may also play some role in RMS that is used by prokaryotes to avoid phages and plasmids, by plasmids to compete with other plasmids, and by plasmids to avoid segregation from the cell. As a result, many phages and plasmids have developed the anti-restriction systems allowing them to avoid degradation of their DNA by bacterial restrictases. The anti-restriction mechanisms include, among others, modification of the phage genome (methylase proteins), transient occlusion of restriction sites, subversion of host RMS activities (activation of host methylation), and direct inhibition of restriction enzymes (antirestriction proteins) (Montange and Batey, 2008). *In silico* analysis by Touchon and Rocha (2010) suggested that CRISPR often targets RMS both in phages and in plasmids and thus may serve as an additional mechanism of RMS, neutralizing the mobile genetic elements that are able to interfere with RMS.

CRISPR loci might also function in other processes not associated with defense against phages. Because the loci literally represent clusters of repetitive sequences, they might allow for homology-driven genome rearrangements. This is supported by finding that many large insertions/translocations are found in the CRISPR loci of two related *Thermotoga* species (DeBoy et al., 2006).

The CRISPR system might also be involved in regulating endogenous cellular processes. The formation of biofilm and swarming motility of the bacterium *P. aeruginosa* was found to be dependent on the activity of CRISPR. Infection of *P. aeruginosa* with bacteriophage DMS3 leads to the loss of biofilm formation and swarming motility, and disruption of CRISPR or several *cas* genes restores it. In *M. xanthus*, fruiting body development upon starvation involves genes of the CRISPR locus that are co-transcribed with the surrounding *cas* genes and repeats.

These additional functions of the CRISPR/*cas* pathway indicate that either the system had these functions in the past and then adapted some additional functions of crRNA-mediated defense against phages or vice versa, and the more ancient defense system became useful in other cellular processes.

CRISPRs in evolutionary context

Evolution of bacterial resistance against viruses and viral counters represents one of the most ancient combats on Earth. Thus, it is not surprising that co-evolution of certain microbial and viral populations resulted in selective sweeps at a CRISPR locus and frequent single-nucleotide polymorphisms in the viral sequence corresponding to all but the most recently incorporated spacers (Heidelberg et al., 2009). Co-evolutionary interactions between bacteria and phages maintain population-wide genetic diversity. Such interactions might indeed be extremely old. For example, CRISPRs from Leptospirillum populations found in acid mine drainage and subaerial biofilms were found to contain 37 distinct CRISPR arrays containing a total of 6,044 spacer sequences, with nearly half of them being unique (Vale and Little, 2010). Most of the analyzed arrays had either the viral or plasmid/transposon origin, suggesting that CRISPR loci might contain records of various interactions encountered in the past (Marraffini and Sontheimer, 2008).

One would predict that CRISPR loci should have an extreme turnover rate. However, the metagenomic analysis of microbial communities and the analysis of CRISPR repeats, leaders, and *cas* genes reveal a more conserved picture. For example, the phylogeny of *cas1*, the most conserved CRISPR-associated gene, suggests co-evolution of *cas* subtypes and their operon organization (Haft et al., 2005; Makarova et al., 2006). A similar clustering pattern is obvious upon comparison of sequence similarity among repeats or the periodicity of repeat units (Haft et al., 2005). Moreover, the analysis of CRISPRs in the *Sulfolobus* genus showed that leader sequences also correlate with their corresponding repeats (Lillestøl et al., 2009). The aforementioned analysis suggests that CRISPR elements, including the sequence of CRISPR loci, the *cas* genes, and the leader sequences have co-evolved as a single functional unit. The *cas* gene complex might be under particular pressure as it is required to retain the interactions among multiple Cas proteins as well as the interactions with non-coding components of the system—repeats and leader sequences. Interestingly, the analysis of the phylogenetic tree of the module's signature polymerase of a different CRISPR system based on the RAMP module showed no correlation with evolution of the core *cas* genes, subtype-specific *cas* genes, and non-coding components. It is possible that the RAMP module, with its associated RNA-targeting effector activity, evolved independently of the CRISPR system itself (Haft et al., 2005; Makarova et al., 2006).

Evolution of CRISPR elements appears to include two components, vertical and horizontal ones. Whereas vertical inheritance includes the acquisition of new sequences from parental population, the horizontal component includes gene transfer from phages and other bacteria. Indeed, the phylogenetic tree of the CRISPR system does not agree with the established bacterial/archaeal taxonomy. Closely related bacterial species might have different CRISPR content, and distant species occasionally carry similar CRISPR loci (Haft et al., 2005; Makarova et al., 2002). A good example of horizontal transfer of a CRISPR locus is the existence of similar CRISPR elements in the bacterium *T. maritima*, and in several archaeal species. CRISPR loci can be transferred via plasmids, phages, and mobile elements. Thus, it is not surprising that there are similarities among genomic CRISPRs found among bacterial species that are hosts to the same plasmid (Haft et al., 2005). The genome of *C. difficule* carries multiple CRISPR loci, with two of them residing on prophages and one in an excisable prophage-like sequence involved in sporulation.

Remaining questions on CRISPRs

CRISPR elements have high specificity: bacterial hosts are no longer protected from phages that acquire a single mutation in spacer-derived sequences. As a consequence, mutated phages are highly infective and should quickly overpower the host. If this is the case, new phage mutants will quickly fix in populations. Are CRISPR loci able to respond fast enough to infection with the acquisition of new phage-derived spacers, or is this process rather slow? Presently, it is unclear how rapidly CRISPR-based resistance is established. It remains to be shown what proportion of host population incorporates viral spacers before lysis occurs.

Another remaining question—how do CRISPRs avoid targeting and degrading their own phage-derived CRISPR spacers? The current model suggests that viral DNA is targeted via direct Watson–Crick pairing, and degradation of viral genome seems to be triggered by mismatches at specific positions between the viral sequence and the repeat sequence flanking the phage-derived spacer (Marraffini and Sontheimer, 2010b). It is also not clear whether every spacer or only those that were incorporated recently (for example, Brouns et al., 2008) are transcribed during this step.

The exact mechanism of sequence recognition and degradation mediated by crRNA is not clear either. Similarly, the mechanism of recognition, selection, and integration of new viral sequences is not well understood. What is the role of Cas proteins in the process? Two of the known Cas proteins—Cas 1 and Cas 2—are present among the most known CRISPR systems. The crystal structure of Cas 1 indicates that it has a metal-dependent DNase activity, so it is thought to be involved in the initial recognition and acquisition of viral motifs (Wiedenheft et al., 2009), whereas Cas 2 has been shown to cleave single-stranded RNA within U-rich regions. A number of Cas proteins have been biochemically or structurally characterized, but many of them are still not assigned to a particular CRISPR function.

Conclusion

The incorporation of new sequences into the bacterial genome is not a rare process as horizontal gene transfer in bacteria occurs frequently during transformation, transduction, and conjugation. The process of acquisition of new spacer elements by bacteria is an amazing biological phenomenon. It suggests that organisms as simple as bacteria are able to adapt to phage infection by modifying their genome in a directed Lamarckian manner. These new sequences are acquired directly from the phage genome and are used for sRNA-mediated inhibition of translation of phage RNAs during the process of infection. It should be noted, however, that the CRISPR is not a panacea for bacteria; a certain percentage of phages can quickly become infectious, suggesting that phages can modify their genomes very efficiently to avoid recognition. The fact that phages are able to adjust almost immediately to the acquisition of new CRISPR loci by bacteria suggests a constant adaptive arms race between bacteria and phages. Therefore, CRISPRs are truly a microbial immune system (Horvath and Barrangou, 2010) allowing strain-specific resistance—a memory of past infections that permits resistance against future encounters—to be acquired while assuring host integrity through self/non-self-discrimination.

Exercises and discussion topics

1. What are the similarities between the CRISPR-Cas system and RNAi system?

2. Describe the differences between the CRISPR-Cas system and RNAi system.

3. What are PAMs, and what is their function?

5. What experimental evidence allowed researchers to outline the function of CRISPR?

4. The evolution of the CRISPR-Cas system represents Darwinian or Lamarckian evolution? Why?

5. Explain vertical and horizontal evolution of CRISPR elements.

6. What are the similarities between function of CRISPR-Cas system and piRNAs?

7. What is a function of the leader sequence in CRISPR and the RAMP module?

8. Why are CRISPR considered bacterial-adaptive immunity? Is it similar to the human immune system?

9. What proteins are associated with the CRISPR-Cas system, and what is their function?

10. What is a possible mechanism of CRISPR integration in case of RNA bacteriophage infection?

References

Barrangou et al. (2007) CRISPR provides acquired resistance against viruses in prokaryotes. *Science* 315:1709-1712.

Bolotin et al. (2005) Clustered regularly interspaced short palindrome repeats (CRISPRs) have spacers of extrachromosomal origin. *Microbiology* 151:2551-2561.

Brouns et al. (2008) Small CRISPR RNAs guide antiviral defense in prokaryotes. *Science* 321:960-964.

DeBoy et al. (2006) Chromosome evolution in the Thermotogales: large-scale inversions and strain diversification of CRISPR sequences. *J Bacteriol* 188:2364-2374.

Deveau et al. (2008) Phage response to CRISPR-encoded resistance in Streptococcus thermophilus. *J Bacteriol* 190:1390-1400.

Haft et al. (2005) A guild of 45 CRISPR-associated (Cas) protein families and multiple CRISPR/Cas subtypes exist in prokaryotic genomes. *PLoS Comput Biol* 1:e60.

Hyman P, Abedon ST. (2010) Bacteriophage host range and bacterial resistance. *Adv Appl Microbiol* 70:217-248.

Ishino et al. (1987) Nucleotide sequence of the iap gene, responsible for alkaline phosphatase isozyme conversion in Escherichia coli, and identification of the gene product. *J Bacteriol* 169: 5429-5433.

Jansen et al. (2002) Identification of genes that are associated with DNA repeats in prokaryotes. *Mol Microbiol* 43:1565-1575

Karginov et al. (2009) The CRISPR System: Small RNA-Guided Defense in Bacteria and Archaea. Mol Cell 37:7-19.

Kunin et al. (2007) Evolutionary conservation of sequence and secondary structures in CRISPR repeats. *Genome Biol* 8:R61.

Lillestøl et al. (2009) CRISPR families of the crenarchaeal genus Sulfolobus: bidirectional transcription and dynamic properties. *Mol Microbiol* 72:259-272.

Makarova et al. (2006) A putative RNA-interference-based immune system in prokaryotes: computational analysis of the predicted enzymatic machinery, functional analogies with eukaryotic RNAi, and hypothetical mechanisms of action. *Biol Direct* 1:7.

Marraffini LA, Sontheimer EJ. (2010a) Self versus non-self discrimination during CRISPR RNA-directed immunity. *Nature* 463:568-571.

Marraffini LA, Sontheimer EJ. (2010b) CRISPR interference: RNA-directed adaptive immunity in bacteria and archaea. *Nat Rev Genet* 11:181-190.

Marraffini LA, Sontheimer EJ. (2008) CRISPR interference limits horizontal gene transfer in staphylococci by targeting DNA. *Science* 322:1843-1845.

Mojica et al. (2009) Short motif sequences determine the targets of the prokaryotic CRISPR defence system. *Microbiology* 155:733-740.

Mojica et al. (2005) Intervening sequences of regularly spaced prokaryotic repeats derive from foreign genetic elements. *J Mol Evol* 60:174-182.

Montange RK, Batey RT. (2008) Riboswitches: emerging themes in RNA structure and function. *Annu Rev Biophys* 37:117-133.

Pourcel et al. (2005) CRISPR elements in Yersinia pestis acquire new repeats by preferential uptake of bacteriophage DNA, and provide additional tools for evolutionary studies. *Microbiology* 151:653-663.

Touchon M, Rocha EP. (2010) The small, slow, and specialized CRISPR and anti-CRISPR of Escherichia and Salmonella. *PLoS One* 5:e11126.

Vale PF, Little TJ. (2010) CRISPR-mediated phage resistance and the ghost of coevolution past. *Proc Biol Sci* 277:2097-2103.

Wiedenheft et al. (2009) Structural basis for DNase activity of a conserved protein implicated in CRISPR-mediated genome defense. *Structure* 17:904-912.

15

Gene silencing—ancient immune response and a versatile mechanism of control over the fate of foreign nucleic acids

RNA interference (RNAi) is an epigenetic mechanism of homology-dependent gene silencing that is induced by double-stranded RNA. Since its discovery in plants, RNAi has been observed in a variety of organisms ranging from fungi to animals. RNAi has evolved as a process involved in defense against viruses, protection against transposons, the regulation of chromatin structure, control over gene expression, specification of cell identity, and the regulation of development of multicellular organisms. This chapter describes in detail the RNAi process, the mechanism of **virus-induced gene silencing** (**VIGS**), and **viRNA (virus-derived siRNA)** function.

15-1. Glossary

RNA interference (RNAi): Homology-dependent gene silencing triggered by double-stranded RNAs.

Co-suppression: Silencing of the introduced transgene along with the corresponding endogenous gene(s).

A Dicer or dicer-like protein: The RNase III-family member enzyme that cleaves double-stranded RNAs to produce siRNAs or miRNAs.

RISC: RNA-induced silencing complex. A multiprotein complex that performs the degradation of homology-dependent mRNA using small RNAs as a guide.

viRNAs: Virus-derived siRNAs. Perfectly complementary small interfering RNAs of 21, 22, and 24 nucleotides produced from long viral dsRNAs.

PTGS: Post-transcriptional gene silencing, a term used for RNAi where gene expression is lost as a result of sequence-specific mRNA degradation. The rate of transcription is typically not altered.

TGS: Transcriptional gene silencing. Originally referred to as suppression of transcription as a result of DNA methylation of promoter sequences following exposure of cells to dsRNAs that are homologous to promoter regions.

History of gene silencing

RNAi known as **post-transcriptional gene silencing** (**PTGS**) in plants and quelling in fungi is an epigenetic mechanism that affects gene expression via transcriptional silencing, RNA degradation, or translation inhibition. First observed in higher plants, it attracted interest from other research groups, and later this phenomenon was discovered in *Caenorhabditis elegans*, *Drosophila melanogaster*, insects, fungi, and vertebrates. For more details on RNAi, TGS, and quelling, see also Chapters 10, "Non-Coding RNAs Across the Kingdoms—Animals," and 11, "Non-Coding RNAs Across the Kingdoms—Plants."

Gene silencing is a widespread mechanism used by mammals, plants, and invertebrates. As early as the 1920s, cross-protection was reported to control plant disease epidemics (McKinney, 1929). **Cross-protection** is a phenomenon in which plants infected with a mild strain of a virus is protected from more virulent strains that can cause symptoms of a more severe disease. Although the mechanism underlying cross-protection has remained enigmatic since its first use nearly half a century ago, pathologists have been aware that an underlying mechanism, now known as gene silencing, was responsible for viral resistance. Then, in the mid-1980s, accidental observations on gene silencing started to emerge, most notably in plants, and the theory of **parasite-derived resistance** (**PDR**, later named **pathogen-derived resistance**) became a reality (Stanford and Johnson, 1985). PDR is the mechanism of generating disease resistance through the introduction of part of the pathogen into host cells. By combining two technologies—*Agrobacterium tumefaciens* plant transformation and tissue culture techniques—Powell-Abel et al. (1986) first implemented PDR *in vivo*. This study showed a high level-resistance to tobacco mosaic virus in tobacco plants harboring the TMV coat protein gene. For half a decade, scientists tried to uncover the mechanism behind PDR.

A hallmark discovery of gene silencing occurred in the 1990s when a group of plant physiologists from Richard Jorgensen's lab obtained a number of unexpected results that were difficult to explain (Napoli et al., 1990). They were trying to enhance the color intensity of the petals in petunia flowers by introducing a gene inducing the formation of red pigment in the flowers. An attempt to overexpress chalcone synthase *(CHS)* transgene resulted in a loss of petal color. Suppression was observed when either sense or antisense copies of the *CHS* transgene were over-expressed. The authors coined the term **co-suppression** to describe their observations referring to the ability of exogenous elements to alter the expression of endogenous genes. The mechanism underlying this phenomenon of co-suppression remained unknown until the recent discovery of gene silencing.

The occurrence of a similar phenomenon was also described by Romano and Macino (1992) in *Neurospora crassa*. When the authors transformed a wild-type strain of *N. crassa* with different portions of the carotenogenic albino-3 (al-3) or albino-1 (al-1) genes, they observed that up to 36% of transformants showed an albino phenotype. The authors coined the term **quelling** and found it to be spontaneously and progressively reversible, leading to wild-type or intermediate phenotypes.

Following the introduction of antisense RNA into plant cells, a similar observation was made in *Caenorhabditis elegans*: injecting either antisense or sense RNAs into the germline was equally effective at silencing the homologous target gene (Guo and Kemphues, 1995).

With the extension of the study in 1995, Mello and Fire (Fire et al., 1998) demonstrated that dsRNA was more than ten times as effective at knocking down mRNA than single-stranded RNA. In 2006, they received the Nobel Prize in Physiology and Medicine for their work on the identification of dsRNA as a trigger of gene silencing.

The importance of a dsRNA intermediate for the initiation of PTGS was demonstrated by Waterhouse et al. (1998). In this elegant study, the authors demonstrated that efficient post-transcriptional silencing occurs only in the population of plants that reconstitute dsRNA after a cross between plants carrying the sense strand and plants carrying the antisense strand of said RNA molecule (Waterhouse et al., 1998) (see Figure 15-1).

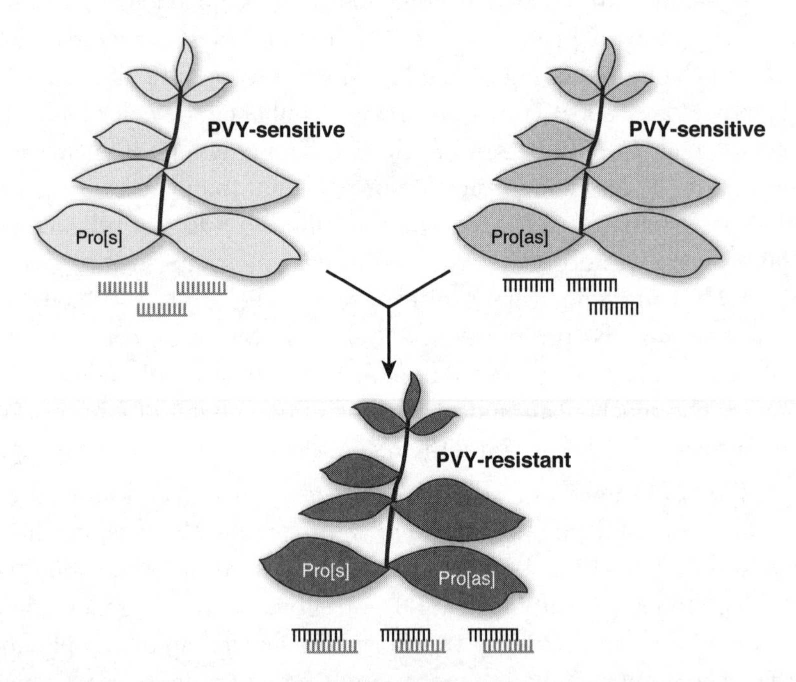

Figure 15-1 Efficient post-transcriptional silencing is achieved through reconstitution of dsRNA against the Pro gene. Silencing is more efficient when both sense- and antisense-producing transgenes are present.

Later on, first in plants (Hamilton et al., 1999) and then in animals (Hammond et al., 2000), it was demonstrated that the occurrence of PTGS is associated with the appearance of short antisense

RNAs complementary to the target mRNA. The development of cell-free *in vitro* biochemical assays for the analysis of RNAi mechanisms enabled substantial progress in understanding of the phenomena (Zamor et al., 2000). This approach made it possible to quickly dissect the biochemical pathways and identify the involved proteins; striking similarities in mechanisms of gene silencing in different organisms were identified (Fagard et al., 2000). Finally, the existence of RNAi in mammals was shown by Svoboda et al. (2000); the authors demonstrated inhibition of various mRNAs by introducing dsRNAs into mouse oocytes. Using cell-free Drosophila extracts, Elbashir et al. (2001) identified and described a new type of small RNAs, known as **short interfering RNAs** (**siRNAs**). The authors showed that chemically synthesized siRNAs are able to trigger efficient RNAi (Elbashir et al., 2001). Injection of these siRNAs into HeLa cells demonstrated their efficiency in triggering RNAi in mammals.

RNAi is a protection mechanism against invading foreign nucleic acids

Viruses and transposable elements are sources of nucleic acids that can disrupt the integrity of host genomes. Co-evolution with viruses forced plants to evolve a sequence-specific protection mechanism—gene silencing—for preventing the spread of the virus. It is possible, however, that plants simply adapted an already existing epigenetic mechanism of RNA degradation for a more specific purpose of degradation of viral RNAs. Gene silencing is triggered by viral dsRNAs to produce two main classes of small RNAs, siRNAs and miRNAs. siRNAs are involved in regulating foreign nucleic acids through PTGS and transcriptional gene silencing (TGS).

PTGS as an antiviral mechanism

PTGS causes mRNA degradation in a sequence-specific manner (see Figure 15-2 and Table 15-1). The first described biological function of gene silencing was an antiviral protection mechanism discovered in plants. There is significant evidence that plant antiviral protection involves PTGS. First, it was discovered that PTGS was induced by a transgene containing a viral sequence that then targeted homologous viral RNAs for silencing to confer virus resistance (Lindbo et al., 1993). Another line of evidence that PTGS is an antiviral mechanism is based on a vast number of studies that used reverse genetics to knock out important components of gene silencing pathways such as RDR3 and its interacting partner SGS3 (see Chapter 11 for details). These studies showed that inactive silencing components rendered plants more susceptible to virus infections. Probably the most convincing piece of evidence supporting the role of PTGS as an antiviral mechanism comes from the experiments that demonstrate that almost all viruses contain silencing suppressors that can influence the immunity of plants against viruses.

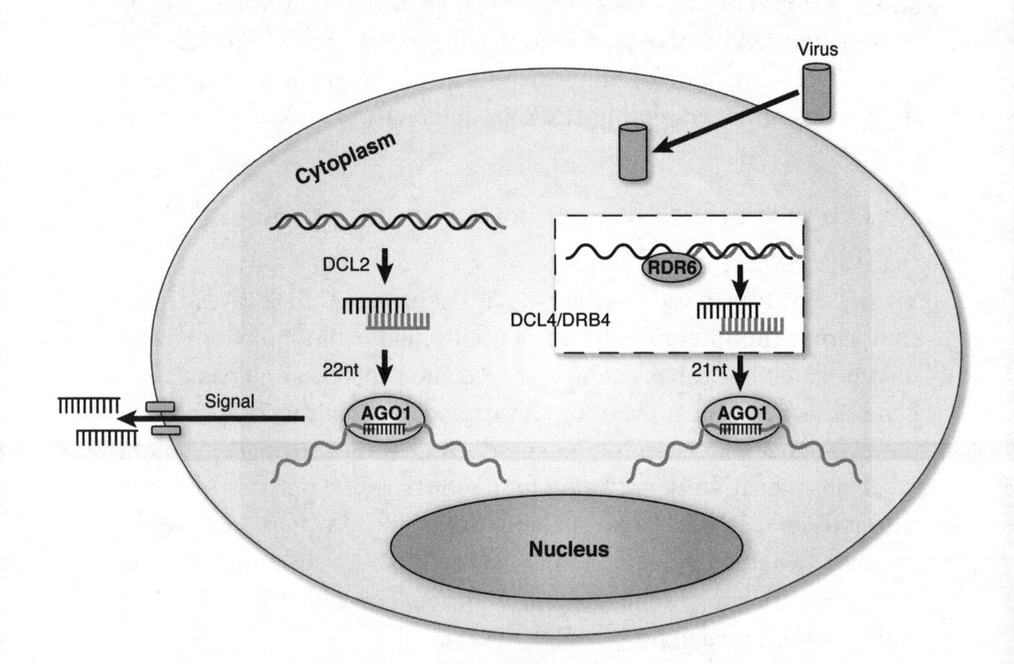

Figure 15-2 Key steps in PTGS. Viral dsRNAs are diced and amplified (dashed box) by RDR6/DCL4/DRB4 producing 21-nt siRNAs. In the absence of DCL4, DCL2 might produce 22-nt siRNA but at a lower efficiency. One strand of the ssRNA is used as a guide to degrade viral and messenger RNA in a sequence-specific manner. Small interfering RNA duplexes might also be involved in a cell-to-cell signal.

Table 15-1 Proteins involved in PTGS and TGS in *Arabidopsis thaliana*

Protein	Abbreviation	Function
PTGS		
Dicer-like 4	DCL4	The main DCL involved in PTGS. Processes 21-nt small RNAs from viral dsDNA.
Dicer-like 2	DCL2	Processes 22-nt small RNAs from viral dsDNA, although the efficiency is lower than that of DCL4. DRB4-independent.
dsRNA-binding protein 4	DRB4	Interacts with DCL4 in the antiviral defense silencing pathway.
Hua enhancer 1	HEN1	An RNA methyltransferase that promotes 2'-O-methylation of 3' overhangs of the sRNA.
Hasty	HST	Exportin-5 homolog: exports the 21-22-nt sRNA to the cytoplasm.
Argonaute 1	AGO1	The main slicer of viral sRNAs. AGO7 and AGO2 might rcplacc AGO1, albeit with lower efficiency. Has higher affinity to siRNAs with 5'- uracil nucleotides.
Argonaute 2	AGO2	Backs up AGO1 function in protection against viruses. Has higher affinity to siRNAs with 5'-adenine nucleotides.
Suppressor of gene silencing	SGS3	Interacts with RDR6 and AGO7.
TGS		
Dicer 3	DCL3	Processes dsRNAs into heterochromatic siRNAs.
Argonaute 4	AGO4	Binds 24-nt siRNAs that guide cleavage or *de novo* methylation.
Argonaute 6	AGO6	Involved in DNA methylation and 24-nt siRNA accumulation.
Hua enhancer 1	HEN1	Stabilizes siRNA by 2'-O-methylation of 3' overhangs.
Methyltransferase 1	MET1	A DNA methyltransferase involved in CG maintenance methylation.
Domains rearranged methylase 2	DRM2	A DNA methyltransferase involved in *de novo* methylation (CG, CNG, CNN).

Virus-induced gene silencing (VIGS)

VIGS employs viral genomes or intermediates of reverse transcription as substrates for the internal silencing machinery, and it also helps plants control viral infections. Smith et al. (1994) found that if they transformed tobacco with an untranslatable version of the potato virus Y coat protein, transcript levels correlated with virus resistance. A similar phenomenon was observed in transgenic tobacco plants containing an untranslatable version of the tobacco etch virus coat protein. The transgenic plants had multiple integrations of this sequence, and virus resistance was mediated by targeted RNA degradation. Later on, it was shown that siRNAs can be produced directly from viral genomes. However, in order to be potent effectors, they undergo a round of amplification in a plant RdRP-dependent manner. This system demonstrates how plants can use their endogenous silencing machinery to fight viral infections; thereby, it confirms the importance of small ncRNAs in the immune system in addition to the prokaryotic **CRISPR-Cas** system.

Because many plant viruses have a dsRNA genome or, in many cases, their replication includes dsRNA formation, these viral dsRNAs can potentially be targeted by DCLs and other components of siRNA processing. It is thus logical that viral dsRNAs serve as substrates for small RNA biogenesis. Indeed, the coexistence and thus co-evolution of plants and viruses resulted in the development of a complex network of defense and counter-defense. Some of these mechanisms include silencing or RNA-mediated antiviral immunity triggered by plants and suppression of RNA silencing by viruses.

Viral genomes are composed of either DNA or RNA, and they can be single-stranded (ss) or double-stranded (ds). Also, ssRNA viruses can have positive sense (+) or negative sense (-) RNAs, and most plant viruses are (+)ssRNA viruses. As an antiviral mechanism, double-stranded RNA is derived from replication of genomes of DNA or RNA viruses. More than 70% of plant viruses have single-stranded (ss)RNA genomes replicated by a virus-encoded RNA-dependent RNA polymerase (RdRP) which produces a dsRNA replicative intermediate. Some viruses have double-stranded (ds) RNAs in which the genome itself is a source of dsRNA. Double-stranded DNA and ssDNA viruses are abundant, and they produce DNA and RNA replicative intermediates. Double-stranded RNA is produced by host-encoded RdRP using a replicative intermediate as a template.

Small RNAs produced in infected plants are referred to as **virus-derived small RNAs (viRNAs)**. Generating viRNAs from ssRNA viruses occurs in two major ways. Because (+)ssRNA viruses carry in the genome their own RNA-dependent RNA polymerase (RdRP). dsRNA intermediates produced upon replication of (+)ssRNA are mainly recognized by DCL4, which dices the intermediates into 21-nucleotide viRNAs. Curiously, the DCL4 mutant is not impaired in viral resistance, although 21-nt viRNAs disappear. Instead, a new pool of viRNAs that are 22 nt in length is generated by DCL2 (refer to Figure 15-2). These data have been confirmed by the fact that the *dcl4-dcl2* double knockout mutant indeed shows an increased viral susceptibility. DCL2 is involved in processing **nat-siRNAs (naturally occurring antisense RNA)**, whereas the aforementioned data suggest that DCL2 might also function as a back-up to DCL4. DCL3 has been shown to be responsible for generating viRNAs only against certain (+)ssRNA viruses. DCL3 is involved in the synthesis of 24 nt-long viRNAs in wild-type plants infected with *Tobacco rattle virus* and *Cucumber mosaic virus* and in *dcl2/dcl4* mutants infected with *Turnip crinkle virus*. Also, triple *dcl2/dcl3/dcl4* mutants have an increased virus titer. The fact that several DCL proteins are involved in processing viRNAs supports the importance of small RNA-mediated antiviral immunity in plant evolution.

The second mechanism of viRNA production from (+)ssRNA and (-)ssRNA genomes includes the function of DCL1 (see Figure 15-3). ssRNAs can fold into complex structures, including dsRNA hairpins recognized by DCL1 and processed to viRNAs (Ding and Voinnet, 2007). The contribution of the miRNA-specific DCL1 to immunity against (+)ssRNA viruses is not dramatic because *dcl2/dcl3/dcl4* and *dcl1/dcl2/dcl3/dcl4* mutants showed similar susceptibility to Cucumber mosaic virus or *Turnip crinkle virus*, and viRNAs produced by DCL1 were barely detectable even in the triple *dcl2/dcl3/dcl4* mutant (Ding and Voinnet, 2007).

The production of viRNAs from DNA viruses is not like that from RNA viruses. For example, in the case of the dsDNA virus, *Cauliflower mosaic virus* (CaMV), mRNA produced from the

35S-driven polycistron folds into hairpin-containing secondary structures. Processing of these hairpins is initiated by DCL1 followed by dicing by DCL4 and DCL3. Another DNA virus—circular DNA geminivirus—produces sense and antisense transcripts upon replication. ViRNAs are produced from dsRNAs generated through sense-antisense pairing of these transcripts. DCL2 might also be involved in viRNA production from CaMV as reduced viRNA accumulation was only observed if all three Dicers—DCL4, DCL2, and DCL3—were disabled (Moissiard and Voinnet, 2006). DNA viruses such as CaMV or *Cabbage leaf curl geminivirus* (CaLCuV) are the exception; all the four DCLs are involved in siRNA biogenesis, as seen in the infected hosts. Double-stranded DNA plant viruses have a long intergenic region (for example, the CaMV 35S promoter region), and therefore, DCL1 plays a role in antiviral defense by dicing these areas that resemble the secondary structure of miRNAs. DCL3 is involved in antiviral mechanisms during a nuclear phase of CaMV virus multiplication.

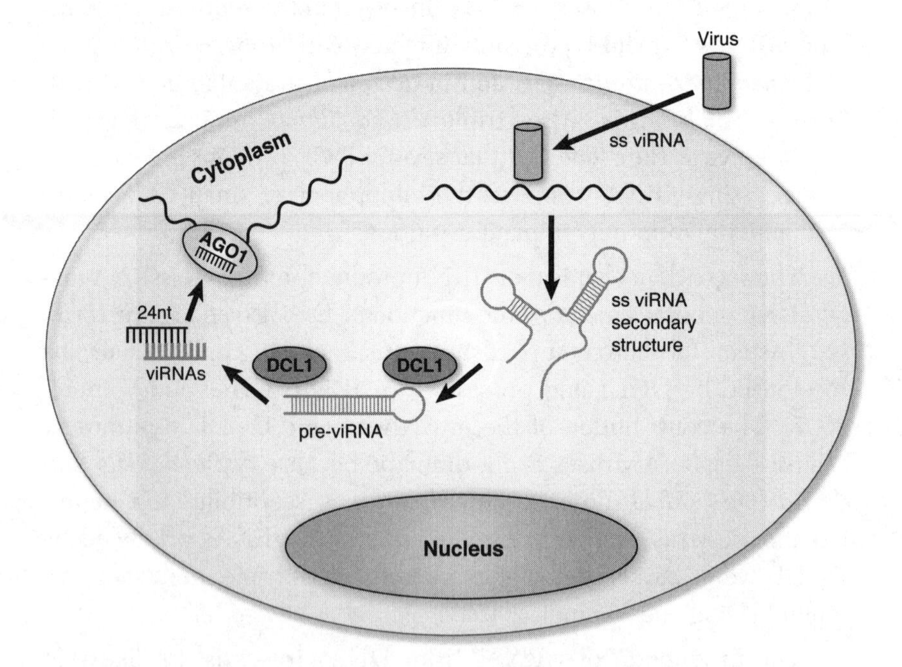

Figure 15-3 Production of viRNAs from (+)ssRNA and (-)ssRNA genomes.

Thus, all the four Dicers are important for providing protection against DNA viruses. More details of the mechanism of plant protection against RNA and DNA viruses are discussed in the review by Ding and Voinnet (2007).

viRNAs generated by DCLs are incorporated into RISCs to initiate cleavage of viral genomes. This serves dual purpose: it reduces the number of functional viral genomes, thus inhibiting viral infection, and it also allows for amplification of new viRNAs. Although the **viRNA RISC complexes** (**vi-RISC**) mostly include AGO1, there apparently exists a substantial redundancy allowing other AGO proteins to participate in the process. Two mechanisms of replication of viRNAs using RdRP are known. If primary viRNAs target viral ssRNA genomes, cleaved ssRNAs can be used by plant RdRP, namely RDR6, to produce dsRNAs, which are further processed by DCL4 (Ding and Voinnet, 2007) (see Figure 15-4A). The process also requires the function of an RNA-helicase, SILENCING DEFECTIVE3 (SDE3), and a putative mRNA export factor, SILENCING DEFECTIVE5 (SDE5). This mechanism allows silencing to spread; whereas initial targeting of viral genomes might have occurred at a specific location, processing through RdRP/DCL4 allows generating new viRNAs to other areas of viral genomes. This process, known as **transitivity**, was first reported to occur in virus-induced gene silencing of transgenes, during which transgene-specific siRNAs from regions absent in the recombinant viral genome accumulate.

An alternative mechanism of generating various viRNAs functions independently of primary viRNAs (see Figure 15-4B). Aberrant transcripts frequently generated by viral transcription, including those that lack 5' cap- or 3' poly-A modifications, might recruit RDR6, SDE3, SGS3, and AGO1 to generate novel dsRNAs. The produced dsRNAs are targeted by DCL4 for the generation of secondary viRNAs.

Double-stranded RNA-binding proteins (DRBs) facilitate DCLs in the dicing process. Compared to DCLs, DRBs do not contain hierarchical redundancy. In *A. thaliana*, DRB4 interacts with DCL4 in the antiviral defense silencing pathway. Interestingly, unlike DCL4, DCL 2 and DCL3 do not require DRBs for processing virus-associated siRNAs.

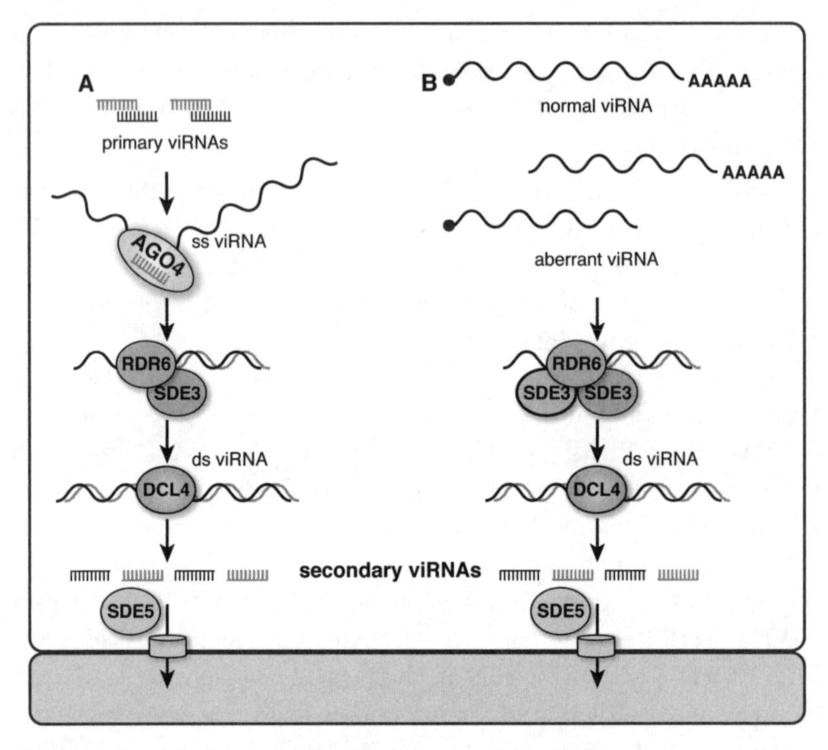

Figure 15-4 RdRP-dependent mechanisms of viRNA replication.

In *A. thaliana*, upon dicing, the sRNA 3' overhanging ends are 2'-O-methylated by the methyltransferase HUA ENHANCER 1 (HEN1), which protects them from degradation. It is known that HEN1 participates in the antiviral RNA silencing pathways because in comparison to wild-type plants, the *Arabidopsis hen1* mutant exhibits an increased susceptibility to *Cucumber mosaic virus* (CMV) and exhibits a fivefold increase in the virus titer. After dicing is completed, the stabilized sRNA duplexes are then retained in the nucleus for TGS at the chromatin level or exported to the cytoplasm for PTGS, possibly via the exportin-5 homolog HASTY (HST). These sRNAs are incorporated into a large ribonucleoprotein complex, the RISC.

Each RISC contains an Argonaute (AGO) protein that has an sRNA-binding PAZ domain and a PIWI domain. The PIWI domain has structural similarity to RNaseH and has the endonucleolytic activity to digest the target RNA using a guide strand, a process called slicing. *A. thaliana* contains the ten predicted family members of AGO (AGO1–AGO10) with established roles for AGO1, AGO4,

AGO6, and AGO7 in sRNA-directed silencing. Slicer activity has been demonstrated for AGO1, AGO4, and AGO7. Although no such activity has been yet demonstrated for AGO2, *ago2* mutant is more sensitive to *Turnip crinkle virus* (TCV) or *Cucumber mosaic virus* (CMV). AGO4 and AGO6 are required in the TGS pathway. AGO1 and AGO7 are required in PTGS. AGO1 is the main slicer for viral RNAs because it has a higher affinity for more compact structures compared to AGO7. AGO1 has additional roles in the miRNA and siRNA pathways. During the slicing process, the two RNA strands in sRNA unwind and become separated. One strand, called the **guide strand**, is preferentially incorporated into the RISC complex to guide the degradation of transcripts or viral genomes, whereas the other one is rapidly degraded. AGO has a preference for the strand that has the weakest base-pairing interaction at the 5' terminus.

Transcriptional gene silencing regulates foreign nucleic acids

Transposable elements (TEs) can be regulated by DNA methylation through the process of TGS (see Figure 15-5). Transposable elements are DNA sequences that have the capacity to excise and insert into a genome. They have been identified in all organisms from prokaryotes to eukaryotes and can contribute substantially to the size of a genome. For example, TEs comprise approximately 12% of the *C. elegans* genome, 37% of the mouse genome, 45% of the human genome, and up to 80% of the plant genome. From bacteria to humans, TEs have accumulated over time and continue to be a main player in genomic evolution. The activity of TEs can affect a genome either positively or negatively. For example, TEs can play a significant role in genomic evolution by promoting gene inactivation, modulating gene expression, or inducing homologous and/or non-homologous recombination. However, TEs are also able to produce various genetic alterations upon insertion, excision, duplication, or translocation, rendering deleterious effects or diseases in the host organism. It is not surprising that organisms have evolved mechanisms that control the translocation of TEs. One of the mechanisms is mediated through TGS, involving DNA methylation. Generally, deactivation of transposons has been shown to occur by methylation, whereas activation of transposons occurs by demethylation.

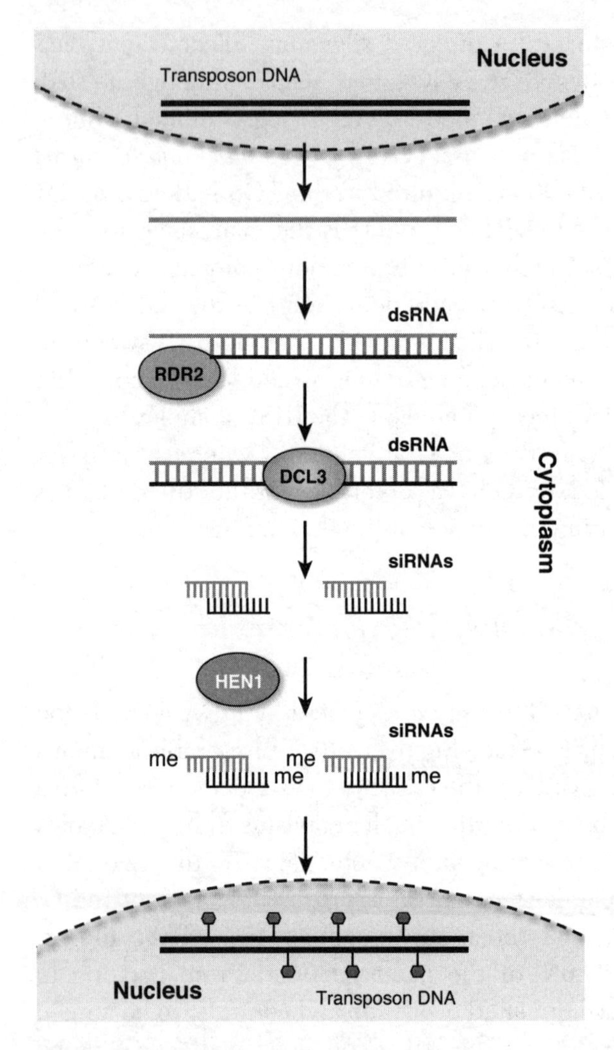

Figure 15-5 siRNA-regulated methylation of transposons.

The heterochromatin-formation pathway produces 24-nt sRNAs (hcRNAs) that mediate TGS through the maintenance of DNA methylation and chromatin structure. In *A. thaliana*, the pathway uses DCL3 to cleave dsRNAs derived from endogenous transcripts. TEs provide dsRNA templates used in TGS through mechanisms that are not fully understood. Double-stranded RNAs derived from endogenous transcripts can also be generated through the pathway involving the RNA-dependent RNA polymerase 2 (RDR2) and 6

(RDR6) (also known as SGS2/SDE1). AGO 4 is the main AGO involved in TGS. It binds 24-nt siRNAs that guide cleavage or *de novo* DNA methylation. AGO6 also functions in DNA methylation and accumulation of siRNAs.

At a site homologous to the 24-nt sRNAs, *de novo* methylation is carried out through **RNA-directed DNA methylation (RdDM)** using various DNA methyltransferases such as DNA METHYLTRANSFERASE1 (MET1), DEFECTIVE IN RNA-DIRECTED DNA METHYLATION2 (DRD2), and others. Other key enzymes involved in TGS are DNA glycosylases, lyases, chromatin-remodeling proteins, and RNA polymerases. *De novo* DNA methylation is methylation of cytosines in all sequence contexts (CG, CNG, and CNN where N is A, T, or C). Promoters, and sometimes coding regions, are targets for DNA methylation. DNA methylation at a promoter prevents binding of factors necessary for transcription.

In Drosophila and vertebrate germlines, TE silencing relies on PIWI/Argonaute proteins and a class of sRNAs known as **PIWI-interacting short RNAs (piRNAs)**. Interestingly, dicer-like proteins are not involved in producing piRNAs. In Drosophila, the majority of piRNAs target TEs, whereas in vertebrates most piRNAs target repetitive sequences with only a minor complementarity to TEs.

Transitive silencing defends the host from invaders

Resistance to viral infection depends locally on the process of **hypersensitive response (HR)**—rapid induction of free radicals that kills the host cells around the pathogen infection site. Systemic protection is achieved through the process of **systemic acquired resistance (SAR)**—induced whole plant resistance against similar and different pathogens (see box 15-2). It is also possible that these signals might be either dependent on viRNA production or might be viRNAs themselves. Both primary and secondary viRNA might be involved in the immunization of non-infected cells by arriving to non-infected cells prior to the arrival of the replicating virus (Ding and Voinnet, 2007). Infected cells can communicate with non-infected cells by sending primary and secondary viRNAs through plasmodesmata (see Figure 15-6).

15-2. HR and SAR

Hypersensitive response is a massive production of free radicals followed by cell death providing a demarcation zone around the place of viral infection. HR allows the previously infected tissue to tolerate future infections. Systemic tissues of locally infected plants also acquire protection. The mechanism of plant defense known as systemic acquired resistance (SAR) depends on spreading of a signal from the site of infection to non-infected tissues. The nature of the signal is not clear, but it might include various volatile molecules such as methyl salicylate and methyl jasomate.

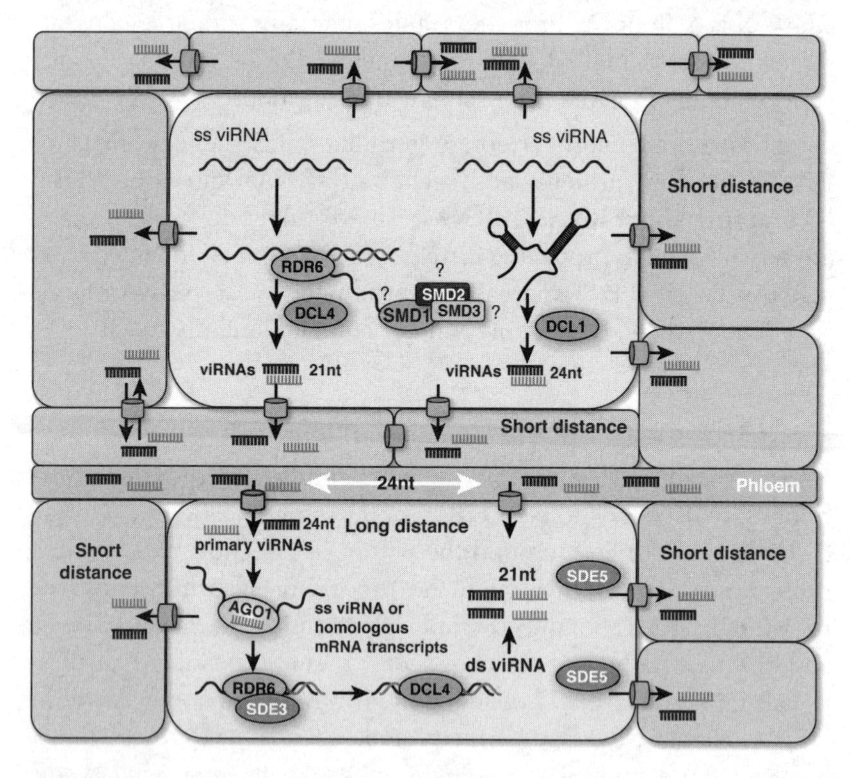

Figure 15-6 Short and long distance movement of viRNAs.

This communication occurs only between 10 to 15 neighboring cells surrounding the infected cells (Dunoyer et al., 2005) and might not require the function of RDR6 and SDE3, but it does require the three *SILENCING MOVEMENT DEFICIENT* genes (*SMD1*, *SMD2*, and *SMD3*) of so far unknown function. Because short-range systemic signaling transports both 21- and 24-nt viRNAs, it depends on the function of both DCL4 and DCL1. In contrast, long-distance signaling by viRNAs occurs through the phloem using 24-nt viRNAs and requires the function of RDR6, SDE3, and possibly SDE5, which convert homologous transcripts into new dsRNAs in the cells receiving short-range signals (see Figure 15-6). These cells use DCL4 to process new dsRNAs into secondary 21-ntsiRNAs that move to another 10 to 15 cells (Ding and Voinnet, 2007). Thus, it can be hypothesized that short- and long-range signaling events are directly related to primary and secondary viRNA synthesis, respectively. Hence, RDR6 is important for amplification of a sufficient number of copies of viRNAs to be distributed throughout the plant. The systemic transport of anti-viral viRNAs inhibits the potential spread of infection and immunizes plants against future inoculations.

Fungi, plants, and nematodes encode the eukaryotic RNA-dependent RNA polymerase (RdRP or RDR) that can generate new sources of dsRNA leading to the amplification of silencing in the organism. Transitive gene silencing occurs in both plants and nematodes. In this process, a virus or a pool of TE-derived RNAs is amplified using RdRPs. This leads to the propagation and spread of the silencing signal beyond the region that was initially targeted for gene silencing. *Arabidopsis* encodes six RdRPs that work with DCLs to control the biogenesis of sRNAs. The function of RDR2 is required for the production of 24-nt siRNAs by DCL3, which are involved mainly in the hcRNA pathway and sometimes in antiviral gene silencing. RDR6 is involved in the production of siRNAs by DCL1, DCL2, or DCL4 and apparently in long-distance signaling. RdRP1 influences virus replication and is involved in antiviral defense in plants. The loss of function of RdRP1 in *Arabidopsis* resulted in the enhanced accumulation of viral RNAs and an increased susceptibility to viral infections (Yu et al., 2003). These studies are consistent with the role of RdRP involved in virus defense.

Gene silencing is associated with cell-to-cell and non-cell autonomous and systemic signals in plants and nematodes. The systemic signal sends a message to distal tissues that renders the entire organism resistant to further infection of the same virus. In nematodes, the systemic signal requires the transmembrane protein SID-1 that efficiently transports dsRNA longer than 100 nt (Winston et al., 2002).

In plants, the cell-to-cell signal is believed to move to 10 to 15 cells through the plasmodesmata, whereas the systemic signal is thought to pass through the phloem. The presence of a systemic silencing signal is supported by grafting experiments that showed that the GUS silencing signal can move from roots to new shoots through GUS-expressing scions, and the spreading of the signal through wild-type tissue was up to 30 cm (Palauqui et al., 1997). Biological evidence for this signal is further supported by the systemic action of silencing suppressors p25 of potato virus X and 2b of cucumber mosaic virus, and the location of sRNAs in phloem tissue.

Many suggestions have been made as to the nature of these signals, although convincing evidence has been difficult to obtain. Some of the candidates are long dsRNAs (that is, precursor siRNAs), dsRNA molecules, or products of dsRNAs that might be produced through a DICER-independent pathway, modified products of methylated target genes, and 21 to 24-nt dsRNAs. Recent studies suggest that 21-nt siRNA duplexes are the mobile signal between plant cells as opposed to their long precursor molecules (Dunoyer et al., 2010). Also, a systemic signal has been identified through grafting experiments supplemented with sRNA high-throughput sequencing (Molnar et al., 2010). The results from these experiments suggest that 24-nt sRNAs produced by the DCL3/AGO4 pathway are systemic signals and might act as signals in the hcRNA pathway.

A molecular arms race between viruses and hosts

RNA silencing is a highly complex system made up of numerous different proteins and processes. The mechanics of infecting host plants with plant viruses relies on the ability of a virus to overcome the plant gene silencing defense mechanisms. RNA silencing provides an ample opportunity for pathogens to develop the ways of avoiding the host's defense machinery. The effectiveness of small RNA-mediated anti-viral immunity has by no means gone unnoticed through viral evolution. A variety of elaborate and sophisticated strategies have been adopted by plant viruses in their attempts to elude these mechanisms. For example, viruses encode suppressor proteins capable of interfering with various steps of the PTGS and TGS pathways (see Table 15-2). **Viral suppressors of RNA silencing (VSRs)** working in counter-defense against silencing is the best described phenomenon. It is estimated that every virus contains at least one silencing suppressor, although for many of them the exact mode of action remains unknown. In many cases, the level at which they act within the gene silencing pathway remains to be elucidated.

Table 15-2 Plant and animal silencing suppressors

Virus genus	Virus name	Silencing suppressor	The outcome of a silencing suppressor
Closterovirus	*Beet yellows virus (BYV)*	P21	An RNA binding protein involved in suppression by sequestering siRNAs
Cucumovirus	*Cucumber mosaic virus (CMV)*	2b	Binds AGO
Nodavirus	*Flock house virus (FHV)*	B2	Binds double-stranded RNAs
Orthomyxovirus	*Influenza virus A*	NS1	Binds double-stranded RNAs
Tombusvirus	*Tomato bushy stunt virus (TBSV)*	P19	Binds 21-nt siRNA and 22-nt siRNAs (with a lower affinity)
Potexvirus	*Potato virus X (PVX)*	P25	Inhibits the systemic and cell-to-cell movement of the silencing signal
Polerovirus	*Beet western yellows virus (BWYV)*	P0	Promotes ubiquitin-dependent proteolysis of AGO1
Caulimovirus	*Cauliflower mosaic virus (CaMV)*	P6	Binds to DRB4 and inactivates DCL4
Geminivirus	*African cassava mosaic virus (ACMV)*	AC4	Competes with AGO by binding single-stranded siRNAs, preventing RISC assembly

The mechanism of VSRs includes disruption of their production and/or an export of plant-produced anti-viral viRNAs. The importance of VSRs is confirmed by the fact that this viral counter-defense mechanism is known to be present in most well-described viruses (Ding and Voinnet, 2007). As a result, 35 families of VSRs defined by specific function and genomic positions have been reported to date. VSRs are dsRNA-binding proteins belonging to the following major families: p14 from *Pothos latent virus* (PLV); p15 from *Clump virus*

(CV); p19 from *Tomato bushy stunt virus* (TBSV), p21 from *Beet yellows virus* (BYV), γB from *Barley stripe mosaic virus* (BSMV); and HC-Pro from *Tobacco etch virus* (TEV), CP from TCV, and 2b from CMV. The main mechanism of VSR functions is the binding of 21-nt dsRNA, the predominantly produced viRNAs (see Figure 15-7). The binding might result in the precipitation of RNAs of this size making them unavailable for targeting viral genomes and priming noninfected cells. Another possible mechanism might include prevention of HEN1-mediated methylation at the 3' end of viRNAs, which makes them unavailable for RISCs. Specific VSRs that bind HEN1 include BYV p21, TBSV p19, TEV, and HC-Pro. In addition, the most commonly known proteins involved in virus replication (*Tobacco Mosaic Virus* (TMV) and *Oilseed Rape Tobamovirus* (ORMV) replicases) also interfere with sRNA methylation.

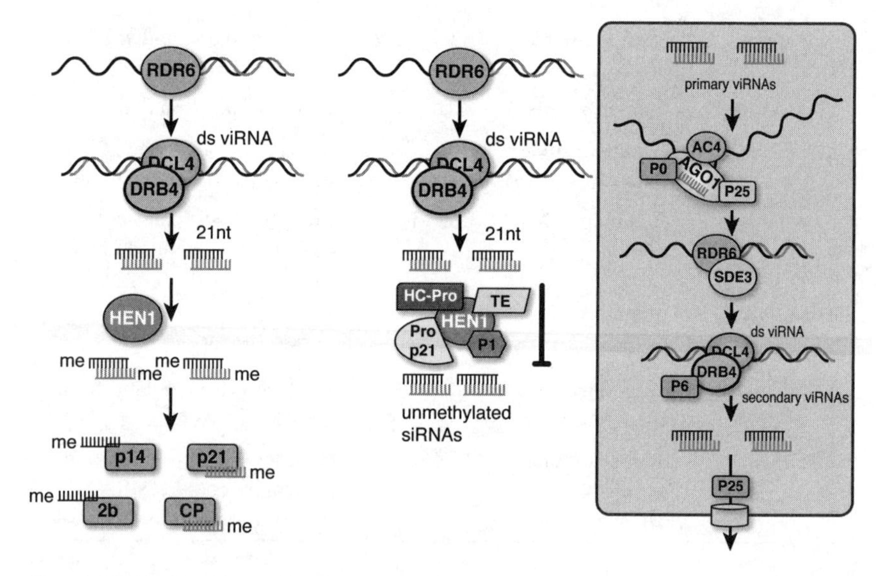

Figure 15-7 Function of VSRs.

Some VSRs, such as PLV p14, BYL p21, TCV CP, and MV 2b, have no preference for the size of RNAs they bind. As a result, they are able to sequester 21-nt viRNAs and bind long dsRNA substrates such as viral hairpins. The *Tombusvirus* P19 is also a dsRNA-binding protein, and what makes it interesting and unique is that this suppressor selects its substrate on the basis of length of the RNA duplex region. P19 binds 21-nt duplex siRNAs with a much higher affinity than 22-nt dsRNA duplexes.

Another group of VSRs is able to inhibit the spread of systemic silencing by preventing the export of viRNAs. The P25 protein is one of such VSRs that prevents systemic transgene silencing in *Nicotiana benthamiana* plants infected with the *Potato virus X* (PVX). P25 from PVX strongly inhibits the production of 24-nt RNAs but only partially inhibits that of 21-nt RNAs upon infection of *Nicotiana benthamiana* plants (Hamilton et al., 2002). One of the first experiments suggesting that P25 inhibits the movement of the signal showed that systemic silencing did not occur if P25 was deactivated, the reverse being demonstrated as well (Voinnet et al., 2000). Later, it was shown that P25 suppression of RNA silencing was required for the cell-to-cell movement of potato virus X (PVX) (Bayne et al., 2005). Similarly, the 2b protein of CMV, an inhibitor of systemic silencing, is able to sequester 24-nt viRNAs, preventing them from reaching the phloem. This model fits well with the observation that only 24-nt siRNAs are involved in the phloem-dependent silencing signal.

In contrast, the CMV 2b protein is able to target the AGO1-loaded RISC complexes preventing their function. The polerovirus P0 protein also interferes with AGO1 activity, but the mechanism of silencing suppression is different. P0 encodes an F-box protein that promotes ubiquitin-dependent proteolysis of AGO1 proteins; such mechanism of suppression avoids inhibiting the systemic signal that is important for this phloem-limited virus. DNA viruses also encode silencing suppressors. For example, cauliflower mosaic virus protein P6 encodes a silencing suppressor that binds to the protein DRB4 in the nucleus of cells. DRB4 is inactivated through binding, and thus is unavailable as a cofactor involved in dicing activities of DCL4. Geminiviruses contain the AC4 protein, a silencing suppressor that competes with AGO proteins by binding to single-stranded siRNAs, thereby preventing RISC assembly.

There is a constant arms race between host plants and foreign nucleic acids produced by either viruses or transpositions of transposable elements, and this phenomenon can be seen across kingdoms.

Does the development of symptoms in infected plants represent one of the off-target effects?

Could the appearance of symptoms in infected plants be the result of action of viral suppressors? As discussed earlier, ncRNAs are critically important for plant development and stress tolerance. The knowledge of the mechanism of action of VSRs allows us to suggest that the development of symptoms in the infected plants is the result of malfunctions of miRNA/siRNA biogenesis. It is not surprising that abnormalities associated with viral infections are similar to those observed in mutants deficient in miRNA-biogenesis (Ding and Voinnet, 2007). How is it then that altering miRNA pathways results in an increase of biotic resistance, as observed for miR-398a and *Pseudomonas* infection of Arabidopsis (Navarro, et al., 2006)? It is possible that the alteration of host small RNA pathways might, in this case, be a deliberate act of counter-defense.

Conclusion

From the initial discovery of the phenomena cross-protection, parasite-derived resistance, quelling, RNAi, and others to more recent exciting findings of VIGS and silencing suppression it became more and more apparent that all these RNAi mechanisms represent a sophisticated immune response system that hosts employ to combat foreign nucleic acids. Over the last 15 years since the discovery of gene silencing, we have gained enormous knowledge on the mechanisms involved in the innate immunity response. However, we are only beginning to understand the mechanisms behind the components of gene silencing and strategies used to overcome the hosts' silencing innate immunity. For example, viroids are relatively small ssRNA plant pathogens of about 350 nt that do not encode any proteins, but they contain an unusually high level of secondary structures. Interestingly, viroids are able to bypass the host defense system responsible for gene silencing. There are still many unanswered questions and exciting discoveries that remain to be made.

Exercises and discussion topics

1. Describe the phenomena of cross-protection, co-suppression, and quelling. Are there any differences in these processes?
2. Describe experiments by Fire and Mello demonstrating RNAi in *C. elegans*.
3. Describe the "reconstitution" experiment by Waterhouse et al.
4. Describe PTGS. List the proteins involved and describe their functions.
5. Describe TGS. List the proteins involved and describe their function.
6. Describe DCL4- and DCL2-dependent mechanisms of generation of siRNAs involved in PTGS.
7. Describe the mechanism of VIGS.
8. Describe two mechanisms of generation of viRNAs.
9. Explain the role of individual AGO proteins in silencing.
10. Describe the mechanisms of generation of primary and secondary viRNAs.
11. Describe the mechanisms of short- and long-distance signaling in VIGS. Describe the experiments supporting the existence of short- and long-distance signaling.
12. Describe HR and SAR phenomena. Are these phenomena epigenetic in nature?
13. Describe the phenomenon of suppression of RNAi by viral suppressors.
14. Describe main mechanisms through which viral suppressors interfere with RNAi.

References

Bayne et al. (2005) Cell-to-cell movement of Potato Potexvirus X is dependent on suppression of RNA silencing. *Plant J* 44:471-482.

Ding SW, Voinnet O. (2007) Antiviral immunity directed by small RNAs. *Cell* 130:413-426.

Dunoyer et al. (2005) DICER-LIKE 4 is required for RNA interference and produces the 21-nucleotide small interfering RNA component of the plant cell-to-cell silencing signal. *Nat Genet* 37:1356-1360.

Dunoyer et al. (2010) Small RNA duplexes function as mobile silencing signals between plant cells. *Science* 328:912-915.

Elbashir et al. (2001) RNA interference is mediated by 21-and 22-nucleotide RNAs. *Genes Dev* 15:188-200.

Fagard et al. (2000) AGO1, QDE-2, and RDE-1 are related proteins required for post-transcriptional gene silencing in plants, quelling in fungi, and RNA interference in animals. *Proc Natl Acad Sci USA* 97:11650-11654.

Fire et al. (1998) Potent and specific genetic interference by double-stranded RNA in *Caenorhabditis elegans*. *Nature* 391:806-811.

Guo S, Kemphues K. (1995) parI, a gene required for establishing polarity in *C. elegans* embryos, encodes a putative Ser/Thr kinase that is asymmetrically distributed. *Cell* 81:611-620.

Hamilton et al. (2002) Two classes of short interfering RNA in RNA silencing. *EMBO J* 21:4671-4679.

Hamilton AJ, Baulcombe DC. (1999) A species of small antisense RNA in posttranscriptional gene silencing in plants. *Science* 286:950-952.

Hammond et al. (2000) An RNA-directed nuclease mediates post-transcriptional gene silencing in Drosophila cells. *Nature* 404:293-296.

Lindbo et al. (1993) Induction of a highly specific antiviral state in transgenic plants: implications for regulation of gene expression and virus resistance. *Plant Cell* 5:1749-1759.

McKinney HH. (1929) Mosaic diseases in the Canary Islands, West Africa and Gibraltar. *J Agric Res* 39:557-578.

Moissiard G, Voinnet O. (2006) RNA silencing of host transcripts by cauliflower mosaic virus requires coordinated action of the four *Arabdiopsis* Dicer-like proteins. PNAS 103:19593-19598.

Molnar et al. (2010) Small silencing RNAs in plants are mobile and direct epigenetic modification in recipient cells. *Science* 328:872-875.

Napoli et al. (1990) Introduction of a chimeric chalcone synthase gene into petunia results in reversible co-suppression of homologous genes in trans. *Plant Cell* 2:279-289.

Navarro et al. (2006) A plant miRNA contributes to antibacterial resistance by repressing auxin signaling. *Science* 312:436-9.

Palauqui et al. (1997) Systemic acquired silencing: transgene specific post-transcriptional silencing is transmitted by grafting from silenced stocks to non-silenced scions. *EMBO J* 16:4738-4745.

Powell-Abel et al. (1986) Delay of disease development in transgenic plants that express the tobacco mosaic virus coat protein gene. *Science* 232:738-743.

Romano N, Macino G. (1992) Quelling: transient inactivation of gene expression in Neurospora crassa by transformation with homologous sequences. *Mol Microbiol* 6:3343-3353.

Smith et al. (1994) Transgenic plant virus resistance mediated by untranslatable sense RNAs: expression, regulation, and fate of nonessential RNAs. *Plant Cell* 6(10):1441-1453.

Stanford JC, Johnson SA. (1985) The concept of parasite-derived resistance-Deriving resistance genes from the parasite's own genome. *J Theor Biol* 113:395-405.

Svoboda et al. (2000) Selective reduction of dormant maternal mRNAs in mouse oocytes by RNA interference. *Development* 127:4147-4156.

Voinnet et al. (2000) A viral movement protein prevents spread of the gene silencing signal in *Nicotiana benthamiana*. *Cell* 103:157-167.

Waterhouse et al. (1998) Virus resistance and gene silencing in plants can be induced by simultaneous expression of sense and antisense RNA. *Proc Natl Acad Sci USA* 95:13959-13964.

Winston et al. (2002) Systemic RNAi in *C. elegans* requires the putative transmembrane protein SID-1. *Science* 295:2456-2459.

Yu et al. (2003) Analysis of the involvement of an inducible *Arabidopsis* RNA-dependent RNA polymerase in antiviral defence. *MPMI* 16:206-216.

Zamore et al. (2000) RNAi: double-stranded RNA directs the ATP-dependent cleavage of mRNA at 21 to 23 nucleotide intervals. *Cell* 101:25-33.

16

Epigenetics of germline and epigenetic memory

In previous chapters, various existing mechanisms of epigenetic modifications, such as imprinting, gene silencing, control over transposon activity, and several other related processes were described, and their roles in the functioning of the cell were explained. The most controversial concepts in this area of research, which still remain difficult to test experimentally, are the existence of **epigenetic memory** and the mechanism of occurrence of **transgenerational changes** in response to environmental stimuli. The epigenetic memory is the transmission of information across generations through changes in DNA methylation, histone modifications, and small RNAs. Epigenetic marks can be transferred from one somatic cell to another through mitotic cell divisions, a process referred to as **epigenetic somatic inheritance**. Mitotic and meiotic epigenetic inheritance involves similar epigenetic mechanisms, but there are some substantial differences.

This chapter focuses on how transgenerational inheritance of epigenetic states occurs through meiotic cell divisions. Multiple reports on various organisms, including plants and animals, suggest that the progeny of exposed organisms exhibit changes in behavior, physiology, genome stability, and various cellular functions that might be largely dependent on a proper function of the epigenetic machinery. In order for epigenetic inheritance to contribute to changes in the progeny, epigenetic modifications must escape reprogramming. In this chapter, some extreme cases of epigenetic memory in ciliates are described and some details of epigenetic modifications occurring in gametes of animal and plant cells are provided. There are also some explanations of how epigenetic memory is formed and how it is passed from one generation to another.

Epigenetic reprogramming in ciliates

Ciliates are perhaps one of the earliest model systems for studying non-genetic modes of inheritance. In Oxytricha, these modes of inheritance include mRNA-mediated DNA rearrangements in mRNA-dependent transmission of spontaneous somatic mutations to the next generation. As suggested by Nowacki and Landweber (2009), the somatic ciliate genome is actually an "epigenome" formed through templates and signals arising from the previous generation.

Cortical inheritance

Cortical inheritance is one of the classical examples of non-Mendelian inheritance described in ciliates. After conjugation, Paramecium cells sometimes fail to separate and produce abnormal doublets that only occasionally revert back to singlets (Nowacki and Landweber, 2009). The doublet character is not encoded by the genetic information from the nucleus or the cytoplasm, but rather by the outer layer of the cell cortex, suggesting that the preformed cell structure might play a critical role in inheritance. Similar evidence of cortical inheritance was also reported for Oxytricha. It cannot be excluded that cortical organization of genes also plays an important role in this type of inheritance because genetic mutations causing altered cortical organization have been reported (reviewed in Nowacki and Landweber, 2009). This epigenetic phenomenon represents one of the cases of environmentally induced (or acquired) phenotypic changes that passed from one organism to another without the involvement of DNA or RNA, although the molecular mechanisms of cortical inheritance are still poorly understood.

Homology-dependent epigenetic inheritance

In addition to sequence-dependent gene silencing occurring in most somatic cells of most eukaryotes, homology-dependent mechanisms in ciliates are also responsible for programming of genome rearrangements during development. In ciliates, two distinct nuclei, the diploid

micronucleus (MIC) and a DNA-rich **macronucleus (MAC)**, decouple the germline and somatic functions. MICs give rise to new micro- and macronuclei, whereas the old MAC is lost and replaced by a new one.

The development of a new MAC involves extensive rearrangements of the germline genome. Rearrangements primarily involve the elimination of transposons, repetitive sequences, and numerous single-copy **Internal Eliminated Sequences (IESs)** from coding and non-coding genomic sequences. The sequence-specific elimination of genomic DNA discards nearly all non-genic DNA, thus resulting in more simple gene-rich genomes. For instance, in Paramecium, approximately 40,000 genes span about 72 Mb (reviewed in Nowacki and Landweber, 2009). Sequence elimination is not the only type of genome unscrambling. In Oxytricha, the remaining DNA segments (**Macronucleus Destined Segments**, or **MDSs**) are reordered through frequent translocations or inversions. You can find more details about RNA-directed DNA elimination and rearrangements in Chapter 9, "Non-Coding RNAs Across the Kingdoms—Protista and Fungi."

In Paramecium, developmentally regulated genome rearrangements are apparently controlled by homology-dependent maternal effects that score the presence or absence of a gene in the macronucleus. It was shown that microinjection of a specific DNA sequence into the parental macronucleus protects this sequence from being eliminated from the somatic genome of the progeny. It has been hypothesized that there is a *trans*-nuclear genome comparison that allows for the elimination of sequences that are not present in MICs.

How does this comparison occur? There must exist a transfer of information between the parental MAC and the developing MAC, which allows a comparison of the encoded information. Because in the majority of cases the non-coding sequences are eliminated, it was initially hypothesized that the entire genome of the parental MAC was transcribed and transcripts were exported to the developing MAC where they regulated rearrangements in a homology-dependent manner (reviewed in Nowacki and Landweber, 2009). Initially two possibilities for controlling IES excision were proposed: Either the maternal transcripts provide templates for restoration of the genome after the constitutive random sequence elimination or the transcripts simply inhibit the excision of homologous sequences. In

Paramecium, the former hypothesis was rejected based on the observation that mutated IES DNA injected into the old MAC was not copied in IES maintained in the new MAC. The protection mechanism is more likely to operate in ciliates. In Paramecium, if a specific transcript in the parental MAC is degraded via RNAi, the corresponding DNA sequences of genes and IESs are eliminated in the developing MAC. A template repair model can be applied to other ciliates, including Oxytricha. DNA rearrangements are more complex in Oxytricha as compared to Paramecium, and it is possible that the maternal transcripts might provide templates for cleavage and repair during the DNA unscrambling process.

RNA template-guided genome unscrambling in Oxytricha

A study by Nowacki et al. (2008) demonstrates that maternal RNA templates trigger extensive genomic rearrangements in Oxytricha. This work suggests that a complete RNA pool is produced in a short window of the development from the maternal somatic genome. These RNAs direct precise DNA rearrangements that include DNA deletion and unscrambling. Injection of artificial templates containing segment translocations reprograms the DNA rearrangement pathway, suggesting that RNA molecules guide the sequence-specific genome rearrangements. The repair template model for these rearrangements was supported by the fact that point mutations introduced into RNA templates were occasionally transferred to the rearranged DNA molecules. It was suggested that RNA molecules serve not only as guides for the sequence-specific DNA elimination, but they also provide a template for proofreading the rearranged DNA sequence. As such, this phenomenon represents rare evidence of RNA-directed spontaneous somatic mutations that are not transmitted to the germline genome.

In Oxytricha, RNA-mediated DNA rearrangements and DNA repair delete 95% of the germline genome and rearrange the remaining DNA sequence. These processes severely fragment their germline chromosomes and then sort and reorder the hundreds of thousands of MDS pieces remaining. A new RNA-mediated pattern of DNA rearrangements and occasional point mutations can be transmitted to the progeny, thereby bypassing the usual DNA mode of inheritance. In this manner, the acquired epigenome can be passed through generations without altering the germline genome.

Non-coding RNA-mediated programming of DNA rearrangements

Two distinct RNAi pathways mediate homology-dependent changes in ciliate DNA. Deletions of genomic DNA can be induced by siRNAs stemming from long dsRNAs; feeding Paramecium cells with *E. coli* encoding gene-specific long dsRNAs triggers the deletion of a target gene. In Tetrahymena, a class of approximately 23 to 24nt RNAs is produced endogenously from several loci, and it is hypothesized that these RNAs direct DNA elimination.

Some of the epigenetic effects, such as maternal inheritance of alternative DNA rearrangements, cannot be explained by the function of siRNAs. An alternative to siRNA-mediated DNA elimination is represented by the function of specifically modified histones. In *Tetrahymena thermophila*, the elimination of micronuclear-specific DNA is directed by methylation of lysine 9 of histone H3. The accumulation of scan RNAs (scnRNAs) in meiotic micronuclei is dependent on the function of meiosis-specific Dicer- and PIWI-like proteins. The research demonstrated that the elimination of DNA in the developing macronucleus is impaired in cells that are deficient for Dicer and PIWI-like proteins. It is thus proposed that DNA rearrangements in the developing MAC of Tetrahymena are determined by scnRNA-directed methylation of histone H3 on lysine positions 9 (H3K9) and 27 (H3K27) and binding of the modified histones to specific DNA sequences.

The elimination of DNA targeted by methylated histones is analogous to heterochromatin-related transcriptional gene silencing in other eukaryotes (reviewed in Nowacki and Landweber, 2009). Also, scnRNAs in Tetrahymena migrate from the parental to zygotic macronucleus and become enriched in micronucleus-specific sequences during conjugation. This observation suggests that scnRNAs function as a template for a *trans*-nuclear comparison. It is further hypothesized that those scnRNAs that find and pair homologous DNA sequences in the parental macronucleus are degraded if the old macronucleus is destroyed. The scnRNAs that find their targets in the germline are exported to the newly developing macronucleus where they direct sequence-specific DNA elimination.

Both the scan RNA model and the maternal transcript model might function in ciliates. A combined model proposed for Tetrahymena and Paramecium suggests that scnRNAs are selected through their interactions with long maternal RNAs. scnRNAs might be compared to an RNA copy of the maternal MAC genome via RNA/RNA binding rather than to MAC DNA via RNA/DNA binding. You can find more details on the function of scnRNAs and possible models for RNA-mediated DNA elimination in Chapter 9.

Epigenetic reprogramming and epigenetic memory in animals

In order to pass any changes in epigenetic makeup to the progeny, epigenetic marks need to avoid reprogramming or resetting that occurs twice in the mammalian lifecycle. The first wave of epigenetic reprogramming takes place during gametogenesis, whereas the second one occurs early during embryogenesis. Reprogramming is a robust process that includes passive and active losses and the reestablishment of epigenetic marks. These marks establish patterns of inheritance that are not normally associated with changes in DNA sequences because they rely on DNA methylation, histone modifications, and the function of small non-coding RNAs. Similar to genetic memory, epigenetic memory requires stable inheritance and should be passed onto the progeny through meiosis (although epigenetic inheritance mainly does not follow the Mendelian laws). There are many mechanisms that are able to support the **non-Mendelian mechanism of inheritance**. They include the genomic imprinting via differential methylation patterns; the prepacking of genes involved in early development in histones, rather than in **protamines;** in sperm cells the retention of the CenH3 proteins required for nucleosome assembly at centromers in mammalian sperm; the inheritance of **PIWI-associated interfering RNAs** (**piRNAs**); and many others. The ability of mammals to pass on epigenetic information to the progeny might not only depend on a proper function of the aforementioned mechanisms but also might require a certain degree of flexibility in response to environmental cues.

Epigenetic reprogramming first occurs in **primordial germ cells (PGCs)** and then after fertilization is characterized by extensive changes in DNA methylation and chromatin modifications. In sperm cells, genes that are involved in early development are associated with histones, whereas those that are required later during embryonic development are associated with protamines. In the embryo, the latter are temporarily repressed using reversible histone marks.

Epigenetic reprogramming in gametes

Differentiated somatic cells have a defined epigenetic status that allows them to be defined as cells belonging to a specific tissue type. The division of these cells occurs through mitosis, which relatively faithfully copies genetic and epigenetic information from parent to daughter cells. In contrast, germline cells acquire their totipotency upon the erasure of parental imprints occurring in PGCs. These cells undergo meiosis and transmit genetic and epigenetic information to the next generation (Combes and Whitelaw, 2010). PGCs are derived from epiblast cells; in male embryos, they enter mitotic arrest until near birth when mitosis of spermatogonial stem cells begins, and they do not enter meiosis until puberty. In females, meiosis begins immediately after the time PGCs enter the ovary in the embryo; these cells enter meiotic arrest in prophase I of meiosis and then continue meiosis during ovulation.

A fertilized single-cell zygote undergoes four consecutive divisions becoming a 16-cell morula. The morula itself consists of external and internal cells. Cells committed to either the external or internal layer begin to express different genes, thus leading to the formation of different tissues. External cells representing a larger group of cells become the trophoblast (trophectoderm) cells that develop into the chorion, the embryonic portion of the placenta. Internal cells develop into an inner cell mass (ICM), which gives rise to the embryo and its associated yolk sac, allantois, and amnion. After the commitment to become either a trophoblast cell or an inner cell mass cell is made, different genes are expressed by cells of these two regions. ICM cells form the lower layer—the hypoblast, developing into the yolk sac—and the upper layer—the epiblast, giving rise to ectoderm, mesoderm, and endoderm. In mice, precursors of PGCs arise very early during epiblast development, at 6.0 to 6.5 days post coitum prior to epiblast differentiation into ecto-, meso-, and endoderm. Around embryonic day 11.5 (E11.5), PGCs migrate into the gonadal ridge where they undergo developmental differentiation to oocytes or spermatocytes in the adult organism.

Epigenetic marks of early PGCs resemble those of somatic cells as both cells have similar **imprinted genes** in which a single copy is expressed according to the parent of origin. The experiments using flow cytometry allowed to demonstrate that imprints are largely intact in migrating PGCs: out of four imprinted genes analyzed, only *Igf2r* was expressed biallelically, whereas *Snrpn*, *Igf2*, and *H19* were expressed from a single allele (Szabo et al., 2002). Although somatic cells maintain these imprints throughout their life cycle and multiple cell divisions, PGCs quickly lose them; in mice, PGCs express imprinted genes biallelically already at E11.5, a time when they arrive at the genital ridge. Methylation analysis revealed that imprints in PGCs are demethylated between E11.5 and E12.5 (Hajkova et al., 2008) (see Figure 16-1). This rapid loss of DNA methylation is paralleled by several other reprogramming activities, including the loss of repressive demethylation of histone 3 lysine 9 (H3K9), and many other histone modifications. Because the maintenance DNA methyltransferase is expressed in PGCs, extensive demethylation occurring over these few cell cycles is likely an active process of reprogramming (Hajkova et al., 2008). The degree of DNA demethylation is fairly significant, with female PGCs exhibiting an approximately 70% decrease in methylation and male PGCs—about a 60% decrease (Popp et al., 2010). It appears that reprogramming is not uniform; some loci are demethylated at extremely high levels, whereas the others are largely intact.

DNA remethylation in male germ cells is quick to occur, and at E15-E16 cells restore methylation patterns; these imprints are maintained through meiosis and in a haploid stage. In contrast, remethylation in the female germline does not occur until after birth during oocyte growth, and methylation marks are fully acquired after the metaphase II stage. The differences in the acquisition of new methylation patterns might be associated with the different timing and stages of cell cycle arrest in male and female gametes. In gametes, the remethylated DNA has a methylation pattern similar to that in somatic cells of the developing organism, with differences in methylation patterns primarily at the **imprinting control regions (ICRs)** (Hemberger et al., 2009).

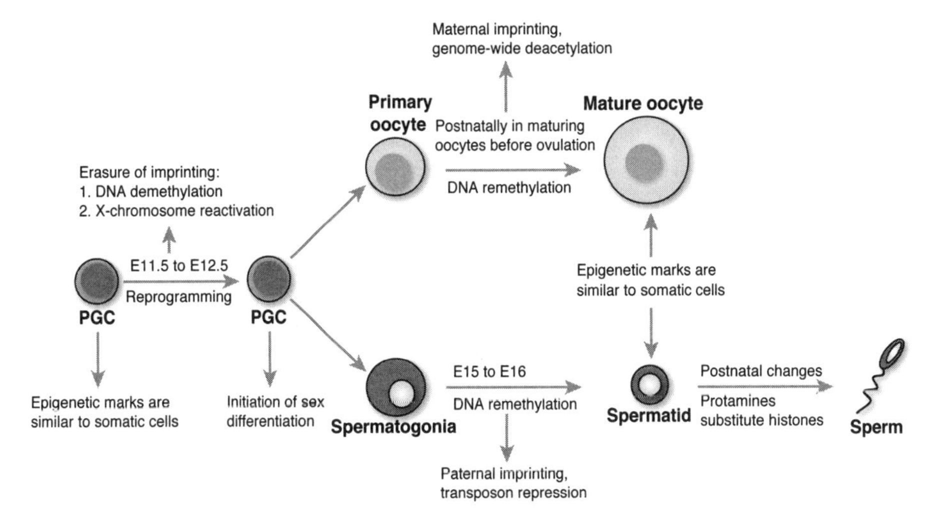

Figure 16-1 First round of epigenetic reprogramming (adapted from Migicovsky and Kovalchuk, 2011 with permission).

The second wave of reprogramming in animals occurs shortly after fertilization. Fertilization is the process of fusion of sperm and ovum cells upon which two parental genomes with completely different epigenetic states unite to form a diploid cell. Because sperm and ovum cells are at different stages of their cell cycle and their haploid genomes carry different epigenetic marks, major reprogramming events occur before the zygote starts transcription and it is ready to divide (see Figure 16-2). Male and female gametes package their genomes in a different manner. Whereas the paternal genome is packaged with special arginine-rich, nuclear proteins called protamines, the maternal genome is enriched in abundantly modified histones (Puri et al., 2010). The replacement of histones by **protamines** occurs during the late stage of spermatogenesis, and it is hypothesized that it provides high levels of condensation that are necessary for the protection of sperm DNA against damage and loss. In humans, all but 4% of the histones are replaced by protamines (PRM1 and PRM2), leaving the genes that are essential for early embryogenesis bound to histones (Balhorn, 2007).

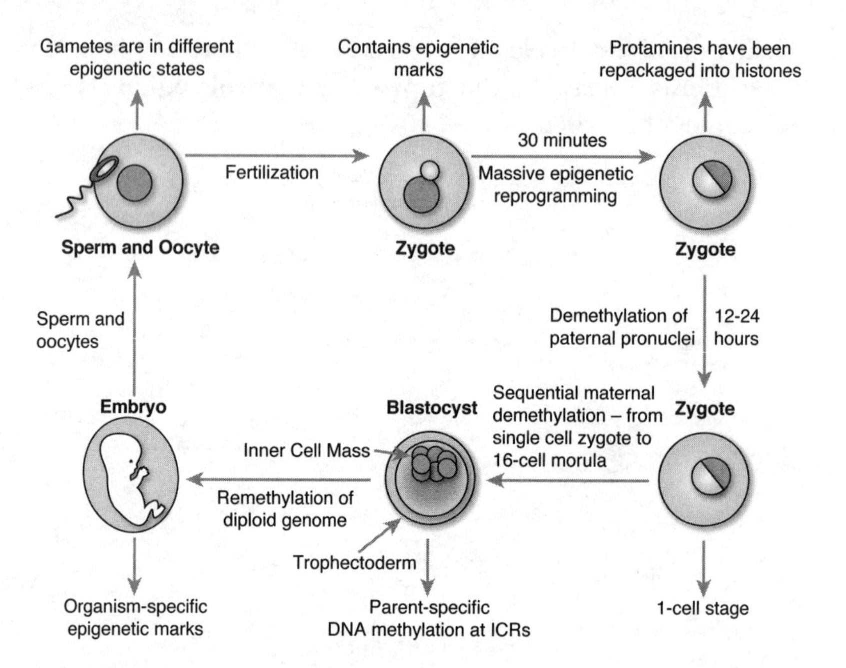

Figure 16-2 Second round of epigenetic reprogramming (adapted from Migicovsky and Kovalchuk, 2011 with permission). See details in the text.

Chromatin remodeling of the paternal and maternal genomes begins when the sperm arrives in the oocyte cytoplasm. Whereas the paternal genome is highly condensed and transcriptionally inert, the maternal genome is transcriptionally active and the oocyte maintains a high capacity for chromatin remodeling. The first steps in the process of decondensation of the paternal genome is the repackaging of nucleosomes associated with protamines into nucleosomes that are made up of DNA bound to histones (Puri et al., 2010). Although protamine-based nucleosomes prevent transcription, chromatin associated with histones is compatible with transcription, replication, and DNA repair processes (Ooi and Henikoff, 2007).

The transition from a protamine-based to a histone-based nucleosome occurs very quickly; in mice, it was shown that most protamines were removed within 30 minutes after fertilization. A recent analysis showed that histone modifications in paternal and maternal genomes are not similar, with the paternal genome having higher levels of histone acetylation as compared to the maternal genome. It is not clear, however, whether this difference is passive due to the fact that the maternal genome is over-represented by de-acetylated histones, or this is a reflection of an active deposition of acetylated histones allowing the more prolific gene expression from the paternal genome.

The second wave of modifications occurs at the level of DNA methylation after histone deposition; the paternal DNA undergoes severe DNA demethylation erasing most of the epigenetic marks. The demethylation process starts within the first 4 to 12 hours following fertilization, depending on species analyzed, and it is thought to be active because it occurs before DNA replication is initiated in the paternal pronucleus (Combes and Whitelaw, 2010). In mice and rats, methylation levels of the paternal genome drop to less than 40% of levels of the maternal genome; however, the degree of demethylation in other species, including humans, is unclear because different reports on different animals suggest variations from 50% to as much as complete demethylation. At this stage, methylation of the maternal genome does not change substantially. It is hypothesized that during the time of the protamine to histone exchange, the paternal DNA is prone to active demethylation, whereas the maternal genome is protected against this process by associating with methylated histones.

During the first zygotic division, the maternal and paternal genomes combine, and each of the daughter blastomeres ends up with equal amounts of methylated DNA. The maternal genome undergoes gradual DNA demethylation with each subsequent division (see Figure 16-3). Similar to the paternal genome, specific loci in maternal genome resist demethylation because they retain repressive histone marks. The exclusion of the maintenance DNA methyltransferase Dnmt1 from the nucleus contributes to the process of maternal DNA demethylation. The demethylation process in the maternal genome is complete by the late morula stage. The levels of residual methylation of the maternal genome might be about 30% higher than those of the paternal genome. It is not clear whether the mechanisms of reprogramming in paternal and maternal genomes after fertilization are similar to those occurring in PGCs. However, it is known that parent-specific DNA methylation is maintained in paternal and maternal genomes of ICRs, whereas methylation differences are erased in PGCs (Edwards and Ferguson-Smith, 2007).

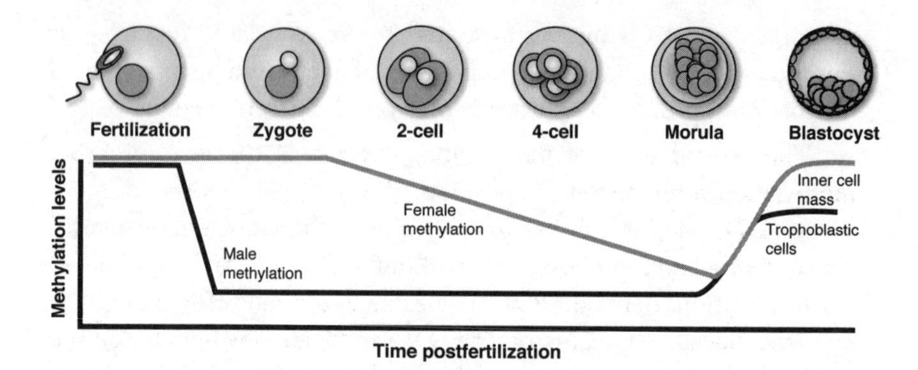

Figure 16-3 Reprogramming of methylation upon fertilization (reproduced from Abdalla H, Yoshizawa Y, Hochi S. (2009) Active Demethylation of Paternal Genome in Mammalian Zygotes. J. Reprod. Dev. Vol. 55, 356-360). Methylation at trophoblast is substantially lower than in inner cell mass.

The process of *de novo* genome-wide DNA methylation occurs shortly after blastocyst implantation in cells committed to the inner cell mass. The newly established methylation patterns that differ by 30% to 50% are then faithfully maintained from that point on through mitotic divisions in somatic lineages (Bonasio et al., 2010).

You can find more details on the process of *de novo* and maintenance DNA methylation in germline and somatic cells in Chapter 4, "DNA Methylation as Epigenetic Mechanism." Epigenetic reprogramming in the early embryo is a fine-tuned mechanism, and the early embryo is very susceptible to internal and external stresses. Errors in this spatially and temporally highly coordinated process of epigenetic regulation are responsible for a high rate of embryo loss after fertilization.

Epigenetic inheritance—an escape from reprogramming

How does epigenetic inheritance occur if paternal and maternal genomes undergo reprogramming associated with a profound loss of DNA methylation and the replacement of specific histone marks? As mentioned earlier, not all epigenetic marks are removed from parental genomes. Thus, acquired epigenetic marks might escape reprogramming, but they must be protected by specific mechanisms. There exist several known mechanisms associated with DNA methylation, protamines, histones, and small RNAs.

Protamine-mediated inheritance

Protamines replace histones during male gametogenesis. The incomplete replacement of histones by protamines might be one of the mechanisms of epigenetic inheritance. The percentage of histones that are retained in the DNA is different in different organisms; in sperm cells of mice, as little as 1% of histones remain bound to DNA, whereas in humans the fraction of nucleosome-bond histones is from 4% to 15%. This fraction of nucleosome-bound histone is not evenly spread across the genome but is concentrated in genes that are involved in early developmental processes. This bias is not surprising. Because chromatin in sperm cells is extremely condensed, it is transcriptionally inert. The activation of transcription requires protamine displacement by active histone marks. In this case, the prepacking of genes involved in early developmental processes with histones prevents their tight condensation by protamines in sperm cells, allowing the activation of transcription in early embryos at a much faster rate (Ooi and Henikoff, 2007). One of the examples of genes associated with histones in sperm cells is the ε- and γ-globin genes in humans. In contrast, the region encoding the β-globin genes are not associated with histones. This difference might be explained by differential involvement of the ε-, γ- and β-globin genes in embryos and adults. The ε- and γ-globin genes are transcribed in the primitive erythroblasts in the embryonic yolk sac, which differentiates at three weeks of gestation. The expression of γ-globin predominates during the fetal period, whereas the β-globin and δ-globin genes are not required in the embryo. On the contrary, the expression of the β-globin and δ-globin genes is predominant after birth in the bone marrow. Thus, the presence of histones rather than protamines at the ε- and γ-globin loci marks them for the early expression in the embryo.

The differential binding of histones or protamines to specific genomic loci might represent one of the mechanisms of epigenetic memory transmission. Histone retention at loci that have novel epigenetic marks, such as hypo- or hypermethylated loci, might allow them to pass these marks to the progeny. Curiously, sperm cells of some animals do not contain protamines. For example, in zebrafish sperm, chromatin is condensed with extra amounts of linker histone H1 rather than protamines (Wu et al., 2011). It is an interesting hypothesis that epigenetic memory might be more efficiently passed in zebrafish rather than in mice or humans.

Methylation-mediated inheritance

DNA methylation might play one of the most active roles in epigenetic inheritance because methylation represents a more stable and thus a more permanent epigenetic mark as compared to DNA association with certain histones. Differentially methylated loci that escaped reprogramming in PGCs might pass their expression status to the progeny. For example, **intracisternal A-particle (IAP)** retrotransposons are protected from demethylation in PGCs, and thus most IAPs escape reprogramming (Popp et al., 2010). The second round of reprogramming that occurs right after fertilization also leaves IAPs untouched. This protection may have a sequence-specific pattern, and IAPs might influence imprinting of neighboring loci; the reinsertion of IAPs in a different genomic position resulted in parental imprinting of endogens located nearby newly inserted IAPs. Stable inheritance of newly acquired imprints gives rise to epialleles. The best known example of epiallele is represented by the *agouti* locus in mice. This locus is linked to the epigenetic status of IAPs and controls coat color in mice.

The insertion of IAP retrotransposon upstream of the *agouti* locus of the *Avy* allele results in the ectopic expression of agouti protein. Consequently, hair follicle melanocytes synthesize a yellow pigment instead of a black pigment (Morgan et al., 1999). Changes in coat color are also linked to diabetes, obesity, and increased incidences of tumors. A yellow coat color is associated with hypomethylation of the *Avy* epiallele; in the *Avy*/*A* genetic background, *Avy* is dominant over the *A* allele, resulting in a yellow coat color. Hypermethylation of the cryptic IAP promoter leads to silencing of the *Avy* epiallele and reversion to a black coat color. The *Avy*/*A* epiallele phenotype is thus converted to a *A*/*A* wild-type phenotype. The progeny of *Avy*/*A* mothers have varying degrees of a yellow coat color, suggesting that the epigenetic mark is not completely erased in the female germline. Thus, it is a clear example of epigenetic inheritance, but it should be noted that the epiallele is not passed to the progeny with 100% penetrance (Morgan et al., 1999). The insertion of IAP often results in an unstable epigenetic state in the germline, leading to a mosaic appearance of coat colors. The analysis of methylation shows that different cells of the same animal have a different epigenetic status.

Some endogenes were also shown to escape the reprogramming after fertilization, resulting in certain loci having epigenetic marks of parental origin (Bartolomei, 2009). Imprints established in the germline during organism development must also escape the subsequent wave of *de novo* methylation. Genomic regions with differential methylation, known as **differentially methylated regions (DMRs)**, are found in many imprinted genes. They serve as crucial epigenetic marks indicating genomic areas that are methylated differently in maternal and paternal alleles. DMRs that derive from sperm and oocytes behave as ICRs. In fact, most of the germline DMRs represent ICRs that are controlled through *cis*- and *trans*-acting factors. The binding of *trans*-acting factors to ICRs prevents active demethylation of one of the parental alleles while allowing another allele to undergo reprogramming (Bartolomei, 2009). One of the *trans*-acting factors, the PGC7/STELLA protein, controls methylation of multiple paternally or maternally methylated ICRs. STELLA-deficient eggs exhibit normal methylation at ICRs; however, zygotes derived from these eggs are severely hypomethylated at multiple maternal and paternal ICRs. STELLA also appears to be important for the protection of the maternal genome against demethylation; Stella $^{-/-}$ zygotes exhibit a premature, global loss of DNA methylation in the maternal pronucleus. Other proteins, RBBP1/ARID4A and RBBP1L1/ARID4B, are potentially involved in the stability of imprints; these protein pairs regulate the tissue-specific expression of the small nuclear ribonucleoprotein polypeptide N (*Snprn*), which is probably involved in tissue-specific splicing of pre-mRNAs (Bartolomei, 2009).

Differential DNA methylation of germline DMRs/ICRs gives rise to the coordinated allele-specific expression of imprinted genes that are often found in clusters of hundreds of kilobases. Currently in the germline of mice, there are 21 known DMRs, with the majority of them having a maternal-specific origin of methylation. Probably, there are several mechanisms involved in locus-specific imprinting. If DMRs/ICRs are located at the promoter, differential methylation results in silencing of only a given methylated allele. If imprinted areas represent entire clusters, the mechanisms of silencing might be associated with establishment of chromatin insulators that cover the imprinted region and/or the function of sequence-specific siRNAs and piRNAs. The establishment of the insulator-dependent imprinting mechanisms is supported by the analysis of *H19* and *Igf2* loci. These genes are closely linked and are part of the imprinted cluster. *H19* encodes long non-coding RNAs with growth-repressive properties, whereas the *Igf2* gene encodes an autocrine growth factor. This gene pair is reciprocally imprinted, with *H19* being expressed from the maternal locus and the *Igf2* gene expressed only from the paternal locus. Parent-of-origin-specific expression of these two genes is due to enhancer activity stemming from the boundary separating the H19 and Igf2 genes (see Figure 16-4). The insulator itself is represented by a differentially methylated region with a methylation pattern similar to that of the H19 gene. The insulator is unmethylated and active on the maternal chromosome but methylated and inactive on the paternal chromosome (Biliya and Bulla, 2010). The CCCTC-binding factor (CTCF) binds the unmethylated region on the maternal chromosome restricting the activity of the enhancer only at the *H19* locus, thus stimulating the expression of only H19. Methylation of the insulator on the paternal chromosome prevents binding by CTCF, thus allowing the enhancer to act beyond its previous boundary, which promotes the expression of Igf2.

Epigenetic reprogramming includes the erasure of the existing imprints at ICRs followed by the establishment of a new set of sex-specific imprints. Subsequently, these imprint marks are maintained from the zygote to all somatic tissues. Importantly, not all sequences in the genome are equally affected by passive (the first wave) and active (the second wave) demethylation steps. It is believed that the erasure of genomic imprints at ICRs occurs only during the second step because, in mice, ICRs retain their methylation marks up to 11.5 days post coitum (dpc), and demethylation at these regions occurs only at 13.5 dpc (Hajkova et al., 2008).

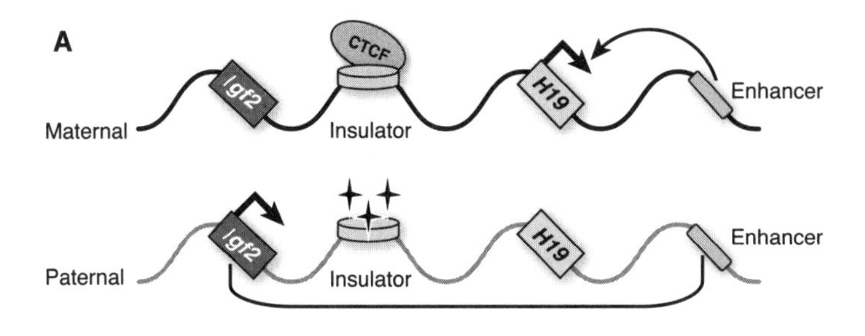

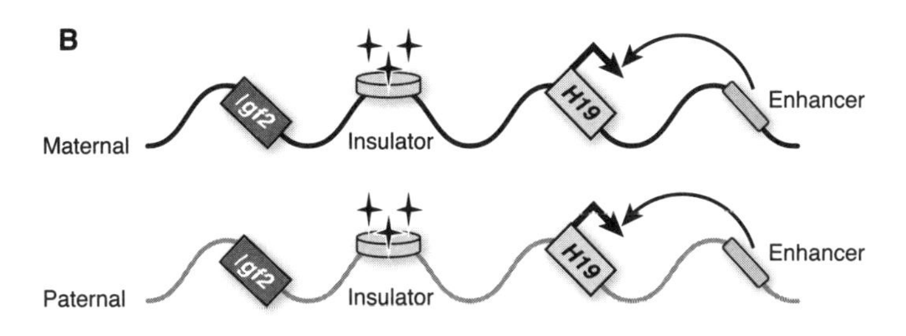

Figure 16-4 Regulation of enhancer activity via methylation of insulator.

For epigenetic inheritance to occur, erased epigenetic marks must be placed on newly replicated DNA. DNA methylation imprints are established through the activity of both the oocyte-specific and somatic forms of Dnmt1; see Chapter 4 for more details on Dnmt enzymes. Changing a balance between these two forms of Dnmt1 might be one of the mechanisms regulating epigenetic inheritance of newly formed epialleles.

The genomic features of maternal and paternal germline DMRs are clearly different. Whereas maternal germline DMRs are CpG-rich and they reside in a promoter region, the paternally methylated germline DMRs are CpG-poor and they are located within intergenic regions. Because CpG islands have low levels of methylation, it is quite unusual that maternal germline DMRs are hypermethylated. DMRs are frequently associated with arrays of tandem repeat motifs. It is hypothesized that these repetitive elements recruit imprinting factors, including Dnmt1. Experimental evidence for the role of tandem repeats in imprint acquisition exists only for the *Rasgrf1* locus but not the *H19* and the *KvDMR* loci.

The maternally expressed *H19* gene is one of the best known imprinted genes in mice. The mouse *H19* gene encodes an RNA that is highly expressed in endoderm and mesoderm. A paternal copy of *H19* is hypermethylated in both somatic tissues and sperm. This pattern of differential methylation is inherited from the gametes and is preserved through embryogenesis, being a clear example of epigenetic memory (Tremblay et al., 1997). Similar to mice, human *H19* is also hypermethylated on the paternal allele.

It is believed that the imprint cycle is initiated by DNA methylation of the germline. However, recent studies suggest that DNA methylation of the germline might not be necessary to start the imprint cycle, and methylation might be acquired at a later developmental stage. The experimental evidence shows that if the *H19* gene is inserted at ectopic positions in the genome, DNA methylation at the paternal DMR is established after fertilization during early development. Also, DNMT3L-deficient oocytes are still present at some ICRs, although in a stochastic manner (Arnaud et al., 2006). Moreover, the *Snrpn* and *Peg3* ICRs were fully maternally methylated in 30% and 14% of the progeny of *Dnmt3L*$^{-/-}$ females, respectively (Arnaud et al., 2006). These reports suggest that some ICRs contain germline-derived (DNA methylation-independent) signatures that can direct methylation of the right parental allele at implantation.

Histone-mediated inheritance

The inheritance of histone marks through both oocyte and sperm might be an additional contribution to reprogramming and epigenetic inheritance. Being the architectural proteins that package DNA into nucleosomal particles, modified histone variants control transcription and perpetuate active or repressive chromatin states. Elevated levels of histone retention observed in mature sperm occur nonrandomly, and sperm cells contribute not only the genome but also specific molecular regulatory factors to the developing embryo. If some histones are not protected and are replaced by protamines, the embryo does not develop correctly (Ooi and Henikoff, 2007).

A well-known example of histone-regulated epigenetic memory is the retention of a special histone H3 variant, CenH3 (CENP-A), in mammalian sperm. In higher eukaryotes, CenH3 is required for determining the location of the centromere and the assembly of centromeric nucleosomes. The replacement of H3 by CENP-A at the centromere ensures correct chromosome segregation during mitosis and meiosis. Thus, centromere identity does not depend on DNA sequence but rather on the presence of CENP-A. The incorporation of newly synthesized CENP-A occurs in the late telophase/early G1 in human cells. CenH3-containing nucleosomes are segregated between daughter DNA strands, which also contain new nucleosomes that are not attached to CenH3, resulting in the dilution of the CenH3 signal during S phase. The dilution of the CenH3 signal leads to a self-propagating epigenetic state, marking the location of the centromere and providing the optimal chromatin configuration for kinetochore positioning and function (reviewed in Migicovsky and Kovalchuk, 2011).

Small RNA-mediated inheritance

Finally, small non-coding RNAs might also regulate epigenetic inheritance through the differential expression of specific ncRNAs in the germline. In mammals, piRNAs are specifically expressed in germ cells to regulate transposon activity (see Chapter 10, "Non-Coding RNAs Across the Kingdoms—Animals"). Half of the human genome consists of transposons, and their derepression occurs during epigenetic reprogramming. The role of piRNAs is to ensure that transposons are not reactivated. In Drosophila, maternally deposited piRNAs are responsible for epigenetic inheritance. Crosses between female flies that were previously not exposed to a specific active transposon and males in which this transposon is silenced cause sterility in the progeny, a phenomenon referred to as **hybrid dysgenesis**. In contrast, the reverse cross does not result in the same outcome, thus supporting the hypothesis that this process is regulated by maternal piRNAs (Brennecke et al., 2008).

piRNAs are also critically important for male fertility in mice as piRNA biogenesis mutants exhibit the degeneration of the germline cells, with somatic cells being relatively unaffected. Although this has not been documented in humans, there is an indication that a Chinese population with genetic variants in some piRNAs is susceptible to spermatogenic failure. Different levels of piRNAs might regulate germline development and alter the control over transposons, resulting in developmental abnormalities and, possibly, phenotypic changes in the progeny. If inherited, the differential expression of piRNAs might alter germline development in the progeny (Jablonka and Raz, 2009).

Because sperm and ovum cells carry a certain epigenetic load of differentially expressed ncRNAs (besides piRNAs), it is also possible that this pool might regulate the development of the embryo or influence genetic and epigenetic make-up of the progeny. It remains to be established whether other types of ncRNAs are involved in passing epigenetic memory across generations.

Epigenetic inheritance of disease

The information presented in previous sections clearly demonstrates that epigenetic marks are necessary for the proper development of the embryo, the establishment of the germline, and parent-of-origin specific gene expression. Dysregulation of imprinting results in severe developmental abnormalities and therefore triggers multiple diseases. Defects in imprint erasure and remethylation cause the establishment of wrong imprints either at the wrong genomic position or on the wrong parental chromosome. As a result, silent alleles are expressed, whereas the alleles that are normally expressed are silenced. An imbalance in the dosage of imprinted genes leads to specific diseases. Dysregulation of imprinting is likely to occur through two main mechanisms. The first mechanism is a primary imprinting defect that occurs due to alterations in methylation/histone association without any changes in the DNA sequence (a primary imprinting defect). A secondary imprinting defect depends on a mutation in a *cis*-regulatory element or a *trans*-acting factor. More details about the role of primary and secondary imprinting defects in the development of diseases are discussed in reviews by Dolinoy et al. (2007) and Horsthemke (2011).

It is not surprising that epigenetic memory is crucial to insure viable offspring. Some recent reports also suggest that the ability to pass new epigenetic information to the progeny is also responsible for disease development. Specifically, transgenerational epigenetic inheritance in humans might result in illnesses associated with epimutations and epialleles that are passed on to the progeny. One of the best-known cases of epigenetic inheritance is an increased risk for hereditary nonpolyposis colorectal cancer (HNPCC), a sporadic colorectal cancer. HNPCC is associated with genetic as well as epigenetic defects in several mismatch repair genes, including *MLH1*, *MSH2*, *MSH6*, *PMS2*, and *MLH3*.

Hypermethylation of the *MLH1* gene triggers neoplastic transformation, and there are reports of germline hypermethylation (Hitchins et al., 2007). As a result of *MLH1* hypermethylation in the germline, the developing embryos carry only one functional allele of the *MLH1* gene. Genetic or epigenetic mutations of the second functional allele in such individuals more frequently trigger cancers typical of HNPCC syndrome. The presence of *MLH1* epimutation in the germline indicates a potential for inheritance from parents to progeny. Although such inheritance is rare, it is still possible. One case of maternal transmission of the epimutation to a son was documented, although the epimutation was not passed on to the next generation as hypermethylation was erased in the son's spermatozoa. In this case, the epimutation of the *MLH1*gene that caused a predisposition to HNPCC in a mother was passed on to a son, thus predisposing him to an increased risk of cancer. Curiously, this epimutation was not passed from a mother to her other children; the epimutation was apparently erased shortly after fertilization during the second wave of reprogramming. Although these results indicate that germline transmission of an epigenetic state that confers disease susceptibility is possible, it is apparently a rare case. Indeed, epimutations that arise in the germline are expected to disappear after fertilization. It is when they escape the reprogramming that epimutations and epialleles are retained and passed on to the progeny; but the occurrence of such situations is probably rare (Hitchins et al., 2007).

Angelman (AS) and Prader-Willi (PWS) syndromes might also be caused by epigenetic inheritance in certain subgroups of patients. Both diseases are caused by the loss of function of imprinted genes in human chromosome 15q11-q13. Analyses of patients with PWS and AS have shown that nearly one third of all patients have the deletion of an **imprinting center (IC)**; and a disease is possibly developed because of an imprinting defect which occurs after fertilization (reviewed in Migicovsky and Kovalchuk, 2011). Whereas in AS patients, the imprinting defect occurred on the chromosome inherited from either the maternal or paternal grandmother, in PWS patients, it occurred in the chromosome inherited from the paternal grandmother. It is possible that at least one mechanism of the development of Prader-Willi and Angelman syndromes is likely to be due to either the failure to erase the maternal imprint during spermatogenesis or incomplete reprogramming.

So far, hereditary nonpolyposis colorectal cancer and Prader-Willi and Angelman syndromes are the only heritable syndromes that are clearly associated with epimutations. It remains to be shown whether they are unique in this respect, or there are other cases when epimutations are responsible for disease development and heritable transmission.

Epigenetic reprogramming and epigenetic memory in plants

Plants do not possess a predetermined germline; rather they maintain the undifferentiated state of germline-like cells. Being gametophyte initials, these cells give rise to megaspores and microspores, which develop into female and male gametophytes, respectively. Two cell divisions of the male gametophyte produce vegetative nucleus (VN) and generative nucleus (GN) that divides into two sperm cells (SC). The female gametophyte consists of an egg cell (EC) that develops into an embryo and the central cell nucleus (CCN) that gives rise to the endosperm. The embryo sac also contains two synergid cells required for pollen tube attraction and three antipodal cells of a yet unknown function. During the fertilization event, two sperm cells reach the embryo sac that binds the central cell and the egg cell, forming an endosperm and an embryo, respectively.

DNA methylation in gametes

The details on DNA methylation control in plant gametes are discussed in Chapter 4. In brief, the level of methylation in VN and CCN decreases, resulting in an increased production of siRNA from repetitive elements and transposons. It is hypothesized that these siRNAs travel from VN and CCN to SC and EC, respectively; there, they reinforce silencing of DNA sequences homologous to these siRNAs.

Demethylation of transposable elements in VN is associated with downregulation of the RdDM pathway and the absence of DDM1. In addition, the DNA methyltransferase MET1 regulates imprinting during male gametogenesis. As a consequence, *FWA* and *FIS2* are exclusively expressed in the endosperm. Thus, the control over imprinting is different in plants as compared to animals because it involves the maintenance of DNA methylation as opposed to *de novo* methylation. The difference in DNA methylation between endosperm and embryo is especially prominent in DMRs; nearly 90% of DMRs exhibited low levels of DNA methylation in the endosperm as compared to the embryo. DMRs corresponding to transposable areas were more than threefold undermethylated in the endosperm.

Changes in histone expression and modifications in gametes

Histone modifications play an essential role in setting apart the chromatin state of both VN and GN as well as EC and CCN. In *Lilium longiflorum*, the levels of histone H1 are maintained in GN but not in VN, whereas the levels of histone H2B are maintained at relatively similar levels. Histone H1 is involved in chromosome condensation during mitosis. Because VN does not divide again (unlike GN), H1 might not be required in VN.

Unlike female cells, male gametic cells contain three novel histones—gH2A, gH2B, and gH3—that are involved in nucleosome formation. Also in Arabidopsis, the germline and mature sperm cells have high levels of the expression of the H3 variant HTR10, whereas this variant is completely replaced by H3.3 variant in the zygote upon fertilization (Ingouff et al., 2007). A substitution of the H3.3 variant is a hallmark of postfertilization development, which is similar in both developing plants and animal embryos. The HTR10 histone variants in sperm cells are completely removed within a few hours before the S phase of the first division of the zygote of the fertilized egg. This indicates that the replacement of HTR10 by H3.3 variants occurs in a replication-independent manner, thus being an active process (Ingouff et al., 2007). A similar situation exists in female gametes; egg cells contain HTR5, whereas CNN contains HTR8 and HTR14 isoforms of H3.3. The process of fertilization eliminates these differences, suggesting that H3.3 isoform-based epigenetic marks are not transmitted to the progeny. Passing epigenetic memory on to the progeny requires that at least some of specific histone variants and differential methylation patterns can escape reprogramming associated with fertilization, allowing maternal or paternal imprints to be formed (Ingouff et al., 2010).

Gene imprinting in plants

The endosperm nourishes the embryo and does not contribute to the genetic or epigenetic makeup of developing offspring. This might be the main reason why imprinting in plants is observed primarily in the endosperm; more than 200 genes are imprinted in the endosperm (Gehring et al., 2011). Some imprinted genes also exist in plant embryos (Jahnke and Scholten, 2009). In maize, the *maternally expressed in embryo 1* (*mee1*) gene is expressed only following fertilization. In contrast, in Arabidopsis, no imprinting in the embryo has been found (Hsieh et al., 2011). The analysis of imprinting in rice showed that out of 56 loci imprinted in the endosperm, only one was imprinted in the embryo (Luo et al., 2011). The locus encodes Os10g0750, an *Ole e 1* homolog, with a possible function of control of the formation of pollen tube (Luo et al., 2011). Unlike in plants, in animals, imprints are established in gametes and are often maintained in the developing embryo (Migicovsky and Kovalchuk, 2011). These imprints are erased and reapplied during gametogenesis.

In plants, no erasure is needed as imprints predominantly are formed in the endosperm. Imprints in the endosperm are possible because of DEMETER (DME) allele-specific activity in CCN where DNA methylation is removed. This results in the maternal-specific expression of imprinted genes. The fertilization event ceases the expression of *DME*, preventing further changes in DNA methylation. *DME* is not expressed in pollen. The expression of DME is essential in CCN, but not in pollen. Plants altered in *DME* in the maternal allele do not complete their development, whereas plants deficient for *DME* in the paternal allele develop normally. As a consequence, only the maternal expression is observed for the majority of imprinted genes. Wolff et al. (2011) identified 52 of 65 imprinted genes that were maternally expressed, whereas Gehring et al. (2011) found 165 of 208 maternally expressed genes.

Passing epigenetic memory to the progeny

As mentioned earlier, the transfer of epigenetic memory of stress encounters between generations requires escaping several rounds of reprogramming that occur during gametogenesis and post-fertilization. As gametes are formed relatively late during development, stress-induced signaling molecules, such as siRNAs and possibly miRNAs, might reach the developing gametes and interfere with the process of methylation/demethylation, histone repositioning, histone variant substitution, and other mechanisms.

The mobility of these small RNAs, together with their ability to establish differential patterns of methylation at distal tissues, was demonstrated by Molnar et al. (2010). Grafting of mutant plants impaired in the function of Dicer-like mutants 2, 3, or 4 onto wild-type plants allowed the authors to demonstrate a strong correlation between silencing at transposons sequences, their methylation, and the production of sequence-specific siRNAs. It was shown that the production of these mobile 24-nucleotide sRNAs required DCL3 and Pol IV (Molnar et al., 2010).

The authors proposed two hypotheses. First, 24-nt siRNAs are possibly involved in the response of meristems to transposon activation in somatic cells, allowing for the reinforcement of transposon silencing in pluripotent cells and giving rise to the next generation. Second, mobile siRNAs might pass the information about stress to gametes or mediate responses to stressors that initiate epigenetic changes in pollen and seeds. Exposure of Arabidopsis plants to several stresses, such as cold, heat, salt, and UVC, leads to transgenerational changes such as global genome hypermethylation, increased genome instability, and increased stress tolerance. The study showed that DCL2 and DCL3 are important for passing the information about stress to the progeny—**transgenerational** changes in the progeny of stressed plants were lost in *dcl2* and *dcl3* mutants (Boyko et al., 2010).

Changes in methylation may play a crucial role in the transmission of epigenetic memory to the progeny. Significant changes in DNA methylation were observed in genetically identical populations of apomictic dandelions exposed to several abiotic and biotic stresses (Verhoeven et al., 2010). The progeny inherited changes in methylation patterns. Moreover, additional methylation changes were also observed in the progeny. Changes in methylation might be sufficient to trigger the development of novel phenotypes. In *Linaria vulgaris*, hypermethylation of the *Lcyc* gene promoter leads to the formation of flowers with radial floral symmetry from bilateral flowers. In this case, changes in DNA methylation represent a true epimutation event that might facilitate co-evolution of plants and pollinators.

These studies and many other studies that are not mentioned here suggest the existence of heritable responses to stress triggered by stress-induced epigenetic changes in plants. Transgenerational inheritance is very important for both vegetatively and asexually propagated plants because the formation of epialleles might be a primary mechanism underlying population variability. Also, for sexually propagating plants, stress-induced epimutations have a chance to escape reprogramming and become epialleles, thus contributing to the diversity in the progeny.

Conclusion

There are many ways for a mammal to pass on epigenetic marks such as inheritance of histones, piRNAs, and methylation signatures. What all these epigenetic modifications that can be passed on as epigenetic memory have in common is their role in maintaining a proper development and their ability to escape near genome-wide reprogramming that occurs in the germline. Epigenetic inheritance plays a crucial role in maintaining important functions, such as the expression of genes involved in early embryo development, imprinting, and transposon silencing. Without epigenetic factors, such as DNA methylation and histone modifications, the development cannot proceed. Epigenetic inheritance might also lead to the inheritance of epimutations, which increase a risk of a disease. Now when we begin to understand the importance of epigenetic memory, it quickly becomes clear that inheritance is not simply a matter of Mendelian genetics, and our understanding of complex life-generating interactions is far from complete.

Exercises and discussion topics

1. Why is the somatic ciliate genome considered to be an "epigenome"?
2. Describe cortical inheritance in Paramecium.
3. Describe the RNA template-guided genome unscrambling in Oxytricha.
4. Give examples of ncRNA-mediated control over DNA rearrangements in ciliates.
5. Describe epigenetic modifications at PGCs.
6. Describe reprogramming of DNA methylation in development of male and female gametes in animals. Explain the first and second waves of the process.
7. What are the imprinted genes? Give examples in animals and plants.
8. Explain how epigenetic inheritance may arise.
9. Describe changes in the type of histones that associate with male and female gametes before and after fertilization in animals.
10. Describe the role of differentially methylated regions (DMRs) in epigenetic inheritance.
11. Explain the role of small RNAs in transgenerational inheritance in animals.
12. List and describe diseases that might have epigenetic nature.
13. Describe changes in DNA methylation in developing gametes in plants.
14. Give examples and describe imprinting in plants.
15. Explain the mechanism of transposon silencing in plant gametes.
16. Explain possible role of small RNAs in transgenerational inheritance in animals.

References

Abdalla et al. (2009) Active demethylation of paternal genome in mammalian zygotes. *J Reprod Dev* 55:356-60.

Arnaud et al. (2006) Stochastic imprinting in the progeny of Dnmt3L-/- females. *Hum Mol Genet* 15:589-98.

Balhorn R. (2007) "The protamine family of sperm nuclear proteins". *Genome Biol* 8:227

Bartolomei MS. (2009) Genomic imprinting: employing and avoiding epigenetic processes. *Genes Dev* 23:2124-2133.

Biliya S, Bulla LA Jr. (2010) Genomic imprinting: the influence of differential methylation in the two sexes. *Exp Biol Med (Maywood)* 235:139-47.

Bonasio et al. (2010) Molecular signals of epigenetic states. *Science* 330:612-6.

Boyko et al. (2010) Transgenerational adaptation of *Arabidopsis* to stress requires DNA methylation and the function of Dicer-like proteins. *PLoS One* 5:e9514.

Brennecke et al. (2008) An Epigenetic Role for Maternally Inherited piRNAs in Transposon Silencing. *Science* 322:1387-1392.

Combes A, Whitelaw E. (2010) Epigenetic reprogramming: Enforcer or enabler of developmental fate? *Develop Growth Differ* 52:483-491.

Dolinoy et al. (2007) Metastable epialleles, imprinting, and the fetal origins of adult diseases. *Pediatr Res* 61:30R-37R.

Edwards CA, Ferguson-Smith AC. (2007) Mechanisms regulating imprinted genes in clusters. *Curr Opin Cell Biol* 19:281-289.

Gehring et al. (2011) Genomic analysis of parent-of-origin allelic expression in Arabidopsis thaliana seeds. *PLoS One* 6(8):e23687.

Hajkova et al. (2008) Chromatin dynamics during epigenetic reprogramming in the mouse germ line. *Nature* 452:877-881.

Hemberger et al. (2009) Epigenetic dynamics of stem cells and cell lineage commitment: digging Waddington's canal. *Nat Rev Mol Cell Biol* 10:526-537.

Hitchins et al. (2007) Inheritance of a cancer-associated MLH1 germ-line epimutation. *N Engl J Med* 356:697-705.

Horsthemke B. (2010) Mechanisms of imprint dysregulation. *Am J Med Genet C Semin Med Genet* 154C:321-8.

Hsieh et al. (2011) Regulation of imprinted gene expression in Arabidopsis endosperm. *Proc Natl Acad Sci USA* 2011 108:1755-62.

Ingouff et al. (2007) Distinct dynamics of HISTONE3 variants between the two fertilization products in plants. *Curr Biol* 17:1032-7.

Ingouff et al. (2010) Zygotic resetting of the HISTONE 3 variant repertoire participates in epigenetic reprogramming in Arabidopsis. *Curr Biol* 20:2137-43.

Jablonka E, Raz G. (2009) Transgenerational epigenetic inheritance: prevalence, mechanisms, and implications for the study of heredity and evolution. *Q Rev Biol* 84:131-176.

Luo et al. (2011) A genome-wide survey of imprinted genes in rice seeds reveals imprinting primarily occurs in the endosperm. *PLoS Genet* 7:e1002125.

Molnar et al. (2010) Small silencing RNAs in plants are mobile and direct epigenetic modification in recipient cells. *Science* 328:872-875.

Morgan et al. (1999) Epigenetic inheritance at the agouti locus in the mouse. *Nat Genet* 23:314-318.

Nowacki et al. (2008) RNA-mediated epigenetic programming of a genome-rearrangement pathway. *Nature* 451:153-158.

Nowacki M, Landweber LF. (2009) Epigenetic inheritance in ciliates. *Curr Opin Microbiol* 12:638-43.

Ooi SL, Henikoff S. (2007) Germline histone dynamics and epigenetics. *Curr Opin Cell Biol* 19:257-265.

Popp et al. (2010) Genome-wide erasure of DNA methylation in mouse primordial germ cells is affected by AID deficiency. *Nature* 463:1101-1105.

Puri et al. (2010) The paternal hidden agenda: Epigenetic inheritance through sperm chromatin. *Epigenetics* 5:386-91.

Szabo et al. (2002) Allele-specific expression of imprinted genes in mouse migratory primordial germ cells. *Mech Dev* 115:157-160.

Tremblay et al. (1997) A 5' 2-Kilobase-Pair Region of the Imprinted Mouse H19 Gene Exhibits Exclusive Paternal Methylation throughout Development. *Mol Cell Biol* 17:4322-4329.

Verhoeven et al. (2010) Stress-induced DNA methylation changes and their heritability in asexual dandelions. *New Phytol* 185:1108-1118.

Wolff et al. (2011) High-resolution analysis of parent-of-origin allelic expression in the Arabidopsis Endosperm. *PLoS Genet* 7:e1002126.

Wu et al. (2011) Genes for embryo development are packaged in blocks of multivalent chromatin in zebrafish sperm. *Genome Res* 21:578-9.

17

Epigenetics of health and disease—cancer

Currently cancer is the second (after cardiovascular disease) leading cause of death in North America, and the number of new cases in the Canadian province of Alberta alone is expected to double to more than 26,000 cases per year by 2025. Worldwide, lung, stomach, liver, colorectal, and breast cancers are the top five types of cancer that cause the most deaths.

What is cancer?

Cancer is a disease involving deregulated cell growth. It arises through a lengthy multistep process that comprises disruption of growth regulatory pathways, resulting in uncontrolled cellular proliferation and altered apoptosis. Carcinogenesis, the process of cancer development, is initiated when a single cell harbors a mutation in an important gene. This mutation-harboring cell is considered to be "initiated," and the mutation-initiated status is largely irreversible. The clonal expansion of the initiated cell stimulated by different promoting agents, such as growth factors and hormones, leads to focal proliferation. This process is known as the promotional stage of carcinogenesis. Notably, the clonal expansion of initiated cells observed at the stage of cancer promotion is reversible if the promotion stimulating agent is removed. Further down the road, clonally expended cells exhibit genome instability, acquire further mutations, and turn into carcinoma cells over time. Further unregulated growth of cancer cells constitutes the progression of cancer. Malignant cells gain the capacity to invade surrounding tissue and metastasize to other tissues and organs. Cancer development is a lengthy process, and often years or decades can pass from the stage of initiation of a single cell to the detection of cancer and metastatic disease.

Multiple "hits" to DNA are necessary to cause cancer. Analyzing the patterns of inherited retinoblastoma in children, Knudson (1971) concluded that the first insult is inherited in the DNA, and the second insult rapidly leads to cancer development. In noninherited retinoblastoma, two "hits" have to take place before a tumor can develop. This theory holds true for all types of cancer and is known as Knudson's "two-hit" cancer hypothesis. Thus, according to Knudson's two-hit theory of carcinogenesis, the loss of function of a tumor suppressor gene requires the complete loss of function of both copies of the gene.

Although the role of genetic changes in cancer has been well studied and established, classic genetics alone cannot apparently account for the differences in cancer susceptibility and cancer manifestation. Indeed, a number of studies have found contradictory results on the incidence of cancer between monozygotic twins who have identical genomes (Poulsen et al., 2007). Currently, cancer is well accepted to be an epigenetic disease, and the role of epigenetic changes in the development of cancer has been extensively studied.

Overall, epigenetics plays an important role in cancer, and heritable changes in DNA methylation, histone modifications, and small RNAs are intricately linked to the initiation, promotion, and progression of cancer.

Importantly, Knudson's two-hit theory of carcinogenesis (Knudson, 1971 and 1996) can also be applied to epigenetic changes (Deng et al., 2010). Indeed, the concept that tumor suppressor genes must be inactivated on both alleles can be interpreted not only in the sense of genetic mutations but also due to epigenetic modifications (see Figure 17-1). For example, DNA hypermethylation and altered chromatin modification of both alleles, or in combination with genetic mutations, have been proven to abolish gene expression, thereby providing growth advantages during cancer development. This chapter discusses the key roles that epigenetic changes play in cancer.

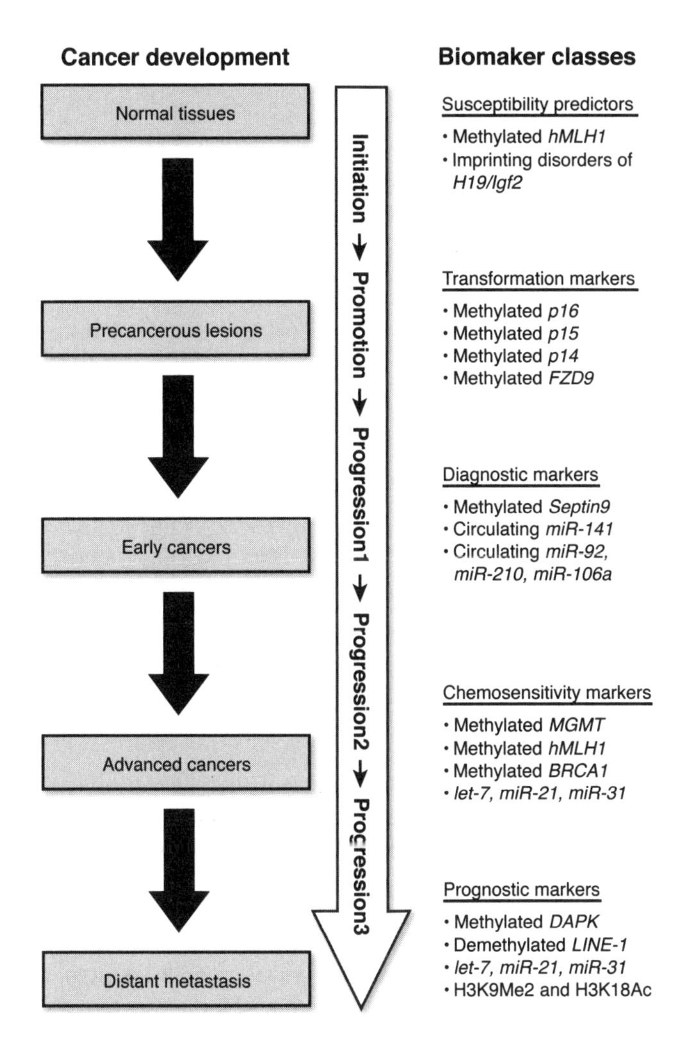

Figure 17-1 Illustration of the multiple stages of cancer development and the epigenetic biomarker classes used in the corresponding stages. (Re-printed with permission from Dajun Deng, Zhaojun Liu, and Yantao Du. Epigenetic Alterations as Cancer Diagnostic, Prognostic, and Predictive Biomarkers. Advances in Genetics, Volume 71, 2010, Pages 125-176.) The clinical and pathologic changes during carcinogenesis are displayed in the rectangles. The long gradient dark arrow indicates that carcinogenesis is a long-term, continuous process. The representative candidates of each class of epigenetic biomarker are listed along the process. Different class biomarkers overlap with each other.

DNA methylation in cancer

DNA methylation is one of the key epigenetic modifications in mammalian genomes and is one of the most studied epigenetic processes (see Chapter 4, "DNA Methylation as Epigenetic Mechanism"). It usually takes place at the 5' position of the cytosine ring within CpG dinucleotides. These are regions of DNA where a cytosine nucleotide occurs next to a guanine nucleotide in the linear sequence of bases along its length. CpG stands for cytosine and guanine separated by a phosphate, which links the two nucleosides together in DNA. Although roughly 70% of CpG residues are methylated, their distribution within the genome is non-random, and the vast majority of the genome is CpG poor (Robertson and Jones, 2000). Interestingly, certain genomic regions possess high frequencies of CpGs and are therefore termed CpG islands. CpG islands commonly occur at the 5' regulatory regions of genes and are usually protected from methylation in normal cells. This condition allows for the important housekeeping genes to be expressed, notwithstanding some of them are methylated in a tissue-specific manner during early development and differentiation, thus controlling the proper gene-expression patterns. In sum, DNA methylation gives rise to normal cell- and tissue-specific gene-expression patterns. Therefore, it regulates important cellular processes, such as cellular differentiation, proliferation, apoptosis, and others (reviewed in Kulis and Esteller, 2010).

In normal noncancerous cells, DNA methylation occurs mainly in repetitive genomic regions and is used to silence and control repetitive satellite DNA and parasitic elements (such as long interspersed nuclear elements (LINEs), short interspersed nuclear elements (SINEs), and other transposable elements and sequences of viral origin). Thus, DNA methylation is important to safeguard genomic integrity. Furthermore, in normal cells, DNA methylation controls imprinting and X-chromosome inactivation in females (Kanwal and Gupta, 2010).

DNA methylation is catalyzed by DNA methyltransferases (DNMTs). DNMT1 is a maintenance DNA methyltransferase that copies the existing methylation patterns following DNA replication, whereas DNMT3A and DNMT3B are *de novo* methyltransferases that interact with unmethylated CpGs to establish new methylation patterns. DNMT-3L enzyme lacks intrinsic methyltransferase activity, but it interacts with DNMTs 3a and 3b to establish DNA methylation of transposable elements (Kanwal and Gupta, 2010).

Methylated DNA is a binding target for methyl-CpG-binding domain (MBD) proteins. These proteins interact with methylated DNA and help recruit histone modifying enzymes and chromatin-remodeling complexes to methylated sites DNA. To date, six methyl-CpG-binding proteins have been identified in mammals: MECP2, MBD1, MBD2, MBD3, MBD4, and Kaiso (Kanwal and Gupta, 2010).

Interestingly, the expression of methyltransferases and MBD proteins is often altered in cancers. A list of methylation machinery proteins affected in cancers is given in Table 17-1.

Table 17-1 Changes in the expression of DNA methyltransferases and Methyl-CpG-binding proteins in human cancers. Modified with permission from Kanwal R, Gupta S. Epigenetics and cancer. J Appl Physiol. 2010;109(2):598-605.

Gene/protein group	Gene/protein name	Type of change	Cancer type
DNA methyltransferases	DNMT1	upregulation, mutation	colon, ovarian
	DNMT3a	upregulation	colorectal, breast, ovarian, esophageal
	DNMT3b	upregulation	breast, hepatocellular-carcinoma, colorectal
Methyl-CpG-binding proteins	MeCP2	upregulation, mutation	prostate
	MBD1	upregulation, mutation	prostate, colon, lung
	MBD2	upregulation, mutation	prostate, colon, lung
	MBD3	upregulation, mutation	colon, lung
	MBD4	upregulation, mutation	colon, gastric, endometrial
	Kaiso	upregulation, mutation	colon, intestinal, lung

For decades, DNA methylation patterns in cancer cells have been recognized to significantly differ from those of normal cells. Both hypo- and hypermethylation events are observed in cancer.

Indeed, cancer cells exhibit **global genome hypomethylation** as well as a concurrent **hypermethylation of the promoter regions** of certain genes, which are usually tumor suppressor genes. Additionally, **genomic imprinting** is affected in cancer cells. The examples, mechanisms, and cellular consequences of the aberrant DNA methylation patterns in cancer cells are discussed throughout this chapter.

DNA hypomethylation in cancer

Some of the initial lines of evidence that the DNA of cancer tissues is significantly hypomethylated were provided by the experiments of Lapeyre and Becker who showed a significant decrease in the levels of 5-methylcytosine in rat hepatocellular carcinomas compared with normal rat liver tumors. Their experiment proved that the level of genome methylation in experimentally induced rodent liver cancer induced by exposure to cancer-causing agents aminofluorene and diethylnitrosamine is significantly lower than that of the non-cancerous liver tissue (Lapeyre and Becker, 1979).

In a key pioneer study, Feinberg and Vogelstin (1983) showed that a substantial portion of CpGs usually methylated in normal tissues is unmethylated in cancer cells. Ehrlich and colleagues further substantiated these studies by showing a reduction in the global 5-me-C content of tumors (Gama-Sosa et al., 1983). Importantly, loss of DNA methylation was seen in every tumor type studied. Later, reviewing the state of knowledge on DNA hypomethylation in cancer, Feinberg and colleagues (2006) stated that “although individual genes vary in hypomethylation, all tumors examined so far, both benign and malignant, have shown global reduction of DNA methylation. This is a striking feature of neoplasia.” Almost all major cancers, such as breast, prostate, liver, lung, colon, gastric, and ovarian cancers, are currently accepted to be characterized by global genome hypomethylation. Overall, cancer cells lose a significant proportion of their 5-methylcytosine content compared with noncancerous cells. The most significant demethylation was observed in the coding regions and introns of genes and in repetitive DNA sequences.

Loss of methylation of repetitive sequences in cancer

In normal cells, pericentromeric areas are highly methylated. Genomic loci such as satellite sequences, long and short interspersed nuclear elements, and other repetitive genomic sequences are also heavily methylated and silenced to prevent their expression and movement, thus ensuring genomic stability. In a variety of tumors, this silencing mechanism is disrupted, and the aforementioned loci are hypomethylated (Kulis and Esteller, 2010). Hypomethylation of repetitive DNA sequences might lead to increased levels of aberrant somatic recombination, reactivation of transposons, and consequently increased mutation, aneuploidy, and general chromosomal and genome instability, which are known and well-established hallmarks of cancer (Pogribny and Beland, 2009).

The research of Erlich and colleagues linked hypomethylation in cancer and chromosomal instability. This hypomethylation was noted in pericentromeric satellite repetitive sequences and was associated with unbalanced chromosomal translocations and breakpoints in the pericentromeric DNA of chromosomes 1 and 16 (James et al., 1999), which were often seen in ovarian and breast cancers and Wilms tumors.

Furthermore, DNA hypomethylation in cancer might reactivate previously silent latent viral sequences in the genome. For example, the hypomethylation and reactivation of human papillomavirus (HPV) were reported in HPV-positive cervix cancer in humans (Badal et al., 2003). Epstein-Barr virus, which is important in lymphoma, may also be controlled by hypermethylation; loss of methylation control might lead to its reactivation and lymphomogenesis (Takacs et al., 2010).

Juxtacentromeric Sat2 and centrameric Satα repeats are known to be hypomethylated in ovarian cancer samples. Hypomethylation is also more prevalent in more aggressive high-state/grade tumors. Similarly, Sat hypomethylation was also reported to occur in breast cancer (Costa et al., 2006).

Hypomethylation of LINE L1 elements is frequently observed in hepatocellular carcinoma, bladder, head and neck, lung, prostate, breast, and esophagus cancers, in uroepithelial and renal cell carcinomas, and other tumors. Loss of L1 methylation and reactivation of L1 have been proposed as surrogate markers for cancer-linked genome demethylation and genome instability (Kulis and Esteller, 2010).

Hypomethylation of individual genes in cancer

Although DNA hypomethylation of individual genes in cancer is relatively uncommon, some single-copy genes might be hypomethylated and therefore aberrantly expressed in cancers. Many of the aberrantly hypomethylated genes are **oncogenes**, the expression of which promotes cell growth and malignant transformation of the cells. Hypomethylation has been reported in the H-ras oncogene in colon and lung cancer. The R-ras oncogene is often hypomethylated and activated in gastric cancers. Demethylation of the 5' end and aberrant expression of the Bcl-2 gene were reported in chronic lymphocytic leukemia (reviewed in (Wilson et al., 2007).

Furthermore, the genes involved in **invasion and metastases** are often found to be hypomethylated and overexpressed in cancer. Similar to oncogenes, these genes are usually methylated and silent in normal cells. Among these genes, urokinase-type plasminogen activator is hypomethylated and expressed in the progression and metastasis of breast and prostate cancers. Aberrant expression and hypomethylation of maspin, another gene involved in tumor invasion and metastasis, have been linked to the progression of thyroid, gastric, gall bladder, colorectal, and pancreatic cancers. In colorectal cancer, maspin expression is the highest at the tumor invasion front (Kulis and Esteller, 2010).

Aberrant hypomethylation might lead to alterations in the tissue specificity of gene expression. For example, gamma synuclein, a gene usually expressed only in the brain, is often demethylated and aberrantly expressed in breast, ovarian, liver, esophageal, cervix, prostate, and other cancers.

Indeed, many of the promoters affected by loss of DNA methylation in cancer belong to tissue-specific genes. Among these, a family of cancer/testis antigens, the expression of which is restricted to adult testicular germ cells under normal conditions, is often aberrantly hypomethylated and activated in various human cancers (Caballero and Chen, 2009). In colon cancer and melanoma, melanoma-associated CT antigens MAGE are hypomethylated and reactivated (reviewed in Pogribny and Beland, 2009). Recent advances in whole genome-based DNA methylation analysis have led to the identification of genes hypomethylated in various tumors and in different stages of carcinogenesis. Aberrant expression of hypomethylated genes gives tumor cells selective advantage over normal cells and leads to the deregulation of cell cycle control, tissue invasion, and metastasis.

Loss of imprinting due to hypomethylation in cancer

One key role of DNA methylation in normal cells is the maintenance of genomic imprinting, a parent-of-origin specific allele inactivation, of a small number of genes (approximately 90 genes in humans). Loss of imprinting (LOI) and loss of monoallelic regulation of the expression of imprinted genes are frequent occurrences in cancer. These epigenetic alterations are currently considered as the most frequent ones in cancer. LOI was shown to affect the insulin-like growth factor-II (IGF2) gene. Initially, IGF2 LOI was attributed to increased DNA methylation, but a number of subsequent studies have proven that DNA hypomethylation is a key reason for IGF2 LOI in many cancers, such as liver, bladder, breast, and colon cancers. Altered imprinting due to hypomethylation has also been observed in the H19 gene in colon and lung cancers (for detailed reviews, see Pogribny and Beland, 2009).

The extent of DNA hypomethylation and the state and grade of cancer have a known association. Hypomethylation is an early event during the early preneoplastic stages of carcinogenesis; it occurs before the malignant transformation of cells and accumulates with cancer progression.

Mechanisms of DNA hypomethylation in cancer

The precise mechanisms of DNA hypomethylation in cancer have not been established. The overall loss of DNA methylation has been attributed to a number of factors. Research evidence strongly suggests that the accurate functioning of DNMTs is absolutely essential for the maintenance of DNA methylation. Reducing the expression of DNMTs either individually or in combination leads to significantly reduced global methylation levels. For example, in mice, the reduction of DNMT1 expression results in the significant hypomethylation of centromeric sequences and endogenous retroviral intracisternal A particle repetitive sequences. Likewise, the loss of DNMT function due to inhibition of its activity leads to the rapid demethylation of DNA.

The integrity of genomic DNA is another important factor that affects DNA methylation. Living organisms are exposed to various endogenous and exogenous factors that damage DNA. The results of several studies have shown that the presence of DNA damage substantially affects the methylation capacity of DNA methyltransferases and leads to DNA hypomethylation. Furthermore, during repair, DNA polymerases incorporate cytosine but not methyl-cytosine. Therefore, the presence of DNA lesions and the activation of DNA repair mechanisms might also contribute to DNA hypomethylation.

Alternatively, the active removal of DNA methyl groups by demethylating enzymes can considerably influence global genomic methylation. Several proteins have been suggested to possess DNA demethylase activity. These are RNA-dependent 5-meC glycosylase, a ribozyme-like demethylase, and MBD2. Two recent studies also suggested that DNMT3A and DNMT3B might have some demethylase activity and act as DNA demethylases. Interestingly, all these putative DNA demethylase proteins are characterized by quite different DNA demethylating mechanisms (reviewed in Pogribny and Beland, 2009). However, despite the loss of DNA methylation in tumors, no actively demethylating enzyme has been unequivocally established.

Recent lines of evidence suggest the possible roles of the SNF2 family of ATP-dependent helicases in maintaining DNA methylation. Indeed, Lsh, a member of the SNF2 family, is important for DNA methylation and is aberrantly expressed in cancer, leading to DNA hypomethylation. Aberrant splice variants of *de novo* methyltransferase DNMT3b have been suggested to play a part in DNA hypomethylation (reviewed in Pogribny and Beland, 2009).

Regardless of the mechanism involved, the loss of genomic DNA methylation can be observed early in tumor development and may be a cause, not merely a consequence, of malignant transformation.

DNA hypomethylation in cancer development

Accumulating evidence suggests that DNA hypomethylation might be important in cancer development and cancer progression. Although DNA hypomethylation is well documented in cancer, its role in cancer development is still being actively investigated. The extent of global DNA hypomethylation and the grade and stage of cancer are strongly associated. DNA methylation has been suggested as an excellent biomarker for the diagnosis and prognosis of cancer. However, a decrease in DNA methylation in cancer might not be a causative factor in tumor development but can merely be a consequence of a transformed state reflecting the undifferentiated state of cancers.

Thus, in providing conclusive evidence that hypomethylation plays a key part in cancer development, the following must be demonstrated:

1. DNA hypomethylation occurs at the very early preneoplastic stages of carcinogenesis.
2. DNA hypomethylation changes seen at preneoplastic stages are also present during later stages of cancer development.
3. Additional DNA hypomethylation changes are acquired during tumor progression.
4. Most importantly, a mechanistic link between the hypomethylation of cellular genomic DNA and cancer development exists (Pogribny and Beland, 2009).

By now, the results of numerous studies have provided solid evidence that DNA hypomethylation occurs at premalignant pathological states or at early preneoplastic stages of carcinogenesis. Furthermore, cancerous tumors have been proven to harbor a much greater degree of hypomethylation compared with preneoplastic lesions and that cancer progression from normal tissue to stage IV of the disease is associated with cumulative methylation change in various types of tumors. For instance, elegant studies using a mouse model of multistage skin carcinogenesis have proven that a significant decrease in the 5-meC content of tumor cells is associated with tumor progression and is correlated with the degree of cancer invasiveness (Fraga et al., 2004).

These lines of evidence suggest that loss of DNA methylation in cancer is not a secondary event but might be a driving force in cancer development and progression.

Furthermore, targeting DNMT1 and LSH results in significant DNA hypomethylation and tumor induction in experimental animal models. These data, together with the evidence on the reactivation of previously silent transposons and genome destabilization upon loss of DNA methylation, provide the mechanistic links between DNA hypomethylation and cancer induction and progression. Recent data on the role of DNA hypomethylation in carcinogenesis were reviewed by Pogribny and Beland (2009).

DNA hypomethylation is also considered to be a mechanism involving drugs, toxins, and viral effects in cancer. Hypomethylation and the aberrant expression of the multidrug resistance protein gene have been shown to result in a significant resistance to anti-tumor agents.

DNA hypermethylation in cancer

Along with global and gene-specific DNA hypomethylation, cancers exhibit a significant degree of DNA hypermethylation in the control regions of genes. DNA hypermethylation might effectively prevent the genes from being expressed, either by directly interfering with the binding of RNA polymerase or via the formation of a repressive chromatin state over methylated promoters.

Mutagenic potential of DNA hypermethylation

DNA hypermethylation is mutagenic. Methylated cytosines (5-mC) are very prone to mutations by deamination. These deamination events lead to C→T transitions. The opposite strand of CpG motifs might also be affected, leading to G→A changes. Thus, CpGs constitute important, often critical, hotspots for mutations. Such G-A transitions have been found in approximately 45% of leukemia and myelodysplasia cases and in about 60% of colon cancers. C→T and tandem CC-TT mutations have been reported to occur in basal cell and squamous cell carcinomas. Overall, approximately one-third of germline point mutations underlying varied human genetic diseases occur at CpGs, and the vast majority of these mutations are C→T mutations.

Oxidation of 5-mC might also lead to C→T transitions at CpG sequences. Indeed, reactive oxygen species, especially oxygen radicals, can react with 5-mC and oxidize the 5, 6-double bond, leading to the formation of 5-mC glycol that can further be deaminated to form thymine glycol. Therefore, 5-mC residues are considered to be "hotspots" for mutations, which can destabilize gene structure and function (Jones and Baylin, 2002).

DNA hypermethylation and gene inactivation

DNA hypermethylation in tumors is considered the key mechanism of epigenetic gene inactivation. The pioneering report on such hypermethylation was made by Baylin and colleagues who proved hypermethylation and silencing of a calcitonin gene in cancer (Baylin et al., 1986).

The first key evidence on the hypermethylation and inactivation of a tumor suppressor gene in cancer was provided by Greger and colleagues whose pioneer work showed a significant hypermethylation of the promoter of retinoblastoma (RB) tumor suppressor gene in retinoblastoma patients (Greger et al., 1989). After this initial discovery, numerous aberrant hypermethylated genes were identified in cancer. Each cancer is characterized by its specific methylome.

Hypermethylation at specific genes typically affects promoter CpG-islands and shuts down the transcription of the affected genes. Hypermethylation is observed at specific CpG islands.

Genes that are hypermethylated and thus transcriptionally inactivated in cancers include those that are involved in key cancer-associated cellular pathways, such as cell cycle control (RB, $p16_{INK4a}$, $p15_{INK4b}$ CDKN2A, APC), Ras signaling (NOREIA, RASSF1A), DNA repair (MLH1, MGMT, BRCA1, WRN), tumor-suppressor genes, apoptotic cell death (death-activated protein kinase, TP73, PYCARD, WIF-1, SFRP1), hormone response (ESR1, ESR2), metastasis (CDH1, CDH13, PCDH10), vitamin response (RARB2, CRBP1), detoxification (GSTP1), and p53 network (p14ARF, p73, HIC-1). Hypermethylation often affects transcription factors (GATA4, GATA5, ID4, RARβ2) and tumor suppressor genes (von-Hippel-Lindau tumor suppressor, e-cadherin), to name a few.

One specific example of methylation altering the cell cycle is the hypermethylation of the cell-cycle inhibitor $p15_{INK4b}$ in leukemias. Another important and frequent methylation-mediated silencing event is that of the cell-cycle inhibitor gene $p16_{INK4a}$ in a number of primary tumors. The cyclin-dependent kinase inhibitor 2A (CDKN2A, $p16_{INK4a}$) is responsible for the maintenance of the protein RB in an active and non-phosphorylated state. The cyclin D-RB pathway-mediated control of the cell cycle is often lost in human cancers. Often, this loss is due to $p16_{INK4a}$ hypermethylation. The most mutated tumor suppressor gene in most human cancers is p53. The protein p53 is a key factor involved in many critical cellular functions, including growth arrest, programmed cell death, and DNA repair. The promoter region of the p53 gene contains 21 CpG dinucleotides, constituting 8.2% of the p53 domain (Laird and Jaenisch, 1996). The high concentration of CpG dinucleotides makes the p53 gene very susceptible to changes in methylation patterning. Many types of tumors exhibit hypermethylation of the p53 gene or methylation-mediated silencing of the $p14_{ARF}$ gene, the protein product of which indirectly inhibits p53 degradation (Robertson and Jones, 1998). Evidence has shown that many other important genes involved in the correct functioning of p53 and normal cell differentiation experience methylation-mediated silencing (Esteller and Herman, 2002). Another hypermethylated gene seen in a number of cancers is p73. The p73 gene encodes a protein similar to p53 and is involved in cell cycle regulation and induction of apoptosis.

Aberrant hypermethylation leads to the silencing of the aforementioned genes and therefore significantly affects the pathways in which they partake. Transcriptional silencing of DNA repair and cell cycle control appears to escalate during the progression toward malignancy. The silencing of genes regulating cell adhesion and motility also enables tumor cells to break away from the primary tumor and to metastasize. Interestingly, however, the loss of such gene products is not always advantageous and might in fact interfere with the tumor's ability to survive in a new location.

Hypermethylation of miRNA genes

Small RNAs, especially microRNAs, have been proposed to play important roles in cancer. Their expression is frequently deregulated in cancers, and they have been extensively studied as potential cancer biomarkers. Expression of several miRNAs can be regulated by DNA methylation. MicroRNAs acting as tumor suppressors might be silenced by hypermethylation. Among these, miR-124a is inactivated by hypermethylation in cancer (Agirre et al., 2009). This hypermethylation mediates the activity of miR-124 on its target, which is CDK6. As a result, RB is inactivated by CDK6-mediated phosphorylation. In bladder cancer, hypermethylation of miR-127 leads to the aberrant expression of oncogenic BCL-6 factor. In breast cancer cells, miR-

34c down-regulation via DNA methylation promotes self-renewal and epithelial-mesenchymal transition. MicroRNAs 34b and 34c are frequently methylated in small-cell lung cancer (Tanaka et al., 2012). In gastric cancer, tumor-specific methylation silences miR-34b and miR-129. MiR-34b is also hypermethylated and repressed in melanoma. Epigenetic DNA methylation-based silencing was shown to affect the miR-200 family (Davalos et al., 2011). The silencing of the miR-200 family promotes epithelial-mesenchymal transition in cancer cells.

In ovarian cancer, tumor suppressor miR-152 is silenced. DNA methylation and silencing miR-152 were consistent with its location at 17q21.32 in intron 1 of the COPZ2 gene, which is also often silenced in endometrial cancer by DNA hypermethylation (Tsuruta et al., 2011). MiR-375 is epigenetically silenced in melanoma. Interestingly, whereas only minimal CpG island methylation was observed in melanocytes, keratinocytes, normal skin, and nevi, very significant hypermethylation and reduced expression of miR-375 were noted in patient tissue samples from primary, regional, distant, and nodular metastatic melanoma. Additionally, ectopic expression of miR-375 inhibited melanoma cell proliferation, invasion, and cell motility, as well as induced cell shape changes. These results strongly suggest that miR 375 might have an important function in the development and progression of human melanomas (Mazar et al., 2011). Furthermore, miR-375 might constitute an important target for the development of therapeutic approaches.

MiR-345 is a methylation-sensitive microRNA involved in cell proliferation and invasion in human colorectal cancer. Other microRNAs affected by DNA methylation include miR-132 (pancreatic cancer), miR-143 (hepatocellular carcinoma), miR-203 (multiple myeloma), miR-9 (gastric cancer), and others. The interplay between DNA methylation and miRNAs might be important in the etiology and pathogenesis of many tumors.

Possible mechanisms of hypermethylation in cancer

Although more and more genes and miRNAs are being identified as hypermethylated and aberrantly silenced in cancer, the mechanisms of DNA hypermethylation remain unclear. Current studies support a hypothesis on the non-random DNA methylation process. Indeed, every cancer type has a unique methylation profile. Some of the hypermethylated genes are seen in many cancers, whereas many of them are cancer- and cell-type specific. Therefore, imagining that aberrant DNA methylation simply occurs randomly would be difficult. Furthermore, most recent studies suggest that there is an interplay between polycomb (PcG) proteins and DNA methylation events, which might lead to the establishment of a specific methylation pattern of target sequences. EZH2, a PRC2 component, has been shown to interact with DNMTs at some gene promoters (for example, TBX3 in breast cancer, HOXD10, SIX3, KCNA1, and MYT1 in osteosarcoma). Thus, PcG proteins might mark some specific sequences for methylation during cancer. In sum, DNA methylation-associated silencing is well established to play a crucial role in tumorigenesis and is a hallmark of all types of human cancer.

DNA methylation as a biomarker

Global and gene-specific DNA methylation levels constitute excellent potential biomarkers for early tumor detection, diagnosis, and, most importantly, prediction of prognosis and treatment responses. Recent studies have proven that DNA methylation signatures affect overall survival, tumor recurrence, and progression (reviewed in Deng et al. 2010).

Cancer methylome

Three key strategies exist for the detection and analysis of DNA methylation:

- Digestion of DNA with methylation-sensitive or insensitive restriction enzymes
- Chemical modification of DNA by bisulphite
- Enrichment and purification of the methylated DNA fraction of the genome using antibodies against methylated cytosine

These approaches may be combined with DNA fingerprinting, array-based profiling, or sequencing. They serve as invaluable tools for the in-depth analysis of cancer methylome. The advent of new technologies has facilitated the generation of comprehensive genome-wide profiles of DNA methylation in cancers. In the future, large-scale comprehensive analyses of cancer methylomes will shed new light on the etiology and pathogenesis of cancer. The field of cancer epigenomics is still at an early stage, and new technologies are still being developed and applied to analyze cancer epigenomes. Colon cancer methylome was the first to be analyzed using the comprehensive high-throughput array-based relative methylation assay. The comprehensive profiling of 13 colon cancers and matched normal mucosa from the same individuals led to the identification of 2707 cancer-specific, differentially methylated regions. Of these, 56% were hypermethylated and 44% were hypomethylated. Comprehensive methylome analyses have been conducted for testicular cancer, hepatocellular carcinoma, melanoma, lymphoma, breast cancer, and several other tumors. These studies facilitated the establishment of novel tumor-specific methylation signatures and the definition of which pathways are epigenetically reprogrammed in tumors. Recent advancements in cancer methylome profiling have been reviewed by Lechner and colleagues (Lechner et al., 2010).

17-1. Changes in DNA methylation in cancer

DNA hypomethylation in cancer

- Overall levels vary within and between cancer types.
- Global DNA hypomethylation may be an early event in carcinogenesis.
- In some cancers (for example, hepatocellular carcinoma), the degree of hypomethylation increases with cancer progression.
- Areas affected by hypomethylation include repeated DNA sequences (satellite DNA, transposable elements).
- Consequences of DNA hypomethylation in cancer include transcriptional activation of promoters of repeat DNA sequences and retroviruses, chromosomal instability, insertional mutagenesis, aberrant recombination, loss of imprinting, altered expression of small RNAs.
- Cancer-causing agents might cause DNA hypomethylation.

Hypermethylation in cancer

- Levels vary between cancers and cancer-specific methylation signatures and may be used as diagnostic tools.
- Hypermethylation leads to inactivation of cell cycle control, DNA repair, and apoptotic processes.
- Hypermethylation might affect microRNAs leading to aberrant expression and function of miRNA targets.
- Hypermethylation in cancer is non-random.

Histone modifications and chromatin in cancer

Histone modifications in normal cells and in cancer cells

Histone modifications—which are comprised of acetylation, methylation, phosphorylation, ubiquitination, and sumoylation—are important in the regulation of transcription and overall genome stability (see Chapter 5, "Histone Modifications and Their Role in Epigenetic Regulation"). Covalent modifications of histone tails regulate numerous cellular chromatin-based processes, such as transcription, DNA repair, and DNA replication. Histone modifications can lead to either transcriptional activation or transcriptional repression, depending on the pattern of modifications to the amino acid residues on histone tails. Overall, histone modifications and chromatin modifiers are responsible for establishing either an "open" permissive chromatin or a "closed" non-permissive one.

Histone modifications control various key cellular processes, such as DNA repair, transcription, DNA replication, and the cell cycle. Deregulation of histone modifications might shift the balance of gene expression, leading to alterations in transcription, DNA repair, proliferation, and apoptosis, resulting in oncogenic transformation and cancer development.

In "normal" cells, genomic regions that include the promoters of tumor-suppressor genes are enriched in histone modifications associated with active transcription, such as the acetylation of H3 and H4 lysine residues (for instance, K5, K8, K9, K12, and K16) and the trimethylation of K4 of H3. Along with those, repetitive DNA areas and other heterochromatic regions are characterized by the trimethylation of K27, the dimethylation of K9 of H3, and the trimethylation of K20 of H4, which function as repressive marks (Sawan and Herceg, 2010).

Changes in histone modifications have been reported in various cancer types. In general, cancer cells are characterized by the loss of the "active" histone marks on tumor-suppressor gene promoters and by the loss of repressive marks, such as the trimethylation of K20 of H4 or the trimethylation of K27 of histone H3 at subtelomeric DNA and other DNA repeats. This leads to a more "relaxed" chromatin conformation in these regions.

So far, relatively little is known about the patterns of histone modification disruption in human tumors. Imbalances between histone acetylation and deacetylation have been detected in various tumors. Whereas hypoacetylation has been associated with the silencing of tumor suppressors, hyperacetylation has been linked to a more active transcriptional state that might be associated with uncontrolled proliferation and the activation of oncogenes. It is widely believed that the abnormal state of histone acetylation is associated with the downregulation of HATs (histone acetyl transferases) and the overexpression of HDACs (histone deacetylases) reported in breast cancer.

In 2005, Fraga and colleagues demonstrated that the loss of histone H4K16 acetylation and H4K20 trimethylation constitutes an important hallmark of human cancers (Fraga et al., 2005).

The levels of the methylation of H3K4, the dimethylation of H3K9, the trimethylation of H3K9, and the acetylation of H3 and H4 are significantly reduced in prostate cancer tissue as compared to normal prostate tissue. Decreases in the levels of the dimethylation of H3K4 and H3K9 or H3K18 acetylation levels were shown to be important molecular markers of pancreatic, lung, and kidney cancers (Kanwal and Gupta, 2010).

Histone modifications play important roles in lung cancer. The significant hyperacetylation of H4K5 and H4K8 and the hypoacetylation of H4K12 and H4K16 have been shown to be a characteristic feature of non-small cell lung carcinoma (Van Den Broeck et al., 2008). Interestingly, the loss of H4K20 trimethylation was found to be more frequent in lung squamous cell carcinoma than in adenocarcinoma. The levels of H4K20 methylation might serve as potential epigenetic biomarkers of squamous cell carcinoma. Levels of histone modifications also constitute important predictive factors in lung cancer. Patients with tumors harboring high levels of histone H3K4 dimethylation or low histone H3K9 acetylation levels have favorable survival prognoses (reviewed in Sawan and Herceg, 2010).

Histone-modifying enzymes in cancer

Several studies have attempted to identify the roles of histone-modifying proteins in cancer, and the aberrant expression and activity of numerous histone-modifying enzymes has been found in cancer cells.

Using a knockdown model of the HDAC2 protein, investigators observed an increased induction of apoptosis in cervical carcinoma cells. This provides evidence that HDAC2 might indirectly suppress apoptosis through deacetylating histone residues, promoting tighter histone and DNA binding. This tighter binding might suppress the transcription of apoptotic proteins and lead to carcinogenesis. Another knockdown model showed that the knockdown of HDAC1 or HDAC3 suppressed the growth and survival of cervical carcinoma cells. This study revealed that HDAC1 and HDAC3 indirectly promote cell growth and increase the survival of these carcinogenic cells (Kanwal and Gupta, 2010).

Altered levels of and mutations in histone acetyltransferases, such as p300, CBP, pCAF, MOZ, and MORF, have been reported in colon, stomach, endometrial, and lung cancers as well as hematological malignancies. The enhancer of zeste homolog 2 (EZH2), an H3K27-specific methyltransferase, provides an important novel connection between altered histone methylation and carcinogenesis. EZH2 is frequently found to be overexpressed in various solid tumors, including prostate, breast, colon, skin, and lung cancer. RNAi-based suppression of EZH2 significantly diminished breast and prostate tumor growth. On the other hand, the overexpression of EZH2 leads to increased invasiveness and immortalization of mammary epithelial cells.

Mixed lineage leukemia (MLL) and EZH2 catalyze the addition of methylation to H3K4 and H3K27, respectively, which are arguably two of the most important histone methylation marks. *MLL* rearrangement and the deregulation of *EZH2* are among the most common mutations in leukemia and solid tumors, respectively. Overall, histone methyltransferases, MLL1, MLL2, MLL3, NSD1, EZH2, and RIZ1, were shown to be altered in multiple malignancies. Furthermore, the gene encoding for the histone demethylase GASC1 is often amplified in squamous cell carcinoma (Kanwal and Gupta, 2010).

The deregulation or mutations of the recently identified H3K4- or H3K27-specific histone demethylases have been observed in solid tumors. However, their involvement in cancer development and the underlying mechanisms are largely unclear. Thus, histone modification proteins might have a significant role in some types of cancers. Most recent studies have begun to dissect the roles of the mis-establishing, mis-reading, and mis-erasing of various histone marks in cancer (reviewed in Sawan and Herceg, 2010).

Polycomb proteins and cancer

Moreover, polycomb group (PcG) and trithorax group (TrxG) proteins have recently become a new focus in cancer biology research. These proteins affect the covalent modification of histone tails, the position or composition of nucleosomes, and DNA methylation, thereby affecting chromatin structure and transcriptional activity. PcG proteins are transcriptional repressors, whereas TrxG proteins are transcription activators (see Chapter 2, "Chromatin Dynamics and Chromatin Remodeling in Animals"). The balance between PcG and TrxG proteins affects the expression of genes that induce cellular senescence, a tumor-suppressive mechanism that opposes cellular proliferation. PcG-TrxG-mediated chromatin dynamics affect the expression of genes that regulate apoptosis, and PcG-TrxG complexes regulate genome stability. The gain of PcG and the loss of TrxG functions were reported in various human cancers.

It has been recently suggested that there might exist a putative "targeting" mechanism for aberrant DNA methylation of certain genes in cancers, and the genes normally repressed by PcG proteins are especially susceptible to aberrant DNA methylation in cancers. This PcG-based model could also be used to explain the tissue-specificity and differences in DNA methylation targets between various cancers, especially considering that PcG targets genes that are also regulated in a tissue-specific manner.

Another important component of promoter activation and inactivation is due to chromatin remodeling, specifically the SWI/SNF complex. The SWI/SNF complex rearranges chromatin by destabilizing the histone-DNA complex. Chromatin remodeling can open or close regions of the genome required in normal cell functioning. However, changes in the location of chromatin compaction can result in the aberrant transcription of proto-oncogenes or the suppression of genes required for normal cellular functioning. Evidence has indicated the important role that the SWI/SNF complex plays in cancer, more specifically the loss of SNF5 in pediatric cancers.

The interconnected epigenetic pathways of miRNA and histone modifications have also been implicated in some cancers. In a recent study, it was identified that the trimethylation of histone H3K27 can induce a loss of control of miR-449a/b (Yang et al., 2009). This might account for the high proliferation rate that is observed in cancer cells. Although some of the important histone modifications identified in cancer progression and the changes in histones involved in the cell cycle have been identified, many modifications still remain to be revealed in order to fully understand the power of histones in cancer cells.

MicroRNAs in cancer

Recent studies have reported the involvement of both genetic and epigenetic mechanisms in miRNA dysregulation that can potentially lead to cancer development (for more details on small RNAs, see Chapters 7, "Non-Coding RNAs Involved in Epigenetic Processes: A General Overview," and 10, "Non-Coding RNAs Across the Kingdoms—Animals"). Genetic mechanisms are usually correlated with chromosomal abnormalities, which can lead to the deletion, amplification, or translocation of miRNAs. It has been suggested that greater than 50% of miRNA loci are at or near fragile genomic sites, which are prone to breakage and rearrangement in cancer cells. For example, miR-15a and miR-16-1, two tumor-suppressor microRNAs, are severely down-regulated in approximately 70% of patients with chronic lymphocytic leukemia (CLL) due to chromosomal deletions or mutations at the 13q13.4 loci where they are situated. It was later found that miR-15 and -16 were responsible for targeting the anti-apoptotic factor B-cell lymphoma 2 (BCL2) and that the loss of miRNA was one of the root causes of BCL2 overexpression observed in most CLL cases. Alternatively, the genomic amplification of oncogenic miR-21 was found in breast cancer specimens (Koturbash et al., 2011).

Of particular interest are epigenetic alterations that can potentiate microRNAome changes. A number of miRNAs have been detected in the vicinity of CpG islands, and it has been found that the methylation status of these CpG islands might have a drastic influence on the expression of miRNAs. In general, about 10% of miRNAs are regulated by cytosine DNA methylation. Extensive studies have documented profound alterations of miRNA expression in all major human cancers. A comprehensive list of miRNAs and their associations with different cancers is shown in Table 17-2.

Table 17-2 Alterations of miRNA Expression Related to Prognosis of Cancers. Modified with permission from Dajun Deng, Zhaojun Liu, and Yantao Du. Epigenetic Alterations as Cancer Diagnostic, Prognostic, and Predictive Biomarkers. Advances in Genetics, Volume 71, 2010, Pages 125-176.

miRNA	Cancer type	Associated characteristics (case number)
let-7	Lung cancers	Downregulation, poor survival (N = 143)
	Lung cancer	Downregulation, poor survival (N = 104)
	Squamous cell lung cancer	Downregulation, poor survival (N = 121)
	Ovarian carcinoma	Downregulation, poor survival (N = 214)
	Ovarian carcinoma	Downregulation, poor survival (N = 72 + 53)
	Ovarian carcinoma	Downregulation, poor prognosis (N = 28)
	Head and neck cancers	Downregulation, poor survival (N = 104)
	Gastric carcinoma	Downregulation, poor survival (N = 353)
miR-15b	Melanoma	Upregulation, poor survival and recurrence (N = 128)
miR-18a	Colorectal cancer	Upregulation, poor prognosis (N = 69)
	Ovarian carcinoma	Upregulation, poor prognosis (N = 28)
miR-21	Nonsmall-cell lung cancer	Upregulation, poor prognosis (N = 48)

Table 17-2 Alterations of miRNA Expression Related to Prognosis of Cancers. Modified with permission from Dajun Deng, Zhaojun Liu, and Yantao Du. Epigenetic Alterations as Cancer Diagnostic, Prognostic, and Predictive Biomarkers. Advances in Genetics, Volume 71, 2010, Pages 125-176.

miRNA	Cancer type	Associated characteristics (case number)
	Nonsmall-cell lung cancer	Upregulation, poor prognosis (*N* = 47)
	Head and neck carcinoma	Upregulation, poor survival (*N* = 169)
	Head and neck carcinoma	Upregulation, poor survival (*N* = 113)
	Squamous cell esophagus cancer	Upregulation, poor prognosis (*N* = 170)
	Colorectal cancer	Upregulation, poor prognosis (*N* = 113)
	Breast cancer	Upregulation, poor prognosis (*N* = 113)
	Pancreas carcinoma	Upregulation, liver metastasis (*N* = 56)
	Diffuse large B-cell lymphoma	Upregulation, good survival (*N* = 103)
*miR-29c** (*miR-29c*)	Malignant pleural mesothelioma	Upregulation, longer survival (*N* = 142)
	Chronic lymphocytic leukemia	Downregulation, poor prognosis (*N* = 110)
miR-31	Colorectal cancer	Upregulation, in advanced cancers (*N* = 12)
	Colorectal cancer	Downregulation, poor differentiation (*N* = 35)
	Breast cancer	Downregulation, more metastasis (*N* = 56)
miR-96	Prostate carcinoma	Upregulation, cancer recurrence (*N* = 79)
miR-146b	Squamous cell lung cancer	Upregulation, poor survival (*N* = 71)
miR-155	Lung cancer	Downregulation, poor survival (*N* = 104)
	Diffuse large B-cell lymphoma	Upregulation, longer relapse-free survival (*N* = 103)
miR-181a	Nonsmall-cell lung cancer	Upregulation, good survival (*N* = 47)
miR-196a-2	Pancreatic cancer	Upregulation, poor survival (*N* = 107)
miR-200	Ovarian carcinoma	Downregulation, poor survival (*N* = 55)
	Ovarian carcinoma	Upregulation, poor prognosis (*N* = 28)
	Colorectal cancer	Upregulation, poor prognosis (*N* = 24)

Table 17-2 Alterations of miRNA Expression Related to Prognosis of Cancers. Modified with permission from Dajun Deng, Zhaojun Liu, and Yantao Du. Epigenetic Alterations as Cancer Diagnostic, Prognostic, and Predictive Biomarkers. Advances in Genetics, Volume 71, 2010, Pages 125-176.

miRNA	Cancer type	Associated characteristics (case number)
miR-205	Head and neck cancers	Downregulation, poor survival (*N* = 104)
miR-210	Breast cancer	Upregulation, poor survival (*N* = 219)
miR-214 and *miR-433*	Gastric cancer	Upregulation, poor survival (*N* = 353)

MicroRNAs as oncogenes

Overexpression and amplification are the main criteria in defining an miRNA as an oncomiR. The most abundant cancer-related oncomiR is miR-21. This miRNA is upregulated in more than 15 different cancers, including some of the most aggressive cancers, such as glioblastoma, lymphoma, pancreatic, and lung cancers. MiR-21 is located on chromosome 17 and is regulated by active STAT3, which is erroneously expressed and activated in many cancer types. MiR-21 can act as an oncogene by regulating the tumor-suppressor genes PTEN and PDCD and suppressing their anti-apoptotic activity. The inactivation of miR-21 in several cell lines resulted in increased cell death by reactivating caspases as well as an increased sensitivity to gemcitabine via the activation of PTEN.

The miR-221/222 tandem, located less that 1 kb from one another on chromosome X, is an example of an oncomiR cluster. The aberrant expression of this cluster has significance in thyroid carcinoma, hepatocellular carcinoma, pancreatic adenocarcinoma, non-small cell lung cancer, and prostate cancer. The possible oncomiR mechanism of miR-221/222 operates through the suppression of p27/Kip, a key mediator of cell cycle progression inhibitors.

The miR-17-92 cluster was among the first miRNA groups discovered to be deregulated in a number of human tumors, including lymphomas, leukemias, and lung, breast, and testicular cancers, among others. The miR-17-92 cluster contains six miRNAs that are transcribed together as a single polycistron. The upregulation of these miRNAs is correlated with increased levels of cellular proliferation and decreased levels of apoptosis (Koturbash et al., 2011).

MicroRNAs as tumor-suppressors

Among the numerous tumor-suppressor miRNA genes, the most remarkable are members of the let-7 family, the miR-15a-16-1 cluster, the miR-34 family, and the miR-143-145 cluster. The special characteristics and mechanisms of the tumor-suppressor activity of the let-7 family, the miR-15a-16-1 cluster, and the miR-34 family were recently unraveled in several excellent reviews.

Less is known about the miR-143-145 cluster. It consists of two miRNAs, miR-143 and miR-145, that are located approximately 1500 kb apart within the fragile site 5q33. Many cancers exhibit the downregulation of these miRNAs (colorectal, ovarian, breast, and lung cancers as well as chronic lymphocytic leukemia, cervical, bladder, and prostate cancers). The only confirmed target for miR-143 is ERK5, which is involved in cell growth promotion and proliferation, and the overexpression of ERK5 has already been detected in several cancers. In addition, a number of oncogenes (MYCN, KRAS, MCF2, AKAP13, FLI1, NOVA1, and RAB7) are among the predicted targets for this cluster. However, this has yet to be validated experimentally.

Other potential tumor-suppressor miRNAs are miR-99, miR-100, miR-125a, miR-125b, miR-126, miR-139, and miR-140. These miRNAs are downregulated in three or more cancers, and experimentally confirmed targets include ERRB2, ERBB3, the vascular cell adhesion molecule VCAM1, and histone deacetylase HDAC4. Additional research is required to determine their effect on these targets.

MicroRNAs with both oncogenic or tumor-suppressor functions

There are a number of miRNAs that are overexpressed in one type of cancer and downregulated in another. For example, miR-205 is upregulated in lung, bladder, and pancreatic cancers. In contrast, it is significantly downregulated in prostate and breast cancers as well as esophageal squamous cell carcinoma. MiR-224 is highly overexpressed in prostate and bladder cancers and in hepatocellular carcinoma, but it is downregulated in ovarian epithelial carcinoma and lung cancer.

Additionally, miR-29b is upregulated in thyroid papillary carcinoma and breast cancer, but significantly downregulated in lung cancer. Notably, miR-29b targets the de novo DNA methyltransferases DNMT3A and DNMT3B, which were previously found to be strongly upregulated in lung cancer (Fabbri et al., 2007). However, the mechanisms of this tumor type-selective miRNA dysregulation have yet to be elucidated.

MicroRNAs as diagnostic and prognostic markers in cancer

Different tumors and tumor subtypes have specific miRNA signatures, which might be useful as diagnostic and prognostic markers. Bloomston and colleagues report 23 miRNAs that significantly distinguish pancreatic cancer from chronic pancreatitis, with several of these miRNAs being capable of predicting overall survival in cancer patients (Bloomston et al., 2007). In stage II colon cancer, miRNA expression profiles were capable of predicting recurrence rates with an accuracy of greater than 80%, suggesting that miRNA profiling can also be used to determine a tumor's aggressiveness. Similarly, miRNA profiling of hepatocellular carcinoma could accurately differentiate between the tumors and matched normal liver cells.

In addition, miRNA expression patterns can help distinguish tumor histopathological subtypes. For example, endometriosis history in ovarian carcinoma can be differentiated from other ovarian cancer histopathological subtypes via upregulated levels of miR-21, miR-203, and miR-205. On the other hand, the downregulation of miR-145 and miR-222 might be a marker for ovarian clear cell and serous carcinomas as well as endometriosis. Tumor specimen profiling can provide an important signature for the stage and grade differentiation of ovarian cancers in epithelial and serous carcinomas. Furthermore, Laos et al. report that miR-9 and miR-223 are significant biomarkers for the recurrence of ovarian cancer. Finally, the increased expression of miR-18a, miR-93, miR-141, the miR-200-family, and miR-429 in combination with the downregulation of miR-let7b and miR-199a characterize ovarian serous cystadenocarcinoma with decreased progression. With the relatively new discovery of the utility of miRNA profiling of plasma and serum, the role of miRNAs might become very important in minimally invasive cancer diagnostics and prognostics (Koturbash et al., 2011).

MiRNAs and metastasis

About 90% of cancer-related deaths are caused by the development of malignant tumors distant from the primary site lesions as a result of metastasis. This is a complex, multistep process that includes the dissemination of cancer cells through the blood/lymph and results in secondary cancer settlement. Recent studies have suggested an important role for miRNAs in metastasis formation. These miRNAs can be divided into two main categories: **metastatic activators** and **metastatic suppressors**. **Metastatic activators** include miR-10b, miR-21, miR-127, miR-199a, miR-210, miR-373, and miR-520c. MiR-10b, downregulated in a number of cancers, is unexpectedly found to be upregulated in about 50% of metastatic breast cancers. Furthermore, these miRNAs display the ability to promote the migration, invasion, and metastasis of non-invasive breast cancer cells *in vitro* and *in vivo*. MiR-10b exerts its invasive and metastatic phenotype by targeting the cell motility regulator homeobox protein D10 (HOXD10), which is a known inhibitor of angiogenesis. Likewise, miR-21 overexpression has been directly correlated with lymph node metastasis, whereas elevated levels of miR-373 and miR-520c lead to markedly increased metastatic risk.

Several miRNAs have been proposed as **metastatic suppressors**: the let-7 family, miR-100, miR-126, miR-218, and miR-335. Reduced levels of miR-126 and miR-335 were found in breast cancers characterized by poor metastatic-free survival, whereas the significantly decreased expression of miR-let7c, miR-100, and miR-218 differentiate metastatic prostate cancer from high-grade localized prostate cancer. Another member of the let-7 family, miR-let-7b, is correlated with metastatic risk in uveal melanoma.

Interestingly, of the metastatic activator miRNAs, only miR-21 is an miRNA with established oncogenic properties. Alternatively, the majority of metastatic suppressor miRNAs found to date—such as miR-let7b, let7c, miR-100, miR-126, and miR-218—are also considered to be putative tumor suppressor miRNAs. These observations suggest that either metastatic activator miRNAs uniquely regulate key sets of genes involved in invasion and migration or that these activator miRNAs might also have other unknown tumorigenic activities (Koturbash et al., 2011).

Perspectives of epigenetic therapy and diagnostics

Unlike genetic changes, epigenetic modifications in cancer cells are reversible. Therefore, when the cellular epigenome is changed, giving rise to a new gene expression pattern, the cell still harbors the potential to revert back to the prior epigenetic status and to the previous gene expression patterns. Such a potential reversibility of epigenetic changes constitutes an important potential target for therapeutic interventions.

Furthermore, it is important to note that in cancers, tumor suppressor genes are very often silenced by aberrant DNA methylation, rather than mutated. Therefore, the use of pharmacological inhibitors of DNA methylation, such as 5-azacytidine and 5-aza-2'-deoxycytidine, has great potential to reactivate genes that have been inappropriately silenced.

DNA methylation changes are often paralleled by histone modifications, and together, these lead to aberrant gene expression in carcinogenesis. Therefore, multiple therapeutic strategies will be needed to accurately reverse aberrant gene expression patterns that are generated via multiple types of epigenetic modifications. To reactivate genes that have been inappropriately silenced, histone deacetylase inhibitors are used often in parallel with DNA methylation inhibitors. Recently, two HDAC inhibitors Zolinza (Vorinostat) and Istodax (Romidepsin), have been approved by the U.S. Food and Drug Administration for the treatment of cutaneous T cell lymphoma. These novel epigenetic medications can induce cellular differentiation, cell cycle arrest, and apoptosis. Clinical trials are underway to determine if HDAC inhibitors can also be used in the treatment of other malignancies.

Furthermore, because these epigenetic changes can be detected prior to tumor development, they might serve as biomarkers for the early detection of disease, prognosis, and response to treatment. Investigations regarding the potential value of epigenomic profiling are currently underway, using samples collected from tissue biopsies or from less intrusive methods, such as blood and urine samples, to evaluate the DNA-methylome and to map histone modifications (Liloglou and Field, 2010).

The importance of epigenetics, especially in our understanding of susceptibility to disease, is reflected in the huge undertaking needed to establish an international human epigenome. This can be used to identify individuals at a higher risk for developing diseases so that preventative strategies can be implemented to reduce the number of incidences and improve patient survival and quality of life.

Conclusion

While for decades, cancer researchers have focused their efforts on uncovering cancer-associated genetic defects, the recent paradigm shift has led to acceptance of cancer as an epigenetic disease. Currently, epigenetic deregulation has become recognized as a hallmark of cancer.

Epigenetic deregulation including aberrant DNA methylation, altered histone modifications, abnormal imprinting, and small RNAome occur early during cancer predisposition and progress and accumulate during cancer development.

New technologies currently allow establishing the epigenome profiles of cancers, and in the future, this epigenome information might be used to generate novel biomarkers that will help guide diagnosis, prognosis, and treatment of cancer. Novel epigenome knowledge-based strategies might be also used for cancer prevention.

Exercises and discussion topics

1. What is a Knudson's hypothesis?
2. What are the key epigenetic changes that are deregulated in cancer?
3. What is the role of DNA hypomethylation in cancer?
4. How is DNA methylation lost in cancer?
5. Which pathways are affected by hypermethylation in cancer?
6. What are the possible mechanisms of DNA hypermethylation in cancer?
7. Are histone modifications altered in cancer cells?
8. Which histone-modifying enzymes are affected in cancer?
9. What are the roles of PcG and TrxG proteins in cancer?
10. What is a cancer epigenome, and why is it important to study it?
11. What are the roles of microRNAs in cancer?
12. What makes microRNAs to be excellent cancer biomarkers?

References

Agirre et al. (2009) Epigenetic silencing of the tumor suppressor microRNA Hsa-miR-124a regulates CDK6 expression and confers a poor prognosis in acute lymphoblastic leukemia. *Cancer Res* 69:4443-4453.

Badal et al. (2003) CpG methylation of human papillomavirus type 16 DNA in cervical cancer cell lines and in clinical specimens: genomic hypomethylation correlates with carcinogenic progression. *J Virol* 77:6227-6334.

Baylin et al. (1986) DNA methylation patterns of the calcitonin gene in human lung cancers and lymphomas. *Cancer Res* 46:2917-2922.

Bloomston et al. (2007) MicroRNA expression patterns to differentiate pancreatic adenocarcinoma from normal pancreas and chronic pancreatitis. *JAMA* 297:1901-8.

Caballero OL, Chen YT. (2009) Cancer/testis (CT) antigens: potential targets for immunotherapy. *Cancer Sci* 100:2014-2021.

Costa et al. (2006) SATR-1 hypomethylation is a common and early event in breast cancer. *Cancer Genet Cytogenet* 165:135-143.

Davalos et al. (2011) Dynamic epigenetic regulation of the microRNA-200 family mediates epithelial and mesenchymal transitions in human tumorigenesis. *Oncogene* Aug 29, 2011. doi: 10.1038/onc.2011.383.

Deng et al. (2010) Epigenetic alterations as cancer diagnostic, prognostic, and predictive biomarkers. *Adv Genet* 71:125-176.

Esteller M, Herman JG. (2002) Cancer as an epigenetic disease: DNA methylation and chromatin alterations in human tumours. *J Pathol* 196:1-7.

Fabbri et al. (2007) MicroRNA-29 family reverts aberrant methylation in lung cancer by targeting DNA methyltransferases 3A and 3B. *Proc Natl Acad Sci USA* 104:15805-15810.

Feinberg et al. (2006) The epigenetic progenitor origin of human cancer. *Nat Rev Genet* 7:21-33.

Feinberg AP, Tycko B. (2004) The history of cancer epigenetics. *Nat Rev Cancer* 4:143-153.

Feinberg AP, Vogelstein B. (1983) Hypomethylation distinguishes genes of some human cancers from their normal counterparts. *Nature* 301:89-92.

Fraga et al. (2005) Loss of acetylation at Lys16 and trimethylation at Lys20 of histone H4 is a common hallmark of human cancer. *Nat Genet* 37:391-400.

Fraga et al. (2004) A mouse skin multistage carcinogenesis model reflects the aberrant DNA methylation patterns of human tumors. *Cancer Res* 64:5527-5534.

Gama-Sosa et al. (1983) The 5-methylcytosine content of DNA from human tumors. *Nucleic Acids Res* 11:6883-6894.

Greger et al. (1989) Epigenetic changes may contribute to the formation and spontaneous regression of retinoblastoma. *Hum Genet* 83:155-158.

Jaenisch R, Bird A. (2003) Epigenetic regulation of gene expression: how the genome integrates intrinsic and environmental signals. *Nat Genet* 33:245-254.

James et al. (1999) Abnormal folate metabolism and mutation in the methylenetetrahydrofolate reductase gene may be maternal risk factors for Down syndrome. *Am J Clin Nutr* 70:495-501.

Jones PA, Baylin SB. (2002) The fundamental role of epigenetic events in cancer. *Nat Rev Genet* 3:415-428.

Kanwal R, Gupta S. (2010) Epigenetics and cancer. *J Appl Physiol* 109:598-605.

Knudson A. (1971) Mutation and cancer: statistical study of retinoblastoma. *Proc Natl Acad Sci USA* 68(4):820-823.

Knudson AG. (1996) Hereditary cancer: two hits revisited. *J Cancer Res Clin Oncol* 122:135-140.

Koturbash et al. (2011) Small molecules with big effects: the role of the microRNAome in cancer and carcinogenesis. *Mutat Res* 722:94-105.

Kulis M, Esteller M. (2010) DNA methylation and cancer. *Adv Genet* 70:27-56.

Laird PW, Jaenisch R. (1996) The role of DNA methylation in cancer genetic and epigenetics. *Annu Rev Genet* 30:441-464.

Lapeyre JN, Becker FF. (1979) 5-Methylcytosine content of nuclear DNA during chemical hepatocarcinogenesis and in carcinomas which result. *Biochem Biophys Res Commun* 87:698-705.

Lechner et al. (2010) Cancer epigenome. *Adv Genet* 70:247-276.

Liloglou T, Field JK. (2010) Detection of DNA methylation changes in body fluids. *Adv Genet* 71:177-207.

Mazar et al. (2011) Epigenetic regulation of microRNA-375 and its role in melanoma development in humans. *FEBS Lett* 585:2467-2476.

Pogribny IP, Beland FA. (2009) DNA hypomethylation in the origin and pathogenesis of human diseases. *Cell Mol Life Sci* 66:2249-2261.

Poulsen et al. (2007) The epigenetic basis of twin discordance in age-related diseases. *Pediatr Res* 61:38R-42R.

Robertson KD, Jones PA. (1998) The human ARF cell cycle regulatory gene promoter is a CpG island which can be silenced by DNA methylation and down-regulated by wild-type p53. *Mol Cell Biol* 18:6457-6473.

Robertson KD, Jones PA. (2000) DNA methylation: past, present and future directions. *Carcinogenesis* 21:461-467.

Sawan C, Herceg Z. (2010) Histone modifications and cancer. *Adv Genet* 70:57-85.

Takacs et al. (2010) Epigenetic regulation of latent Epstein-Barr virus promoters. *Biochem Biophys Acta* 1799:228-235.

Tanaka et al. (2012) Frequent methylation and oncogenic role of microRNA-34b/c in small-cell lung cancer. *Lung Cancer* 76:32-38.

Tsuruta et al. (2011) miR-152 is a tumor suppressor microRNA that is silenced by DNA hypermethylation in endometrial cancer. *Cancer Res* 71:6450-6462.

Van Den Broeck et al. (2008) Loss of histone H4K20 trimethylation occurs in preneoplasia and influences prognosis of non-small cell lung cancer. *Clin Cancer Res* 14:7237-7245.

Wilson et al. (2007) DNA hypomethylation and human diseases. *Biochim Biophys Acta* 1775:138-162.

Yang et al. (2009) miR-449a and miR-449b are direct transcriptional targets of E2F1 and negatively regulate pRb-E2F1 activity through a feedback loop by targeting CDK6 and CDC25A. *Genes Dev* 23:2388-2393.

18

Epigenetics of health and disease—behavioral neuroscience

The new discipline of epigenetics is of special interest to neuroscientists because it can provide a molecular basis for the complex mechanisms underlying cognition and behavior. In recent years, more and more research has been providing evidence that neurological processes are indeed regulated by epigenetic components: DNA methylation, histone modification, chromatin remodeling, and noncoding RNAs. Neurological processes orchestrated by epigenetics include neuron development and function, neuronal plasticity, and memory formation just to name a few.

DNA methylation and brain

During brain development, DNA methylation is believed to be important in regulating the proliferation of neural stem cells and their differentiation into neurons and glial cells (Mattson, 2003). Methylation of DNA in the brain is catalyzed by three main enzymes: DNMT1, DNMT3a, and DNMT3b. DNMTs (DNA methyltransferases) are expressed throughout neural development and promote neuronal survival and plasticity (Mehler, 2008). MBDs (methy-CpG-binding domains) were shown to play an important role in brain development and cognitive functions, such as learning and memory (Chahrour et al., 2008; Mehler, 2008).

A growing body of research extends the function of DNA methylation beyond the developing nervous system. DNA methylation was shown to be important in synaptic plasticity, learning, and memory in adult CNS (central nervous system) neurons (Feng et al., 2010), and CNS repair.

Chromatin remodeling, histone modifications, and the brain

Chromatin remodeling and histone modifications play a key role in the accessibility of the nucleosome to the transcription machinery, and thus in gene silencing. The level of gene expression depends on different factors, including modifications in histone N-terminal tails by methylation, acetylation, phosphorylation, SUMOylation, and ubiquitination. A change in a "histone code" or a specific pattern of histone modifications could result in gene activation or silencing (Jenuwein and Allis, 2001).

Understanding the concept of chromatin remodeling might provide an insight into how stable changes in gene expression in the brain produce long-lasting changes in behavior (Colvis et al., 2005). Increasing evidence indicates that changes in the chromatin structure occur not only during development, but also in mature neurons (Renthal and Nestler, 2008). Neuronal signaling appears to be strongly regulated by addition and removal of histone acetylation and histone and DNA methylation (Renthal and Nestler, 2008). Recent studies showed that the neuronal levels of monoacetylated H4 decrease progressively during aging, whereas the levels of acetylation of non-histone proteins might increase in neurons during development and with age (Mattson, 2003).

It was demonstrated that chromatin remodeling via the posttranslational modification of histone proteins (primarily histone H3 phosphorylation and acetylation) is important for the formation of long-term memory (Peleg et al., 2010; Gupta et al., 2010). Some data suggest that histone methylation is actively regulated in the hippocampus and facilitates the formation of long-term memory, whereas the deregulation of H4K12 acetylation might cause memory impairments in the aging mouse brain (Gupta et al., 2010; Peleg et al., 2010).

HDACs catalyze the removal of an acetyl group from the N-terminal tails of histone proteins, whereas HATs catalyze the reverse reaction (Sleiman et al., 2009). The alteration in balance between

HDAC and HAT functions can lead to cancer as well as to neurodegenerative disease (Hahnen et al., 2008). HDAC inhibitors have shown a therapeutic efficacy in animal models of neurodegenerative diseases (Fischer et al., 2007; Abel and Zukin, 2008), suggesting their neuroprotective role.

Non-coding RNAs and the brain

About 98% of transcribed genomic DNA are non-coding RNAs that do not function as messenger, transfer, or ribosomal RNAs (Yu, 2008). Among those, miRNAs are small non-coding single-stranded RNA molecules (approximately 22 nucleotides long) that can regulate gene expression via an RNA interference pathway by altering mRNA stability and translational initiation. Each miRNA can regulate thousands of different mRNA targets (Nelson, Wang, and Rajeev, 2008). They are involved in many biological processes, such as cellular development, differentiation, proliferation, cell division, apoptosis, onset and progression of cancer, and other diseases. The expression of miRNA molecules depends on different factors and can be regulated by other epigenetic agents, such as DNA methylation and chromatin structure (Saito and Jones, 2006).

The expression of miRNAs is cell- and tissue-specific, and miRNAs are particularly abundant in the brain (Kosik, 2006). Many miRNAs are expressed in a spatially and temporally controlled manner in the nervous system, suggesting that their regulation might be important in neural development and function (Barbato, Ruberti, and Cogoni, 2009).

In the past years, a growing body of scientific evidence has indicated that miRNAs might be a contributing factor to aging-related neurodegenerative diseases (Nelson and Keller, 2007; Colvis et al., 2005; Nelson et al., 2008). Schaefer et al. (2007), for instance, demonstrated that a substantial loss of mature miRNA in the cerebellum of mice with the knocked-out Dicer gene causes progressive neurodegeneration in a mouse model, suggesting that the same mechanism might be relevant in human neurodegenerative disorders (Schaefer et al., 2007).

Overall, it is now well-accepted that epigenetic mechanisms are key players in the health and diseases of nervous system (reviewed in Urdinguio et al., 2009a) (see Figure 18-1).

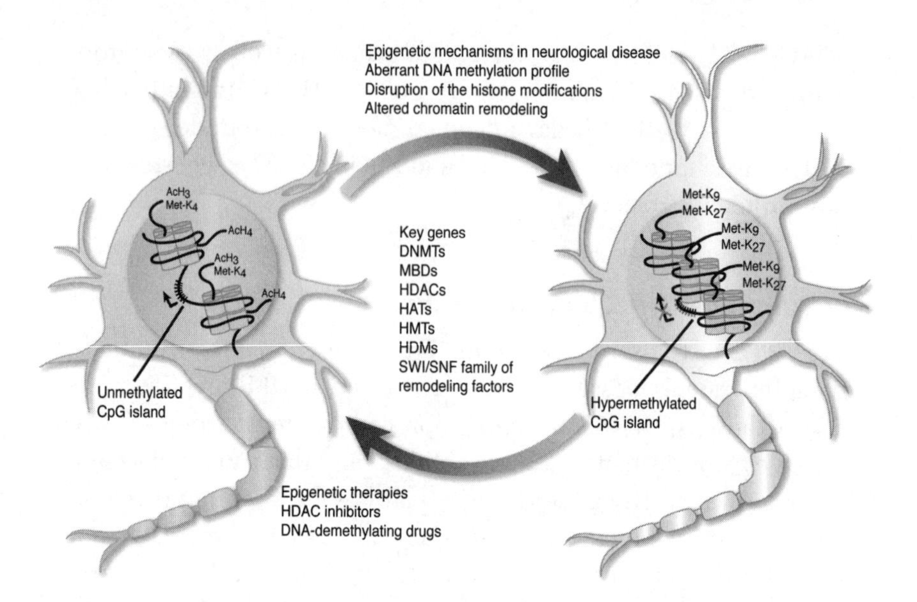

Figure 18-1 Epigenetic mechanisms in the nervous system. (Re-printed with permission from Urdinguio RG, Sanchez-Mut JV, Esteller M. Epigenetic mechanisms in neurological diseases: genes, syndromes, and therapies. Lancet Neurol. 2009; 8(11):1056-72.) In healthy neurons or glia (left), expression of the mRNA of a gene occurs in the presence of an unmethylated promoter CpG island and a set of histone modifications associated with open "chromatin" conformation (for example, hyperacetylation and methylation of lysine 4 of histone H3). Cylinders indicate octamers of histones, consisting of histones H2A, H2B, H3, and H4. These form the nucleosomes, and the double strand of DNA is wrapped around them. A combination of selection and targeted disruption of the DNA methylation and histone-modifier proteins disrupts the epigenetic circumstances in the cell. The aberrant epigenetic inactivation of the disease-associated genes ("closed" chromatin conformation) is associated with dense hypermethylation of the CpG island promoter, the appearance of repressive histone chemical modifications such as methylation of lysines 9 and 27 of histone H3, and nucleosome positioning (right). Epigenetic drugs such as DNA-demethylating drugs and HDAC inhibitors can partially rescue the distorted epigenetic processes and restore gene expression of the neuronal or glial gene by removing chemical modifications (for example, DNA methylation) and inducing the presence of modifications (for example, histone acetylation). Ac=acetylation. DNMT=DNA methyltransferase. HAT=histone acetyltransferase. HDAC=histone deacetylase. HDM=histone demethylase. HMT=histone methyltransferase. MBD=methyl-CpG binding domain protein. Met-K4=methylation of lysine 4. Met-K9=methylation of lysine 9. Met-K27=methylation of lysine 27. SWI/SNF=switching/sucrose nonfermenting chromatin-remodeling complex.

A key evidence that epigenetic mechanisms mediate neurodevelopment—the classical experiments of Meaney and Szyf on rodents

Using a rat model, Michael Meaney and Moshe Szyf and their colleagues performed an excellent set of experiments and demonstrated a direct relation between variations in maternal care and the phenotype of the offspring. The authors not only suggested but provided the experimental evidence that maternal effects on the progeny are mediated by epigenetic mechanisms, notably by DNA methylation. Their data have shown that variations in maternal licking/grooming (LG) behavior influence behavioral and hypothalamic-pituitary-adrenal (HPA) responses to stress in adult offspring through epigenetic mechanisms. LG is a source of tactile stimulation that triggers important endocrine and metabolic responses and regulates somatic growth (Kappeler and Meaney, 2010). It was shown that the progeny of the dams that display high LG behavior had decreased methylation of glucocorticoid receptors (GR) in the hippocampus, which leads to increased GR expression and results in a decreased stress response. This early-life experience has a long-lasting stable effect on the progeny, but it can be reversed by cross-fostering experiments (Francis et al., 1999). Pups born to low LG mothers but reared by high LG mothers show stress responses that are similar to those of the normal offspring of high LG mothers and vice versa. Even more intriguing is the evidence that these effects of maternal care can be transmitted to the next generations (Francis et al., 1999). Individual differences in the expression pattern of some genes in the brain that regulate the stress response (GR in hippocampus, a central benzodiazepine receptor (CBR) in amygdala, and a corticotropin-releasing factor (CRF) in hypothalamus) can be transmitted from one generation to the next through behavior (nongenomic transmission) (Francis et al., 1999). This was demonstrated in the handling experiments; postnatal handling of pups can alter the stress response in the offspring and increase the mother's LG behavior (Francis et al., 1999). The expression patterns of GR, CBR, and CRF in the adult brain of handled offspring of low LG mothers are comparable to those in the offspring of handled or non-handled high LG mothers (Francis et al., 1999).

These pioneer experiments have proved that the quality of maternal care in rodents has a widespread effect on the phenotype that persists into adulthood. Furthermore, they provide an excellent model to study epigenetic mechanisms mediating the influence of early life experiences on health later in life. Furthermore, the aforementioned experiments on rodents provided an important novel insight into how epigenetic mechanisms might regulate neurological processes in the brain and subsequently behavior, thus suggesting that a similar epigenetic regulation might be applied to humans.

As a result, it is currently well-accepted that the quality of parental care significantly influences mental health, including the risk of psychopathology, as well as stress responses, emotional function, learning and memory, and neuroplasticity. All of the aforementioned effects are based on epigenetic reprogramming of gene expression as a function of early care. For example, in rats, it has been shown that variations in maternal care in the first week of life are associated with alterations in DNA methylation and H3K9 acetylation of the *NR3C1* promoter region, and consequently with the expression of the GR17 splice variant of the *NR3C1* gene in the hippocampus of adult offspring.

Notwithstanding, relatively little is known about global large-scale epigenetic changes in the brain that are induced by parental care. The recent landmark study by the research groups of Meaney and Szyf has provided yet another key piece of evidence of the global epigenome changes in the brain as a function of quality of maternal care (McGowan et al., 2011). In this study, the researchers analyzed hippocampal samples from the adult offspring of rat mothers that differed in the frequency of pup licking/grooming in the first week of life (that is, High versus Low LG adult offspring) and performed an analysis of DNA methylation, H3K9 acetylation and gene expression of a contiguous 7 million base pair region of rat chromosome 18 containing the *NR3C1* gene at 100 bp spacing. Their analysis revealed a widespread but patterned response to maternal care among High and Low LG adult offspring. Importantly, they have shown that variations in maternal care are associated with coordinate epigenetic changes spanning 100 kilobase-pairs. The adult offspring of high compared to low maternal care mothers exhibit significant epigenetic changes in promoters, exons, and gene ends associated with higher transcriptional activity across many genes within the locus examined. Indeed, the array data analysis revealed peaks and valleys of H3K9 acetylation and DNA methylation levels throughout a number of regions, suggesting a widespread epigenomic response to variations in maternal care (see Figure 18-2).

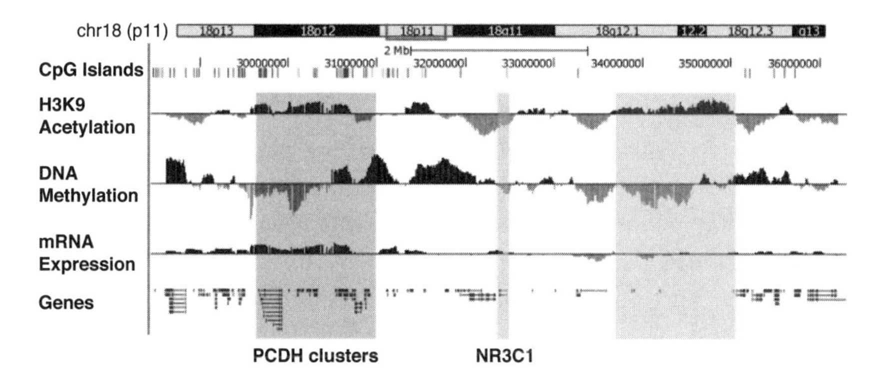

Figure 18-2 The pattern of H3K9 acetylation, DNA methylation, and gene expression among high and low LG adult offspring across approximately 7MB of chromosome 18. (Re-printed with permission from McGowan PO, Suderman M, Sasaki A, Huang TC, Hallett M, Meaney MJ, Szyf M. Broad epigenetic signature of maternal care in the brain of adult rats. PLoS One. 2011; 6(2):e14739.) Tracks show CpG Islands, differences in H3K9 acetylation, DNA methylation and gene expression between High (black) and Low LG (gray) adult offspring (H–L) and the locations of known genes (bottom row) across the chromosomal locus (see Methods S1). Highlighted regions show the location of the NR3C1 gene (center), Protocadherin gene clusters (left), and a large mainly intergenic region (right).

Therefore, the epigenetic response to maternal care is not evenly distributed, with many sequences showing little or no response and clustered regions showing enhanced responses. In total, researchers found significant differential DNA methylation in 1,413 probes and significant differences in H3K9 acetylation in 713 probes out of 44,000 probes that covered the studied region. Variations in epigenetic signaling across the locus appeared within annotated genic regions as well as in the regions where no genes were annotated. Interestingly, the chromosomal region containing the *protocadherin-*α, -β, and -γ (*Pcdh*) gene families implicated in synaptogenesis exhibited the highest differential epigenetic response to maternal care. Notwithstanding, the mechanisms responsible for the coordinated maternal care-induced epigenomic response and its maintenance into adulthood are still unknown and need to be investigated in the future.

Neurological diseases—an epigenetic connection

Initially, the recognition of the important role of epigenetic mechanisms in human diseases started in oncology, but it has recently been extended to other areas of medicine, such as neurological, neurodevelopmental, and neurodegenerative disorders.

Neurodevelopmental disorders

Interestingly, a number of genetic studies have shown that epigenetic deregulation can occur as a result of mutation in genes encoding for key epigenetic regulatory proteins and can lead to neurodevelopmental disorders such as Rett syndrome, ATRX syndrome, mental retardation, autism, and others (reviewed in Urdinguio et al., 2009a).

Rett syndrome

One of the best studied neurodevelopmental diseases is Rett syndrome (RTT) (OMIM 312750). RTT is an X-linked neurological disease caused by mutations in a MBD protein (MeCP2) that affects one girl in every 10,000 to 15,000 live births. MeCP2 was one of the first members of the MBD family discovered. It is absolutely essential for proper brain function and development. Loss of MeCP2 has been shown to delay neuronal maturation and synaptogenesis and cause RTT (reviewed in Urdinguio et al., 2009a). The disease develops gradually; until about 6 to 18 months of age, girls with RTT syndrome develop normally. Later, they start to deteriorate and develop signs of mental retardation, stereotyped hand movements, ataxia, seizures, microcephaly, autism, and respiratory dysfunctions. Mechanistically, MeCP2 is involved in the crosstalk between DNA methylation, chromatin remodeling, and transcription. It binds to methylated DNA together with the Sin3a/HDAC co-repressor complex, thus providing a link between DNA methylation and histone deacetylation. Furthermore, MeCP2 can directly bind to the promoters of a subset of activated genes in association with the transcriptional activator CREB1 (cAMP response element binding protein 1).

Interestingly, originally it was thought that RTT development was linked to MeCP2 deficiency in neurons. Yet, new studies have implicated non-neural cells in the development of the disease.

Recently, it has been shown that MeCP2 is expressed not only in neurons, but also in all glial cell types, including astrocytes, oligodendrocyte progenitor cells, oligodendrocytes, and microglia. Furthermore, the analysis of the RTT mouse model has shown that loss of MeCP2 expression occurs in astrocytes and is associated with low H3K9me3 levels and high enrichment levels of H3Ac.

ATRX syndrome

ATRX syndrome (OMIM 301040) is an X-linked disorder that is caused by mutations in a chromatin-remodeling protein. ATRX is characterized by severe mental retardation, unusual facial appearance, urogenital abnormalities, skeletal abnormalities, and alpha thalassaemia. The ATRX protein is a member of the Snf2 (sucrose non-fermenting 2) family of enzymes. Mutations in the ATRX gene lead to aberrant DNA methylation in repetitive elements, including ribosomal DNA repeats, subtelomeric repeats, and Y-specific satellite repeats. Furthermore, ATRX interacts with chromatin-remodeling proteins of the polycomb group, and mutations negatively influence these important interactions, thus affecting proper cellular functioning.

Rubinstein-Taybi syndrome

Rubinstein-Taybi syndrome (OMIM 180849) is associated with the dysfunction of a histone acetyltransferase (HAT). It is an autosomal dominant disorder that is commonly characterized by mental retardation and numerous physical abnormalities, such as broad radially deviated thumbs and halluces, postnatal growth deficiency often followed by an excessive weight gain in later childhood or puberty, specific facial features, and an increased risk of tumors.

Other neurodevelopmental syndromes

Other neurodevelopmental syndromes include the Coffin-Lowry syndrome, a disease caused by the loss-of-function of a histone phosphorylase and chromatin modifier RSK2. Altered levels of DNA methylation are also seen in Friedrich's ataxia, fragile X syndrome. and others. These too are caused by mutational inactivation of epigenetic regulatory genes.

Epigenetic mechanisms underlying neurodegenerative diseases

Neurodegenerative diseases are diseases in which the nervous system progressively and irreversibly deteriorates. This chapter reviews some of them: Alzheimer's disease (AD), Parkinson's disease (PD), Huntington's disease (HD), and multiple sclerosis (MS). The origin and mechanisms underlying the onset and progression of most neurodegenerative pathologies are largely unknown. Some of them are caused by genetic mutation, for example HD, but the genetic or epigenetic origin of the sporadic forms of AD, PD, or MS is still debatable. A lot of recent studies provide evidence that neurodegenerative diseases might be mediated by aberrant epigenetic mechanisms.

Alzheimer's disease

Alzheimer's disease (AD) is a common form of dementia that is characterized by cognitive and—in the advanced stage—physical impairments. AD is a slowly progressive complex disease of the brain, the development and progression of which depends on many factors, including genetic and environmental components. Although the cause of AD remains unknown, it is associated with the formation of neuritic plaques and neurofibrillary tangles in the brain. These aggregates might affect the neuronal network and synaptic plasticity and might lead to cognitive impairments (Chouliaras et al., 2010). Neurofibrillary tangles are formed by hyperphosphorylation of the tau protein. Tau proteins are axonal proteins that are abundant in neurons and are important for the stability of microtubules. Neuritic plaques are made up of the large fibrils of the amyloid-β protein precursor (APP), which are neurotoxic.

A growing body of evidence suggests that nearly 200 genes might be involved in Alzheimer's disease pathogenesis (Cacabelos, 2007). Recent studies reveal that epigenetic alterations could also play an important role in the onset and progression of AD. Current research on AD pathophysiology is mainly focused on the aberrant processing of APP and tau protein (Chouliaras et al., 2010).

DNA methylation in AD

Altered methylation patterns in the brains of AD patients were first demonstrated in the mid-1990s. Decreased levels of S-adenosylmethionine (-67% to -85%) and its demethylated product

S-adenosylhomocysteine (-56% to -79%) were reported in cerebral cortical subdivisions, hippocampus, and putamen of AD patients as compared to controls (Morrison, Smith, and Kish, 1996). In a study by West et al. (1995), the analysis of human post-mortem brain tissues from AD patients showed that the promoter region of the APP gene was hypomethylated in comparison to the healthy controls (West, Lee, and Maroun, 1995). However, a later study by Barrachina and Ferrer (2009) revealed no differences in the methylation status of the APP promoter between AD and control brain tissues (Barrachina and Ferrer, 2009), suggesting that more research needs to be conducted in order to obtain a more comprehensive picture.

Recent evidence from studies of monozygotic twins (MZ) supports the notion that epigenetic mechanisms might play a role in AD. In a study by Fraga et al. (2005), differences in methylation and acetylation patterns were observed in MZ twins. Interestingly, genomic 5-methyl-C content and acetylation levels of H3 and H4 were significantly different between twins later in life and almost indistinguishable during the early years of life (Fraga et al., 2005).

A high-throughput screening study in human cerebral cortex revealed two genes—S100a2 (S100 calcium binding protein A2) and Sorbs3 (sorbin and SH3 domain containing the 3 cell-adhesion protein)—that are differently methylated in Alzheimer's tissues as compared to controls. Sorbs3 was hypermethylated, whereas the analysis of S100a2 revealed a decrease in DNA methylation in Alzheimer's patients as compared to control cases. An alteration in the methylation pattern of S100a2 and Sorbs3 might play a role in synaptic impairment in AD (Urdinguio, Sanchez-Mut, and Esteller, 2009b). Furthermore, hypermethylation of the promoter region of the Nep (neprilysin) gene in murine cerebral endothelial cells was reported, suggesting its role in amyloid-β accumulation in AD (Chen et al., 2009).

Chromatin remodeling and histone modifications in AD

The role of chromatin remodeling and histone modifications in AD is less studied than DNA methylation; however, it should be noted that changes in the chromatin structure are the prominent pathological features of many neurodegenerative diseases. Post-mortem analysis of human hippocampal neurons showed an increase in phosphorylated histone H3 in neuronal cytoplasm in AD patients in comparison to age-matched controls. Interestingly, this pattern was restricted to the cytoplasm and was not observed in the nucleus of the cell, suggesting that the aberrant cytoplasmic localization of phosphorylated histone H3 leads to neuronal dysfunction and neurodegeneration in AD.

miRNAs in AD

miRNA profiling of human patients with AD identified a number of miRNAs upregulated in peripheral blood mononuclear cells (Schipper et al., 2007). For example, a significant increase in the expression of miR-34a and 181b in mild sporadic AD was observed, as compared to age-matched normal controls (Schipper et al., 2007). However, further investigations still need to be conducted to assess their potential as blood-based diagnostic and prognostic biomarkers.

Various miRNAs were reported to be altered in rodent models of AD (see Table 18-1). Considering the fact that some miRNAs are highly conserved between mammals whereas others are poorly conserved, additional research needs to be undertaken in this area to determine whether all of these miRNAs are active in the human brain and have similar functions.

For a broader review on epigenetic regulation and AD pathophysiology, see Chouliaras et al. (2010).

Table 18-1 miRNAs associated with AD

miRNA	Brain area	Status	The role in disease
miR-9	Hippocampus	Upregulated	Needs to be investigated
miR-128			
miR-15a	Cortex	Downregulated	Candidate binding sites in the 3'UTRs of beta site amyloid precursor protein (APP) cleaving enzyme (BACE)
miR-29b-1			
miR-19b			
miR-9	Cortex	Downregulated	Candidate binding sites in the 3'UTRs of BACE and PSEN1

Table 18-1 miRNAs associated with AD

miRNA	Brain area	Status	The role in disease
101 15a 106b Let-7	Cortex	Downregulated	Candidate binding sites in the 3'UTR of APP
miR-210 miR-181c miR-22 miR-26b miR-363 miR-93	Cortex	Downregulated	Seem to be unrelated to obvious targets
miR-9 miR-132	Cerebrospinal fluid	Downregulation	Correlation to impaired neurogenesis and neuronal differentiation
miR-423	Hippocampus	Upregulated	Modulate IDH2 (isocitrate dehydrogenase 2) expression and IDH2 reduction is involved in oxidative stress in AD prefrontal cortex
miR-98	Cerebellum	Downregulated	
miR-107	Medial frontal cortex	Downregulated	Correlates with BACE1 mRNA increase, miR-107 modulates expression of BACE1 in cell culture

Parkinson's disease

Parkinson's disease (PD) is a progressive neurological disorder characterized by the selective and progressive loss of dopaminergic neurons in the midbrain substantia nigra, which causes a large number of motor and nonmotor impairments. Nearly six million people worldwide have PD, and it affects 1% of the population older than 65 (Abel and Zukin, 2008; Gasser, 2001). The main symptoms of this disorder include tremor, rigidity, bradykinesia, and gait impairment. With the progression of the disease, patients might develop non-motor symptoms, such as depression, sleep disorders, and cognitive impairments (mainly dementia). It is not known what causes PD; however, dementia and motor impairments in this disease are associated with the deposition of Lewy bodies. Lewy bodies are abnormal cytoplasmic eosinophilic aggregates of α-synuclein, parkin, ubiquitin, and other proteins that are found in the dopaminergic neurons in PD and other neurological disorders.

DNA methylation and histone modifications in PD

It has been recently shown that methylation regulates α-synuclein expression and is decreased in the brain of PD patients (Jowaed et al., 2010). Jowaed et al. (2010) found that methylation of α-synuclein intron 1 was reduced in DNA from sporadic PD patients' substantia nigra, putamen, and cortex. The authors demonstrated that methylation of this intron region regulates α-synuclein expression in the luciferase reporter experiments, thus suggesting that hypomethylation of α-synuclein intron 1 could contribute to an increased expression of α-synuclein in the brain of PD patients (Jowaed et al., 2010). Toxicity induced by α-synuclein can be rescued by the administration of histone deacetylase inhibitors in both cell culture and transgenic flies.

Another recent study correlated the plasma levels of APP and α-synuclein and the levels of SAM, SAH (markers of methylation) in patients with PD. A higher SAM/SAH ratio was associated with better cognitive performance.

The role of histone modifications and chromatin remodeling is less studied in PD. However, HDAC inhibitors might play a promising role in PD treatment. It has been recently found that valproate (VPA), an antiepileptic drug, acts as an HDAC inhibitor. VPA was demonstrated to have neurotrophic and neuroprotective effects on dopaminergic neurons in *in vitro* rodent PD models (Chen et al., 2006). In another culture study, it was shown that VPA induces apoptosis of microglial cells, thus reducing the inflammation and attenuating dopaminergic neurotoxicity (Chen et al., 2007). These studies indicate that HDAC inhibitors have potential in the treatment of PD.

miRNAs in PD

miRNAs are necessary for the survival of neurons in PD and other neurological disorders (Hebert and De Strooper, 2007). Experiments in animal models showed that Dicer deletion in the mouse midbrain dopaminergic neurons leads to the cell's death in the substantia nigra (Kim et al., 2007). It was shown in humans that miR-133b, which is specifically expressed in midbrain dopaminergic neurons, is deficient in the brain of PD patients (Kim et al., 2007).

The role of single nucleotide polymorphism (SNP) and the subsequent disruption of miRNA function was also implicated in PD. Genetic analysis revealed that SNP (rs 127202208) within the fibroblast growth factor 20 (FGF20) gene is associated with PD (Wang et al., 2008). FGF20 3'UTR contains a predicted binding site for miRNA-433. Functional assays demonstrated that the risk allele containing rs 127202208 disrupts the binding site of miR-433 and leads to elevated FGF20 translation *in vivo* and *in vitro* (Wang et al., 2008). An increase of FGF20 translation is correlated with the increased α-synuclein expression in PD brains (Wang et al., 2008).

Huntington's disease

Huntington's disease (HD) is an autosomal, dominant, neurodegenerative disease that affects about 3 in 100,000 individuals (Barbato et al., 2009). HD is characterized by incoordination, progressive involuntary movements, neuropsychiatric disturbances, and cognitive decline. HD is caused by polyglutamine expansion in the 5'-coding region of the huntingtin (htt) gene.

DNA methylation and histone modifications in HD

It was hypothesized by Farrer et al. (1992) that the process of DNA methylation might be responsible for the expression of Huntington's disease (Farrer et al., 1992). McCampbell et al. found that histone acetylation was reduced in cells expressing mutant polyglutamine (McCampbell et al., 2001). More recently, it was suggested that inhibition of histone acetyltransferases might be a primary cause of cellular pathogenesis in Huntington's disease and other polyglutamine disorders (Hughes, 2002).

The role of HDAC inhibitors in HD was studied using a Drosophila model, a mouse model, and *C. elegans*. Studies in animal models showed that the treatment with HDAC inhibitors had beneficial effects and an ameliorated disease phenotype. However, more research needs to be done to investigate the efficiency of HDAC inhibition in human HD.

miRNAs in HD

One of the molecular alterations associated with HD is the aberrant activity of the transcriptional repressor RE1-silencing transcription factor (REST). REST silences neuronal gene expression in non-neuronal tissues, whereas in healthy neurons, REST is sequestered in the cytoplasm in part through binding to huntingtin (Zuccato et al., 2003). In the brains of patients with HD, REST is unable to bind to huntingtin and has higher localization in the nucleus (Zuccato et al., 2003). REST interacts with other transcriptional repressors such as CoREST. Recently, it has been shown that miRNA might be involved in the regulation of the REST silencing complex. Profiling of miRNA expressed in the brain of HD patients showed a significant decrease in the expression of miR-9 (targets REST) and miR-9* (targets CoREST) and an increase in miR-132 expression in comparison to healthy controls. However, another study reported the opposite results for miR-132 expression in post-mortem brains of HD patients (Johnson et al., 2008), thus indicating that tissues from more patients need to be analyzed in order to obtain a more comprehensive picture. In general, alterations in the expression of 11 miRNAs were reported in HD: miR-9/9*, miR-124a, miR-132, miR-29b, miR-29a, miR-330, miR-17, miR-196a, miR-222, miR-485, and miR-486. Using Illumina massively parallel sequencing of RNA samples from HD patients, Marti et al. (2010) confirmed the aberrant expression of 9 out of 11 previously reported miRNAs. The deep sequencing analysis just added more complexity to our understanding of the possible role of miRNAs in HD. For more details, see Marti et al. (2010).

Multiple sclerosis

Multiple sclerosis (MS) is an inflammatory autoimmune disorder characterized by damage to myelin and axons in the brain and spinal cord, which is followed by neurodegeneration (Compston and Coles, 2008). The manifestation of the disease could be accompanied by one or several symptoms: fatigue, sensory impairments (numbness and tingling), visual and cognitive impairments, tremors, muscular weakness or rigidity, as well as urinary, intestinal, and sexual dysfunction. MS belongs to a category of multifactorial disorders that is derived from a combination of different factors. The low concordance rate of MS in monozygotic twins (25%) (Willer et al., 2003) suggests some genetic influence, but it also predicts that some epigenetic factors might be involved in the disease's generation and progression. Moreover, discordance in dizygotic twins (only a 5% probability for both members of the pair to develop MS) directs toward the epigenetic role in disease pathogenesis. The interaction of genetic predisposition and environmental factors (which can shape an individual's epigenetic background) is likely to play a major role in the onset of MS, rather than each of the factors alone. However, an attempt to explain disease discordance in twins failed to find evidence for either genetic, epigenetic, or transcriptome differences (Baranzini et al., 2010). In a recent study by Baranzini et al. (2010), genomes, epigenomes, and transcriptomes of CD4+ lymphocytes from MS-affected and unaffected MZ twins were analyzed. T lymphocytes (including CD4+) play a central role in the pathophysiology of MS (Zhang et al., 1992). No reproducible differences in single nucleotide polymorphisms (SNPs), insertion-deletion polymorphisms (indels), or CpG methylation patterns were observed.

DNA methylation and MS

There is some evidence suggesting that alterations in DNA methylation might play a role in the mechanism of MS pathogenesis. Recent data show that DNA isolated from the white matter of MS brains contains only about one-third of the amount of methyl cytosine found in DNA from normal subjects. It has been also shown that DNA methylation of cytosine of the Pad2 promoter in the white matter from MS patients is one-third of the methylation level in normal brain tissues (Mastronardi et al., 2007). Findings by Mastronardi et al. (2007) (Mastronardi et al., 2007) also demonstrated a twofold higher DNA demethylase activity in tissues from MS patients as compared to the white matter of healthy individuals and those with Alzheimer's,

Huntington's, or Parkinson's disease. Additionally, aberrant methylation has been observed in peripheral blood mononuclear cells (PBMCs) and in cell-free plasma of patients with MS (Liggett et al., 2010).

Another important genetic risk factor for MS is the major histocompatibility complex (MHC). Some evidence indicates that the severity of the disease is determined by the presence and interaction of toxic alleles in MHC, a region of the genome that encodes membrane receptors for antigens and peptides, which together with T-cells produce an immune response. In particular, *HLA-DRB1*1501* and *HLA-DRB5* alleles were linked to MS (Gregersen et al., 2006). Using methylation analysis by pyrosequencing, Handel et al. revealed no significant differences in DNA methylation of CpG dinucleotides across *HLA-DRB1*1501* and *HLA-DRB5* in positive benign patients when compared to positive malignant patients (Handel et al., 2010). These results demonstrate that there is no effect of toxic allele-specific DNA methylation on disease severity. However, the authors suggest that there could be time- or tissue-specific effects of DNA methylation.

miRNAs and MS

miRNAs were proposed to be important in the molecular mechanisms implicated in MS, suggesting their role as candidate biomarker targets in MS (Otaegui et al. 2009). The work of Otaegui et al. (2009) showed that miR-18b and miR-599 might be involved at the time of relapse, whereas hsa-miR-96 might play a role in remission in MS patients (Otaegui et al., 2009). The altered levels of microRNAs were also seen in the blood of RRMS patients. The best single miRNA marker—hsa-miR-145—enabled discriminating MS from controls with a specificity of 89.5%, a sensitivity of 90.0%, and an accuracy of 89.7%.

Epigenetics of psychiatric disorders

On the one hand, psychiatric disorders have been proposed to be of a genetic origin, but up to now no mutations in specific genes have been identified to be causatively related to their development. On the other hand, a growing body of research evidence suggests the involvement of epigenetic mechanisms in pathogenesis of major psychiatric disorders such as schizophrenia and bipolar disorder (Pogribny and Beland, 2009). Both hyper- and hypomethylation of DNA have been reported. Hypermethylation of the reelin (RELN) gene was shown to be important for the pathogenesis of schizophrenia. Furthermore, the promoter of the catechol-O-methyltransferase (COMT) gene, the product of which partakes in the degradation of synaptic dopamine in the human brain, is often hypomethylated in schizophrenia and bipolar disorder (Pogribny and Beland, 2009).

Global DNA methylation in white blood cells of patients with schizophrenia was significantly different from that of healthy subjects. Schizophrenia patients exhibited a significant DNA hypomethylation. Interestingly, hypomethylation of the serotonin receptor type-2A gene (HTR2A) was observed in DNA derived from saliva of patients with schizophrenia and bipolar disorder.

A recent large-scale whole methylome analysis has revealed some intriguing global patterns of methylome in normal human brain and in the brains of patients with schizophrenia and depression. A new pioneering study, Methylome DB, provides the first source of comprehensive brain methylome data covering whole-genome DNA methylation profiles of human and mouse brain specimens as well as investigations of schizophrenia and depression methylomes (Xin et al., 2012). Overall, in the future, epigenetic biomarkers might become useful in diagnostics and even staging of psychiatric disorders.

Recently, alterations in histone lysine methylation that affect gene expression were also shown to contribute to changes in brain transcriptomes in mood and psychosis spectrum disorders, including depression and schizophrenia.

Additionally, newly emerging studies suggest the possible key role of miRNAs in psychiatric pathologies (reviewed in Dinan, 2010). miRNA patterns were significantly altered in prefrontal cortex of patients with schizophrenia and bipolar disorder. Moreover, human microRNAs miR-22, miR-138-2, miR-148a, and miR-488 are associated with panic disorder and regulate several anxiety candidate genes and related pathways.

Conclusion

In summary, epigenetic changes are now considered to be the main regulators of normal brain development and function, as well as important players in a wide variety of neurological and psychiatric disorders, thereby constituting important targets for the future of epigenetic therapies.

Exercises and discussion topics

1. What is the role of DNA methylation in brain?
2. What are the roles of histone modifications and chromatic remodeling in the brain?
3. Which epigenetic mechanisms are important for the proper function of nervous system function and why?
4. What were the key conclusions of the Meaney-Szyf experiments?
5. Why was it important to expend the original Meaney-Szyf studies?
6. Which neurodevelopmental disorders result from gene mutations?
7. What is the molecular basis of the Rett syndrome?
8. Describe the epigenetic changes in Alzheimer's disease.
9. Describe the role of epigenetics in Huntington's disease.
10. Describe the epigenetic changes in multiple sclerosis.
11. Which epigenetic changes are important in psychiatric diseases?

References

Abel T, Zukin RSs (2008) Epigenetic targets of HDAC inhibition in neurodegenerative and psychiatric disorders. *Curr Opin Pharmacol* 8:57-64.

Baranzini et al. (2010) Genome, epigenome and RNA sequences of monozygotic twins discordant for multiple sclerosis. *Nature* 464:1351-1356.

Barbato et al. (2009) Searching for MIND: microRNAs in neurodegenerative diseases. *J Biomed Biotechnol* 2009:871313.

Barrachina M, Ferrer I. (2009) DNA methylation of Alzheimer disease and tauopathy-related genes in postmortem brain. *J Neuropathol Exp Neurol* 68:880-891.

Cacabelos R. (2007) Pharmacogenetic basis for therapeutic optimization in Alzheimer's disease. *Mol Diagn Ther* 11:385-405.

Chahrour et al. (2008) MeCP2, a key contributor to neurological disease, activates and represses transcription. *Science* 320:1224-1229.

Chen et al. (2009) The epigenetic effects of amyloid-beta(1-40) on global DNA and neprilysin genes in murine cerebral endothelial cells. *Biochem Biophys Res Commun* 378:57-61.

Chen et al. (2006) Valproate protects dopaminergic neurons in midbrain neuron/glia cultures by stimulating the release of neurotrophic factors from astrocytes. *Mol Psychiatry* 11:1116-1125.

Chen et al. (2007) Valproic acid and other histone deacetylase inhibitors induce microglial apoptosis and attenuate lipopolysaccharide-induced dopaminergic neurotoxicity. *Neuroscience* 149:203-212.

Chouliaras et al. (2010) Epigenetic regulation in the pathophysiology of Alzheimer's disease. *Prog Neurobiol*, 90:498-510.

Colvis et al. (2005) Epigenetic mechanisms and gene networks in the nervous system. *J Neurosci* 25:10379-10389.

Compston A, Coles A. (2008) Multiple sclerosis. *Lancet* 372:150215-17.

Dinan TG. (2010) MicroRNAs as a target for novel antipsychotics: a systematic review of an emerging field. *Int J Neuropsychopharmacol* 13:395-404.

Farrer et al. (1992) Inverse relationship between age at onset of Huntington disease and paternal age suggests involvement of genetic imprinting. *Am J Hum Genet* 50:528-535.

Feng et al. (2010) Dnmt1 and Dnmt3a maintain DNA methylation and regulate synaptic function in adult forebrain neurons. *Nat Neurosci*, 13:423-430.

Fischer et al. (2007) Recovery of learning and memory is associated with chromatin remodelling. *Nature*, 447:178-182.

Fraga et al. (2005) Epigenetic differences arise during the lifetime of monozygotic twins. *Proc Natl Acad Sci USA* 102:10604-10609.

Francis et al. (1999) Nongenomic transmission across generations of maternal behavior and stress responses in the rat. *Science* 286:1155-1158.

Gasser T. (2001) Genetics of Parkinson's disease. *J Neurol* 248:833-40.

Gregersen et al. (2006) Functional epistasis on a common MHC haplotype associated with multiple sclerosis. *Nature* 443:574-577.

Gupta et al. (2010) Histone methylation regulates memory formation. *J Neurosci* 30:3589-99.

Hahnen et al. (2008) Histone deacetylase inhibitors: possible implications for neurodegenerative disorders. *Expert Opin Investig Drugs* 17:169-184.

Handel et al. (2010) No evidence for an effect of DNA methylation on multiple sclerosis severity at HLA-DRB1*15 or HLA-DRB5. *J Neuroimmunol* 223:120-123.

Hebert SS, De Strooper B. (2007) Molecular biology. miRNAs in neurodegeneration. *Science* 317, 1179-1180.

Hughes RE. (2002) Polyglutamine disease: acetyltransferases awry. *Curr Biol* 12:R141-143.

Jenuwein T, Allis CD. (2001) Translating the histone code. *Science* 293:1074-1080.

Johnson et al. (2008) A microRNA-based gene dysregulation pathway in Huntington's disease. *Neurobiol Dis* 29:438-445.

Jowaed et al. (2010) Methylation regulates alpha-synuclein expression and is decreased in Parkinson's disease patients' brains. *J Neurosci* 30:6355-6359.

Kappeler L, Meaney MJ. (2010) Epigenetics and parental effects. *Bioessays*, 32, 818-827.

Kim et al. (2007) A MicroRNA feedback circuit in midbrain dopamine neurons. *Science* 317:1220-1224.

Kosik KS. (2006) The neuronal microRNA system. *Nat Rev Neurosci* 7:911-920.

Liggett et al. (2010) Methylation patterns of cell-free plasma DNA in relapsing-remitting multiple sclerosis. *J Neurol Sci* 290:16-21.

Marti et al. (2010) A myriad of miRNA variants in control and Huntington's disease brain regions detected by massively parallel sequencing. *Nucleic Acids Res* 38:7219-7235.

Mastronardi et al. (2007) Peptidyl argininedeiminase 2 CpG island in multiple sclerosis white matter is hypomethylated. *J Neurosci Res* 85:2006-2016.

Mattson MP. (2003) Methylation and acetylation in nervous system development and neurodegenerative disorders. *Ageing Res Rev* 2:329-342.

McCampbell et al. (2001) Histone deacetylase inhibitors reduce polyglutamine toxicity. *Proc Natl Acad Sci USA* 98:15179-15184.

McGowan et al. (2011) Broad epigenetic signature of maternal care in the brain of adult rats. *PLoS One* 6:e14739.

Mehler MF. (2008) Epigenetics and the nervous system. *Ann Neurol* 64:602-617.

Morrison et al. (1996) Brain S-adenosylmethionine levels are severely decreased in Alzheimer's disease. *J Neurochem* 67:1328-1331.

Nelson PT, Keller JN. (2007) RNA in brain disease: no longer just "the messenger in the middle." *J Neuropathol Exp Neurol* 66:461-468.

Nelson et al. (2008) MicroRNAs (miRNAs) in neurodegenerative diseases. *Brain Pathol* 18:130-8.

Otaegui et al. (2009) Differential micro RNA expression in PBMC from multiple sclerosis patients. *PLoS One* 4:e6309.

Peleg et al. (2010) Altered histone acetylation is associated with age-dependent memory impairment in mice. *Science* 328:753-756.

Pogribny IP, Beland FA. (2009) DNA hypomethylation in the origin and pathogenesis of human diseases. *Cell Mol Life Sci* 66:2249-2261.

Renthal W, Nestler EJ. (2008) Epigenetic mechanisms in drug addiction. *Trends Mol Med* 14:341-350.

Saito Y, Jones PA. (2006) Epigenetic activation of tumor suppressor microRNAs in human cancer cells. *Cell Cycle* 5:2220-2222.

Schaefer et al. (2007) Cerebellar neurodegeneration in the absence of microRNAs. *J Exp Med* 204:1553-1558.

Schipper et al. (2007) MicroRNA expression in Alzheimer blood mononuclear cells. *Gene Regul Syst Bio* 1:263-274.

Sleiman et al. (2009) Putting the "HAT" back on survival signalling: the promises and challenges of HDAC inhibition in the treatment of neurological conditions. *Expert Opin Investig Drugs* 18:573-84.

Urdinguio et al. (2009a) Epigenetic mechanisms in neurological diseases: genes, syndromes, and therapies. *Lancet Neurol* 8:1056-1072.

Urdinguio et al. (2009b) Epigenetic mechanisms in neurological diseases: genes, syndromes, and therapies. *Lancet Neurology* 8:1056-1072.

Wang et al. (2008) Variation in the miRNA-433 binding site of FGF20 confers risk for Parkinson disease by overexpression of alpha synuclein. *Am J Hum Genet* 82:283-289.

West et al. (1995) Hypomethylation of the amyloid precursor protein gene in the brain of an Alzheimer's disease patient. *J Mol Neurosci* 6:141-146.

Willer et al. (2003) Twin concordance and sibling recurrence rates in multiple sclerosis. *Proc Natl Acad Sci USA* 100:12877-12882.

Xin et al. (2012) MethylomeDB: a database of DNA methylation profiles of the brain. *Nucleic Acids Res* 40:D1245-1249.

Yu Z. (2008) Non-coding RNAs in Gene Regulation. In *Epigenetics*, ed. J. Tost, 171-186. Norfolk, UK: Caister Academic Press.

Zhang et al. (1992) Regulation of myelin basic protein-specific helper T cells in multiple sclerosis: generation of suppressor T cell lines. *Cell Immunol* 139:118-130.

Zuccato et al. (2003) Huntingtin interacts with REST/NRSF to modulate the transcription of NRSE-controlled neuronal genes. *Nat Genet* 35:76-83.

19

Epigenetics of health and disease—diet and toxicology, environmental exposures

Health and disease phenotypes have long been thought to be a consequence of an individual's genetic composition. There is a strong foundation for this line of thought, including the discovery of many genetically based diseases such as sickle-cell anemia, albinism, and many others. However, there are many other diseases in existence that cannot be exclusively or fully defined and explained by genetics. In addition, with the complete sequencing of the human genome, the realization that far fewer genes exist than had previously been thought and a surprising degree of gene conservation between humans and other species, there has come to be a greater acceptance of other factors that could be contributing to an organism's phenotype (McKie, 2001). One such factor has been the environment (McKie, 2001). As Dr. Craig Venter, the scientist who started Celera Genomics, so elegantly put it "We simply do not have enough genes for this idea of biological determinism to be right. The wonderful diversity of the human species is not hard-wired in our genetic code. Our environments are critical." (McKie, 2001).

During the past decade, it has become increasingly apparent that epigenetic changes modulate molecular, cellular and organismal responses to the changing environment (Jaenisch and Bird 2003, Wade and Archer 2006, Weidman et al. 2007, Jirtle and Skinner 2007). Every moment, living organisms interact with various aspects of the environment and are exposed to a variety of physical, chemical, and social factors. These interactions may be positive and negative.

The exact combination of exposure to different environmental factors as well as the timing and degree of exposure that each of us experiences varies extensively. All of them may potentially reshape not only the genome but also the epigenome (Kovalchuk 2008).

Linking environment to phenotypic and epigenomic changes

It is known that certain environmental factors can affect our health and influence the appearance and severity of many health-related phenotypes. Examples of this include the health effects of second-hand smoke and socioeconomic status. Second-hand smoke exposure has been clearly linked to the development lung cancer over the past half century, whereas different childhood socio-economic classes have been associated with differences in commonly used health indicators such as body mass index (BMI) and blood pressure (Power et al., 2007; Weaver, 2009; Zhu et al., 2003).

It has also been shown that epigenomic changes can alter gene expression as well as the phenotypes associated with the affected genes. Studies such as those that have been done on the mice agouti locus have managed to establish apparent links between all three: the environment, epigenetics, and phenotypes (Cooney, Dave, and Wolff, 2002). The Agouti locus has been connected to coat color, obesity, and cancer risk in the members of the Muridae family. Changes in diet (specifically in regard to the methyl content) have been shown to foster epigenetic changes at the Agouti locus and provoke phenotypic changes in the aforementioned traits (McEwen, 2008).

As we can see, there does appear to be some basis of support linking the environment and the phenotype to each other through epigenetic mechanisms. And although there are ample environmental factors to look at, this chapter focuses primarily on the prenatal environment, specifically with regard to maternal diet, the psychological environment and stress exposure, air pollution exposure, diet, and chemical and radiation exposure and their effects on health and the epigenome.

The importance of the prenatal environment

The prenatal stage during an organism's development is a highly important and sensitive period (Nafee et al., 2008). The prenatal environment to which the fetus is exposed during this time can produce major repercussions for the developing organism in later life stages (Nafee et al., 2008). During development, the expression of specific genes at specific times and varying levels of expression help determine the normal progression of development. The differential gene expression can be regulated either genetically through the recognition of DNA sequence regulatory elements or epigenetically (Nafee et al., 2008). Several adult diseases, such as hypertension and insulin resistance, have been linked to the conditions present in the prenatal environment (Unterberger et al., 2009). Given the importance of the establishment of DNA methylation patterns early during development, it is easy to see how exposure to epigenetic-altering factors during this period can affect the normal development and long-term health of the fetus (Dolinoy et al., 2007). A well-researched factor that has been found to affect epigenetic modifications in the prenatal environment is maternal diet.

Early work has shown that different nutritional conditions can produce different epigenetic patterns in mouse embryos being cultured *in vivo*. The following studies explore the resulting epigenetic effects that restricted maternal diets have on offspring in primates. The first study by Unterberger and colleagues (Unterberger et al., 2009) investigated general changes to the global DNA methylation landscape of baboons that occurred in parallel to maternal dietary restrictions. In this study, mothers were fed either a control diet or a nutrient-restricted diet (MNR) over the course of their pregnancies (Unterberger et al., 2009). Caesarean sections were performed at two different gestational ages, 0.5G and 0.9G (Term 185 dG), to see if methylation effects were age-dependent (Unterberger et al., 2009). Also, tissue-specific methylation patterns were investigated through the collection and methylation analysis of tissues from the liver, lungs, brain, and heart (Unterberger et al., 2009). The study revealed differences in methylation patterns associated with maternal diet that were tissue and age specific. In general, maternal diet restrictions seemed to have little effect on DNA methylation in the heart; however, a decrease in global methylation was observed in the kidney of 0.5 MNR group, and increases in global methylation were seen in kidneys and frontal cortex of 0.9 MNR group. The tissue specificity suggests that epigenetics might play a role in fetal organ development (Unterberger et al., 2009).

Usually, human studies are bound by much tighter ethical restrictions than animal studies, making it more challenging to study various effects directly in humans. Heijmans and colleagues attempted to overcome challenges by looking at archival data on the Dutch Hunger Winter (1944–45) (Heijmans et al., 2008). This event provided the researcher with a unique opportunity to study the effects of maternal diet restrictions in humans because it was marked by a defined period, food rations were controlled, and an uninterrupted presence of registries and healthcare made it easy to trace individuals exposed prenatally to famine. In this particular study, they looked at the IGF2 gene, a maternally imprinted gene involved in human growth in 60 different people. In each case, unexposed same-sex siblings were used as controls. After analyzing methylation of five different CpG dinucleotides in the IGF2 gene, they found that all but one was significantly less methylated in the prenatally-exposed-to-famine sibling, providing further support for the importance of early mammalian development as a time for establishing epigenetic patterns (Heijmans et al. 2008).

Further study gained even more insight into epigenetic effects of famine. In a follow-up study, Tobi and colleagues analyzed methylation of 15 loci implicated in metabolic and cardiovascular diseases in individuals who were prenatally exposed to famine in 1944–45. Researchers analyzed imprinted loci (GNASAS, GNAS A/B, MEG3, KCNQ1OT1, INSIGF, and GRB10), a putatively imprinted locus (IGF2R), and non-imprinted loci (IL10, TNF, ABCA1, APOC1, FTO, LEP, NR3C1, and CRH) (Tobi et al, 2009).

The study revealed that methylation of *INSIGF* was lower among individuals who were periconceptionally exposed to famine as compared to their unexposed same-sex siblings, whereas methylation of *IL10*, *LEP*, *ABCA1*, *GNASAS*, and *MEG3* loci was higher. Interestingly, methylation of *INSIGF*, *LEP*, and *GNASAS* was sex-specific. Taken together, these data show that persistent changes in DNA methylation constitute an important common result of prenatal famine exposure and that these DNA methylation changes are dependent on sex and gestational timing of exposure (Tobi et al., 2009).

Both primate and human studies associate specific epigenetic changes with altered maternal diets, thus indicating that a relationship between environmental exposures and the dynamic epigenetic landscape might exist. The related findings of these studies also open the door for future studies investigating epigenetics as a possible mechanism to explain some of the associations that have been made between health environment and phenotype.

Psychological environment—stress and health implications

Stress is an encompassing term used to describe factors that elicit physiological and behavioral stress responses in reaction to an existent or illusionary threat to the organism's well-being in a certain environment. The ability to experience stress and respond to it in an appropriate manner is an important and vital aspect of good health. On the other hand, the dysfunction of this system can turn it from something originally designed to protect the organism into something that threatens the integrity of what was meant to protect (de Kloet et al., 2009). Early exposure to stress has been associated with negative effects on growth and intelligence as well as to increased risks of mental illnesses and obesity later on in life (Cirulli et al., 2009). However, the negative effects of stress are not restricted to the early stages of development as is evidenced by the development of post-traumatic stress disorder (PTSD) after exposure to a potentially traumatic event (PTE) (Uddin et al., 2010).

As has been mentioned, the early years of an organism's life are a sensitive period when it is particularly vulnerable to certain environmental exposures (Szyf, 2009). In rats, maternal care has been found to affect the hypothalamic-pituitary-adrenal (HPA) stress response by causing alterations in the epigenome that affects glucocorticoid receptor expression (see Chapter 18, "Epigenetics of Health and Disease: Behavioral Neuroscience"). A study by McGowan and colleagues built on this research and applied it to humans by investigating the epigenetic effect that childhood abuse has on the human NR3C1 neural glucocorticoid receptor promoter, a homolog of the affected glucocorticoid receptor in rats.

Because of the previously established link between a history of childhood abuse and the risk of suicide, this study decided to look at the NR3C1 promoter in post-mortem hippocampus obtained from twelve individuals in each of the following categories: suicide victims with a childhood abuse history, suicide victims with no history of childhood abuse, and individuals who died from causes unrelated to suicide. For each of the samples, they performed quantitative reverse transcription PCR (qRT-PCR) on DNA isolated using a chromatin immunoprecipitation assay to determine expression levels for the NR3C1 receptor and sodium bisulfite mapping to determine changes in cytosine methylation within the NR3C1 promoter region. Through their analysis, they found that the presence of glucocorticoid mRNA was significantly reduced in abused suicide victims in comparison to other two groups. It was also found that there was increased cytosine methylation in the promoter region that corresponds to the known relationship found between hypermethylation of DNA and the repression of gene expression. Furthermore, they sequenced the promoter region of NR3C1 as well as the qRT-PCR primer-binding regions in each sample to eliminate the possibility that a change in the DNA sequence could be accounting for any differences in expression. Sequencing yielded no differences between subjects (McGowan et al., 2009).

To further solidify their findings, researchers wanted to show that hypermethylation at this promoter region can alter the efficiency of transcription factor binding. Therefore, they looked at luciferase expression levels in HEK293 cells transfected with an expression vector containing the NR3C1 promoter region next to a promoterless luciferase construct either in the presence or absence of NGFI-A, a transcription factor known to interact with that regulatory region. In cases where the NR3C1 promoter region had been hypermethylated prior to transfection, transcriptional activity was reduced. Together these results suggest that similar to what was seen in the rat model, parental care can potentially affect stress response in later life through epigenetically regulated alterations in neural glucocorticoid receptors (McGowan et al., 2009).

Stress also has been shown to affect the health and well-being of older individuals. PTSD is a debilitating condition after exposure to a traumatizing event. It is characterized by subsequent abnormal responses to stress brought about by HPA axis dysregulation (Uddin et al., 2010). Previous work has shown that there are particular changes in the immune system that are seen in individuals with PTSD. A study by Uddin and colleagues sought to link this disease and its distinctive immune system changes to modifications in the epigenome. In order to do this, they looked at DNA methylation in 27,000 CpG sites of 14,000 genes using the Illumina methylation microarray (Uddin et al., 2010). The DNA they used for this analysis was obtained from 100 different individual blood samples (23 of which came from individuals diagnosed with PTSD) (Uddin et al., 2010). Their results indicated that certain clusters of functionally related genes in blood samples of PTSD-affected individuals tended to share methylation statuses that were different from those seen in the PTSD-unaffected blood samples. In the case of the immune-related genes under investigation, there appeared to be a general pattern of hypomethylation corresponding to the enhanced immune system response associated with individuals suffering with PTSD. A few other interesting findings included hypermethylation of genes associated with sensory perception of sound, which ties in with one of the major symptoms of PTSD; hypomethylation of genes implicated in DNA methylation; and imprinting-related diseases (Uddin et al., 2010).

In the past few years, new research has begun to look at RNA-mediated epigenetic modifications as a potential source of mediation between stress in the environment and phenotypic changes to health (de Kloet et al., 2009). It has recently been predicted that the 3' UTR region of hippocampal glucocorticoid receptors might contain areas recognizable to several different miRNAs (de Kloet et al., 2009). Further research into some of these particular miRNAs and their implications for the regulation of glucocorticoid receptors in the brain has been promising (Vreugdenhil et al., 2009). However, the role that the environment might have on these interactions remains to be identified (de Kloet et al., 2009).

Epigenetic effects of air pollution

The progression of industrialization and an increase in world population have led to rising levels of air pollution. Over the past few decades, elevated levels of air pollution have emerged as prospective contributing factors for the development of various diseases and health detriments (Wilker et al., 2010). The following section reviews two studies looking into the association of air pollution exposure with the development of elevated blood pressure and asthma as well as epigenetic modifications that might be involved in these relationships.

Black carbon is a combustion derivative that is used as a measure of traffic pollution. It has been associated with cardiovascular morbidity as well as the conditions associated with cardiovascular disease. In separate research study, it has been shown that miRNA is involved in the growth of new blood vessels, a process called angiogenesis. Related to this is the association between improper regulation of miRNA pathways involved and heart disease (Wilker et al., 2010). The study by Wilker and colleagues sought to unify the two. Over the course of 13 years, this study followed 789 white males determined to be free of any chronic medical conditions. All participants lived on average about 17.6 km from a testing site. After the initial screening, subjects visited the study center for further data collection, which included blood samples, measurements of systolic and diastolic blood pressure, BMI, and lifestyle questionnaires. The levels of black carbon were measured hourly for seven days prior to visits and then averaged. After making adjustments to compensate for potentially confounding factors by using two different models, they found that there was an increase in both systolic and diastolic blood pressure for a standard deviation increase in black-carbon levels (0.415ug/m^3). Next, they genotyped 19 single nucleotide polymorphisms (SNPs) from ten different genes. SNPs were selected based on previous studies that linked miRNA processing genes with various diseases, including cardiovascular disease. They considered SNPs found to be homozygous in at least ten or more of the participants for statistical purposes. Using this data, they were able to implicate four SNPs to significantly different relationships between black-carbon levels and blood pressure levels (Wilker et al., 2010). These results suggest the possibility that RNA-mediated epigenetic mechanisms might be mediating the effects of black carbon exposure on blood pressure.

Asthma is a very broad disease that is defined by temporary airway blockages and inflammation. Through past research, it has been established that air pollution levels correlate with exacerbation and onset risks of asthma. Air pollution has also been linked to changes in the epigenome (Nadeau et al., 2010). Furthermore, it has been shown that the activity of regulatory T (or Treg) cells, a factor involved in the progression and development of asthma, may be altered by DNA hypermethylation of the Forkhead box transcription factor 3 (Foxp3). In the study by Nadeau and colleagues, they looked at asthmatic and asthma-free children from both a high pollution area (Fresno, California) and a low pollution area (Stanford, California) to establish a possible connection between ambient air pollution, hypermethylation of the Foxp3 gene, a compromised Treg-cell function, and asthma severity. During the study, air quality and meteorological data were tracked, and blood specimens were collected and used for functional assays, flow cytometry studies, methylation studies, and quantitative PCR (Nadeau et al., 2010).

The study uncovered several interesting correlations and findings to support the association between air pollution, epigenetic changes in the Foxp3 locus, and asthma. Functional assays in the study indicated a larger degree of immune suppression/impairment of Treg cells in children suffering with asthma from the Fresno area (FA group) in comparison to other groups. The range of functional impairment was the lowest in the asthma-free group from Stanford (SNA group). Having looked at Foxp3 transcript numbers and the relative percentage of Foxp3 Treg cells, they found that they were both the lowest in the FA group. They also learned that the severity of Global Initiative in Asthma scores (GINA) for the individuals in the FA and asthma group from Stanford (SA) were related to the reduced percentages of Foxp3 Treg cells. Moreover, the severity was generally worse in the FA group than in the SA group, which was also in accordance with the reduced percentage. Finally, the study looked into the differences in DNA methylation in 8 CpG promoter regions and 13 intronic CpG islands in the Foxp3 locus of Treg cells. Methylation levels were the highest in the FA group, followed by the FNA and SA groups, respectively. The lowest levels of methylation were seen in the SNA group. The trends correlated to each other as was expected, thereby giving credence to the notion that air pollution might drive increases in DNA methylation at the Foxp3 locus leading to the impaired Treg cell function, which has ramifications for the pathogenesis of asthma (Nadeau et al., 2010).

Along with studies of exposed human populations, several studies used a well-established murine model to discern the epigenetic effects of air pollution. A study by Yauk and colleagues analyzed DNA damage and DNA methylation in sperm of mature male C57BL/CBA F1 mice that were exposed to HEPA-filtered or ambient air in Hamilton, Ontario, Canada, near two integrated steel mills and a major highway (Yauk et al., 2008). Air quality data were provided by the Ontario Ministry of the Environment (MOE). For the majority of the time, the enclosures and air samplers were downwind of the steel mills. Exposures were the lowest at the beginning and end of the experiment. Hours of westerly wind, total suspended particles (TSP) (mean 93.8 ± 17.0 µg/m3), and metal (mean 3.6 ± 0.7 µg/m3) concentrations were the highest at week 4. Week 3 had the highest PAH concentration (mean 8.3 ± 1.7 µg/m3). Benzo(b)-fluoranthene, benzo(k)fluoranthene, and indeno(123-cd)pyrene were the three most abundant PAHs. Iron, copper, and manganese were the three most abundant metals (see Figure 19-1).

The analysis revealed an increase in global methylation in sperm from mice exposed to whole air compared with HEPA filtered air, using both the cytosine extension assay (CEA) and methyl-acceptance assay (MAA) (see Figure 19-2a and b). A factorial analysis revealed a significant treatment effect (P = 0.004 and P = 0.001 for the two assays, respectively). There was no significant changes in methylation at three weeks (P = 0.126 and P = 0.162 for the two assays, respectively). Hypermethylation arose in mice exposed continuously to particulate air pollution. Furthermore, hypermethylation persisted in 16-week samples. There might be several important implications of hypermethylation observed. Indeed, DNA hypermethylation may be linked with structural changes in chromatin, decreased gene expression, and decreased rates of transposon movement. Furthermore, increased levels of DNA methylation in gametes might lead to potential problems for progeny, but it is unknown whether methylation changes inherited via mature sperm lead to altered epigenetic reprogramming in the fertilized egg. Global methylation measured here does not necessarily reflect the methylation status of active genes; therefore, the consequences of these changes are unknown. The relationship between methylation and instability of expanded tandem repeat sequences remains to be elucidated. Moreover, alterations in global methylation affecting gene expression, chromatin structure, or genome stability could have an

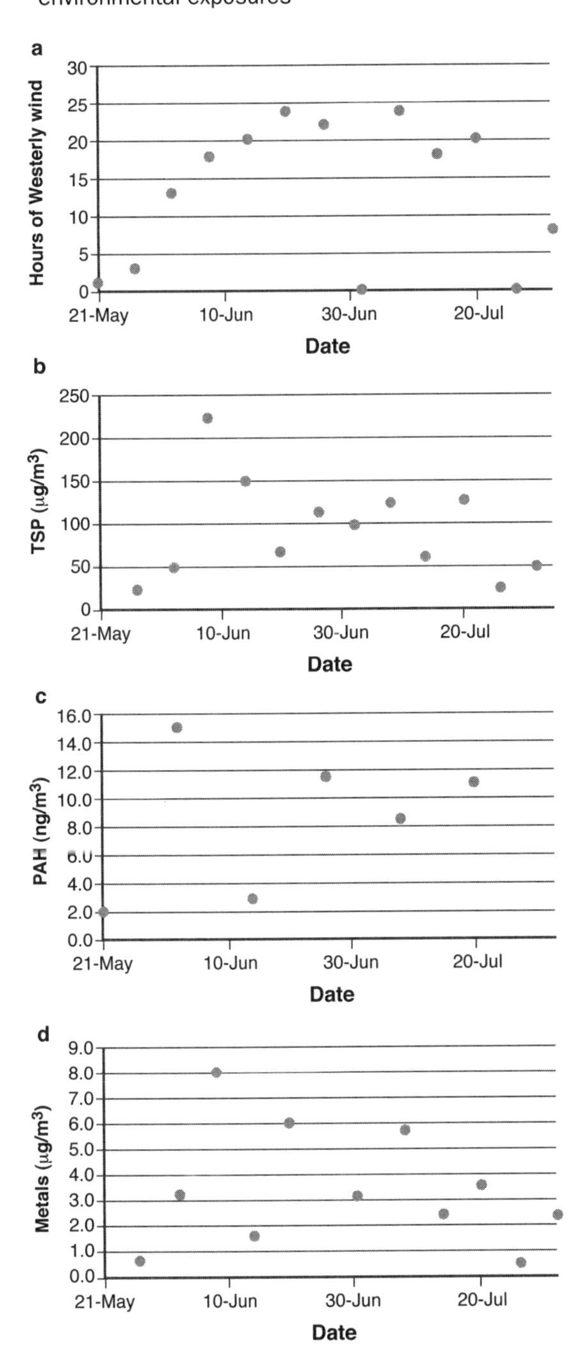

Figure 19-1 Air quality parameters. Air samplers were located approximately 400 m north-east of the mouse sheds. (Adapted with permission from Yauk C et al. Germ-line mutations, DNA damage, and global hypermethylation in mice exposed to particulate air pollution in an urban/industrial location. Proc Natl Acad Sci U S A. 2008;105:605-10.) The data were collected from the MOE database from May 14 to August 1, 2004. (**a**) Hours of westerly wind. (**b**) Total suspended particulate. (**c**) Polycyclic aromatic hydrocarbons. (**d**) Metals.

effect on spermatogonial stem cell function. It is possible that alterations in methylation are transmitted to subsequent generations, providing a persistent epigenetic signal. Future work should target localized regions of DNA and investigate the potential of urban air pollutants to modulate methylation of important genomic sites (Yauk et al., 2008).

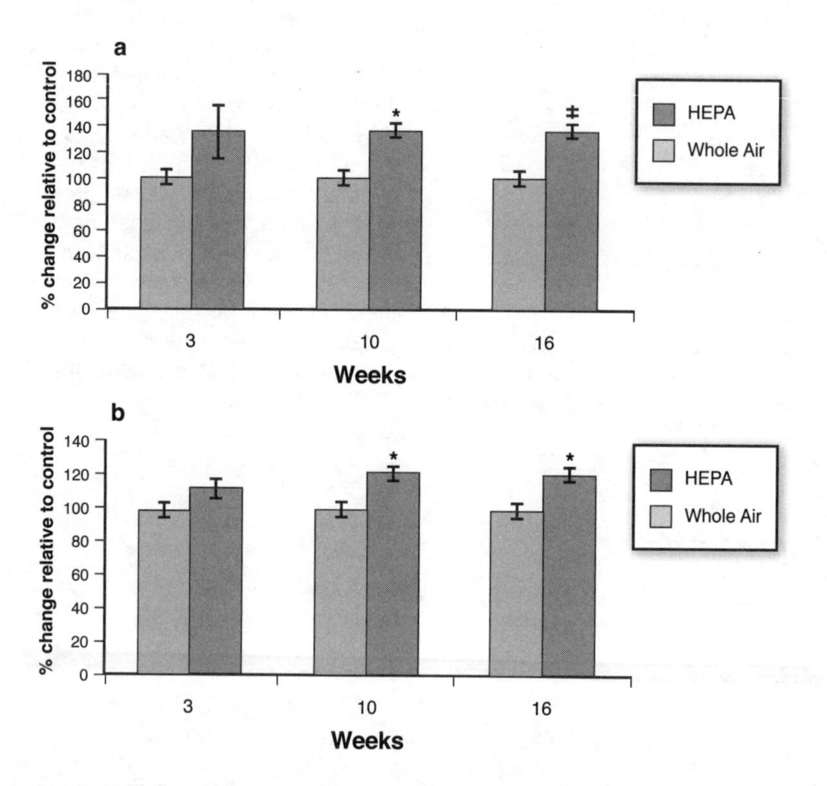

Figure 19-2 Global methylation of male germ-line DNA. (Adapted with permission from Yauk et al., 2008.) (**a**) The percentage of global methylation changes relative to age-matched controls, using the cytosine extension assay. (**b**) The percentage of methylation changes relative to age-matched controls, using the methyl-acceptance assay. Bars indicate standard error. *, $P < 0.05$; [dd], $P = 0.067$, two-tailed Wilcoxon rank test.

Use of twin models to study environmental effects

Monozygotic (MZ) twins provide a unique model for researchers to study the effect environmental factors can have on an organism. The phenomenon occurs if two individual cells or groups of cells arising from a single zygote begin to divide independently of one another, thus leading to the development of two separate organisms known as MZ twins (Singh et al., 2002). Although they are genetically identical, substantial differences in phenotypes between the two might exist. Twin studies have been used as a tool to help investigate what factors other than DNA sequence might contribute to phenotypic discordance and the mechanisms behind it (Fraga et al., 2005; Singh et al., 2002). As we have seen in the aforementioned research, the only mechanism that has been proposed as a means of guiding these differences is epigenetics. An interesting study by Fraga and colleagues sought to investigate whether or not epigenetics has a place as a genuine candidate for this role.

The study looked at 80 Caucasian twins ranging in age from 3 to 74. To obtain a value of phenotypic/environmental differences between the twins, extensive questionnaires were administered and analyzed. Quantification of global histone H3 and H4 acetylation and DNA methylation was done using lymphocyte cells in order to quantify epigenetic differences in twin pairs (Fraga et al., 2005). In both cases—histone modification and DNA methylation—the study found the youngest twins to be most epigenetically similar and the oldest twins to be most epigenetically different, even after taking into account the possible contribution of stochastic events. The level of epigenetic differences between twin pairs also generally corresponded to how much relative time they had spent apart and the degrees of differences existent in their respective natural health-medical histories, with the highest epigenetic differences being among those spending the most time apart and/or having the greatest differences in history. Finally using real-time quantitative PCR expression analyses, they were able to correlate different DNA methylation patterns to different gene expression levels in MZ twin pairs. They found similar results if they did similar analyses using epithelial mouth cells, intraabdominal fat and skeletal muscle biopsies. This study validates epigenetics as a prospective candidate in this area and encourages future studies investigating a possible role of epigenetics in the relationship between the environment and the phenotype (Fraga et al., 2005).

Diet—epigenetic effects of bioactive food compounds

The causal link of diet, toxicology, and epigenetics in human disease development is significant, but studies are still in their infancy (Duthie, 2010b). The majority of analyses on epigenetic mediation of risk have associated DNA methylation and cancer through diet.

Bioactive food components

Folate, choline, methionine, zinc, epigallocatechin gallate (ECGC), diallyl disulfide, resveratrol, sulforaphane, genistein, and vitamin B_{12} are just a few of the bioactive food components (BFCs) and micronutrients receiving support from epidemiological and preclinical studies for roles in modulating cancer risk (Duthie, 2010b; Ross, 2010).

These compounds (see Table 19-1) have been shown to affect epigenetic processes and have positive outcomes, such as controlling proliferation, upregulating apoptosis, reducing inflammation (Duthie, 2010b), and regulating DNA repair, differentiation, hormonal balance, and angiogenesis (Ross, 2010). Most bioactive food compounds, such as folate and flavonoids (genistein and EGCG), are associated with a decreased risk of heart disease and cancers (Duthie, 2010b).

Table 19-1 Selected bioactive food compounds (BFCs) that affect epigenetic status.

Compound	Food source
Sulforaphane (SFN)	Cruciferous vegetables; broccoli sprouts
Diallyl disulfide (DADS)	Garlic
Folate/Folic acid/B_9 micronutrient	Green leaf vegetables, legumes, liver, egg yolks, among others
Genistein	Soy
Epigallocatechin gallate (ECGC)	Green tea
Curcumin	Turmeric

Effects of BFCs on DNA methylation

In an elegant review, Sharon Ross (2010) outlined at least four modes by which bioactive food compounds and nutrients may influence DNA methylation (Ross, 2010):

- By affecting the supply of methyl groups and production of S-adenosylmethionine (SAM).
- By altering DNA methyltransferase activity.
- By affecting DNA demethylation activity.
- The DNA methylation pattern itself might affect the cellular responses to a given nutrient.

Bioactive food compounds in diet have been shown to reactivate genes (for example, tumor suppressor genes) that were epigenetically silenced by DNA methylation in cultured cells (Ross, 2010; Fang et al., 2003; Fang et al., 2007). For example, in studies EGCG and genistein inhibited the activity of DNA methyltransferases and restored methylation patterns and gene expression of tumor suppressor genes in neoplastic cultured cells (Fang et al., 2003; Fang et al., 2007). The epigenetically reactivated genes in human colon, prostate, mammary, and esophageal cancer cell lines included *p16INK4a*, *O^6-methylguanine methyltransferase*, *human mutL homolog1*, *retinoic acid receptor beta*, and many others (Fang et al., 2003, Fang et al., 2007). Importantly, a chronic deficiency in major dietary bioactive methyl group donors such as methionine, choline, folic acid, and vitamin B_{12} can induce the development of liver cancer in rodents (reviewed by Pogribny, 2012).

Furthermore, important evidence connecting diet and DNA methylation stems from an agouti (A^{vy} metastable epiallele) murine model. This mouse model has been successfully used to study how the fetal epigenome and phenotype are affected by maternal nutrition (Dolinoy, 2008; Duthie, 2010b). In general, metastable epialleles are genetically similar alleles but differ in expression due to specific epigenetic modifications that occur early during development (Dolinoy, 2008). In the agouti model, the wild-type allele encodes a paracrine-signaling molecule that produces eumelanin (*a*) (brown) or pheomelanin (*A*) (yellow) fur. In A^{vy} mice, an intracisternal A particle (IAP) retrotransposon is inserted upstream of the transcription start site for the agouti gene. Methylation of the 5' long terminal repeat (LTR) end of the IAP decreases agouti expression and results in a wild-type phenotype (brown coat). Hypomethylation of this LTR results in a yellow coat.

The offspring's yellow coat color is linked to the methylation status of the agouti gene, and this status was found to depend on the amount of bioactive compounds such as polyphenols, phytoestrogens, and methyl group donors (for instance, folic acid, vitamin B_{12}, choline, and so on) in the mother's diet during early stages of fetal development (Dolinoy, 2008; Duthie, 2010b; Ross, 2010). Several recent studies demonstrated that methyl donor supplements in a mother's diet could alter an offspring's phenotype (Ross, 2010); however, which supplement is necessary or sufficient for the epigenetic and phenotypic change is not yet clear (Ross, 2010). The yellow phenotype is also associated with an increased risk of obesity, and susceptibility to cancer and other chronic diseases (Ross, 2010). Importantly, these disease risks decrease with dietary supplementation with bioactive foods, promoting methylation of the IAP promoter.

Among bioactive food compounds, genistein and soy positively affect coat color and DNA methylation in agouti mice. In addition, the resulting DNA hypermethylation at the A^{vy} locus seems to protect against adult-onset obesity, further suggesting that maternal diet could alter disease susceptibility (Ross, 2010).

The agouti mouse model has also served as a biosensor for toxicants in the diet (Ross 2010). Among the most studied, bisphenol A has been used to manufacture polycarbonate plastic and is associated with increased body weight, increased risk of breast and prostate cancer, and changed reproductive function (Dolinoy, 2008). In one study, after maternal feeding of bisphenol A pre-mating and throughout gestation and lactation, the offspring's coat color shifted to yellow, and methylation was decreased within the A^{vy} IAP. Interestingly, in that study supplementation with either methyl donors or genistein counteracted the negative effects of toxicants on DNA methylation. Therefore, even though toxins can change offspring's phenotype, it can be "rescued" with proper supplementation of the maternal diet.

Effects of BFCs on histone modifications

Bioactive dietary compounds also influence histone modification levels. Among those, sulforaphane (SFN), diallyl disulfides, and resveratrol inhibit histone deacetylase activity to increase transcription factor access to DNA and induce gene expression (Duthie, 2010b). SFN was shown to inhibit HDAC activity and increase histone acetylation in HCT116 human colorectal cancer cells, various prostate epithelial cells (BPH-1, LnCaP, and PC-3), and human embryonic kidney 293 cells (Ross 2010). Furthermore, SFN dose-dependently increased the levels of acetylated histone H4 at the *p21* promoter, leading to increased p21 protein expression.

In the *Apc*^{min} mouse model, administration of SFN increased histone acetylation in gastrointestinal cells, and was correlated with suppressed tumor development.

Interestingly, BFC supplementation has shown beneficial results in a practical setting. In healthy humans, one cup of broccoli sprouts inhibited HDAC activity, and increased histone H3 and H4 acetylation, in blood mononuclear cells only three to six hours after eating (Ross, 2010).

Administration of garlic-derived compound DADS was correlated with the increased H4 and/or H3 acetylation in the *CDKN1A* promoter (Druesne-Pecollo, Chaumontet, and Latino-Martel, 2008). Furthermore, this was associated with increased levels of CDKN1A mRNA and p21 protein levels. These changes contributed to reduced proliferation and G2/M phase cell cycle arrest in HT-29 and Caco-2 human cancer cell lines (Ross, 2010; Druesne-Pecollo et al., 2008).

Genistein was found to affect histone modifications differently from SFN and DADS. Chromatin remodeling was altered by reducing histone H3 lysine 9 (H3K9) methylation and deacetylation. In addition, genistein induced *p21* and *p16* expression and decreased cyclin in prostate cancer cells by increasing acetylation of histones H3 (H4K4) and H4.

Effects of BFCs on microRNA

Dietary insufficiency and lack of methyl donor factors methionine, choline, vitamin B_{12}, and folic acid is known to lead to the formation of liver cancer in rodents. This deficiency is associated with global genome and gene-specific hypomethylation, and aberrant expression of epigenetic mediators (DNMTs, methyl CpG binding proteins, and HMTs). Moreover, studies have also linked dietary deficiency to altered levels of microRNAs. Among these affected by dietary deficiency were microRNAs that regulated apoptosis, cell proliferation, and cell-to-cell connections. Expression of *miR-34a*, *miR-127*, *miR-200b*, and *miR-16a* was inhibited, which was correlated with changes in their target proteins (E2F3, NOTCH1, BCL6, ZFHX1B, and BCL2) targets. Furthermore, changes in miRNA expression occurred early during the carcinogenesis process, and the epigenetic effects were persistent and not reversed when rats were fed a methyl-adequate diet.

Curcumin is one BFC shown to alter the miRNA profile in human BxPC-3 pancreatic cancer cells (Ross, 2010; Sun et al., 2008). In one study, 11 miRNAs upregulated 18 miRNAs downregulated by a 72-hour treatment with 10 μmol/L of curcumin. miRNA-22 was upregulated, and its putative targets were the SP1 transcription factor and estrogen receptor 1 (ESR1). Antisense miRNA-22 enhanced the expression of those targets, suggesting that curcumin is a BFC that might play a major role in mediating anticancer effects in pancreatic cells through epigenetic mechanisms (Sun et al., 2008).

Diet epigenetics—perspectives

In sum, the phenotype of an organism is influenced by compounds found in nutrition and in the environment. The compounds discussed here (folate, SFN, DADS, curcumin, genistein, BPA) barely touch the surface of the variety of compounds able to alter the epigenome. For instance, hundreds of studies have elaborated on the effects of folate. Folate itself is a coenzyme critical for methylation and for nucleotide synthesis (Choi and Mason, 2002). Folate has been associated with preventative mechanisms for cancers of the lung, esophagus, brain, pancreas, bone marrow, cervix, and especially colorectum (Choi and Mason, 2002). Many studies (Kim et al., 1996; Duthie, 2010a) have outlined the specifics of this one compound as there are also, although less numerous, reviews discussing SFN, DADS, and the like. The volume of associated literature demonstrates how relevant and abundant the effects of dietary compounds are on biological frameworks.

Although there are many studies, correlation of the results to practical applications is lacking or unclear, especially in relation to humans.

In fact, some epigenetic modulators are currently being put into practice. There has been recent success in using HDAC and DNMT inhibitors for therapeutic intervention of chronic inflammatory disease (Szarc vel Szic et al., 2010; Altucci and Stunnenberg, 2009). For some, the correlations between dietary studies on model systems and humans are based on theoretical principles.

There are a multitude of compounds in certain foods, adding more confounding factors to correlating these studies with practical applications. In addition, some compounds consumed in the diet might also have negative epigenetic effects. The best example is soy. Although soy genistein seems to have wondrous gene expression control implications through altering histone modifications, the other constituent lunasin, also found in soy, might modify chromatin in an opposite manner (de Lumen, 2005). Lunasin is also stated to have cancer prevention ability by inhibiting acetylation of histones and killing actively transforming cells (Ross 2010; de Lumen, 2005). It acts by binding to deacetylated histones, which could upset the mode by which genistein works.

A major limitation to applying these studies to humans is based on compound concentration. In general, the concentrations of the bioactive food components used in many studies were higher than what could be achieved nutritionally (Duthie, 2010b). In consumed foods, there are low amounts of polyphenolic compounds, and the effect of these on DNA methylation in humans is not clear (Ross, 2010). In the agouti mice studies, the diets consisted of many methyl donors and cofactors, implying that a combination of BFCs might be necessary. More so, excessive amounts of polyphenols might cause excessive and unwanted modifications as well (Ross, 2010).

The EGCG levels used in the experimental studies were up to fiftyfold higher than the blood and urine concentrations after drinking tea and can be achieved by drinking 18 or more cups of green tea a day. Similarly, the genistein concentrations used may be three- to tenfold higher than what can be achieved from eating soy products. Confounding factors also come into play. Polyphenols are rapidly metabolized in bodies through mechanisms of glucuronidation, sulfation, and methylation (Duthie, 2010b). These processes might contribute to the low internal availability of BFCs for *in vivo* effects.

Radiation-induced epigenetic changes

All living organisms are exposed to ionizing radiation (IR) on a day-to-day basis. IR is an essential diagnostic and treatment modality, yet it is also a potent genotoxic and cancer-causing agent. In addition to diagnostic and therapeutic medical radiation exposures, most radiation exposures today stem from background radiation, cosmic rays, radioactive waste, radon decay, nuclear tests, and accidents at the Chernobyl, Fukushima, and other nuclear power plants.

Direct IR exposure effects

Direct IR exposure strongly influences epigenetic effectors. Exposure to radiation has been reported to affect DNA methylation patterns (Kalinich et al., 1989; Tawa et al., 1998; Kovalchuk et al., 2004). Acute exposures to X-rays or γ-rays were noted to result in global hypomethylation (Tawa et al., 1998; Kalinich et al., 1989). It was recently shown that the IR exposure leads to the profound dose-dependent and sex- and tissue-specific global DNA hypomethylation. The IR exposure also affects methylation of the promoter of the p16 tumor suppressor in a sex- and tissue-specific manner. The DNA hypomethylation observed after irradiation was related to DNA repair. It also correlated with the IR-induced alterations in the expression of DNA methyltransferases, especially *de novo* methyltransferases DNMT3a and DNMT3b. Most importantly, the radiation-induced global genome DNA hypomethylation appeared to be linked to genome instability in the exposed tissue.

Among the histone modifications that change upon radiation exposure, phosphorylation of histone H2AX is being studied most intensively. Histone H2AX, a variant of histone H2A, is rapidly phosphorylated at Ser139 upon the induction of DNA strand breaks by irradiation, and it can be effectively visualized within repair foci using phosphospecific antibodies (Sedelnikova et al., 2003). Recent studies have also indicated that radiation-induced global loss of DNA methylation might correlate with the changes in histone methylation, specifically with the loss of histone H4 lysine trimethylation (for a detailed review of radiation-induced epigenetic effects, see Ilnytskyy and Kovalchuk, 2011; Kovalchuk and Baulch, 2008).

IR exposure also significantly affects microRNAome of the exposed animal and human tissues. IR-induced microRNAome changes are sex and tissue specific and persist over a long time (reviewed in Dickey et al., 2011).

Epigenetics of indirect IR effects

Even though a significant body of evidence points toward the epigenetic nature of the radiation-induced bystander and transgeneration effects, until recently very few studies addressed the exact epigenetic changes related to the indirect radiation response.

Epigenetics of IR-induced bystander effects

The pioneering work of Kaup et al. has shown that DNA methylation is important for the maintenance of the radiation-induced bystander effect in cultured cells. Using cultured human keratinocytes, they demonstrated that the dysregulation of DNA methylation in naive cells exposed to the medium from the irradiated cells persists for 20 passages (Kaup et al., 2006). A wide range of extensive cell-culture-based studies address the role of phosphorylated histone H2AX in bystander effects (reviewed in Kovalchuk and Baulch, 2008).

Epigenetic changes were also shown to be important in whole-tissue- and whole-organism-based bystander effect models. The reconstituted 3D human tissue model offers an excellent alternative to cell cultures. The recent study by Sedelnikova and colleagues examined bystander effects in two reconstructed human 3D tissue models: airway and full-thickness skin. Following the microbeam irradiation of cells located in a thin plane through the tissue, a variety of biological endpoints were analyzed in distal bystander cells (up to 2.5 mm away from the irradiated cell plane) as a function of post-exposure time (zero hours to seven days). They detected a significant increase in the levels of phosphorylated H2AX in bystander tissues and extensive long-term increases in apoptosis and micronucleus formation, as well as the loss of nuclear DNA methylation (Sedelnikova et al., 2007).

Further insight into the role of epigenetic changes in bystander effects comes from the animal-based studies, where irradiation was shown to induce DNA damage and modulate the epigenetic effectors in distant bystander tissues.

The Kovalchuk and Engelward laboratories pioneered in studies on the role of epigenetic changes in radiation-induced bystander effects *in vivo*. To analyze *in vivo* bystander effects, they developed a mouse model whereby half of an animal body was exposed to radiation, and the other half was protected by a medical grade shield (see Figure 19-3) (Koturbash et al., 2006b). This model was used to monitor the induction and repair of DNA strand breaks in the cutaneous tissue.

In addition to this well-established endpoint, the authors also explored the possibility of epigenetic mechanisms (that is, DNA methylation and alterations in DNA methyltransferases and methyl-binding proteins) in the generation and/or maintenance of a radiation-induced bystander effect in cutaneous tissue. They have shown that radiation exposure to one half of the body leads to elevated levels of DNA strand breaks, and altered levels of key proteins that modulate methylation patterns and silencing in the bystander half of the body at least 0.7 cm from the irradiated tissue.

In a follow-up study, researchers investigated the possibility that the localized X-ray irradiation induces long-term persistent epigenetically modulated bystander effects in distant tissues. Researchers monitored the occurrence of epigenetic changes (that is, DNA methylation, histone methylation, and miRNA expression) in spleen tissue seven months after the localized cranial irradiation. This analysis has revealed that the localized cranial radiation exposure led to the decreased levels of global DNA methylation. It also altered the levels of key proteins that modulate methylation patterns and silencing (that is, *de novo* methyltransferase DNMT3a and methyl-binding protein MeCP2) and contributed to the reactivation of the LINE1 retrotransposon in the bystander spleen, located at least 16 cm from the irradiation site (Koturbash et al., 2007).

Overall, it is now well-accepted that the bystander effect occurs *in vivo* in distant tissue, persists over a long period of time, and is epigenetically regulated in a tissue-specific manner.

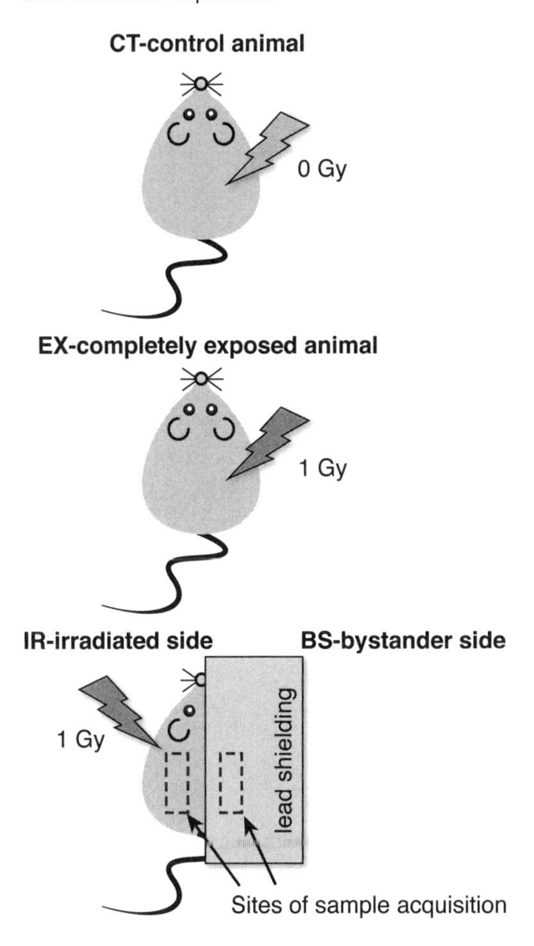

Figure 19-3 Induction of in vivo bystander effect. (Adapted with permission from Koturbash I et al. Irradiation induces DNA damage and modulates epigenetic effectors in distant bystander tissue in vivo. Oncogene. 2006;25:4267-75.) Animals had lead shielding covering half their bodies during exposure to 1 Gy of X-rays. Animals in the unirradiated control cohort were sham treated. Animals in the fully irradiated cohort received 1 Gy of whole-body X-ray exposure. (CT—unirradiated skin; IR—skin from fully irradiated animals; IR ½—skin from the irradiated side of hemi-shielded animals; BS ½—skin from the shielded, bystander side of hemi-shielded animals.) Note that ventral skin was taken from the area adjacent to the thigh, thus bystander samples were taken at least 0.7 cm from the edge of the irradiated side.

Additionally, the IR exposure led to very significant alterations in the microRNA expression profiles in bystander animal tissues. The exact functions of miRNAs in the bystander effect have still to be delineated (reviewed in Dickey et al., 2011).

Epigenetics of transgenerational effects

The occurrence of genome instability and elevated mutation rates in the progeny of exposed parents was attributed to some, yet unknown, mechanisms, possibly epigenetic mechanisms. Relatively few studies have addressed the potential epigenetic alterations in offspring of irradiated parents. The first direct evidence of the epigenetic effectors involvement in transgenerational responses comes from the study by Koturbash and colleagues (Koturbash et al., 2006a). They utilized an *in vivo* mouse model to analyze the role of epigenetic parameters in the transgenerational radiation effects. In this study, the C57Bl/6 mice were exposed to 2.5 Gy of X rays and mated seven days after exposure. Several mating groups were established: maternal exposure only, paternal exposure only, and both parents exposure. Mock-treated mating pairs served as controls. To test whether changes in DNA methylation were observed in the somatic tissue of offspring, global DNA methylation was measured in spleen, thymus, and liver of offspring. Additionally, authors analyzed the levels of phosphorylated H2AX in the aforementioned tissues.

A significant loss of DNA methylation was observed in the thymus of offspring upon paternal and combined parental exposure. The DNA methylation changes were correlated with the alterations in the levels of DNA methyltransferases, methyl-binding proteins, and levels of phosphorylated histone H2AX (see Figures 19-4 and 19-5).

Specifically, DNMT1 expression was dramatically decreased in the thymus tissue of the progeny of exposed males and the progeny with the combined paternal and maternal exposure. The levels of DNMT3a and 3b were also significantly downregulated in the progeny of exposed males and in the combined parental exposure group. The decrease in global cytosine DNA methylation and DNMTs levels observed in the thymus of the progeny upon paternal and combined parental irradiation were correlated with a significant decrease in the level of methyl-binding protein MeCP2 and increased levels of phosphorylated H2AX.

In the recent follow-up study, Filkowski and colleagues analyzed the nature and mechanisms underlying the disruption of DNA methylation and microRNA expression status in the germline and progeny of exposed parents (Filkowski et al., 2010). It was shown that paternal irradiation leads to upregulation of the miR-29 family in the exposed male germline, which causes decreased expression of *de novo* methyltransferase—DNA methyltransferase 3a—and causes significant hypomethylation of long interspersed nuclear elements 1 (LINE1) and short interspersed nuclear elements B2 (SINE B2). Moreover, epigenetic changes in the male germline lead to deleterious epigenetic effects in the somatic thymus tissue from the progeny of exposed animals. Hypomethylation of LINE1 and SINE B2 in the thymus tissue from the progeny was associated with a significant decrease in the levels of lymphoid-specific helicase (LSH) that is crucial for the maintenance of methylation and silencing of repetitive elements. Paternal irradiation leads to significant changes in microRNAone of the offspring tissues.

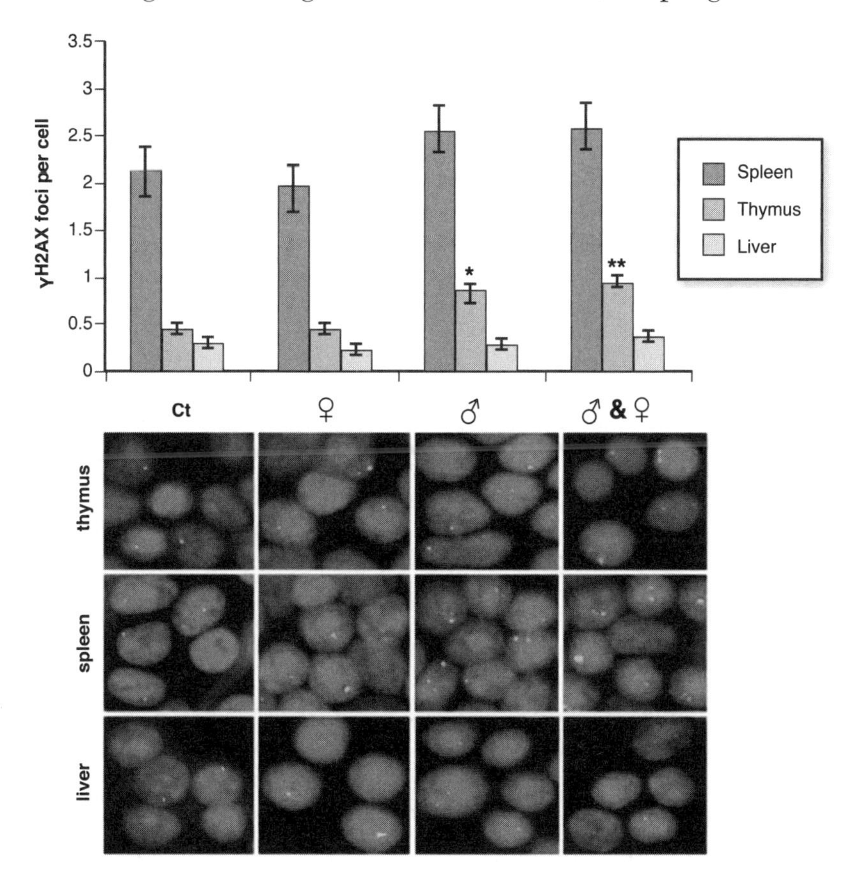

Figure 19-4 Significant accumulation of γH2AX foci in thymus of the progeny of exposed parents. (Adapted with permission from Koturbash et al. Epigenetic dysregulation underlies radiation-induced transgenerational genome instability in vivo. Int J Radiat Oncol Biol Phys. 2006; 66:327-30.) Mean values ± SEM. Significant differences from the control animals are shown: $*p < 0.0125$, $**p < 0.0001$ according to post hoc Bonferroni correction.

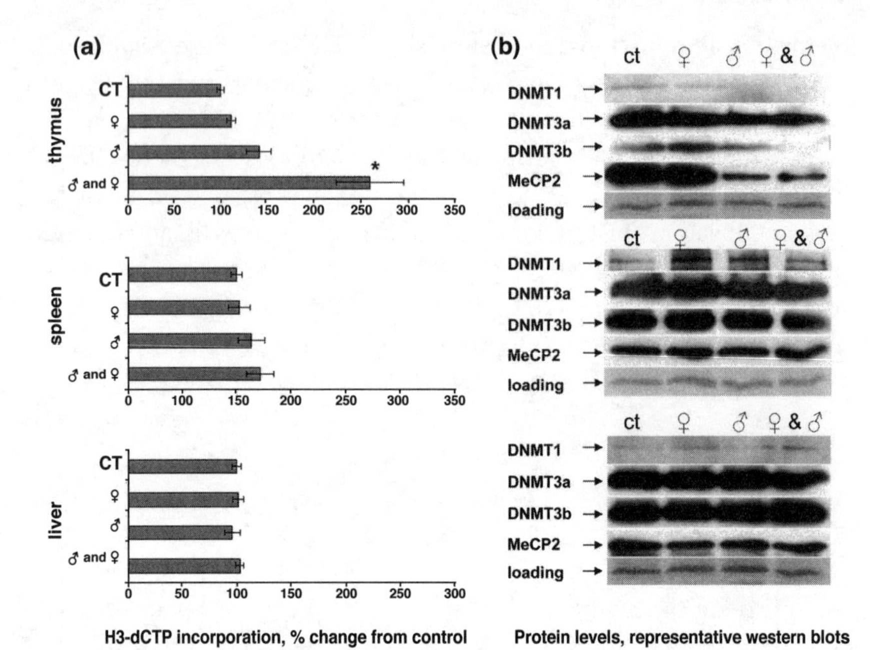

Figure 19-5 Parental exposure leads to global genome hypomethylation and alters the expression of DNA methyltransferases and methyl-binding proteins in thymus of the offspring. (Adapted with permission from Koturbash et al. 2006a.) (**a**) Levels of global genome DNA methylation. The extent of [3H]dCTP incorporation is inversely proportional to the levels of methylation. Mean values ± SEM. Significant differences from the control animals are shown: $*p < 0.0125$ according to post-hoc Bonferroni correction. (**b**) Representative western blots. Each experiment included pooled lysates from five animals for each cohort, with equal representation of each animal. Western blots were repeated at least four times to ensure the reproducibility and robustness of the results. (Ct = progeny of the control cohort; [female] = progeny of the maternal exposure group; [male] = progeny of the paternal exposure group; [male]+ [female] = progeny of the combined parental exposure group.)

In sum, epigenetic changes seem to be the plausible mediators of the direct and indirect radiation effects, including the radiation-induced genome instability, bystander, and transgenerational effects (Kovalchuk and Baulch, 2008).

Epigenetic effects of chemical carcinogens

Environmental exposure to natural and manmade chemical and physical agents is one of the major causes of human cancer (Wild, 2009).

The development of cancer is a complex, multifactorial process characterized by many biologically significant and interdependent alterations. It is clear that cancer, by itself, can trigger epigenetic alterations that reflect the transformed state of neoplastic cells; however, the distortion of the cellular epigenetic status in normal cells might also have an effect on the predisposition to precancer-specific pathological states and cancer development. Recent advances in the field of cancer research indicate that a wide range of chemical carcinogens in addition to genetic changes affect cellular epigenetic landscape. This refines the classical model of multistage carcinogenesis that involves genetic changes to include both genetic and epigenetic changes at each stage of carcinogenic process. In a broad sense, carcinogenesis can be induced through either genotoxic or non-genotoxic mechanisms. Genotoxic carcinogens are agents that interact directly or after metabolic activation with DNA, causing mutations and leading to tumor formation (Shuker, 2002). Non-genotoxic carcinogens are a diverse group of chemical compounds that are known to cause tumors by mechanisms other than directly damaging DNA (Lima and Van der Laan, 2000). Nonetheless, accumulating evidence suggests that despite different mechanisms of action with regard to DNA reactivity, both classes of agents lead to prominent epigenetic abnormalities in tissues that are susceptible to carcinogenesis as a result of exposure. A recent review article by Koturbash and colleagues provides an excellent overview of the epigenetic alterations induced by various chemical carcinogens and toxicants (Koturbash, Beland, and Pogribny, 2011).

The aromatic amine 2-acetylaminofluorene (2-AAF), a potent carcinogen and toxicant is known to cause epigenetic deregulation. 2-AAF exposure leads to aberrant hypermethylation of numerous genes, altered microRNAome expression and deregulation of histone modification levels. Exposure to a well-established liver carcinogen, phenobarbital, results in hypermethylation and altered expression of numerous genes, including p53 and p16 tumor suppressors (reviewed in Koturbash et al., 2011).

Tobacco smoke is the most widespread carcinogen that contains numerous cancer-causing agents such as *N*-nitrosamines, aromatic amines, polycyclic aromatic hydrocarbons, aldehydes, and benzene, just to name a few. Exposure to tobacco smoke causes significant alterations in global and gene-specific DNA methylation and histone modification patterns in the tissue of exposed individuals (Koturbash et al., 2011).

Benzene is one of the key components of tobacco smoke. Benzene exposure leads to global genome hypomethylation, hypermethylation of *p15*, *MSH3*, and *SEMA3C* genes, and hypomethylation of *MAGE-1* and *RUNX3* (*AML2*) genes (Koturbash et al., 2011).

Chronic environmental exposure to metals and metalloid elements, such as nickel, cadmium, chromium, and arsenic, also significantly affects cellular epigenetic profiles (reviewed in Koturbash et al., 2011).

Conclusion

The relationship between certain environmental factors and health are known; however, what is responsible for driving these interactions is not always apparent, especially in cases where there is a combined effect of genetics and environmental exposure. Epigenetics is a field that looks at elements responsible for changes in gene expression in the absence of corresponding changes in DNA sequence. It is interactions with the genome that make it a promising candidate to explain the gene-environment-phenotype relationship. This chapter has examined only a few of the many studies that use epigenetics as a mechanism to explain the associations observed between the environment and the phenotype as well as some of the approaches that are used to do this. As research continues and technology improves the ability to explore this field further, our knowledge of the emerging area of epigenetics will expand.

Exercises and discussion topics

1. Which environmental factors influence epigenetic status of the cells and tissues?
2. In a primate model, how does maternal dietary restriction influence the epigenetic status of the fetus?
3. How did famine influence epigenetic phenomena in the periconceptionally exposed individuals?
4. Which epigenetic changes are associated with PTSD?

5. Describe the roles of epigenetic changes in air pollution induced asthma and inflammation.
6. How did exposure to contaminated urban air influence murine sperm DNA methylation? What might these changes lead to?
7. Why monozygotic twins are good models for epigenetic studies?
8. Which bioactive food compounds might influence cellular epigenetic status?
9. How can bioactive food compounds and nutrients influence DNA methylation?
10. Which bioactive food compounds influence microRNAome?
11. Which epigenetic parameters are affected by ionizing radiation?
12. What is bystander effect and how can this phenomenon be studied *in vivo*?
13. Describe transgenerational radiation-induced epigenetic effects.
14. Which chemicals can influence epigenetic changes?

References

Altucci L, Stunnenberg HG. (2009) Time for Epigenetics. *Int J Biochem Cell Biol* 41:2-3.

Choi S-W, Mason JB. (2002) Folate Status: Effects on Pathways of Colorectal Carcinogenesis. *J. Nutr.* 132:2413S-2418S.

Cirulli et al. (2009) Early life stress as a risk factor for mental health: Role of neurotrophins from rodents to non-human primates. *Neurosci Biobehav Rev* 33:573-585.

Cooney et al. (2002) Maternal methyl supplements in mice affect epigenetic variation and DNA methylation of offspring. *J Nutr* 132:2393s-2400s.

de Kloet et al. (2009) Glucocorticoid signaling and stress-related limbic susceptibility pathway: About receptors, transcription machinery and microRNA. *Brain Research,* 1293:29-141.

de Lumen, BO. (2005) Lunasin: A Cancer-Preventive Soy Peptide. *Nutr Rev* 63:16-21.

Dickey et al. (2011) The role of miRNA in the direct and indirect effects of ionizing radiation. *Radiat Environ Biophys* 50:491-499.

Dolinoy DC. (2008) The agouti mouse model: an epigenetic biosensor for nutritional and environmental alterations on the fetal epigenome. *Nutr Rev* 66:S7-S11.

Dolinoy et al. (2007) Epigenetic gene regulation: Linking early developmental environment to adult disease. *Reprod Toxicol* 23:297-307.

Druesne-Pecollo N, Chaumontet C, Latino-Martel P. (2008) Diallyl disulfide increases histone acetylation in colon cells *in vitro* and in vivo. *Nutr Rev* 66:S39-S41.

Duthie S. (2010a) Folate and cancer: how DNA damage, repair and methylation impact on colon carcinogenesis. *J Inherit Metab Dis* 1-9.

Duthie, SJ. (2010b) Epigenetic modifications and human pathologies: cancer and CVD. *Proc Nutr Soc* 1-10.

Fang et al. (2007) Dietary Polyphenols May Affect DNA Methylation. *J Nutr* 137:223S-228S.

Fang et al. (2003) Tea Polyphenol (—)-Epigallocatechin-3-Gallate Inhibits DNA Methyltransferase and Reactivates Methylation-Silenced Genes in Cancer Cell Lines. *Cancer Res* 63:7563-7570.

Filkowski et al. (2010) Hypomethylation and genome instability in the germline of exposed parents and their progeny is associated with altered miRNA expression. *Carcinogenesis,* 31:1110-1115.

Fraga et al. (2005) Epigenetic differences arise during the lifetime of monozygotic twins. *Proc Natl Acad Sci USA* 102:10604-10609.

Heijmans et al. (2008) Persistent epigenetic differences associated with prenatal exposure to famine in humans. *Proc Natl Acad Sci USA,* 105:17046-17049.

Ilnytskyy Y, Kovalchuk O. (2011) Non-targeted radiation effects—an epigenetic connection. *Mutat Res* 714:113-25.

Jaenisch RA, Bird A. (2003) Epigenetic regulation of gene expression: how the genome integrates intrinsic and environmental signals. *Nat Genet* 33Suppl:245-254.

Jirtle RL, Skinner MK. (2007) Environmental epigenomics and disease susceptibility. *Nat Rev Genet* 8:253-262.

Kalinich et al. (1989) The effect of gamma radiation on DNA methylation. *Radiat Res* 117:185-197.

Kaup et al. (2006) Radiation-induced genomic instability is associated with DNA methylation changes in cultured human keratinocytes. *Mutat Res* 597:87-97.

Kim et al. (1996) Dietary folate protects against the development of macroscopic colonic neoplasia in a dose responsive manner in rats. *Gut* 39:732-740.

Koturbash et al. (2006a) Epigenetic dysregulation underlies radiation-induced transgenerational genome instability in vivo. *Int J Radiat Oncol Biol Phys* 66:327-330.

Koturbash et al. (2011) Role of epigenetic events in chemical carcinogenesis—a justification for incorporating epigenetic evaluations in cancer risk assessment. *Toxicol Mech Methods* 21:289-297.

Koturbash et al. (2007) Role of epigenetic effectors in maintenance of the long-term persistent bystander effect in spleen in vivo. *Carcinogenesis* 28:1831-1838.

Koturbash et al. (2006b) Irradiation induces DNA damage and modulates epigenetic effectors in distant bystander tissue in vivo. *Oncogene* 25:4267-4275.

Kovalchuk O. (2008) Epigenetic research sheds new light on the nature of interactions between organisms and their environment. *Environ Mol Mutagen* 49:1-3.

Kovalchuk O, Baulch JE. (2008) Epigenetic changes and nontargeted radiation effect—is there a link? *Environ Mol Mutagen* 49:16-25.

Kovalchuk et al. (2004) Methylation changes in muscle and liver tissues of male and female mice exposed to acute and chronic low-dose X-ray-irradiation. *Mutat Res* 548:75-84.

Lima BS, Van der Laan JW. (2000) Mechanisms of nongenotoxic carcinogenesis and assessment of the human hazard. *Regul Toxicol Pharmacol* 32:135-143.

McEwen BS. (2008) Understanding the potency of stressful early life experiences on brain and body function. *Metabolism* 57:S11-S15.

McGowan et al. (2009) Epigenetic regulation of the glucocorticoid receptor in human brain associates with childhood abuse. *Nat Neurosci* 12:342-348.

McKie R. (2001) Revealed: the secret of human behaviour. Environment, not genes, key to our acts. *The Observer*, Retrieved from http://www.guardian.co.uk/science/2001/feb/11/genetics.humanbehaviour.

Nadeau et al. (2010) Ambient air pollution impairs regulatory T-cell function in asthma. *J Allergy Clin Immunol* 126:845-852 e10.

Nafee et al. (2008) Epigenetic control of fetal gene expression. *BJOG* 115:158-168.

Power et al. (2007) Life-course influences on health in British adults: effects of socio-economic position in childhood and adulthood. *Int J Epidemiol* 36:532-539.

Ross SA. (2010) Diet and Epigenetics. In *Bioactive Compounds and Cancer*, eds. JA Milner & DF Romagnolo, 101-123.

Sedelnikova et al. (2007) DNA double-strand breaks form in bystander cells after microbeam irradiation of three-dimensional human tissue models. *Cancer Res* 67:4295-4302.

Sedelnikova et al. (2003) Histone H2AX in DNA damage and repair. *Cancer Biol Ther* 2:233-235.

Shuker DE. (2002) The enemy at the gates? DNA adducts as biomarkers of exposure to exogenous and endogenous genotoxic agents. *Toxicol Lett* 134:51-56.

Singh et al. (2002) Epigenetic contributors to the discordance of monozygotic twins. *Clin Genet* 62:97-103.

Sun et al. (2008) Curcumin (diferuloylmethane) alters the expression profiles of microRNAs in human pancreatic cancer cells. *Mol Cancer Ther* 7:464-473.

Szarc vel Szic et al. (2010) Nature or nurture: Let food be your epigenetic medicine in chronic inflammatory disorders. *Biochem Pharmacol* 80:1816-1832.

Szyf M. (2009) The early life environment and the epigenome. *Biochim Biophys Acta* 1790:878-885.

Tawa et al. (1998) Effects of X-ray irradiation on genomic DNA methylation levels in mouse tissues. *J Radiat Res (Tokyo)* 39:271-278.

Tobi et al. (2009) DNA methylation differences after exposure to prenatal famine are common and timing- and sex-specific. *Human Mol Genet* 18:4046-4053.

Uddin et al. (2010) Epigenetic and immune function profiles associated with posttraumatic stress disorder. *Proc Natl Acad Sci USA* 107:9470-9475.

Unterberger et al. (2009) Organ and gestational age effects of maternal nutrient restriction on global methylation in fetal baboons. *J Med Primatol* 38:219-227.

Vreugdenhil et al. (2009) MicroRNA 18 and 124a Down-Regulate the Glucocorticoid Receptor: Implications for Glucocorticoid Responsiveness in the Brain. *Endocrinology* 150:2220-2228.

Wade PA, Archer TK. (2006) Epigenetics: environmental instructions for the genome. *Environ Health Perspect* 114:A140-141.

Weaver ICG. (2009) Shaping Adult Phenotypes Through Early Life Environments. *Birth Defects Res C Embryo Today* 87:314-326.

Weidman et al. (2007) Cancer susceptibility: epigenetic manifestation of environmental exposures. *Cancer J* 13:9-16.

Wild CP. (2009) Environmental exposure measurement in cancer epidemiology. *Mutagenesis* 24:117-125.

Wilker et al. (2010) Black Carbon Exposures, Blood Pressure, and Interactions with Single Nucleotide Polymorphisms in MicroRNA Processing Genes. *Environ Health Perspect* 118:943-948.

Yauk et al. (2008) Germ-line mutations, DNA damage, and global hypermethylation in mice exposed to particulate air pollution in an urban/industrial location. *Proc Natl Acad Sci USA* 105:605-610.

Zhu et al. (2003) Secondhand smoke stimulates tumor angiogenesis and growth. *Cancer Cell* 4:191-196.

20

Epigenetics and technology—hairpin-based antisensing

RNA interference, known as RNAi, is a fascinating phenomenon that enables us to understand how organisms control expression of endogenous and exogenous RNAs. Moreover, RNAi represents a powerful research and therapeutic tool for the regulation of gene expression in cells. The use of RNAi in functional genomics makes it possible to analyze the function of a specific gene. Its use in agriculture represents the knockdown of specific genes and the pathways for either removing an undesirable trait or adding a new, desirable one. The landmark discovery of the hairpin RNAi construct as an efficient sequence-dependent trigger of gene silencing provided a foundation for a new RNAi technology. This chapter explains the molecular mechanisms of RNAi and discusses practical applications of hairpin-based RNAi in plants and animals.

20-1. Glossary

Antisense technology: Refers to the use of RNAs produced by overexpressing the antisense strands of specific genes.

A hairpin RNAi construct: A transgene construct consisting of the self-complementary sense and antisense fragments with the stuffer sequence (often as an intron) forming a hairpin structure upon transcription. Hairpin structure is recognized by the RNAi machinery producing small RNAs for silencing of endogenous sequences.

hpRNA: A hairpin RNA pre-miRNA-like structure with perfect homology between arms.

ahpRNA: An artificial hpRNA refers to artificially designed hpRNA with arms varying between 50 and 1,000 nt, usually 300 to 600 nt.

shRNA: A short hpRNA; a small hpRNA with arms of 19 to 29 nt in length. The shRNAs are normally used for research and practical applications in animals/humans.

siRNAs: Small-interfering RNAs are perfectly complementary to small RNAs (21 to 25 nucleotides) generated from the long dsRNA by Dicer or Dicer-like enzymes.

ddRNAi: DNA-directed RNAi is an approach utilizing mini-genes that encode siRNAs.

RNAi is one of the rare examples of scientific discoveries that became extremely useful not only for basic science but also for many practical applications. Along with the clarification of the RNA interference pathway and the discovery of its participation in essential biological events, a branch of science has grown to utilize the RNA interference pathway as a biotechnology tool for both basic and applied research. In eukaryotes, this has commonly been accomplished by the expression of self-complementary fragments of transgenes, known as hairpins, that are cloned to express **double-stranded RNA** (**dsRNA**), which promotes silencing of the complementary endogenous sequences. The use of hairpin RNAi has many benefits over conventional mutagenesis, particularly that it can be targeted to a sequence of choice and is time and cost effective. Its execution is limited only by the availability of target gene sequences. The latest discoveries in the field of the hairpin RNAi action have led to the development of powerful techniques for studying gene function and modifying the organism's phenotype.

Hairpin-induced RNAi: How does it work?

RNAi pathways in eukaryotes are complex and multifunctional; therefore, they play roles in development and in protection against pathogens such as viruses. These pathways are of ancient evolutionary origins, with many features conserved between plants and animals.

In mammalian cells, nematodes, and flies, RNAi can be triggered by the direct introduction of dsRNAs, whereas in plants it is usually induced by the transformation with a construct that produces hairpin RNAs. A common point in these processes is the degradation of dsRNA into **small interfering RNAs (siRNAs)** of the two size classes: short siRNA, which is 21 to 22 nt in length produced upon sequence-specific degradation of the target mRNAs, and long siRNAs, which are 24 to 26 nt in length that are involved in triggering systemic gene silencing (see Figure 20-1).

dsRNAs are produced in plants in a variety of ways from endogenous and exogenous precursors. They derive from viral genomes, transcripts from overlapping complementary genes, and endogenous imperfect hairpin transcripts (**primary miRNAs**) and can also be produced by the **RNA-dependent RNA polymerase (RdRP)** activity. As was mentioned earlier, small RNAs are produced by DCL proteins that are partially redundant in their function. In Arabidopsis, DCL2 primarily cleaves natural antisense transcripts and viral RNA into a single 24-nucleotide **naturally occurring antisense siRNA (nat-siRNA)**. This single 24-nt nat-siRNA then leads one of the *cis*-antisense gene pair transcripts to degradation. As a result, the cleaved RNA molecule is converted to dsRNA by RdRP; DCL3 is involved in the slicing of transcripts from endogenous repeats into 24-nt siRNAs, which then direct chromosomal methylation to silence transcription; DCL4 processes a viral RNA and a small group of miRNA-cleaved transcripts into 21-nt transacting siRNAs. It is noteworthy to mention that hairpin dsRNA is processed in the same way as viral RNA. For instance, the Fusaro's group showed that hairpin dsRNAs are cleaved into siRNAs primarily by DCL4. In the absence of DCL4, the processing is taken by DCL2. In *dcl2/dcl4* double mutants, silencing by hairpin RNAi and repression of viral replication do not occur. An additional proof of the commonality between hairpin RNAi and viral defense is the inhibition of both by viral-encoded suppressors of silencing (Dunoyer et al., 2004).

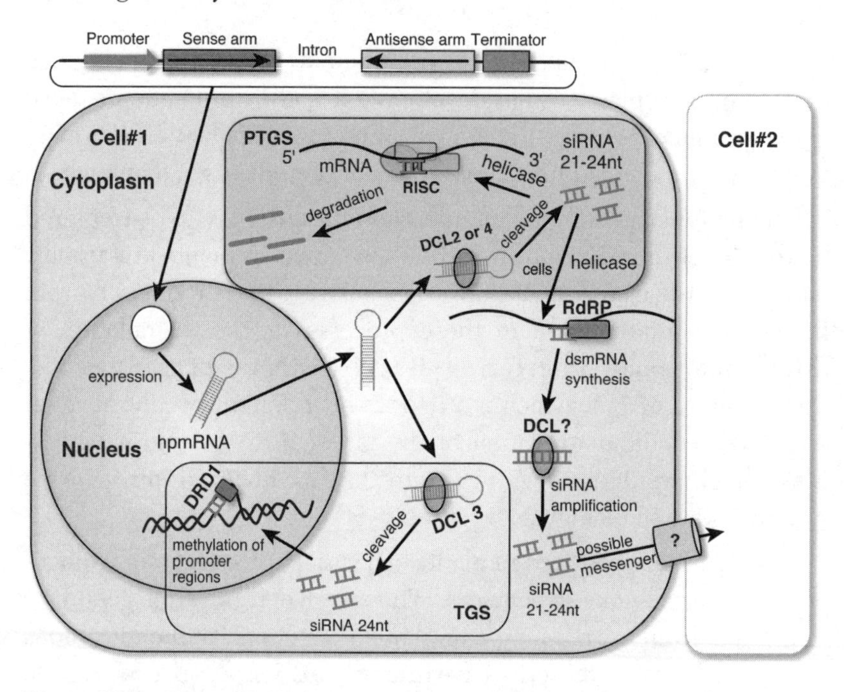

Figure 20-1 The mechanism of the RNAi triggered by the hairpin construct.

After the cleavage, siRNAs serve as guides for a multiprotein complex referred to as the **RNA-induced silencing complex (RISC)** in which the degradation of target mRNAs occurs. One strand is transferred to RISC as a guide to direct cleavage activity. Thus, not all hairpin siRNAs will be of the correct strand to target an mRNA. Consequently, this factor should be taken into account as potentially cross-silencing sequences can be included in constructs.

Hairpins targeting conserved sequences

Both untranslated regions and coding sequences can be successfully used in hairpin RNAi constructs. Untranslated regions are even more useful in designing gene-specific hairpin-silencing constructs because of their variability within highly related gene families. On the other hand, introns seem to be poor targets for PTGS.

An example of using conserved sequences for gene targeting was shown by Reiser et al. (2004). Two closely related ATP/ADP transporters in Arabidopsis—*AtNTT1* and 2—were silenced using a 418-nt hairpin insert from *AtNTT1*, which has 92% homology to *AtNTT2*. Phenotypic analysis and Northern blots clearly demonstrated silencing in both hairpin-transformed lines. Miki et al. (2005) observed a similar effect of simultaneous silencing of seven members of the rice *OsRac* gene family by utilizing highly conserved sequences. The qRT-PCR analysis revealed that the level of silencing of each target generally correlated with its level of sequence homology to the hairpin sequence. However, the *OsRac1* hairpin construct poorly silenced *OsRac5* (the transcript level was on average reduced to 60% of wild-type levels), although the two sequences have 73% homology; similarly, the *OsRac5* construct incompletely silenced *OsRac1* (73% homologous) and *OsRac7* (78% homologous) to about 35% of wild-type levels. Thus, a conserved hairpin approach can be very effective for knockdown of a gene family, but strong silencing is only likely for highly conserved gene families.

Chimeric hairpins to target unrelated genes

When the level of homology between target genes is not satisfactory for the use in conserved hairpins, chimeric hairpin RNAi constructs present an alternative approach. It has been shown (Helliwell et al., 2002) that it is efficient to clone inserts from two genes in a single hairpin vector such that they are driven by a single promoter and expressed as a chimeric hairpin mRNA. An extension of this methodology was demonstrated by Miki et al. (2005). Using gene-specific hairpin inserts (of 221 to 367 nt) derived from the highly divergent 3' untranslated regions of seven members of the *OsRac* gene family in rice, the authors constructed chimeric hairpins. The effectiveness of the constructs decreased as more inserts were included. Thus, double and triple chimeric hairpins effectively silenced their targets, and weak silencing was observed only when four or more inserts were introduced. Using the results of Northern blot analysis, the authors correlated this effect with the decreased amounts of siRNAs derived from each insert sequence.

Designing the hairpins

The application of RNAi in plants and animals improved dramatically with the introduction of the concept of minigenes encoding siRNAs. This approach was initially called DNA-directed RNAi or ddRNAi. Svoboda et al. (2001) constructed a plasmid vector containing the SV40 small intron, EGFP, for the detection of expression and a hairpin formed from inverted repeats of mos mRNA. They showed that the injection of this hpRNA construct into mouse oocytes resulted in RNAi that is similar in its effects to the effects achieved upon injection of dsRNA.

Standard hpRNA constructs include the following required components:

- A promoter
- A sense arm
- An intron
- An antisense arm
- A terminator (Smith et al., 2000)

Transcription from this construct results in the formation of hpRNA. The commonly used sets of promoters, intron spacers, and terminators are

- The *Cauliflower Mosaic Virus* 35S promoter (35S), the intron-2 of the *Pdk* gene of *Flaveria* (767 bp), the Octopine synthase terminator (OCS)
- The napin promoter (Napin), intron 1 of the *Arabidopsis thaliana Fad2* gene (1,147 bp), and the nopaline synthase terminator (NOS) (see Figure 20-2)

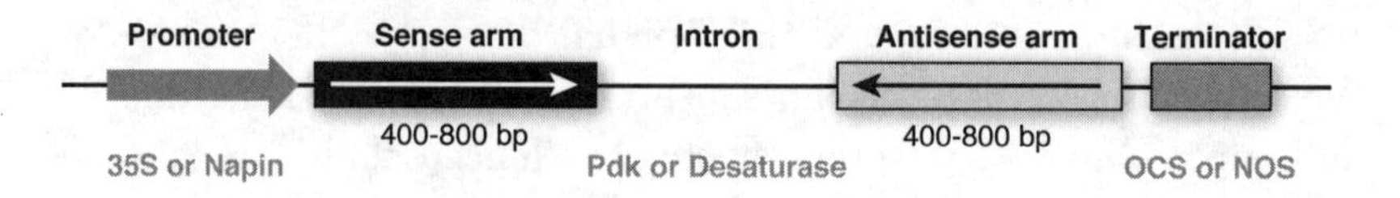

Figure 20-2 Structure of standard hpRNA construct.

It appears that the biogenesis of **artificial hpRNAs** (**ahpRNAs**) and pre-miRNAs is different despite structural similarities of small RNAs involved. Whereas pre-miRNAs are processed by DCL1, ahpRNAs are apparently processed through DCL2 and DCL4 generating 22-nt and 21-nt siRNAs. It is hypothesized that ahpRNAs are processed through the plant viral defense pathway in which plants use DCL4, and to a lesser extent DCL2, for the production of viRNAs. The 21- and 22-nt siRNAs generated from an ahpRNA precursor are capable of being transmitted as silencing signals over a long distance. Further similarities to the biogenesis of viRNAs include the fact that a virus-derived suppressor protein inhibits ahpRNA-induced systemic silencing. Thus, it is possible that a perfect complementarity between sense and antisense arms of ahpRNA triggers DCL4/DCL2-mediated processing similar to the one characteristic for long viral dsRNAs. Various ahpRNA can thus be efficiently used in plants for generating systemic enhanced viral defense or for systemic silencing of specific endogens.

In addition to the considerations of sequence similarities mentioned earlier, several other factors can influence hairpin design. One of these factors is the selection of an insert (**spacer**). A spacer fragment between the arms of the inverted repeat has two advantages. Because *Escherichia coli* does not tolerate well hairpin constructs, it is useful to introduce a spacer for increasing vector stability. Additionally, a new GT115 strain was designed for handling such constructs during cloning steps (Nishikawa and Sugiyama, 2010). Also, utilizing a splicable intron as a spacer has been shown to dramatically increase the frequency of strong silencing phenotypes (Smith et al., 2000). The means by which intron splicing increases the effectiveness of hairpin RNAi is unknown.

Published reports suggest that hairpin spacers of 50 to 1,000 nt can effectively trigger silencing in stably transformed plants (Helliwell and Waterhouse, 2003). The recommended length of the arm is between 300 and 600 bp derived from the coding region or even from the 3' or 5' untranslated regions of a target gene (Wesley et al., 2001). It should be noted that the arms as short as 50 nt are sufficient to trigger significant gene silencing. However, short hairpin inserts should be avoided as they might cause weaker and less frequent silencing. Additionally, longer hairpins are more manageable to clone and manipulate.

Tissue-specific hairpin RNAi

One of the features of the siRNA-mediated RNAi is systemic spread of its silencing signal. Originally, it was proposed that small RNAs associated with PTGS might act as mobile signals that spread silencing throughout the organism (Hamilton and Baulcombe, 1999). On the other hand, these signals can be precursors of small RNAs. In any case, it is not surprising that systemic spread of silencing triggered by hairpins has been observed.

Nevertheless, there are also reports of tissue-specific hairpin RNAi when hairpins are expressed under the control of tissue-specific promoters. For instance, Byzova et al. (2004) made an attempt to modify flower architecture by affecting the expression of MADS genes that control floral organ identity. The *APETALA1* promoter, which is specific to sepals and petals, was used to control expression of a hairpin construct against *APETALA3,* which is normally expressed in petals and stamens. Conversion of petals to sepals in Arabidopsis and *Brassica napus* transformants confirmed that *APETALA3* was downregulated in the organ normally producing petals. Conversely, the development of stamens without any physiological alterations showed that the gene was not silenced in the neighboring organs. This is in contradiction to the hairpin RNAi-mediated spread of silencing signals that might be explained simply by the low transcriptional activity of the floral-organ-specific promoter used.

Inducible hairpin RNAi

Compared to conventional mutagenic methods, the hairpin RNAi technique offers some benefits; one of them is its ability to be induced almost at each developmental stage of an organism. This can be achieved by utilizing inducible promoters that are potentially useful in characterization of gene-silencing phenotypes where constitutive downregulation results in seedling lethality or strong pleiotropic phenotypes. Several systems that successively trigger silencing under particular conditions have been described. Thus, an alcohol-inducible hairpin construct has been shown to activate its transcription under the *alcA* promoter in the presence of ethanol (Chen et al., 2003). A major drawback of this approach is the low rate of hairpin induction following ethanol treatment. Only less than 33% of transformants have a silencing phenotype following induction. Another example of an inducible promoter was reported by Masclaux et al. (2004). The introduced hairpin RNAi construct was driven by the heat-inducible *HSP18.2* promoter of Arabidopsis that was activated in response to 2h treatment at 37°C. Nevertheless, the authors were faced with the problem of some promoter leakage at 23°C and the appearance of silencing phenotypes that were still observed several days following heat treatment. These observations, apparently, can be explained by self-amplification of siRNAs by RdRP.

Guo et al. (2003) described another interesting approach that utilizes the *Cre/loxP* recombination system. Following induction, the *Cre/loxP* recombination system excises a fragment of the transgene sequence that separates a hairpin construct from a constitutive promoter. However, given the irreversible nature of the recombination, it is not possible to discontinue hairpin expression following induction.

A glucocorticoid-inducible vector derived from pHELLSGATE was reported by Wielopolska et al. (2005). The inducible pOp6 promoter got activated through a constitutively expressed, synthetic glucocorticoid-responsive transcription factor (LhGR) in the presence of dexamethasone. As it was shown by Northern blots, induction of silencing in Arabidopsis was observed within one day of transfer to media containing dexamethasone, and recovery of expression began within one day after transfer to plain media.

Applications of hairpin RNAi in plants

Plant hairpin expression vectors that are usually used for triggering RNAi have an inverted repeat of a fragment of the gene of interest behind an appropriate promoter. Additionally, binary vectors should include plant selectable markers that are cloned within the borders of a T-DNA binary vector for *Agrobacterium*-mediated transformation.

To simplify hairpin cloning, a variety of specifically designed hairpin expression vectors are available. Two major concerns in vector selection include the choice of appropriate promoters and selectable markers. For example, the 35S promoter is suitable for high-level constitutive expression in dicots, whereas the maize ubiquitin promoter is preferable for the expression in monocot species.

To reduce time and efforts that are required to clone a hairpin construct, a number of binary vectors have been designed. For instance, Wesley et al. improved the hairpin cloning strategies by introducing a pair of hairpin vectors for conventional cloning of PCR fragments into hairpin expression cassettes (pHANNIBAL and pKANNIBAL) and a third vector (pHELLSGATE) for high-throughput cloning using the Gateway recombination system. All three vectors were designed for the efficient production of intron-spliced hairpin RNAs.

The conventional hairpin cloning vectors pHANNIBAL and pKANNIBAL contain a 35S-driven inverted repeat cassette with the **multiple cloning sites** (**MCSs**). A fragment of the gene of interest is amplified using PCR primers containing restriction sites matching those in the hairpin vectors and cloned into the hairpin cassette on the either side of a *PDK* intron. The pHELLSGATE Gateway cloning vector series are binary vectors that use Gateway recombination sequences instead of MCSs for efficient directional insertion of gene fragments into the hairpin cassette. Because of the extremely high efficiency of the Gateway recombination system, pHELLSGATE was designed to facilitate the construction of large numbers of hairpin-expression binary vectors in the short period of time.

In addition to the aforementioned constructs, versions of these vectors suitable for the use in monocot plants are now available. All these vectors can be obtained from the Commonwealth Scientific and Industrial Research Organisation (CSIRO, Australia, see www.pi.csiro.au/rnai). Vectors developed by other groups include pANDA, a binary vector that utilizes Gateway recombination sites and expresses hairpins from the maize ubiquitin promoter, pKNOCKOUT, and pFGC, the conventional cloning vectors with hairpin expression under the control of the 35S promoter.

The high efficacy of using hairpin RNAi to silence specific plant genes has led to its widespread use. Published reports on hairpin RNAi in plants and animals show a remarkable diversity in the problems to which it has been applied.

Reverse genetics and functional genomics

Compared to a mutagenic approach to studying gene function, hairpin RNAi has a lot of benefits, and primary among these are its simplicity, specificity, and effectiveness. Hairpin RNAi has been demonstrated to be strongly effective against a range of target genes in plants, including transcription factors, biosynthetic genes, and viruses. It effectively triggers PTGS in stably transformed lines as well as in transient transformation systems using Agrobacterium infiltration.

With the continuously growing accessibility of plant genomic sequence information and improving hairpin RNAi binary vectors, the application of these vectors in plant functional genomics becomes possible. To demonstrate a high efficiency of the hairpin RNAi approach for functional genomics in plants, the Arabidopsis Genomic RNAi Knockout Line Analysis (AGRIKOLA) group cloned gene-specific sequence tags that represent all known Arabidopsis genes into the hairpin-silencing constructs. Additionally, they showed the efficiency of their constructs (pAgrikola) by producing a few hundred independent knockdown lines that are now publicly available.

Removal of undesirable traits in crops

The information obtained from reverse genetics and functional genomics can be efficiently used to produce new transgenics with improved agricultural traits. The use of the hairpin RNAi for this purpose is one of the possible approaches. Reported examples of the successful use of hairpin RNAi include silencing of polyphenol oxidase to prevent enzymatic browning of potato, targeting of ACC oxidase that catalyzes the production of the ripening compound ethylene in tomato—resulting in prolonged shelf-life—and downregulation of the Mal d 1 family of allergenic proteins in apple.

A particularly interesting and valuable example was provided by Sunilkumar et al. (2006). Tissue-specific expression of a hairpin transgene was used to reduce the level of gossypol in cottonseed. Gossypol is a terpenoid, which is toxic to humans; therefore, its accumulation in seeds limits the use of cottonseed protein. Nevertheless, gossypol is a beneficial insecticidal protein elsewhere in the plant, and its systemic elimination by conventional breeding techniques results in strongly increased susceptibility to insect attack. The authors have successfully used the hairpin construct expressed from the seed-specific *a-globulin B* promoter to disrupt gossypol biosynthesis in cottonseed tissue by interfering with the expression of the δ-cadinene synthase gene. As a result, the level of seed gossypol was reduced by up to 99%, but levels elsewhere in the plant were not measurably changed.

Another example of using hpRNA technology for the elimination of undesired traits is the manipulation of patatin synthesis. Potato tubers contain patatins, which belong to a family of glycoproteins and represent up to 40% of solute proteins. Patatins function as fatty acid esterases, lipid acyl hydrolases, and acyltransferease. Due to a high concentration of patatins in soluble proteins, it is difficult to use potato plants for the production of various essential therapeutic proteins, such as vaccines, antibodies, glycoproteins, and so on. Potato cultivar was transformed by Agrobacterium carrying the hpRNA vector that included the 35S promoter, a PCS terminator, and the sense/antisense patatin gene *pat3*-k-I (653 bp) separated by Pdk intron (Kim et al., 2008). The levels of patatin and its mRNA were decreased by approximately 99%; however, phenotypic changes were not observed. Low levels of patatin allowed to obtain essential glycoproteins from transgenic plants with the higher concentration and purity. Therefore, patatin-knockdown potato tubers have a potential to be an ideal host for the production of human-therapeutic glycoproteins.

Manipulation of biosynthetic pathways

Engineering of metabolic pathways is another potential use for hairpin RNAi. The most interesting examples are the modification of fatty acid profiles in cottonseed oil and the manipulation of the alkaloid profiles in opium poppy.

It was shown that the manipulation of fatty acid biosynthetic genes by RNAi can result in different oil profiles in seeds of Arabidopsis and cotton (Liu et al., 2002). Cottonseed oil typically contains a relatively high proportion of an unhealthy saturated fatty acid—palmitic acid (26%). Conversely, cottonseed oil is low in the desirable oleic and stearic fatty acids (15% and 2%, respectively). Liu et al. (2002) demonstrated that hairpin RNAi against two fatty acid desaturase genes, which use stearic acid and oleic acid as substrates, can strongly increase the proportion of these fatty acids in cottonseed oil. Thus, silencing of the two genes resulted in an increase of the oleic and stearic fatty acid content to 78.2% and 39.8%, respectively.

Another example of the regulation of plant metabolism is the manipulation of brassinosteroid (BR) content. BRs are synthesized from campesterol (CR) and are plant steroid hormones involved in embryogenesis, cell elongation, vascular differentiation, fertility, and senescence. Key enzymes in BR synthesis in Arabidopsis are Sterol methyl transferease 2 (SMT2), DE-ETIOLATED2 (DET2), and DWARF4 (DWF4). Arabidopsis plants were transformed by Agrobacterium that carried the hpRNA construct containing sense/antisense sequences of *DET2* (455 bp—not a full length), *DWF4* (361 bp), and *SMT2* (289 bp) separated by a 1,352-bp *Chs*A intron from the petunia *Chalcone synthase A* gene (Chung et al., 2010). In transformed plants, the expression of *DET2*, *DWF4*, and *SMT2* was simultaneously decreased, and a semi-dwarf phenotype was observed. Because of the triple knockdown, an intermediate in RB-specific biosynthesis was reduced, whereas an up to 420% increase in CR content was detected. Because CR has been suggested to have cholesterol-lowering effects and anticancer properties, consumption of such plants might be beneficial for human health.

Engineering of metabolic pathways with multiple branches does not always lead to predictable results. This was the case when Allen et al. (2004) used a chimeric hpRNA interference construct to silence the seven-member family of codeinone reductase (COR) enzymes that catalyze the final step of morphine biosynthesis in opium poppy. Silencing of *COR* genes was achieved; however, the reduction in COR enzyme activity vastly changed the profile of alkaloids produced in poppy in an unexpected way. Downregulation resulted in a variety of alkaloids that do not normally accumulate to significant levels.

Pathogen resistance

hpRNAs are also efficiently used in plants to confer pathogen resistance. The efficiency of PTGS is increased by nucleic acids which contain hairpins, inverted repeats, and other high secondary structures. Antisense and hairpin RNA technologies often rely on these secondary structures for the improvement of their products. In addition, viruses are often used in **virus-induced gene silencing (VIGS)** because they are efficient in triggering gene silencing. VIGS relies on the use of viral vectors carrying a transgene that can trigger PTGS, causing the degradation of its homolog within the plant. *Tobacco rattle virus* (TRV) has been used in VIGS to elucidate the mechanisms of floral scent production in petunia and to facilitate the dissection of the flavonoid biosythesis pathway. DNA viruses are also used in VIGS; for example, *tomato golden mosaic virus* (TGMV) was used to silence a meristematic gene called the proliferating cell nuclear antigen (PCNA) gene in *Nicotiana benthamiana*.

In 2010, Di Nicola-Negri et al. showed that hairpin RNA constructs were able to confer resistance to a broad range of *Plum pox virus* (PPV). PPV causes the most detrimental viral disease of plum, apricot, and peach crops. Di Nicola-Negri et al. (2010) generated transgenic *Nicotiana benthamiana* plants carrying in their genome several hpRNA constructs (h-UTR/P1, h-P1/HCPro, h-HCPro, and h-HCPro/P3) that contain the 35S promoter, the Nos terminator, and sense and antisense arms (approximately 700 bp each) separated by 707 bp of partial Dof Affecting Germination (DAG1) introns. Analysis of viral titer showed that transgenic *N. benthamiana* plants carrying aforementioned constructs were resistant to broad range of PPV.

Another example is provided of how wheat can resist infection with *wheat streak mosaic virus* (WSMV). Fahim et al. (2010) used the nuclear inclusion protein 'a' (*Nia*) gene coding for the virus proteinase to generate a hpRNA vector containing the polyubiquitin promoter, the OCS terminator, WSMV NIa sense, and antisense (696 bp each) separated by Pdk intron. Expressing this construct, transgenic wheat showed no disease symptoms upon infection with WSMV. Moreover, ELISA readings from most of the infected transgenic plants were similar to those from uninfected plants. The authors were able to detect 21- and 24-nt hpRNA-derived siRNAs in infected plants, suggesting that DCL4 and DCL3 might be involved in protecting infected plants against viral infection.

Viroids are also targets for hpRNA technology. The *Potato spindle tuber viroid* (PSTVd) causes one of the major diseases of potato and tomato plants. Mature viroid RNA molecules are thought to be resistant to degradation by RNAi due to their highly ordered secondary structure. Schwind et al. (2009) generated the hpRNA (hpRSTVd) vector containing the 35S promoter, the OCS terminator, and the sense/antisense PSTVd gene separated by Pdk intron and showed that transgenic tomato plants expressing this construct were resistant to PSTVd. Again in this case, the hpRNA-derived siRNAs of 21 and 24 nt in length were detected.

Another interesting phenomenon that can be utilized for commercial purposes is virus-induced gene silencing (VIGS). VIGS depends on the incorporation of viRNAs into RISC complexes. If viRNAs have sufficient homology to endogenous plant transcripts, these transcripts are cleaved and degraded. As a result, viral infection results in knockdown mutations of the targeted endogenous gene. This mechanism is utilized by researches to knockdown the function of specific genes. The first well-described report on VIGS included downregulation of several plant mRNAs that carried homology to the 35S leader of CMV (Moissiard and Voinnet, 2006). Hundreds of host transcripts with substantial homology to various viral genomes might exist and be targets of VIGS. It remains to be shown whether this mechanism is employed either by plants or by viruses, and whether it includes the mechanism of fighting infection or breaking plant resistance, respectively. It might be a case of both, depending on viral and host factors.

Using hairpin RNAi in animals

ahpRNAs used in mammalian cells are shorter than ahpRNAs used in plants. Therefore, hpRNAs used in mammalian cells are called **short hpRNAs (shRNAs).** The advantage of using shRNAs in mammalian cells is that shRNAs provide a long-lasting silencing effect. Unlike in plants, shRNAs are processed like pre-miRNAs by the same miRNA biogenesis machinery. The final products from shRNAs are called siRNAs, which have perfect complementarity to the target RNA. Effective shRNAs have been developed for the use in mammals. The vectors contain Pol II promoters, typically the H1 promoter of RNase P or the U6 small nuclear RNA (snRNA) promoter (Olejniczak et al., 2010). Ideally, shRNA contains a short 19- or 29-bp stem and a loop of various sizes and sequences. In addition, a high level of shRNA expression inhibits the natural pre-miRNA pathway. Therefore, it is important to consider achieving optimal expression levels of shRNA.

Comparing to plants, the use of long extragenous hairpin RNAi vectors or dsRNAs in mammals is problematic because they can trigger profound physiological reactions that lead to the induction of interferon synthesis. As a result, degradation of dsRNAs bigger than 30 bp occurs without triggering RNAi. Therefore, a variety of vector-based approaches that express siRNAs as short hairpin shRNAs have been developed to permit their delivery through viral vectors. Designing such shRNA in the context of a naturally occurring RNA polymerase (pol) II-driven microRNA transcript (miR-shRNA) increases the flexibility of this approach without triggering the interferon-dependent response. Conversely, the use of shRNAs may bring inconsistent RNAi as well as an off-targeting effect. Thus, in many cases, it is hard to obtain reliable and predictable RNAi in the animal model by using hairpin constructs. An alternative approach that is widely used to trigger RNAi in animals is the delivery of siRNAs or miRNAs directly into cells. One of the main differences between siRNA and miRNA is that the former causes degradation of a target mRNA to which it is perfectly complementary, whereas the latter inhibits translation of a target mRNA to which it is imperfectly base-paired.

Applications of hpRNAs in mammals

Hairpin technology is commonly used in research and therapeutic applications in animals, although the process seems to be less efficient and less understood as compared to plants.

The use of hpRNA technology in research

Hairpin technology is effectively used in various areas of research with animals. This section provides only a few examples such as the use of hpRNAs for studying apoptosis and cell renewal.

In apoptosis, cytochrome *c* release from mitochondria is triggered by the accumulation of reactive oxygen species (ROS). The voltage-dependent anion channel (VDAC) is known as mitochondrial porin, which lies in the outer mitochondrial membrane (OMM) and facilitates the exchange of adenine nucleotides, Ca^{2+}, other metabolites, and ions. Furthermore, it has been suggested that VDAC might be involved in mitochondria-induced ROS production and promotion of cytochrome *c* release. Because the oxidation state of VDAC is thought to affect apoptosis, cysteine residues, which are the major target in oxidation/reduction reactions, might play a role in apoptosis. Two VDAC1 cysteine residues (Cys^{127} and Cys^{232}) were examined for apoptosis requirements using shRNA (Aram et al., 2010). The shRNA vector containing 19 bp of sense and antisense hVDAC2 coding sequences separated by a short spacer was generated and transfected into cells. The results showed that the level of hVDAC1 expression was low (10% to 20%) in hVDAC1-shRNA cells. However, hVDAC1-shRNA cells in a native or mutated cysteine-less rat model undergo apoptosis either by treatment with ROS-producing agents or by over-expression of VDAC1. Therefore, these findings clarified that VDAC1 cysteine residues were not required for apoptosis.

The self-renewal process of cells can be also studied using shRNAs. Many regulators such as proteins of the transforming growth factor (TGF)-β subfamily are involved in human embryonic stem cells (hESCs). The TGFβ family includes TGFβs, activins, inhibins, nodal, myostatin, bone morphogenic proteins (BMPs), growth differentiation factors, and anti-Müllerian hormone that controls hESC self-renewal and cell fate. Based on recent understanding, TGF/activin/nodal are important for maintaining self renewal of hESCs. SMAD proteins are receptor proteins that are phosphorylated resulting in the activation of TGFβ family signaling. Phosphorylated SMAD2 and SMAD3 interact with co-SMAD4, which is a common component in TGF/activin/nodal. When TGF/activin/nodal receptor signaling is inhibited, the differentiation of hESC occurs. Thus, the suppression of SMAD4 protein has a potential to lead to the same outcome. The *SMAD4* protein was knocked down by the shRNA-SMD4 vector containing approximately 20 bp of SMAD4 coding sequences oriented in sense and antisense separated by a spacer. Cells transfected with shRNA-SMAD4 showed more than 85% reduction of SMAD4 RNA levels (Avery et al., 2010). Although *SMAD4* knockdown was successful, it did not promote the differentiation; instead, it destabilized hESCs. Therefore, it was shown that SMAD4 stabilizes the cell, yet it is not necessary for maintaining the undifferentiated state of hESCs.

The use of shRNAs for therapeutic purposes

hpRNA technology can be efficiently used for therapeutic purposes. The first such examples were attempts of controlling HIV progression. Lee et al. (2002) demonstrated that injection of ddRNAi constructs carrying shRNAs with homology to HIV indeed is able to inhibit viral infection. Later on, Anderson et al. (2003) demonstrated that ddRNAi constructs carrying multiple shRNA effector motifs separated by spacers could produce independent shRNAs targeting independent phases of HIV metabolism. Permanent expression of shRNAs was achieved by infection CD34(+) hematopoietic progenitor cells using an HIV-based lentiviral vector carrying an anti-Rev siRNA construct (Banerjea et al., 2003). Transduced progenitor cells were injected into SCID-hu mouse thy/liv grafts, and a higher level of viral resistance was obtained. Treatment of hematopoietic cells with a vector carrying multiple HIV targets was attempted by Li et al. (2006). Targeting several genes, including the HIV Tat protein and the chemokine receptor CCR5 used by HIV for cellular entry, the authors showed a substantial reduction in HIV replication in primary hematopoietic cells.

The inhibition of viral amplification using shRNAs was attempted by Pongratz et al. (2010). However, a stable expression of antiviral shRNAs was unable to inhibit virus amplification since a rapid emergence of viral mutants. Therefore, the identification of multiple strong inhibitory sequences targeting different or conserved regions of the viral genome was examined. Pongratz et al. generated the first randomized shRNA library against the human immunodeficiency virus type 1 (HIV-1) genome. Later, such libraries were designed by digesting DNA of HIV-1 using DNase I in order to generate diverse overlapping cDNA fragments, which were then inserted into RNAi vectors containing the H1 promoter. Cells expressing the cDNA of Herpes Simplex Virus thymidine kinase (HSV-TK) fused with HIV-1 nucleotide sequences were transfected with shRNA vectors, and powerful HIV-1-specific shRNAs were selected by ganciclovir treatments. More than 50% of 200 selected shRNAs that had little or no toxicity to host cells inhibited more than 70% of an HIV-1 based luciferease reporter assay. This study presented a non-toxic and highly efficient HIV-1-specific approach that can be used for gene therapy (Pongratz et al., 2010).

shRNA technology also showed a promise for the prevention of allogeneic transplant rejection. Immune rejection is one of the major obstacles following organ transplantation. Although several immunosuppressive therapies can prevent graft rejection, long-term immunosuppression results in opportunistic infections and cancer in some recipients. Because the platelet-endothelial cell adhesion molecule-1 (PECAM-1) is an immunoglobulin-like glycoprotein involved in white-blood-cell migration, cellular adhesion, and signal transduction, silencing of PECAM-1 expression was hypothesized to suppress the transplant rejection process (Jaimes et al., 2010). Lentiviral-based vectors encoding shRNA sequences specific for PECAM-1 transcripts were generated and then transduced into monocytic and endothelial cell lines. This resulted in an 80% decrease of PECAM-1 mRNA and protein. Moreover, T-cell cytotoxicity assays showed a significant reduction of T-cell cytotoxicity against PECAM-1 in hpRNA-transduced monocytic cells. T-cell proliferation was also reduced in the transduced cells. These findings suggested that permanent silencing of OECAM-1 expression is possible, and it inhibits the cytotoxic T-cell response to allogeneic target cells, tissues, or organs.

Drawbacks and limitations of hairpin-mediated RNAi

A major limitation to the high-throughput application of hairpin RNAi in plants and animals is the time-consuming nature of transformation. Another issue is potential cross-silencing from hairpin constructs using full-length transcripts. It was found that 68.7% of Arabidopsis transcripts may cause cross-silencing if perfect 21-nt matches are used (Xu et al., 2006). Additionally, 79.9% of gene family members may trigger cross-silencing of other family members if full-length transcripts are used as hairpin inserts. Also, usually it is unnecessary to use full-length open reading frames as hairpin inserts; these results suggest the importance of sequence homology searches as a part of hairpin insert design. The lack of reports of off-target effects provides some confidence that the process is primarily gene-specific. On the other hand, off-target effects have been commonly reported from siRNAs used in animal systems. This appears to arise from as little as 6-nt match between mRNA and the 5' end of miRNA that is required to trigger RNAi.

Conclusion

Despite the lack of a complete picture of its action, the RNAi effect is widely used nowadays in basic research as a method to investigate gene function, and in applied studies as a tool for improving plant traits. RNAi can be artificially triggered by transcriptome of sense/antisense copies of genes, small RNAs, and hairpin RNAi constructs. A better understanding of the components and mechanisms of the RNAi pathway is a future goal shared by many laboratories. Knowledge accumulation in this field will make it possible to improve methods and techniques for the alteration of gene expression making it more specific and thus triggering fewer side effects for the target cell/organism.

Exercises and discussion topics

1. Define RNAi.
2. Explain the structure of a hairpin construct.
3. Explain the benefits of hairpin-based antisensing over regular mutagenesis.
4. Describe ddRNAi.
5. Describe the mechanism of RNAi triggered by introduced hairpin constructs.
6. Describe examples of inducible hairpin RNAi.
7. Explain the differences in RNAi in plants and animals.
8. What are the possible drawbacks of RNAi?

References

Allen et al. (2004) RNAi-mediated replacement of morphine with the non-narcotic alkaloid reticuline in opium poppy. *Nat Biotechnol* 22:1559-1566.

Anderson et al. (2003) Bispecific short hairpin siRNA constructs targeted to CD4, CXCR4, and CCR5 confer HIV-1 resistance. *Oligonucleotides* 13:303-312.

Aram et al. (2010) VDAC1 cysteine residues: topology and function in channel activity and apoptosis. *Biochem J* 427:445-454.

Avery et al. (2010) The role of SMAD4 in human embryonic stem cell self-renewal and stem cell fate. *Stem Cells* 28:863-873.

Banerjea et al. (2003) Inhibition of HIV-1 by lentiviral vector-transduced siRNAs in T lymphocytes differentiated in SCID-hu mice and CD34+ progenitor cell-derived macrophages. *Mol Ther* 8:62-71.

Byzova et al. (2004) Transforming petals into sepaloid organs in Arabidopsis and oilseed rape: implementation of the hairpin RNA-mediated gene silencing technology in an organ-specific manner. *Planta* 218:379-387.

Chen et al. (2003) Temporal and spatial control of gene silencing in transgenic plants by inducible expression of double-stranded RNA. *Plant J* 36:731-740.

Chung et al. (2010) Simultaneous suppression of three genes related to brassinosteroid (BR) biosynthesis altered campesterol and BR contents, and led to a dwarf phenotype in Arabidopsis thaliana. *Plant Cell Rep* 29:397-402.

Di Nicola-Negri et al. (2010) Silencing of Plum pox virus 5'UTR/P1 sequence confers resistance to a wide range of PPV strains. *Plant Cell Rep* 29:1435-1444.

Dunoyer et al. (2004) Probing the microRNA and small interfering RNA pathways with virus-encoded suppressors of RNA silencing. *Plant Cell* 16:1235-1250.

Fahim et al. (2010) Hairpin RNA derived from viral NIa gene confers immunity to wheat streak mosaic virus infection in transgenic wheat plants. *Plant Biotechnol J* 8:821-834.

Fusaro et al. (2006) RNA interference-inducing hairpin RNAs in plants act through the viral defence pathway. *EMBO Rep* 7:1168-1175.

Guo et al. (2003) A chemical-regulated inducible RNAi system in plants. *Plant J* 34(3):383-392.

Hamilton AJ, Baulcombe DC. (1999) A species of small antisense RNA in posttranscriptional gene silencing in plants. *Science* 286:950-952.

Helliwell et al. (2002) High-throughput vectors for efficient gene silencing in plants. *Funct Plant Biol* 29:1217-1225.

Jaimes et al. (2010) Silencing the expression of platelet endothelial cell adhesion molecule-1 prevents allogeneic T-cell cytotoxicity. *Transfusion* 50:1988-2000.

Kim et al. (2008) Development of patatin knockdown potato tubers using RNA interference (RNAi) technology, for the production of human-therapeutic glycoproteins. *BMC Biotechnol* 8:36.

Lee et al. (2002) Expression of small interfering RNAs targeted against HIV-1 rev transcripts in human cells. *Nat Biotechnol* 20:500-505.

Li et al. (2006) RNAi in combination with a ribozyme and TAR decoy for treatment of HIV infection in hematopoietic cell gene therapy. *Ann N Y Acad Sci* 1082:172-179.

Liu et al. (2002) High-stearic and High-oleic cottonseed oils produced by hairpin RNA-mediated post-transcriptional gene silencing. *Plant Physiol* 129:1732-1743.

Masclaux et al. (2004) Gene silencing using a heat-inducible RNAi system in Arabidopsis. *Biochem Biophys Res Commun* 321:364-369.

Miki et al. (2005) RNA silencing of single and multiple members in a gene family of rice. *Plant Physiol* 138:1903-1913.

Moissiard G, Voinnet O. (2006) RNA silencing of host transcripts by cauliflower mosaic virus requires coordinated action of the four Arabidopsis Dicer-like proteins. *Proc Natl Acad Sci USA* 103:19593-19598.

Nishikawa Y, Sugiyama A. (2010) shRNA library constructed through the generation of loop-stem-loop DNA. *J Gene Med* 12:927-933.

Olejniczak et al. (2010) Sequence-non-specific effects of RNA interference triggers and microRNA regulators. *Nucleic Acids Res* 38:1-16.

Pongratz et al. (2010) Selection of potent non-toxic inhibitory sequences from a randomized HIV-1 specific lentiviral short hairpin RNA library. *PLoS One* 5:e13172.

Reiser et al. (2004) Molecular physiological analysis of the two plastidic ATP/ADP transporters from Arabidopsis. *Plant Physiol* 136:3524-3536.

Schwind et al. (2009) RNAi-mediated resistance to Potato spindle tuber viroid in transgenic tomato expressing a viroid hairpin RNA construct. *Mol Plant Pathol* 10:459-469.

Smith et al. (2000) Gene expression - Total silencing by intron-spliced hairpin RNAs. *Nature* 407:319-320.

Sunilkumar et al. (2006) Engineering cottonseed for use in human nutrition by tissue-specific reduction of toxic gossypol. *Proc Natl Acad Sci USA* 103:18054-18059.

Svoboda et al. (2001) RNAi in mouse oocytes and preimplantation embryos: effectiveness of hairpin dsRNA. *Biochem Biophys Res Commun* 287:1099-1104.

Waterhouse et al. (1998) Virus resistance and gene silencing in plants can be induced by simultaneous expression of sense and antisense RNA. *Proc Natl Acad Sci USA* 95:13959-13964.

Wesley et al. (2001) Construct design for efficient, effective, and high-throughput gene silencing in plants. *Plant J* 27:581-590.

Wielopolska et al. (2005) A high-throughput inducible RNAi vector for plants. *Plant Biotechnol J* 3:583-590.

Xu et al. (2006) Computational estimation and experimental verification of off-target silencing during posttranscriptional gene silencing in plants. *Plant Physiol* 142:429-440.

Index

B

C

D

E

H

I

J–K–L

M

O

P

Q–R

S

U–V

W–Z